Computational Plasticity in Powder Forming Processes

Computational Plasticity in Powder Forming Processes

Amir R. Khoei

Associate Professor

Civil Engineering Department
Sharif University of Technology
P.O. Box 11365-9313
Tehran
Iran

2005

ELSEVIER

Amsterdam Boston Heidelberg London New York Oxford Paris San Diego
San Francisco Singapore Sydney Tokyo

ELSEVIER B.V.
Radarweg 29
P.O. Box 211, 1000
AE Amsterdam
The Netherlands

ELSEVIER Inc.
525 B Street, Suite 1900
San Diego, CA 92101-4495
USA

ELSEVIER Ltd
The Boulevard, Langford Lane
Kidlington, Oxford OX5 1GB
UK

ELSEVIER Ltd
84 Theobalds Road
London WC1X 8RR
UK

First edition 2005

ISBN: 0080446361

Printed and bound in Great Britain by MPG Books Ltd, Bodmin, Cornwall

To my wife and son

Azadeh and Arsalan

CONTENTS

PREFACE

The application of computer-aided engineering is essential in modern metal forming technology. Process modeling for the investigation and understanding of deformation mechanics has become a major concern in research, and the finite element method has assumed increased importance, particularly in the modeling of powder forming processes. The finite element method emerges as the preferred approach and its flexibility leads to the development of powerful program packages capable of treating wide range of problems. The FE analysis provides detailed information on conditions within the processed material, which is often more complete than can be obtained even from elaborate physical experiments, and the numerical simulation makes it possible to examine a range of designs, or operating conditions economically. The main purpose of this book is to present the fundamentals and applications of finite element plasticity in powder forming analysis and technology.

The book is primarily written for graduate students and researchers (Masters and PhD students in Mechanical, Materials, Civil and Chemical Engineering). However, it will be useful to practicing engineers who have a good background in FEM and who are interested in applying this technique to the analysis of metal forming processes.

The book begins with a general background on the subject in Chapter 1. A general description of the governing equations for large deformation process of metal powder forming is given in Chapter 2 based on the Lagrangian and arbitrary Lagrangian-Eulerian formulation. Chapter 3 describes the mechanical behavior of powder materials using classical and modern constitutive theories. A combination of a Mohr-Coulomb and hardening cap model, a generalized double-surface plasticity and a three-invariant single plasticity theory are developed in this chapter in the concept of the generalized plasticity formulation. In Chapter 4, the general problem of formulating continuum models for frictional phenomena and of developing computational methods for analyzing these phenomena is introduced. Chapter 5 illustrates the applicability of the cap plasticity in simulating powder forming processes. Chapter 6 is devoted to the application of adaptive FEM strategy in the analysis of powder forming processes. In chapter 7, a new approach is developed based on an endochronic density-dependent plasticity model for describing the isothermal deformation behavior of powder materials. An application of the multi-surface plasticity theory is presented for pressure-independent and pressure-dependent materials in this chapter. In chapter 8, the numerical modeling of powder forming process is simulated by a finite strain plasticity based on hypoelasto-plastic formulation. Chapter 9 presents 2D and 3D numerical modeling of powder forming process using advanced plasticity models. In chapter 10, a computational algorithm is presented for dealing with softening plasticity due to material instability. Finally, the powder compaction software (PCS_SUT), which is designed for pre- and post-processing for computational simulation of the process compaction of powder is presented in Chapter 11.

Sincere thanks are due to a number of individuals. I wish to express my sincere gratitude to Professor Roland W. Lewis, University of Wales Swansea, who helped me to devote my careers to research in powder compaction. I am indebted to the predecessors of this research, in particular Professor D.T. Gethin, Dr. D.V. Tran, and Dr. R.S. Ransing, University of Wales Swansea, whose work formed the basis of the new developments reported here. I wish to thank my colleagues and former graduate students in the Department of Civil Engineering, Sharif University of Technology, who have contributed to the advances in the application of computational plasticity to powder forming; Professor M. Mofid, Dr. A. Bakhshiani, A.R. Azami, S.A. Gharehbaghi, S. Azizi, N. Jamali, A.R. Tabarraie, A. Riahi and M. Anahid. I also thank S. Azizi and A.R. Azami for their contribution in the preparation of this manuscript.

Amir R. Khoei

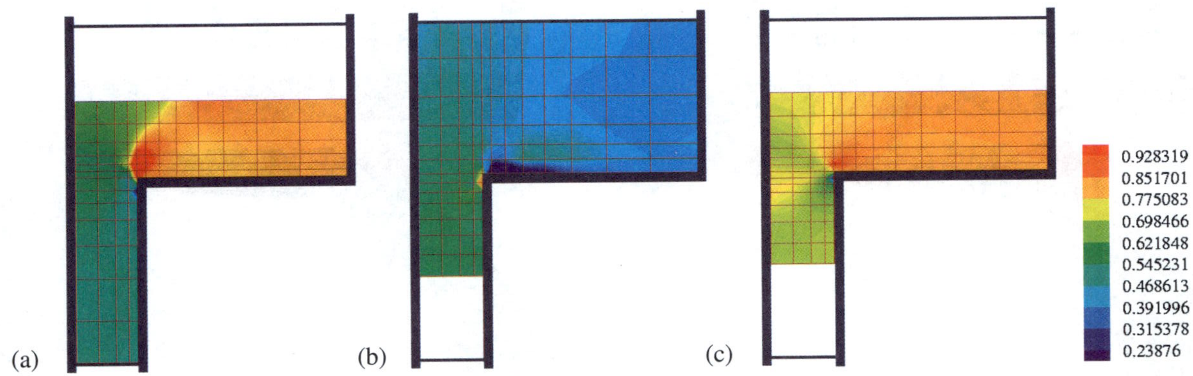

Figure 5.42. Relative density distributions at different movements of punches using ALE formulation;
a) $d_t = 6.06$ mm, b) $d_b = 7.70$ mm, c) $d_t = 6.06$ mm and $d_b = 7.70$ mm.
(see page 160)

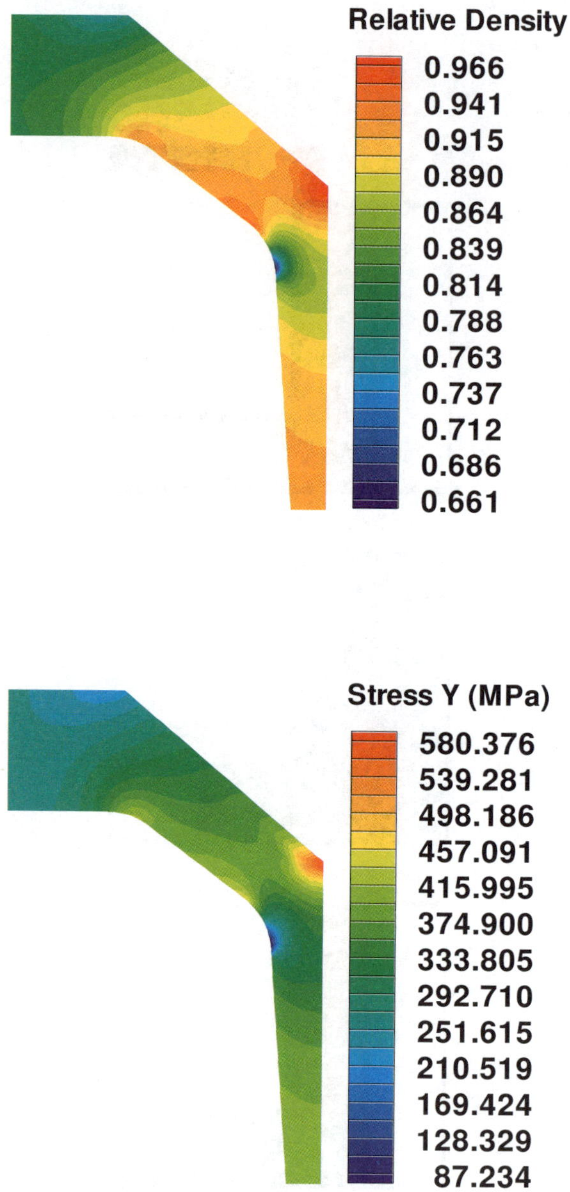

(a)

(b)

Figure 5.54. An automotive component; a) The predicted relative density contour and; b) The normal stress σ_y distribution at the final stage of compaction (see page 168)

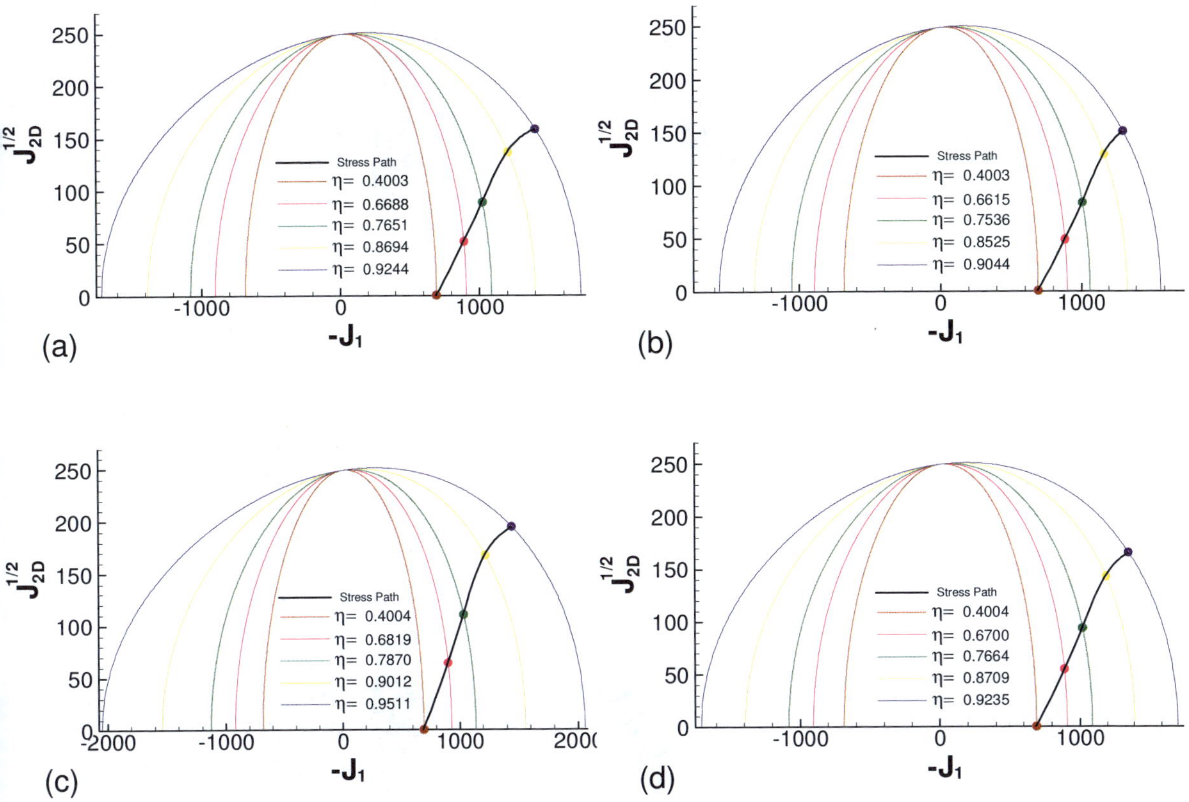

Figure 5.56. An automotive component; The stress paths with their associated yield surfaces at different Gauss points; a) A(30.2, -23.0), b) B(30.0, 0.0), c) C(25.6, 17.7), d) E(28.4, 24.4)
(see page 170)

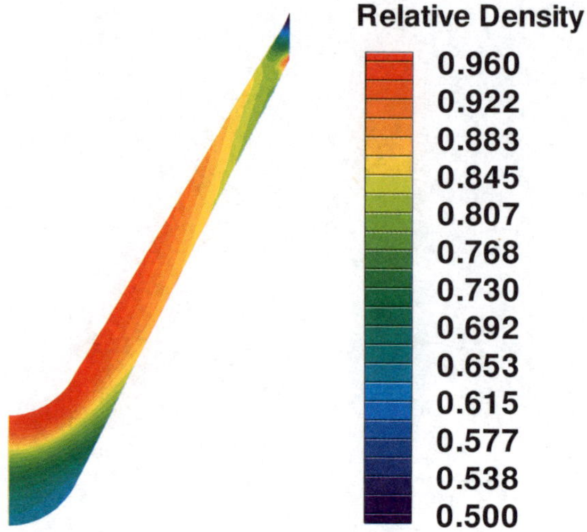

(a)

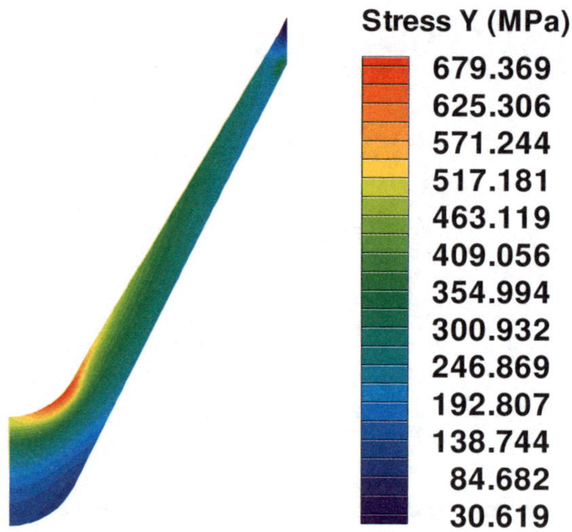

(b)

Figure 5.58. A conical shaped-charge liner; a) The predicted relative density contour and; b) The normal stress σ_y distribution at the final stage of compaction (see page 172)

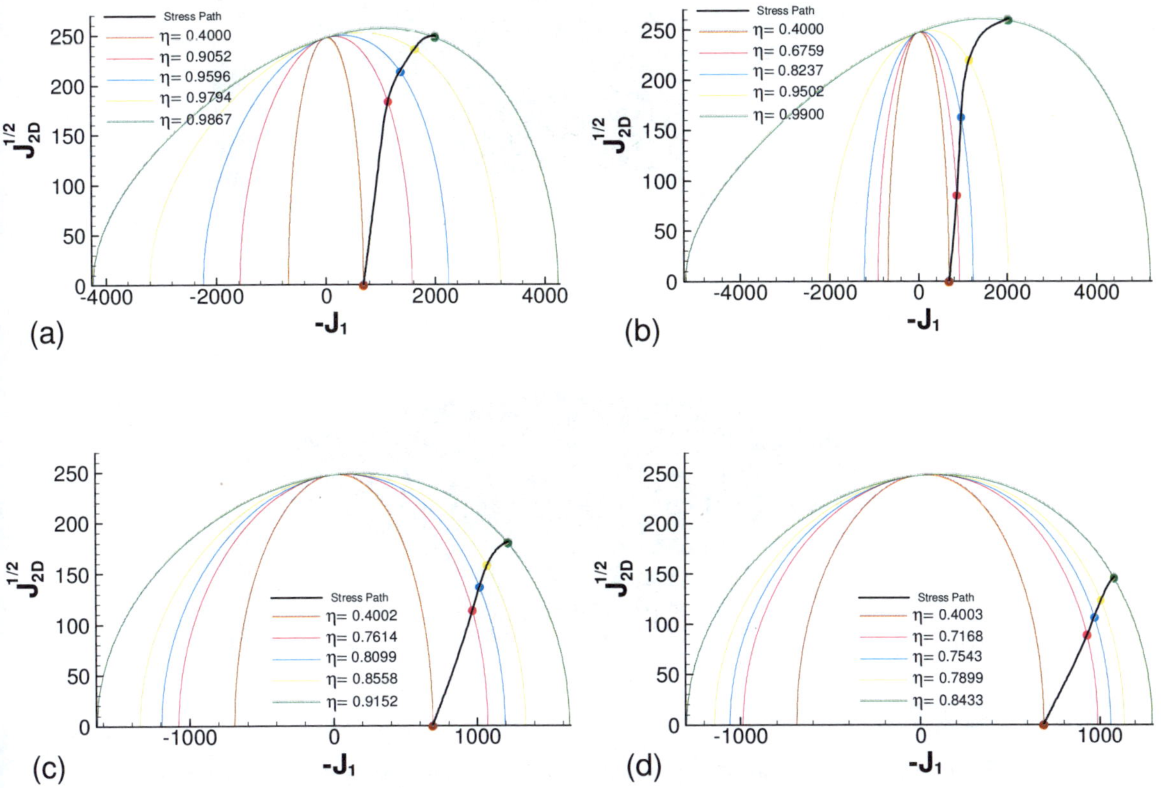

Figure 5.60. A conical shaped-charge liner; 2D stress paths with their associated yield surfaces at different Gauss points; a) A(5.1, 21.5); b) B(21.7, 35.6); c) C(5.6, 11.4); d) D(1.3, 9.1) (see page174)

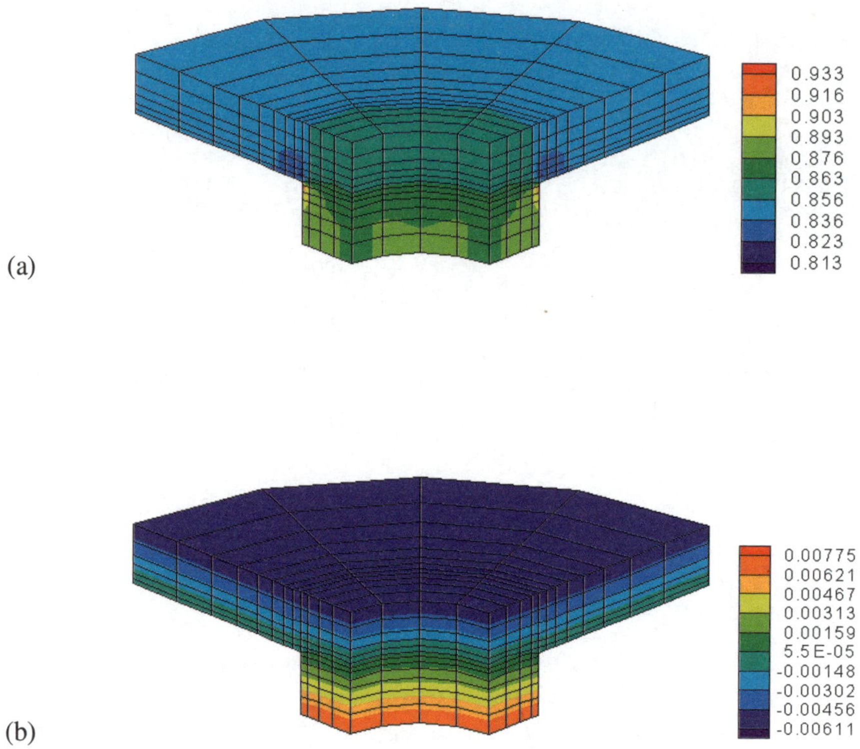

Figure 9.37. A rotational flanged component; a) Relative density contour and, b) Displacement contour at the final stage of powder compaction $d_t = 6.06$, $d_b = 7.7$ mm (see page 333)

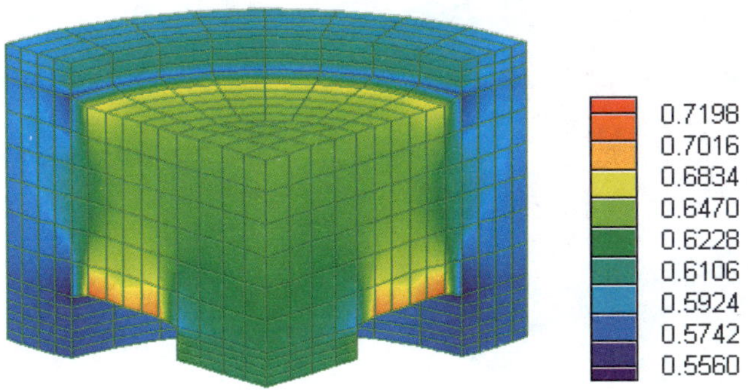

(b)

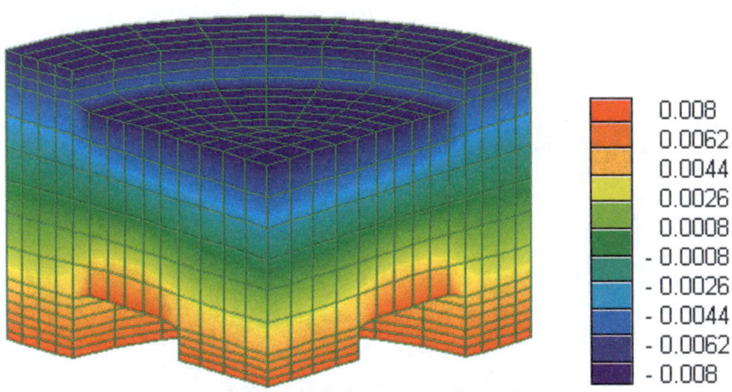

Figure 9.39. A multi-level component; a) Relative density contour and, b) Displacement contour at the final stage of powder compaction $d_t = 8.0$, $d_b = 8.0$ mm.
(see page 335)

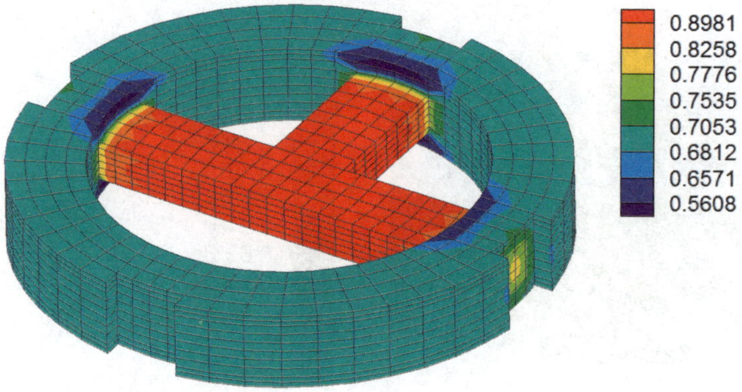

(a)

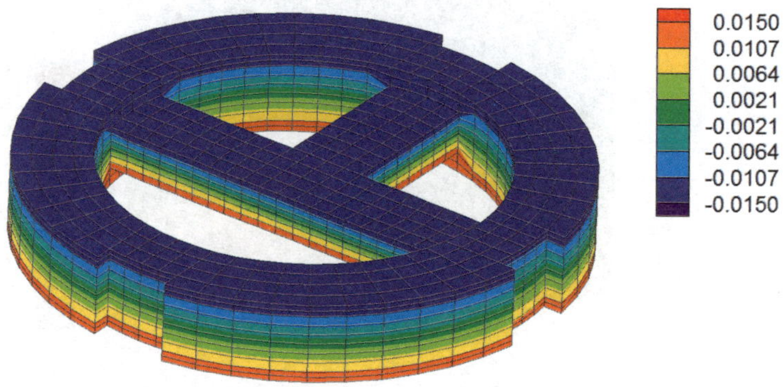

(b)

Figure 9.40. A shaped seal-disk component; a) The relative density contour and, b) The displacement contour at the final stage of compaction.(see page 336)

(a)

(b)

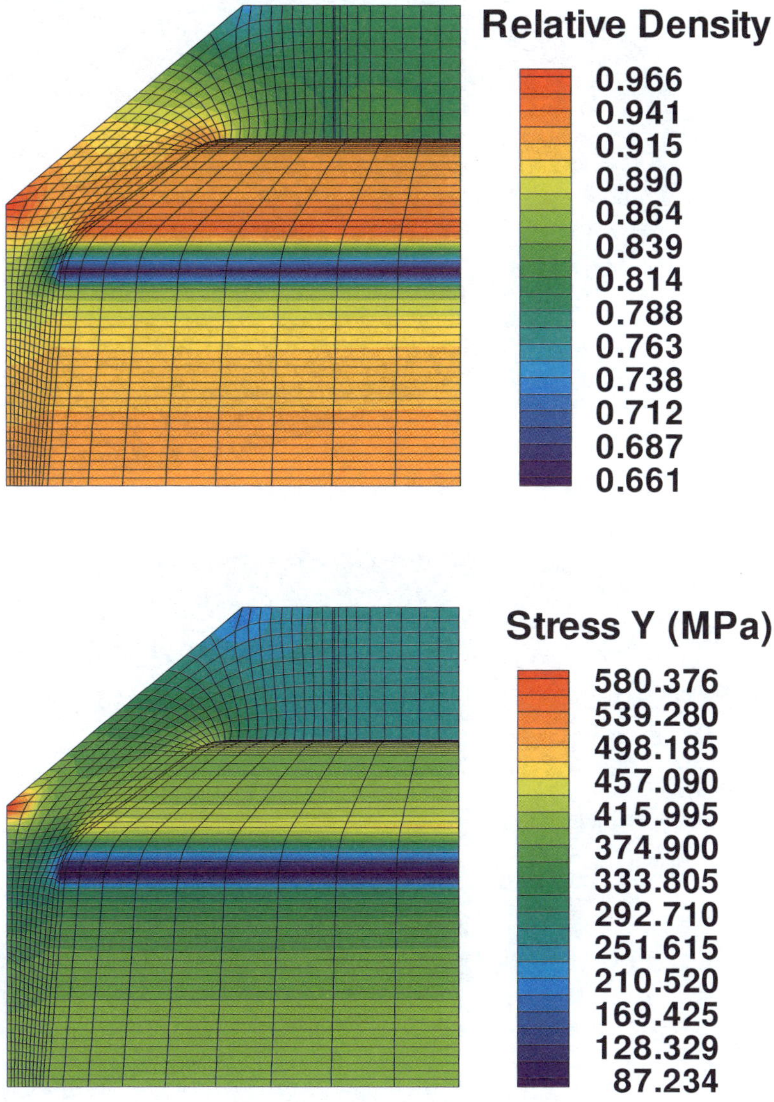

Figure 9.44. An automotive component; a) 3D relative density contour and, b) 3D distribution of normal stress σ_y at the final stage of compaction.(see page 340)

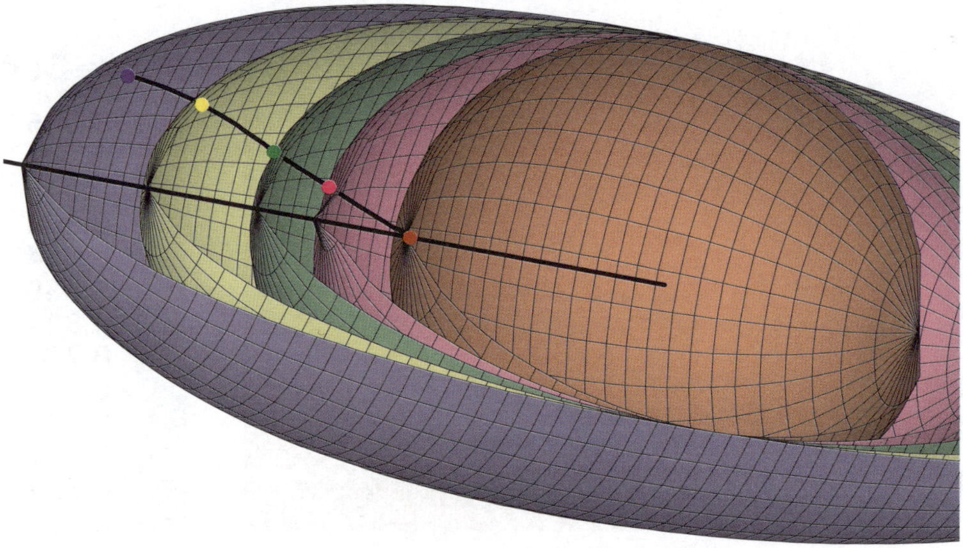

(a)

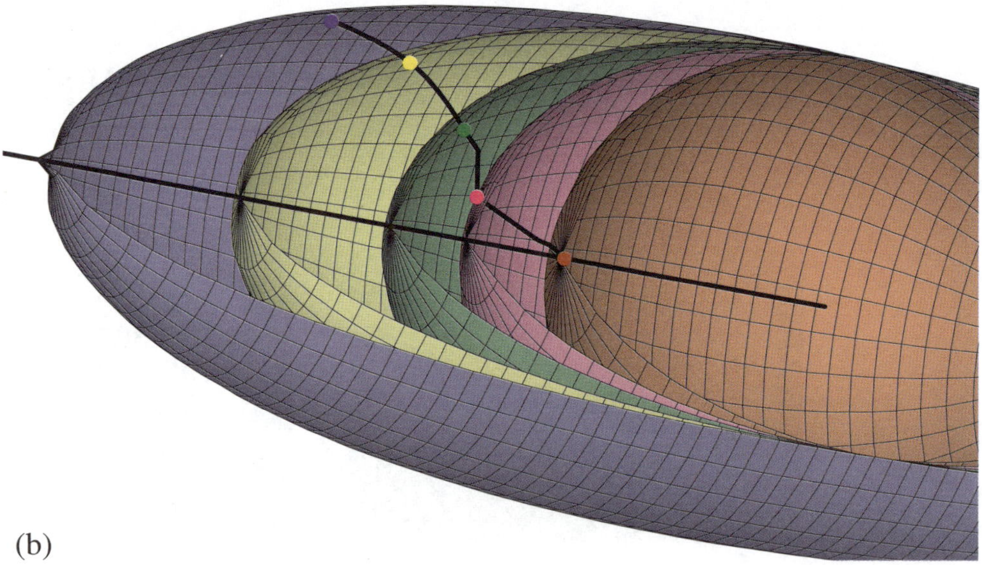

(b)

Figure 9.45. An automotive component; 3D representation of stress paths together with the associated yield surfaces at different Gauss points; a) A(30.2, -23.0); b) C(25.6, 17.7).
(see page 341)

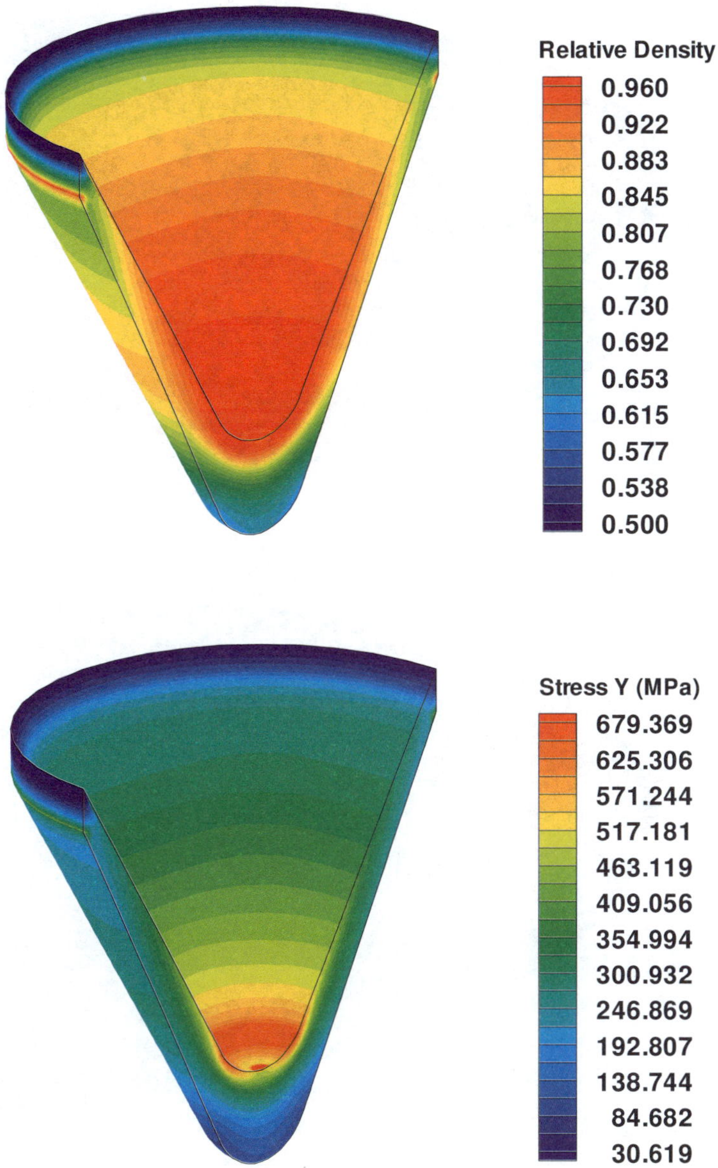

(a)

(b)

Figure 9.49. A conical shaped-charge liner; a) 3D relative density contour and, b) 3D distribution of normal stress σ_y at the final stage of compaction

(see page 345)

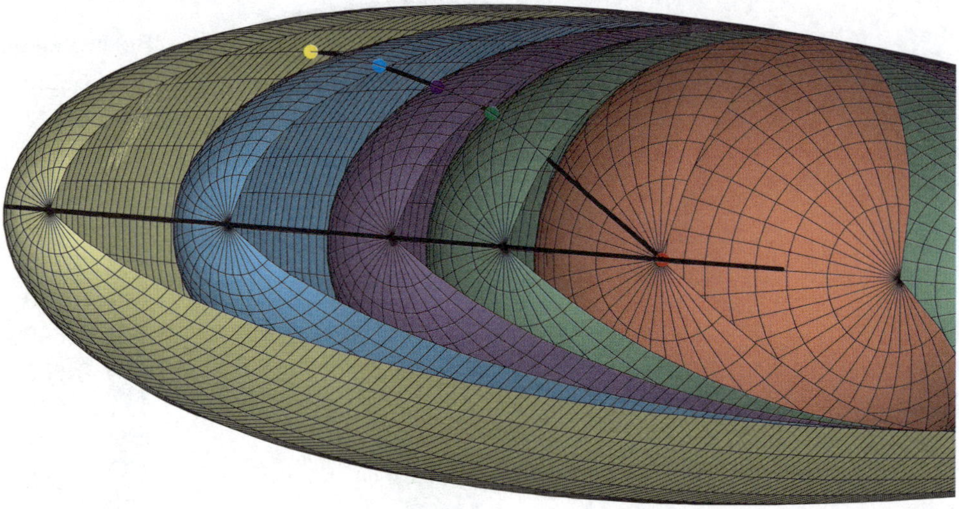

(a)

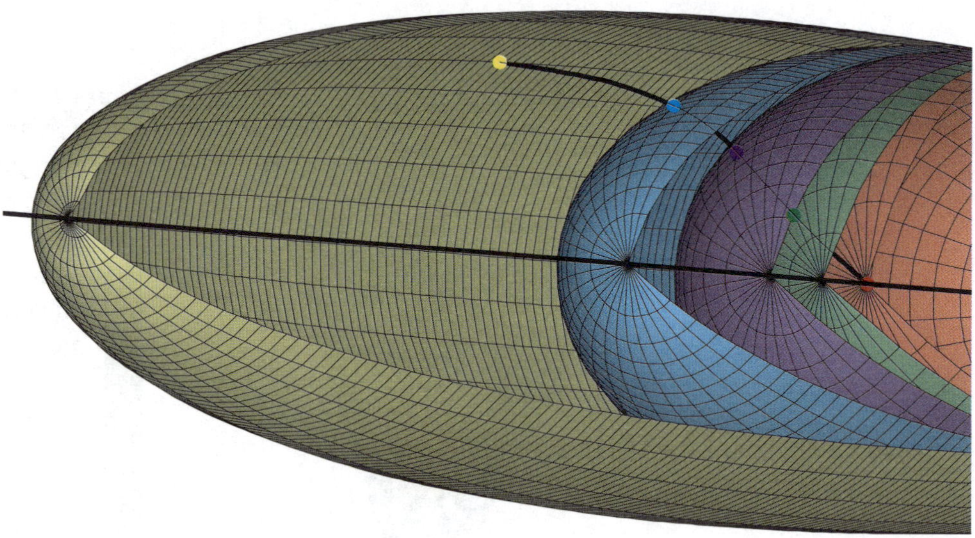

(b)

Figure 9.50. A conical shaped-charge liner; 3D representation of stress paths together with the associated yield surfaces at different Gauss points; a) A(5.1, 21.5), b) B(21.7, 35.6)
(see page 346)

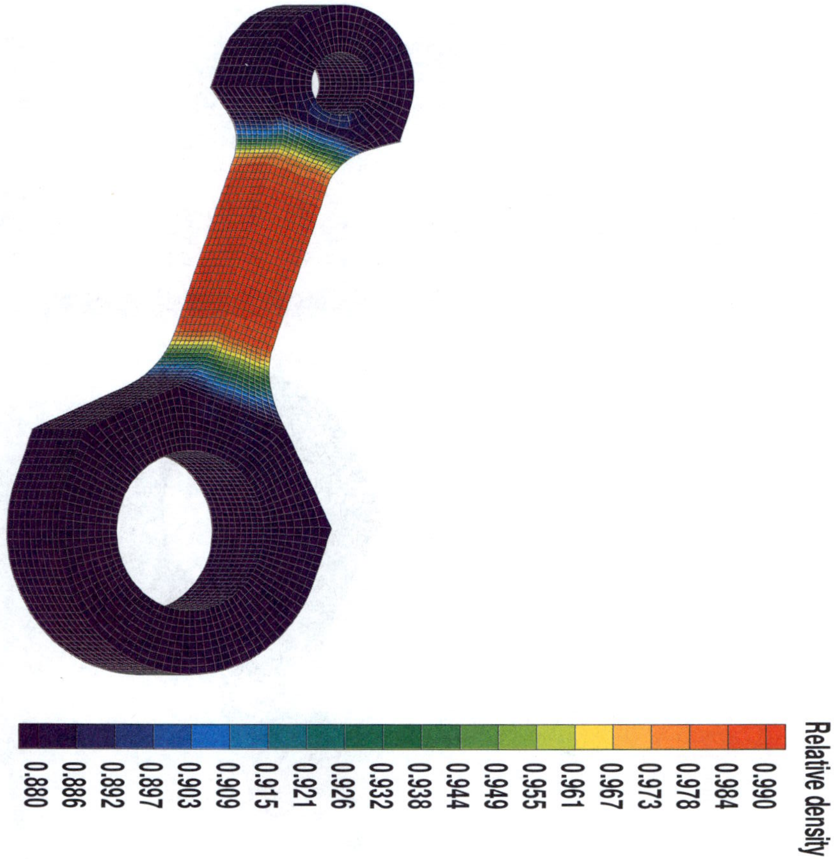

Figure 9.55. A connecting-rod; The evolution of the relative density fields at final stage of compaction (see page 351)

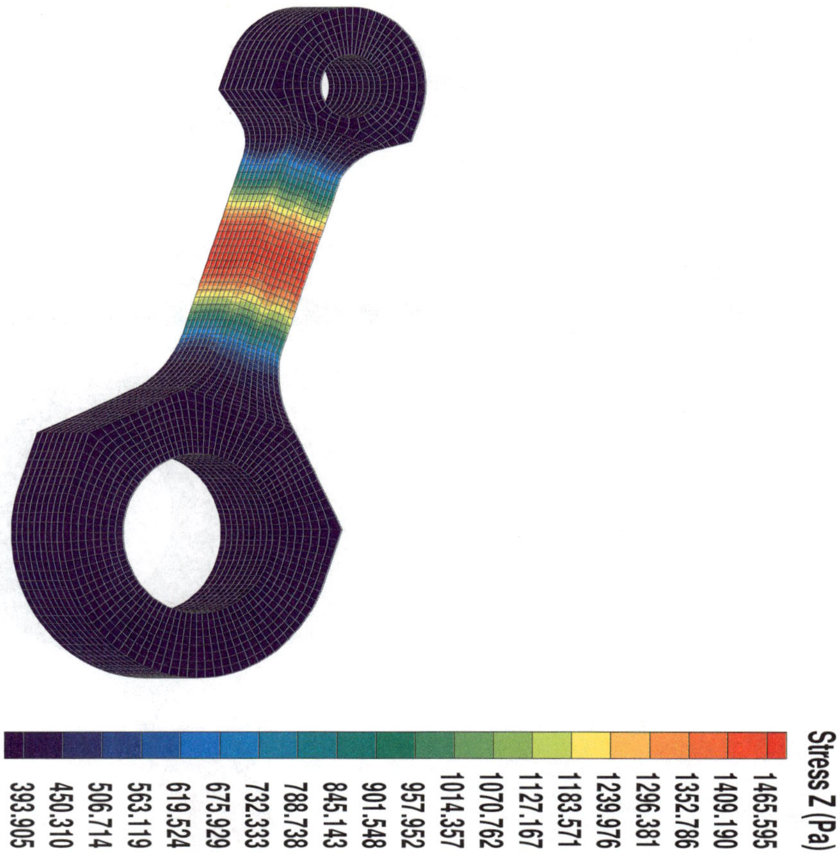

Figure 9.56. A connecting-rod; The normal stress contour at the final stage of compaction
(see page 352)

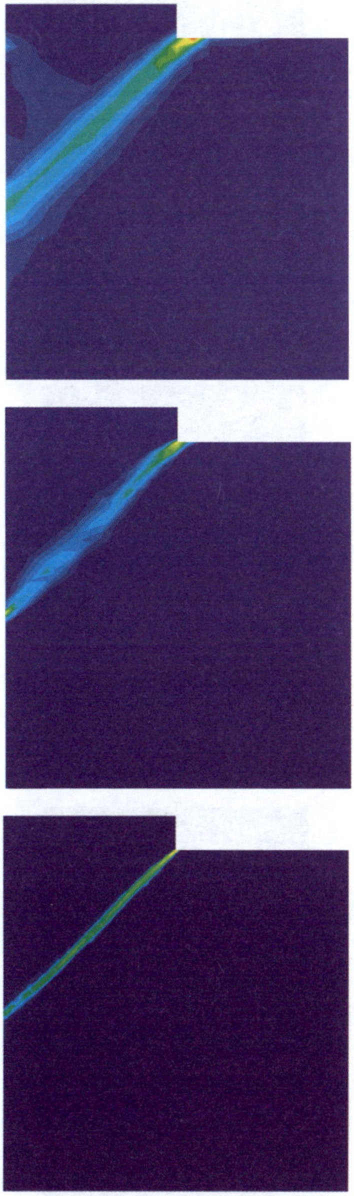

Figure 10.12. A vertical-cut; The effective plastic strain contours for initial and adapted meshes at $\delta = 25$ mm (see page 384)

Figure 10.16. A strip in tension; The effective plastic strain contours for initial and adapted meshes at $\delta = 30$ mm (see page 387)

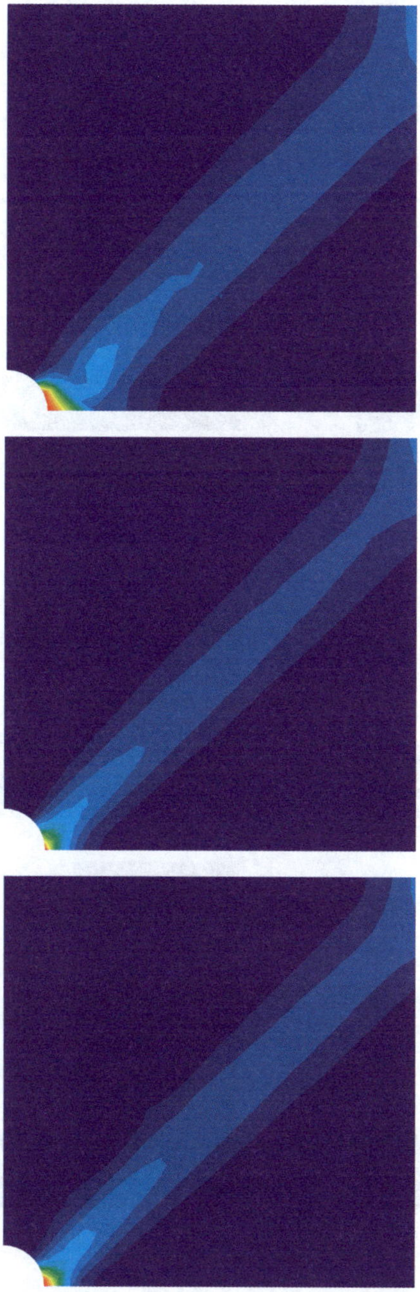

Figure 10.19. A plate with a hole; The effective plastic strain contours for initial and adapted meshes at $\delta = 40$ mm (see page 390)

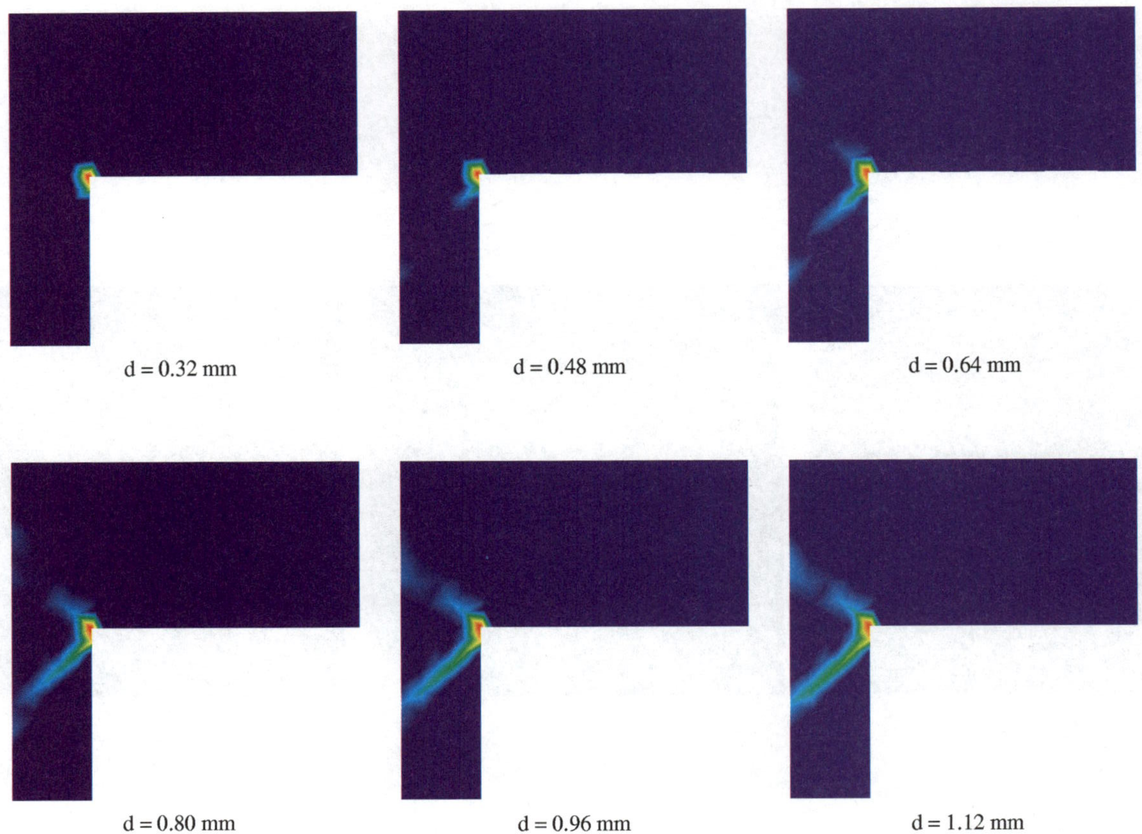

Figure 10.25. Effective plastic strain contours at different displacements using T6C/3C elements (H = -5000) (see page 396)

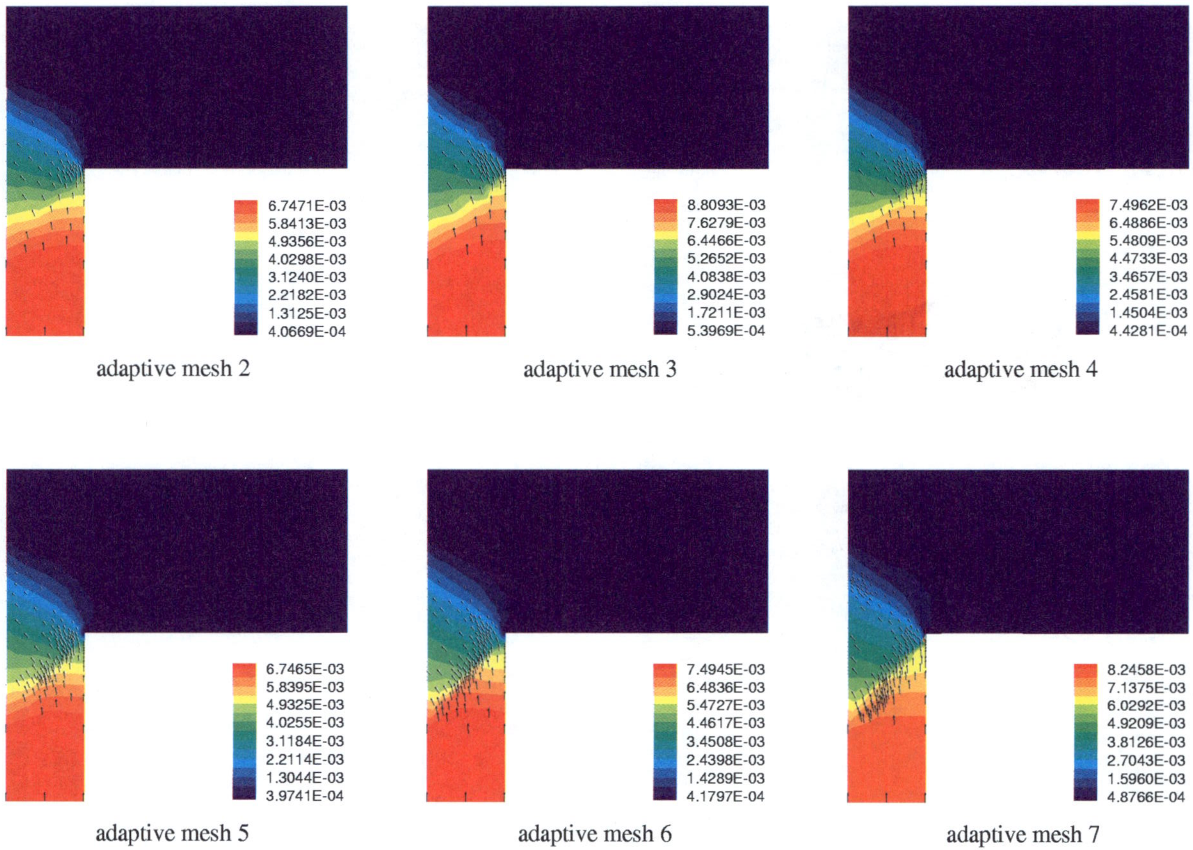

Figure 10.27. Final displacement contours along with their vectors for different adapted meshes (H = -5000)
(see page 398)

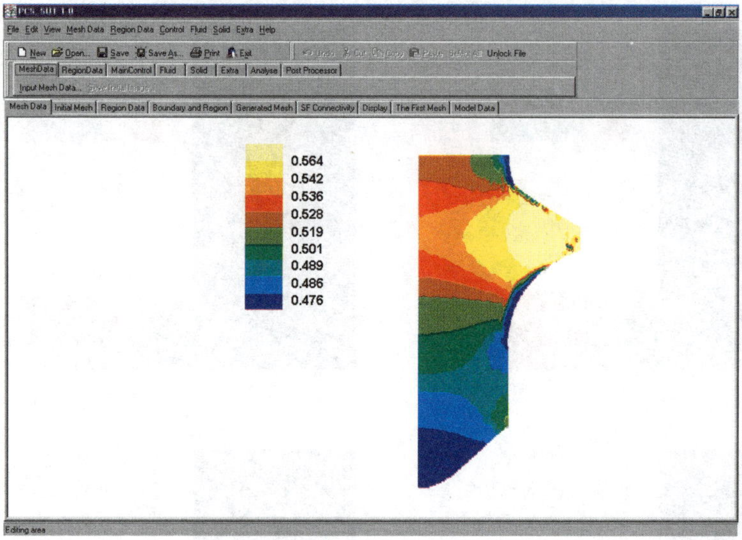

Figure 11.12. Density contour for an axisymmetric shaped tip component in PCS_SUT (see page 416)

1 INTRODUCTION

More and more sectors of the metal forming industry are beginning to recognize the benefits that finite element analysis of metal-deformation processes can bring in reducing the lead time and development costs associated with the manufacture of new components. The research and use of finite element methods in the area of material processing has increased rapidly during the last few decades. If we take the number of published papers as a measure of the research activity in this field, unusual progress can be seen.

A goal in manufacturing research and development is to determine the optimum means of producing sound products. The optimization criteria may vary, depending on product requirements, but establishing an appropriate criterion requires thorough understanding of manufacturing processes. In powder forming technology, proper design and control requires, among other things, the determination of deformation mechanics involved in the processes. Without the knowledge of the influences of variables such as friction conditions, material properties, and workpiece geometry on the process mechanics, it would not be possible to design the dies and the equipment adequately, or to predict and prevent the occurrence of defects. In the mechanical behavior of metal powder forming, the focus of attention has been mainly on two points, the choice of suitable constitutive models and the frictional algorithm for describing the interaction at the powder-die interface. Thus, process modeling for computer simulation is a major concern in modern powder forming technology. In this chapter, we first present an introduction to the application of finite element analysis to metal forming problems and then, a review of the theoretical aspects of large elasto-plastic deformation in powder forming processes.

1.1. Metal forming processes

Metal forming is an ancient art and was the subject of closely guarded secrets in antiquity. In many respects the old craft traditions have been retained until the present time, even incorporating empirical rules and practices in automated production lines. Such techniques have been successful when applied with skill, and when finely adjusted for specific purposes. Unfortunately, serious problems arise in commissioning a new production line or when a change is made from one well known material to another whose characteristics are less familiar. The current trend towards adaptive computer control and flexible manufacturing systems calls for a more precise definition and understanding of the processes, while at the same time offering the possibility of much better control of product dimensions and quality. Practical tests to determine the best tool shapes and forming conditions can be very expensive and wasteful of scarce tool and workpiece material. It is therefore important to provide guidance from theoretical models which can easily be modified.

The use of numerical simulation and computational techniques plays an important role in engineering, especially in simulating forming processes. The forming of industrial products can be accomplished by many different processes all of these involving large and often permanent deformation. By using the method, various forming conditions of loading, friction, boundary geometry, incompressibility and temperature can be considered and unsteady forming processes can be simulated up to very large deformation. The analysis undoubtedly plays an important part in design on the tool and billet geometry, material properties and the processing conditions, which otherwise rely largely on experience. A large variety of processing methods are known for metal forming, each of which has its counterpart in the simulation and numerical analysis. Figure 1.1 illustrates some typical examples of forming such as forging and upsetting, extrusion and drawing, rolling, cutting, sheet metal forming, casting, and metal powder forming.

Forging and upsetting: Forging is the working of metal into a required shape by different methods such as hammering, dropping, upsetting, rolling or pressing forging, respectively. It is the oldest of the metal working arts, having its origin with the primitive blacksmith of biblical time. Most forging operations are carried out hot, although certain metals can be cold-forged. In hot forging operations, the temperature of the workpiece materials is higher than that of the dies. Metal flow and die filling are largely determined by the flow stress and formability, the friction and heat transfer at the die-material interface and the complexity of the forging shape. For a given metal, both flow stress and formability in forging are influenced by the metallurgical structure, temperatures, strains, strain-rates and stresses that occur during deformation. The main objective of the forging process design is to ensure adequate flow of the metal in the dies so that the desired finished part geometry can be obtained without defects and with prescribed properties.

Extrusion: Extrusion is the process of letting a block of metal, the billet, be reduced in its cross section by forcing it to flow through a die orifice under high pressure. Metal flow during extrusion varies considerably, depending on the material, friction, and the shape of the extruded section. Because of the high forces needed to extrude the material, most extrusion processes are done with the material in a hot state. Some metals such as aluminum, lead and tin may be extruded cold. Cold extrusion is a special type of forging process in which cold metal is forced to flow plastically under a compressive force into a variety of shapes. These shapes are usually axisymmetric with relatively small nonsymmetrical features, and, unlike impression die forging, the process does not generate flash. The terms 'cold forging' and 'cold extrusion' are often used interchangeably and refer to well known forming operations such as extrusion, upsetting or heading, coining, ironing and swaging. In cold extrusion, the tool stresses are quite high and hence, the prediction of the forming load and stresses is quite important for die design and machine selection.

Drawing: Drawing is one of the oldest metal forming operations and has major industrial significance. Products that cannot be made with conventional rolling can be processed by drawing. This process allows excellent surface finishes and closely controlled dimensions to be obtained in long products that have constant cross sections. In drawing, a previously rolled, extruded, or fabricated product with a solid or hollow cross section is pulled through a die at a relatively high speed. The die geometry determines the final dimensions, the cross-sectional area of the drawn product, and the reduction in area. Drawing is usually conducted at room temperature using a number of passes or reductions through consecutively located dies. The deformation is accomplished by a combination of tensile and compressive stresses that are created by the pulling force at the exit from the die, by the back-pull tensile force that is present between consecutive passes, and by the die configuration. In the cold drawing of shapes, the basic contour of the incoming shape is established by cold rolling passes that are usually preceded by annealing. After rolling, the section shape is refined and reduced to close tolerances by cold drawing.

Rolling: In the rolling process a metal passed between rolls is plastically deformed. This is a widely used industrial process because it makes possible a high production rate and a close control of the final product. Most engineering metals, such as aluminum alloys, copper alloys and steels, are first cast into ingots and are then further processed by hot rolling into 'blooms', 'slabs' and 'billets'. These are known as semi-finished products because they are subsequently rolled into other products such as plate, sheet, tube, rod, bar and structural shapes. The primary objectives of the rolling process are to reduce the cross section of the incoming material while improving its properties and to obtain the desired section at the exit from the rolls. The process can be carried out hot, warm, or cold, depending on the application and the material involved. Rolling processes on the bases of the complexity of metal flow during the process and the geometry of the rolled product can be classified into the following categories: uniform reduction in thickness with no change in width, uniform reduction in thickness with an increase in width, a moderately nonuniform reduction in cross section and a highly nonuniform reduction in cross section.

Cutting: Metal cutting is one of the most important methods of removing unwanted material in the production of mechanical components. Cutting operations involve the removal of macroscopic chips in the form of ribbons or particles. There is essentially no flow of metal at right angles to the direction of chip flow. Cutting processes are unusually complex largely due to the fact that two basic operations occur simultaneously in close proximity with strong interaction; large strain plastic deformation in a zone of concentrated shear and material transport along a heavily loaded region of relative motion between chip and tool. In general, several simplified models which emphasize different aspects of the problem such as thermal, material and surface considerations are operative simultaneously with varying degrees of importance depending on specific machining conditions. There are basically three widely used cutting operations; 'turning' which is a process using a single point tool which removes unwanted material to produce a surface of revolution, 'milling' which is a process for producing flat and curved surfaces using multipoint cutting tools and 'drilling' which is a complex cutting tool used to produce rough holes.

Sheet metal forming: The products made by sheet metal forming processes include a large variety of shapes and sizes, ranging from simple bends to double curvatures with shallow or deep recesses. Thus, the scope of sheet metal forming is very broad. The basic mechanisms of sheet metal forming can be classified based on the governing deformation mechanisms into stretching, drawing and bending. In most sheet metal parts, the shape is formed almost entirely in the first operation. Subsequent processes are generally trimming, piercing, flanging, and some minor restriking of detail. Depending upon the shape and the relative dimensions of the blank and the tool, one or more basic mechanisms is predominantly involved in sheet metal deformation. The limits of sheet metal forming are determined by the occurrence of defects, such as wrinkles and ruptures in the blank. Wrinkling may form either in the flange over the die surface or in the blank around the die shoulder. Wrinkling over the flat face of the die can be avoided by applying the blank holding force, and the possibility of wrinkling of the blank around the die shoulder can be reduced by taking the radius of the die corner large enough compared with the blank thickness.

Casting and moulding: A metal casting is a shape obtained by pouring liquid metal into a mould or cavity and allowing it to freeze and thus to take the form of the mould. This is the fastest and often the most economical method for obtaining a part of any desired composition. Because of the basic simplicity of the casting process, almost any metal or metal alloy can be cast. The first moments of creation of the new casting are an explosion of interacting events, the release of quantities of thermal and chemical energy. The liquid metal attacks and is attacked by its environment, exchanging alloys, impurities and gas. The surging and tumbling flow of the melt through the running system can introduce clouds of bubbles. During melting, porosity in castings is often produced by solution of gases in liquid metal which are less soluble in the solid metal and therefore precipitate as bubbles, leading to holes. During freezing, liquid surges through the dendrite forest to feed the volume contraction on solidification, washing off branches, cutting flow paths and polluting regions with excess solute, forming segregates. The solidified casting strives to contract whilst being resisted by the mould. The mould suffers and may crush and crack and the casting suffers, being stretched as on a rack. These events, which characterizes the first seconds and minutes of the casting's life, can be difficult to assume good quality products.

Metal powder forming: In powder compaction forming, a specific amount of mixed powder is placed into the compacting die. Powders are commonly pressed into dies to produce near-to-net-shape 'green' bodies prior to final sintering. Pressed products are fabricated from a mixture of powders and additives using the rapid and robust powder compaction process. For instance, powders may be tabletted as medicine, pressed as compacts or mechanical parts, pelletized as feed, or packaged as products for consumption. Compaction is achieved through pressure from above and below on the die. Many factors dictate the amount of pressure that will be used to compact the powder, including: the powder distribution in the die cavity, the lateral flow of the powder mixture, and the specific tolerance needs of the finished part. Once the compaction portion of powder metal is completed, the green compact component is ejected from the die and prepared for the next stage, sintering. In the sintering process, the green compact is placed on a moving belt and pulled through a furnace. The part is heated to a temperature just below the melting point of the primary elements in the compact. This allows the elemental particles to fuse together into a solid mass. Sintering changes the mechanical bonds in the compact into metallurgical bonds, which lead to a finished product that with similar physical properties to wrought or cast items with the same chemical composition. Clearly, powder compaction forming is a key operation of significant economic importance in several industries, such as: agriculture, ceramic, chemical, food, pharmaceutical, powdered metal, and refractory.

1.2. Design of forming process

In metal forming, an initially simple part is plastically deformed between tools or dies to obtain the desired final configuration. Thus, a simple part geometry is transformed into a complex one, in a process whereby the tools store the desired geometry and impart pressure on the deforming material through the tool-material interface. The physical phenomena constituting a forming operation are difficult to express with quantitative relationships. The metal flow, the friction at the tool-material interface, the heat generation and transfer during plastic flow, and the relationships between microstructure/properties and process conditions are difficult to predict and analyze. Often, in producing discrete parts, several forming operations are required to transform the initial 'simple' geometry into a 'complex' geometry, without causing material failure or degrading material properties.

Consequently, the most significant objective of any method of analysis is to assist the forming engineer in the design of forming and/or performing sequences. For a given operation (performing or finish-forming), such design essentially consists of

- establishing the kinematic relationships (shape, velocities, strain-rates and strains) between the undeformed part (billet or blank) and the deformed part (product); i.e., predicting metal flow during the forming operation. This objective includes the prediction of temperatures and heat transfer, since these variables greatly influence local metal flow conditions;
- establishing the limits of formability or producibility; i.e., determining whether it is possible to perform the forming operation without causing any surface or internal defects (cracks or folds) in the deforming material; and

- predicting the stresses, the forces, and the energy necessary to carry out the forming operation. This information is necessary for tool design and for selecting the appropriate equipment, with adequate force and energy capabilities, to perform the forming operation.

For the understanding and quantitative design and optimization of metal forming operations it is useful to consider a metal forming process as a system and to classify these processes in a systematic way. A metal forming system comprises all the input variables relating the geometry and material of the billet or blank, the geometry and material of the tooling, the conditions at the tool-material interface, the mechanics of plastic deformation, the equipment used, the characteristics of the final product, and finally the plant environment in which the process is being conducted. The 'systems approach' in metal forming allows the study of the effects of process variables on product quality and process economics. The key to a successful metal forming operation, i.e., to obtaining the desired shape and properties, is the understanding and control of metal flow. The direction of metal flow, the magnitude of deformation, and the temperatures involved greatly influence the properties of the formed components. Metal flow determines both the mechanical properties related to local deformation and the formation of defects such as cracks or folds at or below the surface. The metal flow is in turn influenced by the process variables, as described below;

Material variables: For a given material composition and deformation/heat-treatment history (microstructure), the flow stress or effective stress, and the workability or formability in various directions (anisotropy), are the most important material variables in the analysis of a metal forming process. For a given microstructure, the flow stress is expressed as a function of strain, strain-rate, and temperature. In order to determine the actual functional relationship, it is necessary to conduct torsion, plane-strain compression, and uniform axisymmetric compression tests. Workability or formability is the capability of a material to deform without failure; it depends on conditions existing during deformation processing, such as: temperature, rate of deformation, stress/strain history, and material variables, such as; composition, voids, inclusions and initial microstructure. In hot forming processes, temperature gradients in the deforming material also influence metal flow and failure phenomena.

Tooling and equipment: The selection of a machine for a given process is influenced by the time, accuracy, and load-energy characteristics of that machine. Optimum equipment selection requires consideration of the entire forming system, including: lot size, conditions at the plant, environmental effects, and maintenance requirements, as well as the requirements of the specific part and process under consideration. The tooling variables include the design and geometry, surface finish, stiffness, and mechanical and thermal properties under conditions of use.

Friction: The mechanisms of interface friction are very complex. One way of expressing friction quantitatively is through a friction coefficient, or a friction shear factor. There are various methods of evaluating friction, i.e., estimating the value of friction coefficient and shear factor. Experimental tests most commonly used are the ring and spike tests for massive forming and the plane-strain-draw and stretch-draw tests for sheet forming.

Deformation mechanics: In metal forming, material is deformed plastically to generate the shape of the desired product. Metal flow is influenced mainly by tool geometry, friction conditions, characteristics of the stock material, and thermal conditions existing in the deformation zone. The details of metal flow influence the quality and properties of the formed product and the force and energy requirements of the process. The mechanics of deformation, i.e., the metal flow, strains, strain-rates, and stresses, can be investigated by process modeling.

Product properties: The macro- and micro-geometry of the product, i.e., its dimensions and surface finish, are influenced by process variables. The processing conditions, i.e., temperature, strain and strain-rate, determine the micro-structural variations taking place during deformation and often influence final product properties. Consequently, a realistic system approach must include consideration of the relationships between properties and micro-structure of the formed material, and the quantitative influences of process conditions on metal flow and resulting micro-structures.

1.3. Finite element analysis in metal forming

The application of the finite element method to metal-forming problems began as an extension of structural analysis technique to the plastic deformation regime, which was mainly based on the plastic stress-strain matrix developed from the Prandtl-Reuss equations. Hydrostatic extrusion, compression, and indentations were analyzed using this matrix and the infinitesimal variational formulations. An analysis method in the area of metal-forming application, in many cases, can be justified only by its solution reliability and computational efficiency. This realization has led to the development of numerical procedures based on the flow formulation. Initial applications of the rigid-plastic finite element method to metal-forming processes were mainly in the analysis of compression and other simple processes. Since those early days, many developments of the numerical techniques have occurred, as well as the continuous growth in the field of applications. Although advances have been made in recent years, the application of solid formulation to metal forming problems is limited. On the other hand, flow formulation has found applications for a wide variety of important forming problems.

A most important improvement was the inclusion of the effects of strain-rate and temperature in material properties and of thermal coupling in the solution. This development has extended the finite element analysis into the warm and hot working range. A further important step in the development of analysis procedures was the development of a user-oriented general-purpose program. The natural course of development of the technique has been in the analysis of two-dimensional and axially symmetric problems, and most recent developments emphasize the application of the finite element method to three-dimensional problems. A further development is the unique application of the finite element method to perform design in metal forming.

Since the 1970's much literature has been published on forming analysis by using the finite element method (Armen and Jones 1978, Pittman *et al.* 1984, Kobayashi *et al.* 1989, Hartley *et al.* 1992). The classical numerical solution for plastic problems of metal forming was originally proposed by Goon *et al.* (1968) and the first application to the flow of hot metals was made by Cornfield and Johnson (1973). Almost, simultaneously Lee and Kobayashi (1973) and Zienkiewicz and Godbole (1974) generalized the procedures to the solution of plastic and visco-plastic problems. The first solution of the flow problem, in which the temperature variation of properties is taken into account, was derived by Oden *et al.* (1973) and Zienkiewicz *et al.* (1978). It has been assumed that large progressive deformation takes place under plastic conditions and that the elastic strain was considered negligible. A further important step in the development of flow procedures was their extension to include iteratively elastic strain effects (Dawson and Thompson 1978).

There are basically two different approaches to the finite element analysis of forming processes, the 'flow' and 'solid' approaches. The 'flow' approach, which takes the material as non-Newtonian fluid and neglects elasticity effects, was first adopted by Lee and Kobayashi (1973), Shah *et al.* (1974), Zienkiewicz and Godbole (1974) and Zienkiewicz *et al.* (1978). With this approach, one can achieve quite large increments in strain and rotation and hence obtain effective solutions to many forming problems. The 'solid' approach which treats the material as an elasto-plastic solid is based on the 'updated Lagrangian' method first defined by McMeeking and Rice (1975). Early research on forming analyses by using this method were described by Lee *et al.* (1977), Nagtegaal and Jong (1981) and Chandra and Mukherjee (1984). In this approach, the boundary value problem is stated such that, in addition to the current states of the body, the internal distribution of the stress is also supposed to be known and the boundary conditions are prescribed in terms of velocity and traction-rate. Distributions of velocity and stress-rate (or displacement and stress-increment) are the solutions to the problem. With the solid approach the constitutive equations can be handled precisely. However, owing to its complicated equations and the strict limitations on the time increment, the method is rather expensive and experiences convergence difficulties.

1.4. Powder compaction processes

The powder compaction process can be described as the forming of any type of powder material by compaction in a container to a desired component shape. This is followed by heating to a temperature which is usually below the melting point to perform sintering of the *green compact.* As the feed stock for this process is in the form of fine particles, or powder, the parts produced in this way are then usually assigned to the *powder metallurgy* process.

Historically, the artifact of production, processing and characterization of metallic powder dates back to 4500-2500 BC in India. Manifestly, the monument of the Delhi iron pillar is a cornerstone in the hallmark of powder technology. The Delhi pillar is one of the earliest applications of iron powder forging techniques. The most significant aspect of the pillar is that there is no sign of any corrosion, in spite of the fact that it has been exposed to the atmosphere for about 1600 years. Modern advancement in the use of powder engineered parts comes mainly from the second world war period. The efforts were mainly restricted to manufacturing cylindrical bushes and plain bearings, using the conventional route of compaction known as die compaction for automotive applications. The more advanced powder processes, called hot isostatic pressing, was originated in 1950 at Battelle's Columbus laboratories (US) and developed as a means of cladding nuclear fuel elements by diffusion bonding.

Over the years the compaction forming process has been the main route of manufacturing engineering parts and used as a competitive method of producing self lubricating bearings and cutting tools. Since this early development, a wide range of structural components, largely automotive, have been produced using modern manufacturing methods. These components have been made form iron powder for structural parts, bronze powder for bearings and filters, carbon powder for electrical products and ceramic powders for wear and corrosion resistant applications. Recently, innovative applications of compaction have been used in the field of food processing.

In contrast to other metal forming techniques, P/M parts are shaped directly from powders while castings are formed from metal that must be melted, and wrought parts are shaped by deformation of hot or cold metal, or by machining. The powder metallurgy industry is a little known industry on the leading edge of a new technology. There are several advantages associated with the powder metallurgy manufacturing route which can be itemized as follows;

- A cost-effective method of forming precision metal components that allows for more efficient design of consumer and industrial products of any metal or alloy.

- The source of new high performance materials and alloys, such as super alloys, tool steels, P/M wrought aluminum alloys, dispersion strengthened metals, and rapid solidification rate materials and metal matrix composites.
- On the leading edge of new manufacturing processes for improved product quality and productivity.
- Saves valuable raw materials through recycling and elimination of costly secondary machining through net and near-net shape design.
- Improves industrial productivity by eliminating manufacturing steps and provides automation, precision and special properties such as self-lubrication and controlled filtration.
- Only way of forming vital metals into products, such as tungsten carbide, dispersion-strengthened materials and self-lubricating bearings.
- Allows a more efficient design of complex parts for products, such as: automobile engines, ride-on lawn movers, business copying machines, power tools and oil/gas well drilling equipment.

Despite these various merits of the P/M production route there are many difficulties which exist in the compaction forming process of powders. These include a compacted part which will have a non-homogenous density distribution. The non-homogeneity in density has wide ranging effects on the final performance of the P/M part.

1.5. Powder metallurgy technology

Powder metallurgy considers the methods of producing commercial products from metallic powders by pressure. Powder compaction forming is a material processing technique utilized to achieve a coherent near-to-net shape industrial component with or without the application of heat. It covers a diversity of processing which includes metal stream atomization, spray forming, compaction, injection moulding and sintering. By taking the advantage of near-to-net shape technology, powder consolidation is traditionally achieved via the cold die compaction and hot isostatic pressing processes. The application of heat during the process is known as sintering and results in bonding the particles together. The general sequence of operations involved in the powder metallurgy process for producing conventional density parts are based on three main operations; mixing, forming and sintering.

Mixing: In the first stage, the element or pre-alloyed metal powders are mixed with lubricants or other alloy additions to produce a homogeneous blend of ingredients. The stock powder is generated by different methods, such as: atomization is the case of metals, or spray drying in the case of ceramics. The process starts with material preparation by mixing the powders, lubricants and binders in a certain proportion. The main structural powders may include more than one type to provide appropriate finished part characteristic. For example, copper is usually used with graphite to improve electrical conductivity and strength of electrical bushes. Lubricant is important to reduce the friction at the die surface. Other additives are optional depending on the P/M parts manufacturer, which can result in reducing sintering time, improvement of the microstructure, and machinability. The initial mixing is performed by the metal powder producer, or the P/M parts manufacturer. The proper mixing of powder is vital to provide a uniform blend of materials for subsequent process stages and is central to achieving final part consistency.

Forming: In the forming stage of the P/M process, the loose powder is pressed into engineered shape in order to achieve the required part with a specific density and density distribution. The forming stage can be established via two different routes, broadly classified as; 'cold' compaction and 'hot' compaction. The cold compaction forming stage can be accomplished using different techniques, Such as: die compaction, cold isostatic pressing, rolling, injection molding and slip casting. Essentially, these various techniques differ from the mode of application of the external force and the compaction is accomplished without the application of heat. The most conventional and widely used is uniaxial die compaction using mechanical or hydraulic press systems. This can be achieved by single or multi-level acting punches depending on part complexity. In the case of a single acting system, the top punch closes the die and compacts the powder to the required shape, followed by punch release. For multi-level components, multiple punches are used which act independently, as shown in Figure 1.2. Alternatively, the hot compaction mode of forming P/M parts can be classified into hot isostatic pressing, hot extrusion, hot die compaction and spraying, among others. In the hot compaction forming process both the external force of pressure and the heat are applied simultaneously in the compacting process.

Sintering: In a typical sintering step, the green compact, placed on a wide mesh endless belt, slowly moves through a controlled-atmosphere furnace. The parts are heated to below the melting point of the base metal, held at the sintering temperature, and cooled. Basically, sintering develops the metallurgical bonds between the powder particles, providing the P/M part's primary functional properties. During sintering, shrinkage of the green compact normally occurs due to metallurgical fusion. P/M parts are generally ready for use after sintering. However, to provide special properties, the parts can be repressed, impregnated, machined, tumbled, plated, or heat treated.

In the **cold die compaction** forming process a controlled amount of mixed powder is automatically gravity-fed into a precision die and is compacted, usually at room temperature, at pressures as low as 138 N/mm^2 or as high as 827 N/mm^2 depending on the density requirements of the part. Higher pressures limit tool life, especially with thin or fragile tool components. Compacting the loose powder produces a coherent piece called 'green' compact. The green compact has a shape of the finished part, when ejected from the die and also has sufficient strength for in-process handling and transport to a sintering furnace. The resulting green compact has very little cohesion and for any further applications the green compact has to be sintered. Owing to low applied pressures and the fact that powder slides relative to the die wall, the density distribution in the green compact is non-homogenous. The typical density attainment during die compaction is limited to 60 to 70 percent of the theoretical density of the fully dense metal.

In the **hot isostatic pressing** (HIP) processes the compaction is typically performed in a gaseous (usually inert argon or helium) atmosphere contained within the pressure vessel. The HIP process consists of the powder to be compacted is canned by a steel container of the desired shape which is subsequently vacuumed. The canned powder mass is compacted by the simultaneous application of hydrostatic pressure and heat. Common pressure levels which extend up to 103 N/mm^2 are combined with temperatures up to 1230 ^{o}C. Processing volumes currently attain sizes up to 1.2 m diameter by 2.7 m long. It is important to note that the end products from the HIP process do not need any further treatment and are ready for use. Owing to high operating pressure and temperature values, and lack of die-tool setup during HIP, the end part results in a more uniform density distribution. In the HIP process the achieved density of the end part is equal to the theoretical density of the metal.

1.6. Mechanical behavior of powder

The compaction of powders has been described qualitatively for a long time and it is well established that the process shows three complex densification mechanisms. Firstly, at low pressure, particle sliding occurs leading to particle re-arrangement. The second stage involves both elastic and plastic deformation of the particles via their contact areas leading to geometric hardening (i.e. plastic deformation and void closure). Lastly, at very high pressure, the flow resistance of the material increases rapidly due to material strain hardening. The research in powder forming is focused on two main manufacturing operations: compaction of the material and the process of sintering. The success of these processes is dependent on achieving a homogeneous density distribution within the compact and also obtaining a finished part of the desired shape. However, in reality, the realization of these aims is difficult to meet. It is with respect to these aspects and to understand the mechanisms of densification, that the predictive capabilities of the finite element method can offer a valuable contribution.

Typically, the main raw material in the compaction forming process is a mass consisting of finely grained powder. The densification, or compaction of a powder body is dependent upon a number of powder characteristics. On the one hand are the structural features e.g. hardness, response to plastic deformation and surface characteristics, which feature respectively during deformation, work hardening and adhesion of the particles. In contrast are the geometrical features e.g. particle size, shape and distribution, and the effect of lubricant additions that govern inter-particle movement and interlocking during pressing. The main stages of densification during cold compaction can be itemized as: particle rearrangement, plastic deformation and fragmentation. That is, the initial transition with pressurization is from a loose array of particles to a closer packing, subsequently, the point contacts deform as the pressure increases leading to plastic deformation, finally to fragmentation of the powder body.

One of the main difficulties which exists in the compaction forming process for powders includes a non-homogenous density distribution which has wide ranging effects on the final performance of the compacted part. The variation of density results in cracks and also in localized deformation in the compact, producing regions of high density surrounded by lower density material, leading to compact failure. The lack of homogeneity is primarily caused by friction, due to inter-particle movement, as well as relative slip between powder particles and the die wall. The die geometry and the sequence of movement result in a lack of homogeneity of density distribution in a compact. In summary, the success of compaction forming depends on the ability of the process in imparting a uniform density distribution in the engineered part. This is largely dependent on process variables, such as: the tool and the part geometry, friction between the powder and the tooling as well as friction between the particles themselves, besides the response of powder to external pressure.

It becomes clear from the above description that the analysis of powder compaction is important because the compact green density influences the further densification of the samples during sintering and hence the performance of the final part. These variations in density of the green compact will cause shape distortion after sintering along with regions of elevated stress and stress concentration. Thus, the control of compact density is very important for the manufacture of reliable products, particularly where they are subjected to a significant level of stress in their duty cycle. In order to perform such analysis, the complex mechanisms of compaction process must be drawn into a mathematical formulation with the knowledge of material and frictional behavior. This task forms a theoretical basis for powder compaction modeling which is practically impossible to solve analytically for realistic engineering components. Once the mathematical model has been incorporated into a computer program, the next requirements are the geometry and material parameters which must be determined accurately from experimental tests.

Clearly, the influence of various variables in powder compaction process can be obtained by numerical simulation on the computer rather than actually building the prototype of the system, for economical reasons.

It is in this respect that the predicting capabilities of the finite element method can make a valuable contribution to the design of the compaction forming processes. In fact, the expensive and time consuming steps of manufacturing pressing tools, trial pressing, sintering, and evaluation of the obtained compact could be replaced, or at least minimized, by having a computational model integrated in the process development stage. Such a procedure directly contributes to the economics of the P/M manufacturing route.

1.7. Numerical simulation of powder forming processes

The numerical simulation of the compaction process is central to an understanding of the mechanics of powder behavior and when it is coupled with experimental inputs the simulation can be considered as an alternative tool to achieve a more economic enterprise. The computational models developed for the modeling of compaction forming processes can be basically classified into two approaches, i.e. 'micro-mechanical' and 'macro-mechanical' approaches. The micro-mechanical approach considers the discrete nature of powder particles. Such a framework encompasses the local behavior between the particles, such as particle contact, sliding, crushing, rolling, segregation *etc* (Turner 1994, Haggblad and Li 1995 and Ransing *et al.* 1998). On the other hand the macro-mechanical, or continuum approach, characterizes the overall behavior of the powder mass by idealizing the powder mass as an equivalent continuum material or a solid continuum due to the characteristic of dilatancy (Chenot *et al.* 1990, Haggblad and Oldenburg 1994 and Lewis and Khoei 1998).

The **micro-mechanical** model focuses on inter-particle behavior where the state of the material is described by many factors, which include, the average number of contacts, the volume fraction of particles, the orientation and distribution of contacts, the contact area and the length of the links connecting the particle centroids. However, the discrete nature of the powder gives rise to difficulty in applying this model as well as the fact that a representative part will comprise many millions of particles. Thus, simplifications are necessary in any micro-mechanical study. Most researchers assume that the powder comprises spherical single size particles arranged in a regular manner (Figure 1.3). However, for application to the simulation of realistic problems, this model is impractical since it is not capable of representing the global behavior. Furthermore, the unit cell model is limited to the representation of powder during crushing only, rather than the complete process of restacking, rearrangement and crushing.

The **macro-mechanical** model has become well established in simulating many other forming processes and has also been adopted in the studies of powder compaction modeling. This type of method is believed to be the most reliable to simulate the powder compaction process since it has the capability of providing useful information on the macroscopic behavior of the powder assembly, such as density distribution, stress state and the compact shape during and after the process. A knowledge of the macroscopic behavior is of primary importance because the current technique of powder forming particularly die compaction produces heterogeneous densification which can cause failure or fracture in subsequent processes. As mentioned before, the macro-mechanical models involve the mathematical theories of plasticity which can be divided into two approaches; a 'flow approach' which considers a succession of infinitesimal increments of distortion, where the instantaneous stress is related to the increment of strain rate and a 'solid approach' which makes use of a total or updated description of motion.

Both the micro-mechanical and macro-mechanical approaches have advantages and disadvantages in the modeling of powder compaction. Nonetheless, from an industrial viewpoint the macro-mechanical approach has a definitive edge over the micro-mechanical approach in that the gross behavior of the powder mass can be modeled and simulated on an industrial scale. It is believed that the micro-mechanical studies still have an important role in providing the intricate micro-mechanical characteristics of powder behavior. However, for application to the simulation of more real problems, this model is impractical since it is not capable of representing the global behavior. Furthermore, the unit cell model is limited to the representation of powder in crushing rather than the complete process of restacking, rearrangement and crushing. Thus, a finite element model based on the macro-mechanical approach is adopted here. In the numerical simulation of the mechanical behavior of metal powder forming, the focus of attention has been mainly on two points, the choice of suitable constitutive models and the frictional algorithm for describing the interaction at the powder-die interface.

A number of numerical solutions have been developed by many authors for simulating the compaction process of metal powders. Much of the relevant literature is devoted to the development of empirical material models which can be used to predict the density distribution of very simple shaped compacts. Morimoto *et al.* (1982) applied the variable elasticity model in which the solid continuum variables, such as Young's modulus and Poisson's ratio, are expressed in terms of the compact density and empirical parameters. Plasticity theories and the corresponding yield criteria are also developed mostly for sintered metal powders in forging problems. An elasto-plastic model by Duffy (1988) and a visco-plastic model by Chenot *et al.* (1990) have been proposed, which are based on a modified von-Mises yield criterion for forging metal powder. A mixed flow formulation was applied to powder compaction problems by Jinka *et al.* (1991) and Lewis *et al.* (1993).

In the process of compaction, the behavior of the powders is assumed to be that of a rate-independent elasto-plastic material. If elastic strains are neglected, the powder may be treated as a fluid and a flow formulation can be applied (Jinka *et al.* 1991). In this case, the powder is treated as a viscous fluid of the non-Newtonion kind. This type of treatment is usually for rate dependent forming process where velocity and pressure are the main variables. However, this formulation experiences difficulty when it comes to dealing with the compressible and hardening behavior of the powder. For this reason, an elasto-plastic formulation for powder compaction material, which takes both the elastic and plastic deformation into account, has been developed by Brekelmans *et al.* (1991), Tran *et al.* (1993) and Gethin *et al.* (1995). An implicit, quasistatic finite element formulation by Brown and Weber (1988) and an explicit time integration method by Haggblad and Oldenburg (1994) have been performed in the modeling and simulation of metal powder die pressing. A constitutive model, based on large strain plasticity, has also been developed by Oliver *et al.* (1996) for simulation of powder compaction processes.

The compaction forming of metal powder is a process involving large deformations, large strain, nonlinear material behavior and friction. Consequently, the numerical analysis of such a highly nonlinear process is a formidable computational problem. For the successful modeling of the powder compaction process several requirements have to be met (Khoei and Lewis 1998, 1999):

- As the compaction process involves a very large reduction in volume, the formulation adopted must be capable of representing this physical process. Thus, the process can be described by a large displacement finite element method based on a 'total' and 'updated' Lagrangian formulation.
- Since powder is a frictional and compressible material, the densification is dependent on the compaction stage and the material model needs to reflect the frictional and yielding characteristics of the powder. A cap plasticity model which reflects the stress state and degree of densification can be applied to describe the constitutive model of powder materials.
- During compaction, powders exhibit strain or work hardening, the volume reduces and the material becomes harder. In this case, the yield function is considered to be a moving surface which is

dependent on the deviatoric and hydrostatic stress state. A hardening rule is used to define the dependence of the yield surface on the degree of plastic straining.

- The involvement of two different materials, which have contact and relative movement in relation to each other must also be considered. Therefore, the interfacial behavior between the die and powder is modeled by a Coulomb friction law and a plasticity theory of friction, in the context of an 'interface' element formulation, is employed.
- The phenomenon of spring-back occurs during unloading. The material properties and yield surface function must consider this phenomenon into account since it may affect the redistribution of density and stresses.

1.7.1. The mathematical model

As the forming process in powder compaction problems is a large deformation process, an elasto plastic analysis often utilizes a large strain model. However, in many forming processes where displacement are large, this is not sufficient and must be coupled with an updating of the calculation region mapped by the elements. The Lagrangian description can be applied to describe the material behavior with respect to either the original domain in the case of a 'total' Lagrangian frame or the domain at the previous calculation step in the case of an 'updated' Lagrangian scheme. In the total Lagrangian description, the element geometry changes are determined by the deformation of the material itself. In other words, we look at the position of a particle and compare this with its initial position. If the initial coordinate is updated every time, following the motion of material, the description becomes updated Lagrangian which always refers the position of a particle to its previous position in the updated coordinate.

An advantage of the Lagrangian description is that the deformation history dependent variables (strain hardening) can be easily taken into account and the changing shape of the formed product can be followed. However, one of the significant limitations of this approach is the progressive distortion which can sometimes lead to entanglement of the finite element mesh. The distortion can cause large approximation errors, or make the Jacobian determinant negative, and make further analysis impossible. In order to solve this problem, an adaptive FEM strategy is applied into the large displacement finite element formulation in Chapter 6 by employing an error estimator and applying an adaptive mesh refinement procedure, which will not only control the discretization error, but automate the simulation process. This consists in defining a new appropriate finite element mesh within the updated, deformed geometry and interpolating (mapping) the pertinent variables from one mesh to another in order to continue the simulation.

In order to control the spatial discretization error, an error estimation for the finite element solution and a consequent adaptive analysis are necessary. For this purpose, a-posteriori error estimator developed by Zienkiewicz and Zhu (1991) using an *h*-refinement adaptive procedure have been employed in the analysis of powder forming processes. For an error estimation, the L_2 norm of stress by a recovery procedure in which the improved value of the finite element solution is the base of an adaptive procedure, is proposed. This adaptive procedure is applied as follows; an error and mesh control component, remeshing of the deformed body, mapping of history dependent variables from the old to the new mesh and a restart of the simulation.

The adaptive mesh refinement is computationally expensive and information must be interpolated from the old mesh to new one. For this purpose, the numerical modeling of powder forming process is simulated by a large strain plasticity in Chapter 9. The hypoelasto-plastic formulation is presented in the context of finite deformation problem involving large strain. The constitutive model is stated in the unrotated frame of reference, in which it is simple to achieve incremental objectivity and also in the unrotated reference frame all constitutive models are cast regardless of finite rotations. While the explicit treatment of both the integration of constitutive equations and the solution of the equilibrium equation is a common approach in the highly nonlinear behavior of powder compaction simulation, an implicit scheme is employed to present the efficiently and accurately the method for this type of nonlinear problems.

1.7.2. The constitutive model

Many studies of powder compaction have been made, most of them concerned with the relationship between compacting pressure and density. Basic equations, such as yield criteria were proposed by means of which it is possible to predict the response of pressed powders. Densification studies have provided an understanding of the basic mechanisms of the compaction process and the nature of green strength. Over the last 70 years, empirical models have been extracted from experimental results expressing pressure-density relationships for different powder types and mixtures. This has lead to theoretical compaction equations for powder materials which describes the relationship between the compact green density and the applied external pressure. The limitation of this category of model is that it cannot predict the local density, stress state and shape of compacts with any accuracy. In addition, there is no general equation which is suitable for every condition, particularly where the load applied starts at a very low level and increases to a high compacting pressure. However, this type of investigation is still necessary to understand the micro-level densification and more importantly to compare compaction performance of various powders.

In order to adequately reproduce powder material behavior under cyclic loading, as a sequence of loading/unloading/reloading, modifications have to be introduced to the classical constitutive models. The cap plasticity model, based on a combination of a convex yield surface consisting of a failure envelope and a hardening elliptical cap is developed for the nonlinear behavior of powder materials in the concept of a generalized plasticity formulation for the description of cyclic loading. This model reflects the yielding, frictional and densification characteristics of powder along with strain and geometrical hardening which occur during the compaction process. The model comprises two surfaces, one to reflect shear failure and the second to capture densification. As double-surface plasticity consists of two different yield functions, special treatment must be made to avoid numerical difficulties in the intersection of two yield surfaces. For this purpose, a three-invariant single cap plasticity with an isotropic hardening rule is developed, which can be compared with some common double-surface plasticity models proposed for powders.

In the analysis of powder forming problems, the classical plasticity models suffer from a serious deficiency when the stress-point reaches a yield surface. In the flow theory of plasticity, the transition from an elastic state to an elasto-plastic state appears more or less abruptly. For powder material it is very difficult to define the location of yield surface, because there is no distinct transition from elastic to elastic-plastic behavior. Thus, the classical plasticity theory cannot deal with such materials and an advanced constitutive theory is necessary. For this purpose, a new approach is developed based on an endochronic density-dependent plasticity model for describing the isothermal deformation behavior of metal powder at low homologous temperature. Constitutive equations are established based on coupling between deviatoric and hydrostatic behavior to the endochronic theory. Although the concept of yield surface has not been explicitly assumed in endochronic theory, it has been shown that the cone-cap plasticity and elliptical density-dependent yield surfaces, which are two most commonly used plasticity models for describing the behavior of powder material can be easily derived as special cases of the proposed endochronic theory.

1.7.3. The frictional model

The proper characterization of friction is central to the success of simulation procedures, especially in the compaction forming process. This is due to the recognized behavior that density variations in components made via the compaction route are mainly, attributed to friction. The friction phenomenon in compaction forming emanates mainly from two sources; inter-particle friction and powder-die wall friction. The inter-particle friction, which is the relative sliding between the powder particles and helps in achieving homogeneity during the pressing operation, has been described in previous section. The powder-die wall friction is the gross sliding between the powder mass and the tooling and occurs at the interface between the powder and the die wall. Modeling of this phenomenon by a standard finite element technique would lead to numerical ill-conditioning due to the high aspect ratios present. In this case, the finite element approach adopted is characterized by the use of 'interface' elements in which a plasticity theory of friction is incorporated to simulate sliding resistance at the powder-tool interface.

The plasticity theory of friction can be achieved by an analogy between plastic and frictional phenomena. In order to formulate such a theory of friction several requirements have to be considered. These requirements, which are similar to the requirements which have to be considered in the theory of elasto-plasticity, are stick (or adhesion) law, stick-slip law, wear and tear rule, slip criterion and slip rule. The constitutive modeling of the frictional behavior of the metal powder is modeled by Coulomb's friction law and the plasticity theory of friction in the context of an interface element formulation.

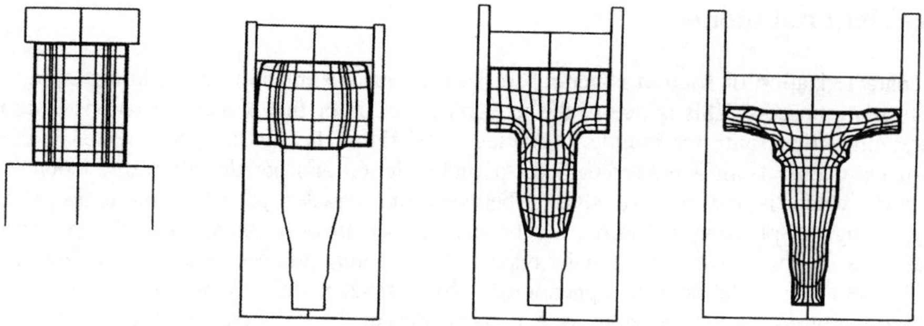

Forging (Rodic and Owen 1989)

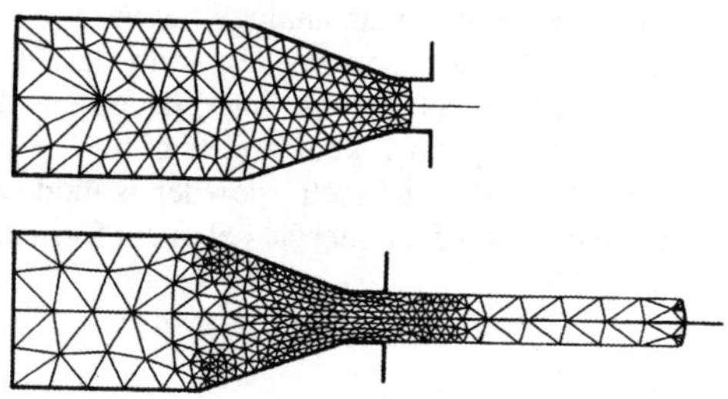

Extrusion (Zienkiewicz *et al.* 1990)

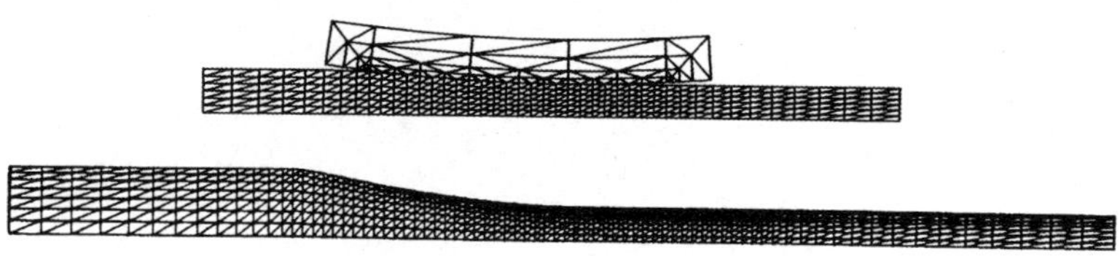

Rolling (Thompson and Berman 1984)

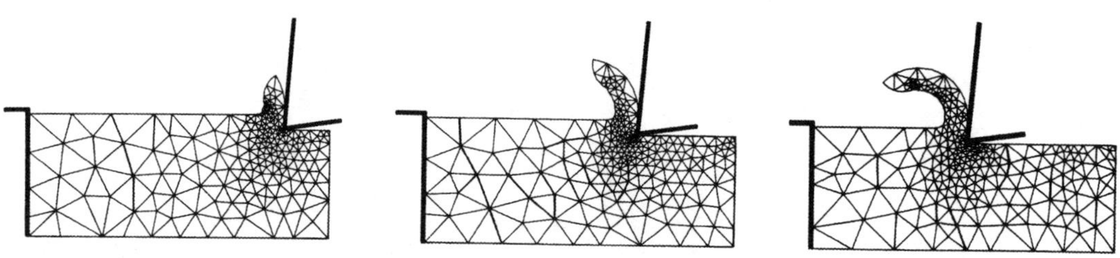

Cutting (Sekhon and Chenot 1993)

Figure 1.1. Some typical forming processes

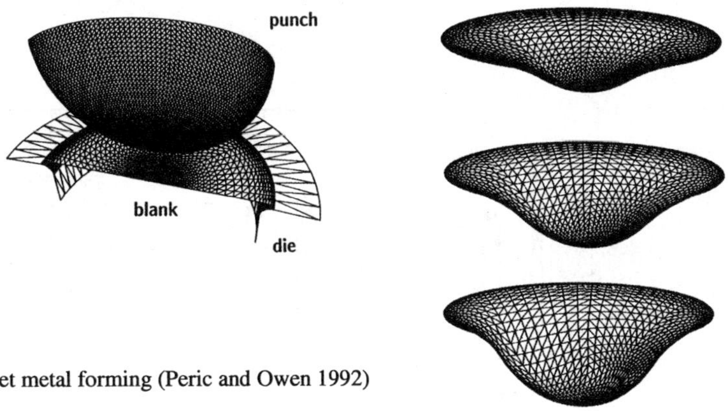

Sheet metal forming (Peric and Owen 1992)

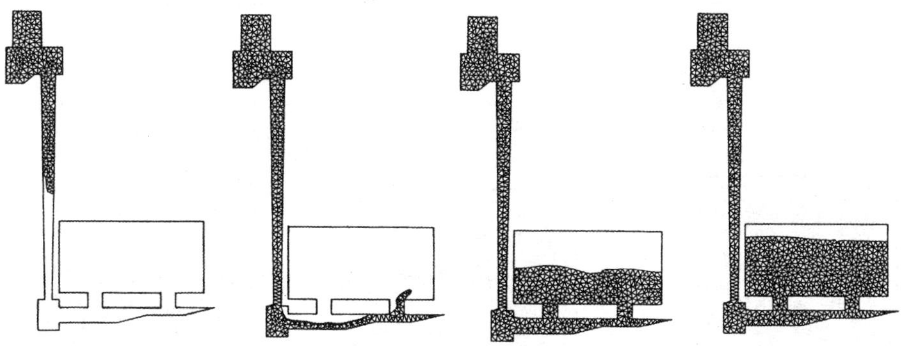

Casting and moulding (Navti 1996)

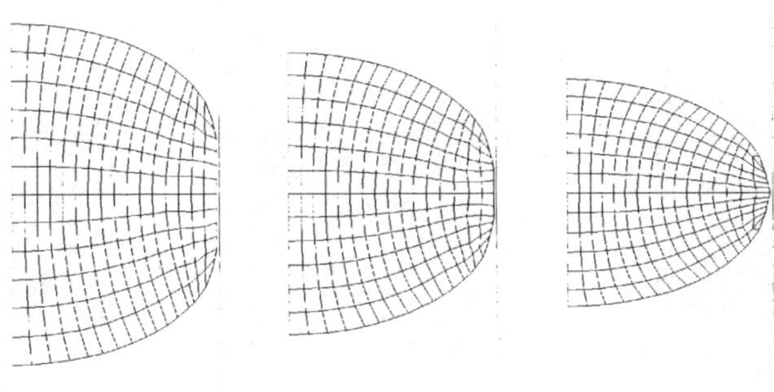

Powder compaction (Khoei and Lewis 1998)

Figure 1.1. Some typical forming processes (continued)

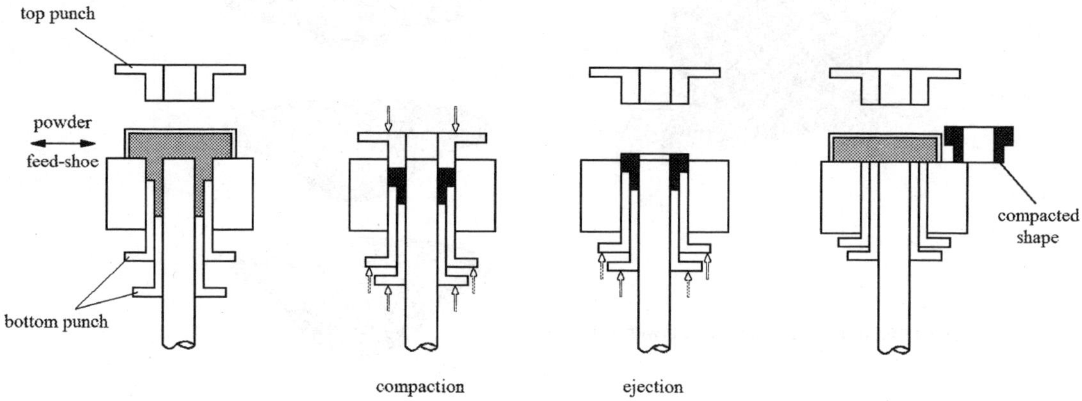

Figure 1.2. The multiple punches for producing multi-level shape component

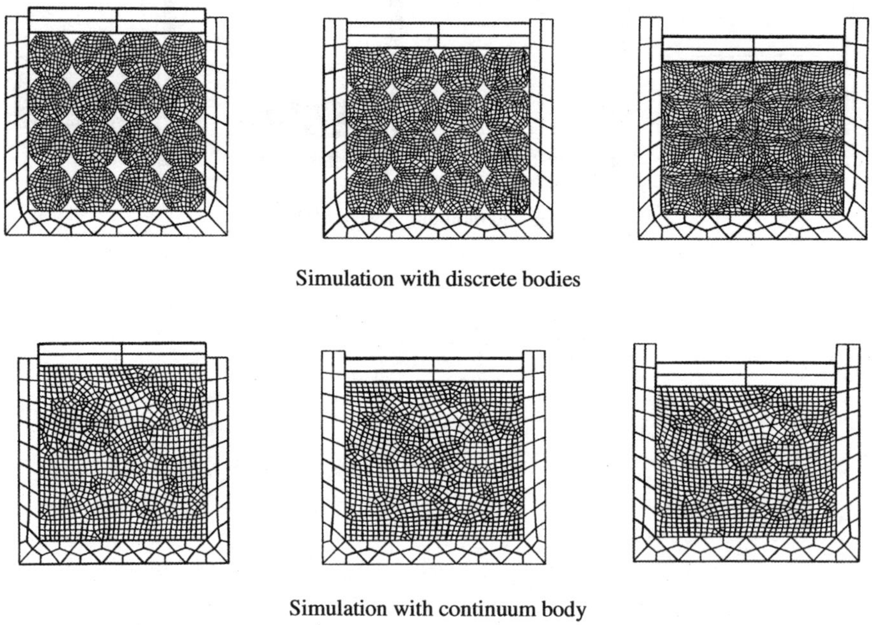

Figure 1.3. A comparison of the macro- and micro-mechanical models in a simple die container (Ransing *et al.* 1998)

2 FINITE ELEMENT MODEL

Metal powder forming processes often cause nonlinear stress behavior in materials and also geometrically nonlinear elastic behavior. Therefore, both material and geometric nonlinear effects should be considered. In the present chapter, the treatment of geometric nonlinear behavior will be discussed via a large finite element deformation based on Lagrangian and arbitrary Lagrangian-Eulerian techniques, and the nonlinear behavior of material by a similar path will be followed in the next chapter.

The numerical simulation of the compaction process is central to an understanding of the mechanics of powder behavior. Since powder compaction forming is a process involving large deformations, large strain, nonlinear material behavior and friction, the numerical analysis of such a highly nonlinear process is a formidable computational problem. In this chapter, the large deformation analysis of metal powder during the cold compaction process is simulated by the finite element method based on a 'total' and 'updated' Lagrangian formulation. As the compaction process involves a very large reduction in volume, the formulation adopted must be capable of representing this physical process. The process is therefore described by a large displacement finite element formulation for the spatial discretization. A generalized Newmark scheme is used for the time domain discretization and then the final nonlinear equations are solved by a Newton-Raphson procedure. Despite the extensive application of the Lagrangian formulation in large deformation problems, it leads to highly distorted and usually useless meshes, which can be particularly observed in higher order elements. Because of severe distortion of elements, a combined technique of Eulerian and Lagrangian approaches which captures the advantages of both methods is then developed. The arbitrary Lagrangian-Eulerian (ALE) technique is proposed based on an arbitrary motion of the mesh so that it optimizes the shape of mesh and elements undergo no distortion

2.1. Introduction

An analysis method in the area of metal powder forming application, in many cases, can be justified only by its solution reliability and computational efficiency. Before embarking upon any analytical study of a powder compaction operation it is neccessary to ask what is required from the eventual solution. In general, most forming processes are of such a complex character that an exact solution is unattainable. Therefore what is neccessary for, and what is of value from, an analytical study?

During the last few decades, a number of finite element analyses for modeling metal powder forming processes have appeared in the literature. Broadly speaking, these analyses may be grouped into two main categories, depending on whether the constitutive laws assumed are of an 'elasto-plastic' type (Brekelmans *et al.* 1991, Haggblad and Oldenburg 1994 and Oliver *et al.* 1996) or a 'rigid-plastic' type (Osakada *et al.* 1982 and Mori and Osakada 1987). These analyses are also known as the 'solid' formulation where the material is considered to behave as an elasto-plastic, or elasto-viscoplastic solid, and secondly the 'flow' formulation where the material behaves as a rigid-plastic or rigid-viscoplastic material where elastic strains are neglected.

A rigid-plastic analysis is suitable if the requirement is a reasonable estimate of the load or pressure necessary to carry out a forming operation, or if it is required to enquire how the load is affected by changing certain process parameters. At the same time it is also possible to obtain a good representation of the deformation modes imposed by certain working processes by the finite element method. Lee and Kobayashi (1973) pioneered an iterative method for solving such nonlinear equations and demonstrated its applicability in two in-plane deformation analyses involving hole expansion and flange drawing of planar anisotropic sheets. For the finite element analysis of metal forming processes, Osakada *et al.* (1982) proposed a method based on the equilibrium of nodal forces by assuming the deforming metal to be a slightly compressible rigid-plastic material. Mori and Osakada (1987) developed a finite element method for a rigid-plastic analysis of plastic deformation of sintered powder metals. In this method, the variational principle of a rigid-plastic porous metal was derived on the basis of the plasticity theory of porous metals and also the effect of density change within the material was taken into consideration.

An elasto-plastic analysis may be necessary if the determination of residual stresses is the aim. For this type, the usual assumption of a linear relation between stress rates and strain rates leads naturally to a system of incrementally linear equations amenable to direct solution methods for calculating nodal displacement rates and element stress/strain rates. An integration of these rate quantities yields their total quantities as a function of an appropriate process variable, such as time. Since metal powder forming processes generally involve both material and geometric nonlinearities, the step size required for an accurate integration is necessarily small. This type of material model could also be of value if the onset of internal (or surface) cracks, or the domains of high strain concentration, or the development of zones of intense strain localization in bulk forming processes have to be assessed accurately. For these types of problems it is essential that the method of solution be capable of establishing a current distribution of stress as a function of the deformation history, which, in turn, is governed by such factors as the geometry of the tooling and the prevailing frictional conditions. In cases where the deformation is ultimately a steady-state one the solution should converge to display this feature.

In the finite element analysis of both the elasto-plastic and rigid-plastic types, stepwise calculation for incremental deformation is repeated from the stress-free state to the final state. In each incremental deformation, it is necessary to estimate the physical values at the end of the increment from the values at the beginning. The simplest method for the estimation is to stretch the formulation for infinitesimal deformation by adding the products of the time interval and the rates or velocities at the beginning to the initial values. However, because of the non-linearity arising from the changes in shape and material property during each incremental deformation, a simple extension of infinitesimal deformation will cause error to some degree, and the error accumulates as the number and size of increments increase. In a metal powder forming analysis which takes account of a change of shape and the nonlinear material property, the rigid-plastic finite element analysis based on infinitesimal deformation provides the stress at each stage without referring to the stress in the previous stages, and thus, the accuracy is usually good enough for practical purposes, but this is not always true as demonstrated by Osakada *et al.* (1982). In the elasto-plastic analysis of large deformation problems by the finite element method, a consideration of finite deformation is essential because the error in incremental stress accumulates significantly.

In the above, we have discussed the applications of the elasto-plastic and rigid-plastic methods to the analysis of metal powder forming processes. Now, a question arises as to the selection of the approach - i.e. a 'flow' approach or a 'solid' approach. A significant contribution to the solution of this question was made by Kudo and Matsubara (1979), in which an attempt was made to examine the comparative merits of various numerical methods. The results were compiled for upsetting of circular solid cylinders under specific conditions, and revealed the importance of certain parameters used in computation, such as mesh systems and the size of a displacement increment. It was shown that the solid formulation needed improvement, particularly in terms of predicting the phenomenon of folding. For elasto-plastic materials, the constitutive equations relate strain rate to stress rates, instead of to stresses. Consequently, it is convenient to write the field equation in the boundary value problem for elasto-plastic materials in terms of the equilibrium of stress rates. Furthermore, in the analysis of a powder forming process, the flow formulation experiences difficulty when it comes to dealing with the compressible and hardening behavior of the powder. For this reason, an elasto-plastic formulation for powder compaction material, which takes both the elastic and plastic deformation into account, is recommended. In this study, an elasto-plastic material model is therefore selected and the constitutive model is formulated in terms of stress rates and strain rates.

2.2. Governing equations for powder forming analysis

As the forming process in powder compaction problems is a large deformation process, an elasto plastic analysis often utilises a large strain model. However, in many forming processes where displacement are large, this is not sufficient and must be coupled with an updating of the calculation region mapped by the elements. Basically there are two types viz. 'Lagrangian' and 'Eulerian' descriptions.

Lagrangian (or material) description describes the material behavior with respect to either the original domain in the case of a 'total' Lagrangian frame or the domain at the previous calculation step in the case of an 'updated' Lagrangian scheme (Hibbitt *et al.* 1970 and Zienkiewicz and Taylor 1989). In the total Lagrangian description, the element geometry changes are determined by the deformation of the material itself. In other words, we look at the position of a particle by comparing with its initial position. If the initial coordinate is updated each time step by following the motion of material the description becomes updated Lagrangian which always refers the position of a particle to its previous position in the updated coordinate. However, it has been shown by Brekelmans *et al.* (1991) that for the more complex shaped parts, the elements can become unacceptably distorted during the process.

An **Eulerian (or spatial) description** is therefore used by the authors which describes the material behavior with respect to fixed position in space in which the physical quantities are expressed as a function of time and position vector in the geometrical space. This implies that we look at the position in space and find out which particle is now occupying that position. In this alternative procedure, it is found that the mesh distortion problem can be reduced since the numerical configuration is not influenced by the material displacement. The original element mesh is sufficiently accurate for any changes in physical quantity and material properties to be captured accurately. However, instability may occur where there are two material properties in one element, especially when the calculation becomes highly non-linear.

More recent formulations are based on an **Arbitrary Lagrangian Eulerian (ALE)** formulation, which is a standard approach in large strain solid mechanics to keep mesh distortion and element entanglement under control (Liu *et al.* 1986, Huetink *et al.* 1990 and Huerta and Casadei 1994). The basic idea of the ALE formulation is the use of a referential domain for the description of motion, different from the material domain (Lagrangian description) and the spatial domain (Eulerian description). When compared to fluid dynamics, where the ALE formulation originated, the main difficulty of ALE solid mechanics is the path dependent behavior of plasticity models. The relative motion between the mesh and the material must be accounted for in the treatment of the constitutive equation.

In this chapter, an updated Lagrangian method is employed in which the coordinates of the finite element mesh must be updated at every load Increment and the solution proceeds as a standard finite element calculation. Having done that difficulty may still arise in analyses, which involve large displacements. The element shapes can become highly distorted leading to an unreliable solution due to mesh degeneration. However, the advantage of using an updated Lagrangian coordinate is that it will not introduce convection terms as an Eulerian coordinate would. When updated Lagrangian coordinates are used, the description of the deformation is in terms of displacement increment (or velocity instead of displacement). This simplifies the formulations desribed in Lagrangian coordinates, since choosing a small time increment can reduce the degree of the nonlinearity. In order to avoid the mesh distorsion and improve the solution quality, an error estimation and the adaptive remeshing schemes are applied in Chapter 6.

2.3. Lagrangian formulation

The nonlinearities in metal powder forming analyses arise from two distinct sources; constitutive nonlinearities and geometric nonlinearities, the latter being due to large displacements. The most commonly used constitutive law is the elasto-plastic formulation, and for this type the linearity of the incremental stress-strain law forms the basis of the equations. The most direct application is in the incremental type solution, where the solution is built up as a series of linear increments.

Geometric nonlinearities were first included by means of an incremental geometric stiffness (initial stress matrix) and the earlier results were obtained on the basis of equilibrium at nodes (Turner *et al.* 1960). The derivation of the initial stress stiffness matrix was finally placed on a firm basis by the use of the Lagrangian or Green's strain by Martin (1966). The latter analysis was established by consideration of additional terms, which took the form of an initial displacement matrix in the incremental solution (Mallett and Marcal 1968). This large displacement analysis has often been approached through an updated local coordinate system, which is usually defined for each element in the structure. Such coordinate systems are of use only when the assumption of small strains is made. Moreover, because of the possibility of finite rotation within the element, this is strictly true only for constant strain elements.

In the motion of the body, three successive configurations can be considered (Figure 2.1); the initial (unstressed, undeformed) configuration γ^0, the current configuration γ and the incremented configuration $\bar{\gamma}$. The Lagrangian description requires a known reference configuration to formulate the equilibrium equations; the total Lagrangian description (TLD) uses the initial unstressed configuration of the body γ^0 and the updated Lagrangian description (ULD) uses the current deformed configuration γ (incidentally, it may be mentioned that an Eulerian approach would use $\bar{\gamma}$ as its reference configuration). However, any intermediate configuration may be used as a Lagrangian reference. Cescotto *et al.* (1979) presented a generalized Lagrangian description (GLD) where the equilibrium equations are referred to an arbitrary known deformed configuration γ^R. In the two limit cases where γ^R coincides with γ^0 and γ, the TLD or ULD, respectively, will be recovered by the generalized Lagrangian description (GLD).

Whether the displacements, or strains, are large or small it is imperative that the equilibrium conditions between the internal and external forces have to be satisfied. Thus, the dynamic equilibrium equation of a body in a deformed configuration γ can be written in a standard form as

$$\sigma_{ij,j} + \rho b_i - \rho \ddot{u}_i = 0 \tag{2.1}$$

where u_i is the current (deformed) displacement, b_i is the body force accelaration, ρ is the current density of powder and σ_{ij} is the total (Cauchy) stress. The constitutive law with respect to the incremental stress can be defined as

$$d\sigma_{ij} = D_{ijkl}(d\varepsilon_{kl} - d\varepsilon^0_{kl}) + \sigma_{ik} d\omega_{kj} + \sigma_{jk} d\omega_{ki} \tag{2.2}$$

where $d\varepsilon_{ij}$ and $d\omega_{kl}$ are the incremental values of the strain and rotation, respectively. The last two terms account for the Zaremba-Jaumann rotational stress changes (negligible generally in small displacement computation). In equation (2.2), D_{ijkl} is a tangential matrix, defined by suitable state variables and the direction of the increment, and ε^0_{ij} refers to strains caused by external actions such as temperature changes.

For geometrically nonlinear elastic behavior, we can select either a total or an updated Lagrangian coordinate system. If the initial undeformed position of a particle of material is $\mathbf{x_0}$ and the total displacement vector is $\mathbf{u}$ then the coordinates of the particle are $\mathbf{x} = \mathbf{x_0} + \mathbf{u}$. A general definition of strains, which is valid whether the displacements or strains are large or small, was introduces by Green and St. Venant. The matrix of Green's strains in a total Lagrangian formulation is given as

$$\boldsymbol{E} = \frac{1}{2}(\boldsymbol{J}_D^T \boldsymbol{J}_D - \boldsymbol{I}) \tag{2.3}$$

where $\boldsymbol{J}_D$ is the deformation Jacobian matrix. Based on Green's strain tensor, the non-linear strain displacement relationship can be defined in terms of the infinitesimal and large displacement components as

$$\boldsymbol{\varepsilon} = \boldsymbol{\varepsilon}_L + \boldsymbol{\varepsilon}_{NL} \tag{2.4}$$

where $\boldsymbol{\varepsilon}_L$ and $\boldsymbol{\varepsilon}_{NL}$ are the linear and nonlinear strains. For a set of virtual displacements, the corresponding virtual Green's strains are given as

$$d\boldsymbol{\varepsilon} = d\boldsymbol{\varepsilon}_L + d\boldsymbol{\varepsilon}_{NL} \tag{2.5}$$

where the linear strain is defined as

$$\boldsymbol{\varepsilon}_L = \{\varepsilon_x,\ \varepsilon_y,\ \gamma_{xy}\}^T = \left\{\frac{\partial u}{\partial x},\ \frac{\partial v}{\partial y},\ \left(\frac{\partial u}{\partial y} + \frac{\partial v}{\partial x}\right)\right\}^T \quad \text{for plane stress/strain problems} \tag{2.6a}$$

$$\boldsymbol{\varepsilon}_L = \{\varepsilon_r,\ \varepsilon_z,\ \gamma_{rz},\ \varepsilon_\theta\}^T = \left\{\frac{\partial u}{\partial r},\ \frac{\partial w}{\partial z},\ \left(\frac{\partial u}{\partial z} + \frac{\partial w}{\partial r}\right),\ \frac{u}{r}\right\}^T \quad \text{for axisymmetric problems} \tag{2.6b}$$

$$\boldsymbol{\varepsilon}_L = \{\varepsilon_x,\ \varepsilon_y,\ \varepsilon_z,\ \gamma_{xy},\ \gamma_{yz},\ \gamma_{zx}\}^T = \left\{\frac{\partial u}{\partial x},\ \frac{\partial v}{\partial y},\ \frac{\partial w}{\partial z},\ \left(\frac{\partial v}{\partial x} + \frac{\partial u}{\partial y}\right),\ \left(\frac{\partial w}{\partial y} + \frac{\partial v}{\partial z}\right),\ \left(\frac{\partial u}{\partial z} + \frac{\partial w}{\partial x}\right)\right\}^T \quad \text{for 3D problems} \tag{2.6c}$$

and the nonlinear terms of equations (2.4) is expressed as

$$\boldsymbol{\varepsilon}_{NL}=\begin{Bmatrix} \frac{1}{2}\left(\frac{\partial u}{\partial x}\right)^2+\frac{1}{2}\left(\frac{\partial v}{\partial x}\right)^2 \\ \frac{1}{2}\left(\frac{\partial u}{\partial y}\right)^2+\frac{1}{2}\left(\frac{\partial v}{\partial y}\right)^2 \\ \frac{\partial u}{\partial x}\frac{\partial u}{\partial y}+\frac{\partial v}{\partial x}\frac{\partial v}{\partial y} \end{Bmatrix} \quad \text{for plane stress/strain problems} \tag{2.7a}$$

$$\boldsymbol{\varepsilon}_{NL}=\begin{Bmatrix} \frac{1}{2}\left(\frac{\partial u}{\partial r}\right)^2+\frac{1}{2}\left(\frac{\partial w}{\partial r}\right)^2 \\ \frac{1}{2}\left(\frac{\partial u}{\partial z}\right)^2+\frac{1}{2}\left(\frac{\partial w}{\partial z}\right)^2 \\ \frac{\partial u}{\partial r}\frac{\partial u}{\partial z}+\frac{\partial w}{\partial r}\frac{\partial w}{\partial z} \\ \frac{1}{2}\left(\frac{u}{r}\right)^2 \end{Bmatrix} \quad \text{for axisymmetric problems} \tag{2.7b}$$

$$\boldsymbol{\varepsilon}_{NL}=\begin{Bmatrix} \frac{1}{2}\left(\frac{\partial u}{\partial x}\right)^2+\frac{1}{2}\left(\frac{\partial v}{\partial x}\right)^2+\frac{1}{2}\left(\frac{\partial w}{\partial x}\right)^2 \\ \frac{1}{2}\left(\frac{\partial u}{\partial y}\right)^2+\frac{1}{2}\left(\frac{\partial v}{\partial y}\right)^2+\frac{1}{2}\left(\frac{\partial w}{\partial y}\right)^2 \\ \frac{1}{2}\left(\frac{\partial u}{\partial z}\right)^2+\frac{1}{2}\left(\frac{\partial v}{\partial z}\right)^2+\frac{1}{2}\left(\frac{\partial w}{\partial z}\right)^2 \\ \frac{\partial u}{\partial x}\frac{\partial u}{\partial y}+\frac{\partial v}{\partial x}\frac{\partial v}{\partial y}+\frac{\partial w}{\partial x}\frac{\partial w}{\partial y} \\ \frac{\partial u}{\partial y}\frac{\partial u}{\partial z}+\frac{\partial v}{\partial y}\frac{\partial v}{\partial z}+\frac{\partial w}{\partial y}\frac{\partial w}{\partial z} \\ \frac{\partial u}{\partial z}\frac{\partial u}{\partial x}+\frac{\partial v}{\partial z}\frac{\partial v}{\partial x}+\frac{\partial w}{\partial z}\frac{\partial w}{\partial x} \end{Bmatrix} \quad \text{for 3D problems} \tag{2.7c}$$

In small displacement theory, the general first order linear strain approximation is obtained by neglecting the quadratic terms.

In equation (2.4), the nonlinear terms of strain $\boldsymbol{\varepsilon}_{NL}$ can be defined as $\boldsymbol{\varepsilon}_{NL}=\frac{1}{2}\boldsymbol{A}_\theta\boldsymbol{\theta}$, with $\boldsymbol{\theta}$ denoting the displacement gradient and $\boldsymbol{A}_\theta$ a suitably defined matrix operator which contains displacement derivatives and can be derived from the definition of $\boldsymbol{\varepsilon}_{NL}$. Thus, the nonlinear strain can be conveniently rewritten as

$$\boldsymbol{\varepsilon}_{NL}=\begin{bmatrix}\frac{\partial u}{\partial x} & \frac{\partial v}{\partial x} & 0 & 0\\ 0 & 0 & \frac{\partial u}{\partial y} & \frac{\partial v}{\partial y}\\ \frac{\partial u}{\partial y} & \frac{\partial v}{\partial y} & \frac{\partial u}{\partial x} & \frac{\partial v}{\partial x}\end{bmatrix}\begin{bmatrix}\frac{\partial u}{\partial x} & 0 & \frac{\partial u}{\partial x}\\ \frac{\partial v}{\partial x} & 0 & \frac{\partial v}{\partial x}\\ 0 & \frac{\partial u}{\partial y} & \frac{\partial u}{\partial y}\\ 0 & \frac{\partial v}{\partial y} & \frac{\partial v}{\partial y}\end{bmatrix}=\frac{1}{2}\boldsymbol{A}_{\theta}\boldsymbol{\theta} \tag{2.8}$$

for plane stress and plane strain problems, whereas for axisymmetric problems

$$\boldsymbol{\varepsilon}_{NL}=\begin{bmatrix}\frac{\partial u}{\partial r} & \frac{\partial w}{\partial r} & 0 & 0 & 0\\ 0 & 0 & \frac{\partial u}{\partial z} & \frac{\partial w}{\partial z} & 0\\ \frac{\partial u}{\partial z} & \frac{\partial w}{\partial z} & \frac{\partial u}{\partial r} & \frac{\partial w}{\partial r} & 0\\ 0 & 0 & 0 & 0 & \frac{u}{r}\end{bmatrix}\begin{bmatrix}\frac{\partial u}{\partial r} & 0 & \frac{\partial u}{\partial r} & 0\\ \frac{\partial w}{\partial r} & 0 & \frac{\partial w}{\partial r} & 0\\ 0 & \frac{\partial u}{\partial z} & \frac{\partial u}{\partial z} & 0\\ 0 & \frac{\partial w}{\partial z} & \frac{\partial w}{\partial z} & 0\\ 0 & 0 & 0 & \frac{u}{r}\end{bmatrix}=\frac{1}{2}\boldsymbol{A}_{\theta}\boldsymbol{\theta} \tag{2.9}$$

and for three-dimensional problems

$$\boldsymbol{\varepsilon}_{NL}=\begin{bmatrix}\frac{\partial u}{\partial x} & \frac{\partial v}{\partial x} & \frac{\partial w}{\partial x} & 0 & 0 & 0 & 0 & 0 & 0\\ 0 & 0 & 0 & \frac{\partial u}{\partial y} & \frac{\partial v}{\partial y} & \frac{\partial w}{\partial y} & 0 & 0 & 0\\ 0 & 0 & 0 & 0 & 0 & 0 & \frac{\partial u}{\partial z} & \frac{\partial v}{\partial z} & \frac{\partial w}{\partial z}\\ \frac{\partial u}{\partial y} & \frac{\partial v}{\partial y} & \frac{\partial w}{\partial y} & \frac{\partial u}{\partial x} & \frac{\partial v}{\partial x} & \frac{\partial w}{\partial x} & 0 & 0 & 0\\ 0 & 0 & 0 & \frac{\partial u}{\partial z} & \frac{\partial v}{\partial z} & \frac{\partial w}{\partial z} & \frac{\partial u}{\partial y} & \frac{\partial v}{\partial y} & \frac{\partial w}{\partial y}\\ \frac{\partial u}{\partial z} & \frac{\partial v}{\partial z} & \frac{\partial w}{\partial z} & 0 & 0 & 0 & \frac{\partial u}{\partial x} & \frac{\partial v}{\partial x} & \frac{\partial w}{\partial x}\end{bmatrix}\begin{bmatrix}\frac{\partial u}{\partial x} & 0 & 0 & \frac{\partial u}{\partial x} & 0 & \frac{\partial u}{\partial x}\\ \frac{\partial v}{\partial x} & 0 & 0 & \frac{\partial v}{\partial x} & 0 & \frac{\partial v}{\partial x}\\ \frac{\partial w}{\partial x} & 0 & 0 & \frac{\partial w}{\partial x} & 0 & \frac{\partial w}{\partial x}\\ 0 & \frac{\partial u}{\partial y} & 0 & \frac{\partial u}{\partial y} & \frac{\partial u}{\partial y} & 0\\ 0 & \frac{\partial v}{\partial y} & 0 & \frac{\partial v}{\partial y} & \frac{\partial v}{\partial y} & 0\\ 0 & \frac{\partial w}{\partial y} & 0 & \frac{\partial w}{\partial y} & \frac{\partial w}{\partial y} & 0\\ 0 & 0 & \frac{\partial u}{\partial z} & 0 & \frac{\partial u}{\partial z} & \frac{\partial u}{\partial z}\\ 0 & 0 & \frac{\partial v}{\partial z} & 0 & \frac{\partial v}{\partial z} & \frac{\partial v}{\partial z}\\ 0 & 0 & \frac{\partial w}{\partial z} & 0 & \frac{\partial w}{\partial z} & \frac{\partial w}{\partial z}\end{bmatrix}=\frac{1}{2}\boldsymbol{A}_{\theta}\boldsymbol{\theta} \tag{2.10}$$

The above system, along with the definitions linking stresses to strains and rotations in equation (2.2) and the boundary conditions, can be discretized and solved numerically. The boundary conditions for the problem are prescribed as $u_i = \bar{u}_i$ on Γ_u for the displacements and $t_i = \sigma_{ij} n_{ij} = \bar{t}_i$ on Γ_t for the tractions.

2.4. Finite element discretization

In order to develop a finite element formulation for powder compaction problems, we need to solve equation (2.1) numerically for both spatial and time discretization. Following the standard procedure of the finite element method, the domain Ω is divided into subdomains (elements). If the displacements within an element are prescribed in the usual manner by a finite number of nodal values, we can obtain the necessary equilibrium equations using the virtual work principle. Thus, equation (2.1) can be written in the weak form as

$$\int_\Omega \delta \boldsymbol{\varepsilon}^T \boldsymbol{\sigma}\, d\Omega - \int_\Omega \delta \mathbf{u}^T \rho \mathbf{b}\, d\Omega + \int_\Omega \delta \mathbf{u}^T \rho \ddot{\mathbf{u}}\, d\Omega - \int_{\Gamma_t} \delta\, \mathbf{u}^T \bar{\mathbf{t}}\, d\Gamma = 0 \tag{2.11}$$

Applying the standard finite element Galerkin discretization process to equation (2.11) with the independent approximations of $\mathbf{u}$ defined as $\mathbf{u} = \mathbf{N}_u^T \bar{\mathbf{u}}$, we will arrive at

$$\int_\Omega \bar{\mathbf{B}}^T \boldsymbol{\sigma}\, d\Omega + \mathbf{M}\ddot{\bar{\mathbf{u}}} - \mathbf{f}_u = 0 \tag{2.12}$$

where the mass matrix, $\mathbf{M}$, and the load vector, $\mathbf{f}_u$, are defined as

$$\begin{aligned} \mathbf{M} &= \int_\Omega \mathbf{N}_u^T \rho\, \mathbf{N}_u d\Omega \\ \mathbf{f}_\mathrm{u} &= \int_\Omega \mathbf{N}_u^T \rho\, \mathbf{b}\, d\Omega + \int_{\Gamma_t} \mathbf{N}_u^T \bar{\mathbf{t}}\, d\Gamma \end{aligned} \tag{2.13}$$

and $\bar{\mathbf{B}}$ is the well known strain matrix relating the increments of strain and displacement (i.e. $d\boldsymbol{\varepsilon} = \bar{\mathbf{B}}\, d\bar{\mathbf{u}}$). The bar suffix is added as, if displacements are large, the strains depend non-linearly on displacements and the matrix $\bar{\mathbf{B}}$ is dependent on $\bar{\mathbf{u}}$. Taking the variation from $\boldsymbol{\varepsilon}_{NL} = \frac{1}{2} \boldsymbol{A}_\theta \boldsymbol{\theta}$, we have

$$d\boldsymbol{\varepsilon}_{NL} = \frac{1}{2} d\boldsymbol{A}_\theta \boldsymbol{\theta} + \frac{1}{2} \boldsymbol{A}_\theta\, d\boldsymbol{\theta} = \boldsymbol{A}_\theta\, d\boldsymbol{\theta} \tag{2.14}$$

and we can determine $\boldsymbol{\theta}$ in terms of the shape function $\mathbf{N}_u$ and nodal parameter $\bar{\mathbf{u}}$ as $\boldsymbol{\theta} = \mathbf{G}\bar{\mathbf{u}}$, thus

$$d\boldsymbol{\varepsilon}_{NL} = \boldsymbol{A}_\theta \mathbf{G}\, d\bar{\mathbf{u}} \tag{2.15}$$

where $\mathbf{G}$ contains the Cartesian shape function derivatives defined by

$$[\mathbf{G}_i]^T = \begin{bmatrix} \frac{\partial N_i}{\partial x} & 0 & \frac{\partial N_i}{\partial y} & 0 \\ 0 & \frac{\partial N_i}{\partial x} & 0 & \frac{\partial N_i}{\partial y} \end{bmatrix} \quad \text{for plane stress/strain problems} \tag{2.16a}$$

$$[\mathbf{G}_i]^T = \begin{bmatrix} \frac{\partial N_i}{\partial r} & 0 & \frac{\partial N_i}{\partial z} & 0 & \frac{N_i}{r} \\ 0 & \frac{\partial N_i}{\partial r} & 0 & \frac{\partial N_i}{\partial z} & 0 \end{bmatrix} \quad \text{for axisymmetric problems} \tag{2.16b}$$

$$[\mathbf{G}_i]^T = \begin{bmatrix} \frac{\partial N_i}{\partial x} & 0 & 0 & \frac{\partial N_i}{\partial y} & 0 & 0 & \frac{\partial N_i}{\partial z} & 0 & 0 \\ 0 & \frac{\partial N_i}{\partial x} & 0 & 0 & \frac{\partial N_i}{\partial y} & 0 & 0 & \frac{\partial N_i}{\partial z} & 0 \\ 0 & 0 & \frac{\partial N_i}{\partial x} & 0 & 0 & \frac{\partial N_i}{\partial y} & 0 & 0 & \frac{\partial N_i}{\partial z} \end{bmatrix} \quad \text{for 3D problems} \tag{2.16c}$$

Thus, we can write $\overline{\mathbf{B}}$ in terms of the linear and nonlinear strain-displacement matrix, from equation (2.5), as

$$\overline{\mathbf{B}} = \mathbf{B}_L + \mathbf{B}_{NL}(\overline{\mathbf{u}}) \tag{2.17}$$

where $\mathbf{B}_L$ is the same matrix as in a linear infinitesimal strain analysis and only $\mathbf{B}_{NL}$ depends on the displacement and can be obtained from equation (2.15) as

$$d\overline{\mathbf{B}} = d\mathbf{B}_{NL} \tag{2.18}$$

and from equation (2.14), we will have

$$\mathbf{B}_{NL} = \mathbf{A}_\theta \mathbf{G} \tag{2.19}$$

In order to obtain the tangential stiffness matrix, the finite element Galerkin discretization formulation (2.12) can be rewritten as a nonlinear function of nodal displacement $\overline{\mathbf{u}}$,

$$\boldsymbol{\Psi}(\overline{\mathbf{u}}) = \int_\Omega \overline{\mathbf{B}}^T(\overline{\mathbf{u}})\,\boldsymbol{\sigma}\,d\Omega + \mathbf{M}\ddot{\overline{\mathbf{u}}} - \mathbf{f}_u = 0 \tag{2.20}$$

Taking appropriate variations of equation (2.20) with respect to $d\overline{\mathbf{u}}$, we have

$$d\boldsymbol{\Psi} = \int_\Omega d\overline{\mathbf{B}}^T\boldsymbol{\sigma}\,d\Omega + \int_\Omega \overline{\mathbf{B}}^T d\boldsymbol{\sigma}\,d\Omega + \mathbf{M}\,d\ddot{\overline{\mathbf{u}}} = \overline{\mathbf{K}}_\mathbf{T}\,d\overline{\mathbf{u}} + \mathbf{M}\,d\ddot{\overline{\mathbf{u}}} \tag{2.21}$$

Substituting the constitutive law definition with respect to the incremental stress, i.e. $d\boldsymbol{\sigma} = \mathbf{D}_\mathbf{T}\,d\boldsymbol{\varepsilon}$, and relation (2.18) into equation (2.21), we will have

$$d\boldsymbol{\Psi} = \int_\Omega d\mathbf{B}_{NL}^T\boldsymbol{\sigma}\,d\Omega + \mathbf{K}_\mathbf{T}\,d\overline{\mathbf{u}} + \mathbf{M}\,d\ddot{\overline{\mathbf{u}}} \tag{2.22}$$

where

$$\mathbf{K}_\mathbf{T} = \int_\Omega \overline{\mathbf{B}}^T\mathbf{D}_\mathbf{T}\overline{\mathbf{B}}\,d\Omega = \mathbf{K}_L + \mathbf{K}_{NL} \tag{2.23}$$

where $\mathbf{D_T}$ is the consistent tangential stiffness matrix which is a function of the displacement and can be obtained by performing a full differentiation on the internal force term. $\mathbf{K}_L$ represents the usual, small displacement stiffness matrix. The matrix $\mathbf{K}_{NL}$ is due to the large displacements and both of these are defined as follows

$$\mathbf{K}_L = \int_\Omega \mathbf{B}_L^T \mathbf{D_T} \mathbf{B}_L \, d\Omega \tag{2.24}$$

$$\mathbf{K}_{NL} = \int_\Omega (\mathbf{B}_L^T \mathbf{D_T} \mathbf{B}_{NL} + \mathbf{B}_{NL}^T \mathbf{D_T} \mathbf{B}_{NL} + \mathbf{B}_{NL}^T \mathbf{D_T} \mathbf{B}_L) d\Omega \tag{2.25}$$

where $\mathbf{K}_{NL}$ is variously known as the 'initial displacement matrix', 'large displacement matrix' *etc.* and contains only terms which are linear and quadratic in $\bar{\mathbf{u}}$. It will be found that this is a matrix which would alternatively be obtained by using an infinitesimal strain approach but adjusting element co-ordinates in the computation of the stiffness.

The first term of equation (2.22) can generally be written as

$$\int_\Omega d\mathbf{B}_{NL}^T \boldsymbol{\sigma} \, d\Omega \equiv \mathbf{K}_{\boldsymbol{\sigma}} d\bar{\mathbf{u}} \tag{2.26}$$

where $\mathbf{K}_{\boldsymbol{\sigma}}$ is a symmetric matrix dependent on the stress level. This matrix is know as 'initial stress' matrix or 'geometric' matrix. Thus, we can define the total tangential stiffness matrix $\overline{\mathbf{K}}_{\mathbf{T}}$, used in the first term of equation (2.12), as

$$\overline{\mathbf{K}}_{\mathbf{T}} = \mathbf{K}_L + \mathbf{K}_{NL} + \mathbf{K}_{\boldsymbol{\sigma}} \tag{2.27}$$

To complete the total tangential stiffness matrix it is necessary only to determine the initial stress matrix $\mathbf{K}_{\boldsymbol{\sigma}}$. Substituting equations (2.18) and (2.19) into (2.26), we have

$$\mathbf{K}_{\boldsymbol{\sigma}} d\bar{\mathbf{u}} = \int_\Omega d\mathbf{B}_{NL}^T \boldsymbol{\sigma} \, d\Omega = \int_\Omega \mathbf{G}^T d\mathbf{A}_\theta^T \boldsymbol{\sigma} \, d\Omega \tag{2.28}$$

where

$$d\mathbf{A}_\theta^T \boldsymbol{\sigma} = d\mathbf{A}_\theta^T \begin{Bmatrix} \sigma_x \\ \sigma_y \\ \tau_{xy} \end{Bmatrix} = \begin{bmatrix} \sigma_x \mathbf{I}_{2\times2} & \tau_{xy} \mathbf{I}_{2\times2} \\ sym. & \sigma_y \mathbf{I}_{2\times2} \end{bmatrix} d\boldsymbol{\theta} \tag{2.29}$$

for plane stress and plane strain problems, whereas for axisymmetric problems

$$d\mathbf{A}_\theta^T \boldsymbol{\sigma} = d\mathbf{A}_\theta^T \begin{Bmatrix} \sigma_r \\ \sigma_z \\ \tau_{rz} \\ \sigma_\theta \end{Bmatrix} = \begin{bmatrix} \sigma_r \mathbf{I}_{2\times2} & \tau_{rz} \mathbf{I}_{2\times2} & 0 \\ & \sigma_z \mathbf{I}_{2\times2} & 0 \\ sym. & & \sigma_\theta \end{bmatrix} d\boldsymbol{\theta} \tag{2.30}$$

and for three-dimensional problems

$$dA_\theta^T \boldsymbol{\sigma} = dA_\theta^T \begin{Bmatrix} \sigma_x \\ \sigma_y \\ \sigma_z \\ \tau_{xy} \\ \tau_{yz} \\ \tau_{zx} \end{Bmatrix} = \begin{bmatrix} \sigma_x \mathbf{I}_{3\times3} & \tau_{xy}\mathbf{I}_{3\times3} & \tau_{xz}\mathbf{I}_{3\times3} \\ & \sigma_y \mathbf{I}_{3\times3} & \tau_{yz}\mathbf{I}_{3\times3} \\ sym. & & \sigma_z \mathbf{I}_{3\times3} \end{bmatrix} d\boldsymbol{\theta} \tag{2.31}$$

where $\mathbf{I}$ is the identity matrix. Equations (2.29), (2.30) and (2.31) can be conveniently rewritten as

$$dA_\theta^T \boldsymbol{\sigma} = \mathbf{M}_\sigma \mathbf{G}\, d\overline{\mathbf{u}} \tag{2.32}$$

where $\mathbf{M}_\sigma$ is a 4×4 matrix of the three stress components for plane stress/strain problems, 5×5 matrix of the four stress components for axisymmetric problems and 9×9 matrix of the six stress components for three-dimensional problems arranged as shown in equations (2.29), (2.30) and (2.31), respectively. Substituting equation (2.32) into (2.28), thus we will have

$$\mathbf{K}_\sigma = \int_\Omega \mathbf{G}^T \mathbf{M}_\sigma \mathbf{G}\, d\Omega \tag{2.33}$$

All the ingredients necessary for computing the large displacement of metal powder forming problem are now available. As a first step displacements $\overline{\mathbf{u}}^0$ are found according to the small displacement solution. This determines the actual strains by considering the nonlinear contribution defined by equations 2.7(a), (b) and (c) together with the appropriate linear contributions. Corresponding stresses can be found by the elastic expressions and $\boldsymbol{\Psi}^0$ determined according to equation (2.20). For successive iterations $(\overline{\mathbf{K}}_\mathbf{T})_n$ is found from equation (2.27). A full discussion of nonlinear iterative solution is given in Section 2.6.

2.5. Time domain discretization

In order to complete the numerical solution, it is necessary to integrate the differential equation (2.12) in time by one of the many available schemes. Various forms utilize the finite element concept in the time domain but here we will introduce the two simplest single step schemes, i.e. the *SSpj* and the *GNpj* methods. The *SSpj* method, based on a weighted residual formulation, was introduced by Zienkiewicz *et al.* (1980, 1984) and Wood (1984). The *GNpj* method, which is based on a generalization of the Newmark's scheme, i.e. the finite difference approach, and called the *Beta-m* method, was presented by Katona and Zienkiewicz (1985) and Katona (1985).

The *SSpj* method stands for Single Step p^{th} order interpolation for j^{th} order differential equation and the *GNpj* method stands for Generalized Newmark p^{th} order interpolation for j^{th} order differential equation. As may be expected, p cannot be less than j for any meaningful solution of the differential equation.

Assuming that the values of the function $\mathbf{u}$, together with appropriate time derivatives, are known at the time station t_n, we seek to determine the values of these at the next time station $t_{n+1} = t_n + \Delta t$. In single step schemes, this is done with the information of the time station t_n only, whereas in multistep schemes, previous values are also used. However, the formula for linear multistep scheme can become very complex when the time step is not constant. On the other hand, a change of time step has no major effect on the single step schemes, so the single step scheme is more suitable for variable time stepping.

2.5.1. $SSpj$ time stepping scheme

The expansion of the unknown vector $\bar{\mathbf{u}}$ will be taken as a polynomial of degree *p*, with known values of $\bar{\mathbf{u}}_n, \dot{\bar{\mathbf{u}}}_n, \ddot{\bar{\mathbf{u}}}_n$, *etc.*, up to $\bar{\mathbf{u}}_n^{p-1}$ at the beginning of the time step Δt. We note that in order to obtain a consistent approximation to all the derivatives, $p \geq 2$ is necessary for the full dynamic equation and $p \geq 1$ is necessary for the first-order equation. Indeed, the lowest approximation is $p = 1$. Here, we consider a second-order system in which the necessary initial conditions require the specification of two quantities, $\bar{\mathbf{u}}_n$ and $\dot{\bar{\mathbf{u}}}_n$. This algorithm, the $SS22$ scheme, is therefore basic for dynamic equations and we present it here in a full form. Using the truncated Taylor series expansion at time t_n, we have

$$\bar{\mathbf{u}} = \bar{\mathbf{u}}_n + \dot{\bar{\mathbf{u}}}_n \tau + \frac{1}{2} \boldsymbol{\alpha}_n \tau^2 \tag{2.34}$$

and the link between the successive values at t_{n+1} and t_n becomes

$$\bar{\mathbf{u}}_{n+1} = \bar{\mathbf{u}}_n + \dot{\bar{\mathbf{u}}}_n \Delta t + \frac{1}{2} \boldsymbol{\alpha}_n \Delta t^2 \tag{2.35}$$

where the only unknown is the vector $\boldsymbol{\alpha}_n$ which represents some average value of the second derivative occurring in the interval Δt (i.e. $\boldsymbol{\alpha}_n = \ddot{\bar{\mathbf{u}}}$).

The recurrence algorithm will now be obtained by inserting $\bar{\mathbf{u}}$ and $\ddot{\bar{\mathbf{u}}}$ obtained by differentiating equation (2.34) into equation (2.12) and satisfying the weighted residual equation with a single weighting function $W(t)$. This gives

$$\int_0^{\Delta t} W(t) \left(\mathbf{K}_\mathbf{T} \bar{\mathbf{u}} + \mathbf{M} \ddot{\bar{\mathbf{u}}} - \mathbf{f}_u \right) d\tau = 0 \tag{2.36}$$

or

$$\int_0^{\Delta t} W(t) \left(\mathbf{K}_\mathbf{T} \left(\bar{\mathbf{u}}_n + \dot{\bar{\mathbf{u}}}_n \tau + \frac{1}{2} \boldsymbol{\alpha}_n \tau^2 \right) + \mathbf{M} \boldsymbol{\alpha}_n - \mathbf{f}_u \right) d\tau = 0 \tag{2.37}$$

where the tangential stiffness matrix $\mathbf{K}_\mathbf{T}$ is defined in equation (2.23). Introducing Θ_n as the weighting parameter given by

$$\Theta_n = \frac{1}{\Delta t^n} \frac{\int_0^{\Delta t} W \tau^n d\tau}{\int_0^{\Delta t} W d\tau} \tag{2.38}$$

we can immediately write

$$\mathbf{K_T}\left(\overline{\mathbf{u}}_n + \dot{\overline{\mathbf{u}}}_n \Theta_1 \Delta t + \frac{1}{2}\boldsymbol{\alpha}_n \Theta_2 \Delta t^2\right) + \mathbf{M}\boldsymbol{\alpha}_n - \overline{\mathbf{f}}_u = 0 \tag{2.39}$$

where $\overline{\mathbf{f}}_u$ represents an average value of $\mathbf{f}_u$ given by

$$\overline{\mathbf{f}}_u = \frac{\int_0^{\Delta t} W\, \mathbf{f}_u \, d\tau}{\int_0^{\Delta t} W \, d\tau} \tag{2.40}$$

From equation (2.39), collecting all the known terms on the right hand side, we have

$$\left(\mathbf{M} + \frac{1}{2}\Theta_2 \Delta t^2 \mathbf{K_T}\right)\boldsymbol{\alpha}_n = \overline{\mathbf{f}}_u - \mathbf{K_T}\left(\overline{\mathbf{u}}_n + \dot{\overline{\mathbf{u}}}_n \Theta_1 \Delta t\right) \tag{2.41}$$

After an evaluation of $\boldsymbol{\alpha}_n$ from equation (2.41), the values of $\overline{\mathbf{u}}_{n+1}$ are determined simply from equation (2.35). This completes the algorithm which is of much practical value in the solution of dynamic problems. In many respects it resembles the Newmark algorithm which we shall discuss in the next section and which is widely used in practice. Indeed, its stability properties turn out to be identical with the Newmark algorithm, i.e. $\Theta_1 = \gamma$ and $\Theta_2 = 2\beta$ and for unconditional stability $\Theta_1 \geq \Theta_2 \geq \frac{1}{2}$, where γ and β are conventionally used Newmark parameters (Zienkiewicz and Taylor 1989).

2.5.2. *GNpj* time stepping scheme

Following the line of the well-known generalized Newmark scheme (*GNpj* method), an unconditionally stable direct solution procedure will be presented. In this method, instead of satisfying the equation in a mean sense over the time interval Δt, the scheme attempts to satisfy the equation at the time station t_{n+1}. Thus, we shall write a recurrence relation linking known values of $\overline{\mathbf{u}}_n$ and $\dot{\overline{\mathbf{u}}}_n$ available at time t_n with the values of $\overline{\mathbf{u}}_{n+1}$ and $\dot{\overline{\mathbf{u}}}_{n+1}$, which are valid at time $t_n + \Delta t$ and are the unknowns. If we assume that equation (2.12) has to be satisfied at each discrete time and $\ddot{\overline{\mathbf{u}}}_n$ is added to the known conditions at t_n with $\ddot{\overline{\mathbf{u}}}_{n+1}$ remaining as the unknown, we require that

$$\left(\int_\Omega \mathbf{B}^T \boldsymbol{\sigma}\, d\Omega\right)_{n+1} + \mathbf{M}\ddot{\overline{\mathbf{u}}}_{n+1} - (\mathbf{f}_u)_{n+1} = 0 \tag{2.42}$$

or

$$(\mathbf{K_T})_{n+1}\overline{\mathbf{u}}_{n+1} + \mathbf{M}_{n+1}\ddot{\overline{\mathbf{u}}}_{n+1} - (\mathbf{f}_u)_{n+1} = 0 \tag{2.43}$$

with appropriate approximations for the values of $\overline{\mathbf{u}}_{n+1}$ and $\ddot{\overline{\mathbf{u}}}_{n+1}$. Applying the *GN22* scheme for displacement, the link between the successive values is provided by a truncated series expansion similar to equation (2.34) for the function $\overline{\mathbf{u}}$, taken in the simplest case as

$$\dot{\overline{\mathbf{u}}}_{n+1} = \dot{\overline{\mathbf{u}}}_n + (1-\gamma)\Delta t \ddot{\overline{\mathbf{u}}}_n + \gamma \Delta t \ddot{\overline{\mathbf{u}}}_{n+1} = \dot{\overline{\mathbf{u}}}^p_{n+1} + \gamma \Delta t \ddot{\overline{\mathbf{u}}}_{n+1} \tag{2.44}$$

$$\overline{\mathbf{u}}_{n+1} = \overline{\mathbf{u}}_n + \Delta t\dot{\overline{\mathbf{u}}}_n + (\frac{1}{2} - \beta)\Delta t^2\ddot{\overline{\mathbf{u}}}_n + \beta\,\Delta t^2\ddot{\overline{\mathbf{u}}}_{n+1} = \overline{\mathbf{u}}^p_{n+1} + \beta\,\Delta t^2\ddot{\overline{\mathbf{u}}}_{n+1} \tag{2.45}$$

where $\dot{\overline{\mathbf{u}}}^p_{n+1}$ and $\overline{\mathbf{u}}^p_{n+1}$ stand for values that are predicted from known parameters at time t_n. The parameters γ and β are chosen in the range of 0–1, but for unconditional stability of the recurrence scheme, it is required that $\gamma \geq \frac{1}{2}$ and $\beta \geq \frac{1}{4}$ (Zienkiewicz and Taylor 1989).

In equations (2.44) and (2.45), we have effectively allowed for a polynomial of second degree plus a Taylor series remainder term in each of the expansions for the function and its derivatives, which can be chosen to give good approximation properties to the algorithm. Substituting $\overline{\mathbf{u}}_{n+1}$ from relation (2.45) into equation (2.43), we have

$$(\mathbf{K}_\mathrm{T})_{n+1}(\overline{\mathbf{u}}^p_{n+1} + \beta\,\Delta t^2\ddot{\overline{\mathbf{u}}}_{n+1}) + \mathbf{M}_{n+1}\ddot{\overline{\mathbf{u}}}_{n+1} - (\mathbf{f}_u)_{n+1} = 0 \tag{2.46}$$

and collecting all the known terms on the right hand side

$$\left[\mathbf{M}_{n+1} + (\mathbf{K}_\mathrm{T})_{n+1}\beta\,\Delta t^2\right]\ddot{\overline{\mathbf{u}}}_{n+1} = (\mathbf{f}_u)_{n+1} - (\mathbf{K}_\mathrm{T})_{n+1}\overline{\mathbf{u}}^p_{n+1} \tag{2.47}$$

For linear problem equation (2.47) can be solved directly for an exact value of $\ddot{\overline{\mathbf{u}}}_{n+1}$. When a nonlinear situation occurs the Newton-Raphson procedure needs to be modified for nonlinear iteration. In the next section, the solution strategy of equation (2.47) will be discussed.

2.6. Nonlinear iterative strategy

The application of the finite element method to nonlinear problems often results in a large set of nonlinear algebraic equations which must be solved for the state variables. Numerous solution methods have been described by many authors in the standard texts and relevant literature (Owen and Hinton 1980, Zienkiewicz and Taylor 1989 and Bathe 1996 *etc.*). Basically, these address two major topics; the solution of the set of linear algebraic equation, and nonlinear iteration.

The method in solving linear algebraic equation can be subdivided into two categories; the direct solution and iterative methods. One of the methods most commonly used is Newton's method and its variations. The computational cost needed to solve the linearized system of equations at each Newton iteration by direct methods increases quadratically as the number of degrees of freedom increases. Therefore, for large-scale problems it is sensible to use an iterative method even for the linearized system of equations. Iterative methods do not usually require the explicit assembly and storage of the global matrix. It must be emphasized that the need for such an alternative iterative strategy, instead of a direct solver, is far more pressing in an analysis of coupled or 'mixed' system because of the coupling of various types of degrees of freedom which tends to increase the size of the problem dramatically. Particularly in a 3D or adaptive remeshing analysis of such a coupled system the cost of direct solution will grow at an alarming rate.

In the nonlinear case, the iteration within one time step can be performed by the Newton-Raphson method, the modified Newton method or the quasi-Newton method. They are introduced in their logical order of development and, indeed, a chronological order. For simplicity, the nonlinear equation system (2.47) can be rewritten as

$$\mathbf{G}(\mathbf{x}) = 0 \tag{2.48}$$

where $\mathbf{x} = \ddot{\mathbf{u}}_{n+1}$. At each time step, the equation needs to be solved by an iterative process until some prescribed tolerance is reached. Assuming that the solution of the ith iteration $\mathbf{x}_i$ is known, and we seek $\mathbf{x}_{i+1} = \mathbf{x}_i + \delta\mathbf{x}_i$, which satisfies

$$\mathbf{G}(\mathbf{x}_{i+1}) = 0 \tag{2.49}$$

Expanding (2.49) in Taylor series and neglecting the second order term, results in

$$\mathbf{G}(\mathbf{x}_i) + \left(\frac{\partial \mathbf{G}}{\partial \mathbf{x}}\right)_{x=x_i} \delta\mathbf{x}_i = 0 \tag{2.50}$$

or

$$\mathbf{J}_i\, \delta\mathbf{x}_i = -\mathbf{G}(\mathbf{x}_i) \tag{2.51}$$

where $\mathbf{J}_i = \left(\frac{\partial \mathbf{G}}{\partial \mathbf{x}}\right)_{x=x_i}$ which is known as the Jacobian matrix. In equation (2.51), the Jacobian matrix can be approximated as

$$\mathbf{J} = \left[\mathbf{M}_{n+1} + (\mathbf{K}_\mathrm{T})_{n+1} \beta\, \Delta t^2\right] \tag{2.52}$$

Substituting the approximation (2.52) in equation (2.51), we will arrive at

$$\left[\mathbf{M}_{n+1} + (\mathbf{K}_\mathrm{T})_{n+1} \beta\, \Delta t^2\right]_{n+1}^{i} \left(\delta\left(\ddot{\mathbf{u}}_{n+1}\right)\right)^{i} = -\left(\mathbf{G}_u\right)_{n+1}^{i} \tag{2.53}$$

where $\mathbf{G}_u$ stands for the vector that can be evaluated from the known initial values. To start a solution, an initial Jacobian matrix $\mathbf{J}_0$ is computed using simply the elastic stiffness matrix $\mathbf{K}^e$. If one retains the $\mathbf{J}_0$ matrix in the whole solution process then only a single matrix decomposition is required and the computational cost can be greatly reduced.

An exact or full Newton-Raphson procedure has to be used when convergence is difficult to achieve. In this method the Jacobian matrix is updated at each step of the iterative process. However, the full Newton-Raphson method is generally very expensive. In contrast to the full Newton-Raphson method, an iterative method called the modified Newton-Raphson method requires a reduced number of assembly and factorization operations of the global tangent. For example, the matrix can be retained after being updated at the beginning of each time step. However, this method often leads to very slow convergence, if not total divergence, particularly when the degree of nonlinearity is high.

An alternative procedure is represented by a class of algorithms called quasi-Newton, which involve Rank-one and Rank-two updates of the coefficient matrix to force it to satisfy a global secant equation at each iteration. This procedure often yields better search directions than does the modified Newton-Raphson iteration, but at the expense of additional storage demands and computing effort associated with matrix updating. Another advantage using quasi-Newton method is that the Jacobian matrix is updated in a simple way after each iteration to avoid the numerical difficulties of the non-symmetric tangential stiffness matrix arising from non-associated plastic models.

The basic requirement of the quasi-Newton methods is that the updated matrix should satisfy the secant condition

$$\mathbf{G}(\mathbf{x}_i) - \mathbf{G}(\mathbf{x}_{i-1}) = \mathbf{A}_i(\mathbf{x}_i - \mathbf{x}_{i-1}) \tag{2.54}$$

where the new matrix $\mathbf{A}_i$ is used sequentially to find $\mathbf{x}_{i+1}$ by solving $\mathbf{A}_i\,\delta\mathbf{x}_i = -\mathbf{G}(\mathbf{x}_i)$ and then set $\mathbf{x}_{i+1} = \mathbf{x}_i + \delta\mathbf{x}_i$.

There are various choices such that the matrix will satisfy the quasi-Newton equation (2.54). Among the popular ones are the Davidon method or 'Rank-one formula' and the Broyden-Fletcher-Goldfarb-Shanno (BFGS) method or 'Rank-two formula' (Fletcher 1980). In particular, their inverse updates are most useful as only one decomposition of the matrix $\mathbf{A}_0$ is required. Theoretically, $\mathbf{A}_0$ can be any full rank matrix, but in practice the intial Jacobian matrix $\mathbf{J}_0$ is often used as a good approximation. In what follows, the inverse updates of the Davidon method and the BFGS method are presented.

The inverse matrix for Davidon's Rank-one, which satisfies equation (2.54), can be written as (Geradin *et al.* 1980)

$$\mathbf{A}_i^{-1} = \mathbf{A}_{i-1}^{-1} + a_i\mathbf{v}_i\mathbf{v}_i^T \tag{2.55}$$

where

$$\mathbf{v}_i = \delta_i - \mathbf{A}_{i-1}^{-1}\,\mathbf{r}_i \qquad \text{and} \qquad a_i = \frac{1}{\mathbf{v}_i^T\mathbf{r}_i} \tag{2.56}$$

where $\mathbf{r}_i = \mathbf{G}(\mathbf{x}_i) - \mathbf{G}(\mathbf{x}_{i-1})$ and $\delta_i = \mathbf{x}_i - \mathbf{x}_{i-1}$.

Rank-two formulas are also often proposed. The most widely used are DFP update and BFGS update (Dennis and More 1977). Both two Rank-two updates preserve matrix symmetry and positive definitness. The inverse update matrix for BFGS's Rank-two, which satisfies equation (2.54), can be written as

$$\mathbf{A}_i^{-1} = (\mathbf{I} - b_i\delta_i\mathbf{r}_i^T)\,\mathbf{A}_{i-1}^{-1}\,(\mathbf{I} - b_i\mathbf{r}_i\,\delta_i^T) + b_i\delta_i\delta_i^T \tag{2.57}$$

where

$$b_i = \frac{1}{\delta_i^T\mathbf{r}_i} \tag{2.58}$$

The relative simplicity of the Davidon method is very attractive. As for the more involved rank two update, it has been pointed out that it does not yield an important improvement of the convergence rate in solving the nonlinear finite element equations of dynamic problems (Geradin *et al.* 1980). Hence, the Davidon Rank-one update should be preferred due to its lower cost.

The nature of the update does not guarantee the preservation of the matrix sparsity. For this reason it is convenient at every iteration to return to the original (sparse) matrix $\mathbf{A}_0$, used in the first iteration, and to reapply the multiplication of equation (2.55) through all previous iterations and their successive multiplications. The ith update of the Davidon inverse method can be easily written in the following recursive form

$$\mathbf{A}_i^{-1} = \mathbf{A}_0^{-1} + \sum_{k=1}^{i} a_k \mathbf{v}_k \mathbf{v}_k^T \tag{2.59}$$

with

$$\mathbf{v}_i = \delta_i - \mathbf{A}_0^{-1}\mathbf{r}_i - \sum_{k=1}^{i-1} a_k \mathbf{v}_k \mathbf{v}_k^T \mathbf{r}_i \tag{2.60}$$

In this manner only the decomposed $\mathbf{A}_0$ matrix, the vectors $\mathbf{v}_i$ and the coefficients a_i need to be stored. It has been observed that in most cases a large number of quasi-Newton updates may lead to an ill-conditioned iteration matrix. It is thus adviseable to restart periodically the iteration procedure either using the initial Jacobian matrix $\mathbf{J}_0$ or by calculating a new Jacobian matrix.

2.7. Arbitrary Lagrangian-Eulerian method

A common feature of powder compaction simulation with solid mechanics constitutive models is the use of a Lagrangian kinematics formulation. This approach has shown to be adequate for problems that do not exhibit large mass fluxes among different parts of the sample. But in practical problems, as those that appear in realistic design processes, the Lagrangian approach leads to highly distorted and usually useless meshes. This difficulty can be particularly observed in higher order elements. Because of severe distortion of elements, the determinant of Jacobian matrix may become negative at quadrature points, aborting the calculations or causing numerical errors. In order to solve these problems, the mesh adaptive strategies have been employed in Chapter 6. However, it is computationally expensive and information must be interpolated from the old mesh to the new mesh.

In fluid mechanics problems, Lagrangian methods basically are inappropriate for high velocity flows of the material and Eulerian meshes are well suited. In Eulerian method, the elements are fixed in space and the material flows through the elements. Therefore, the elements undergo no distortion even though the material experiences large deformation. However, an important disadvantage of Eulerian method is that it is less suited for path dependent material due to the convection of material through the elements and problems with free boundary motion. Hence, it can be concluded that neither Lagrangian nor Eulerian formulation alone is well suited for simulation of processes involving large deformation and path dependent material.

For this purpose, a combined technique of Eulerian and Lagrangian approaches which captures the advantages of both methods has been developed. The arbitrary Lagrangian-Eulerian (ALE) technique is proposed to take the advantages of both Lagrangian and Eulerian methods while minimizing the disadvantages. The term of 'arbitrary' describes that the ALE method is specified based on an arbitrary motion of the mesh so that it optimizes the shape of mesh and elements undergo no distortion.

2.7.1. Basic ALE kinematics

In ALE technique, the reference configuration is applied to describe the motion instead of material configuration in Lagrangian and spatial configuration in Eulerian formulation. This formulation introduces convection terms in finite element equations and consists of two phases. In Lagrangian phase, the mesh and material movements are identical. In Eulerian phase, it is allowed that the mesh to have an arbitrary motion, independent of material motion, keeping the mesh regular. In this study, each time step is analyzed according to Lagrangian phase until required convergence is attained. Then, the Eulerian phase is applied to keep mesh configuration regular. Because of relative displacement between mesh and material, all dependent variables, such as stress and strain are converted through the Eulerian phase.

Let us consider three configurations; the material domain Ω_0, spatial domain Ω and reference domain $\hat{\Omega}$, which is called ALE domain. The material motion is defined by $\mathbf{x}^{\mathrm{m}} = \boldsymbol{f}(\mathbf{X}, t)$, with $\mathbf{X}$ denoting the material coordinates and $\boldsymbol{f}(\mathbf{X}, t)$ a function which maps the body from the initial or material configuration Ω_0 to the current or spatial configuration Ω. The initial position of material points is denoted by $\mathbf{x}^{\mathrm{g}}$ called the reference or ALE coordinate in which $\mathbf{x}^{\mathrm{g}} = \boldsymbol{f}(\mathbf{X}, 0)$. The reference domain $\hat{\Omega}$ is defined to describe the mesh motion and is coincident with mesh points so it can be denoted by computational domain. The mesh motion is defined by $\mathbf{x}^{\mathrm{m}} = \hat{\boldsymbol{f}}(\mathbf{x}^{\mathrm{g}}, t)$. The material coordinate can be related to ALE coordinate, as shown in Figure 2.2 by

$$\mathbf{x}^{\mathrm{g}} = \hat{\boldsymbol{f}}^{-1}(\mathbf{x}^{\mathrm{m}}, t) = \hat{\boldsymbol{f}}^{-1}(\boldsymbol{f}(\mathbf{X}, t), t) = \boldsymbol{g}(\mathbf{X}, t) \quad \text{or} \quad \boldsymbol{g} = \hat{\boldsymbol{f}}^{-1} \circ \boldsymbol{f} \tag{2.61}$$

The material motion can be expressed by a combination of the mesh motion and function $\boldsymbol{g}$ as

$$\mathbf{x}^{\mathrm{m}} = \boldsymbol{f}(\mathbf{X}, t) = \hat{\boldsymbol{f}}(\boldsymbol{g}(\mathbf{X}, t), t) \quad \text{or} \quad \boldsymbol{f} = \hat{\boldsymbol{f}} \circ \boldsymbol{g} \tag{2.62}$$

The mesh displacement can be defined by

$$\mathbf{u}^{\mathrm{g}}(\mathbf{x}^{\mathrm{g}}, t) = \mathbf{x}^{\mathrm{m}} - \mathbf{x}^{\mathrm{g}} = \hat{\boldsymbol{f}}(\mathbf{x}^{\mathrm{g}}, t) - \mathbf{x}^{\mathrm{g}} \tag{2.63}$$

It must be noted that the mesh motion can be simply obtained from material motion replacing the material coordinate by ALE coordinate. The mesh velocity can be defined as

$$\mathbf{v}^{\mathrm{g}}(\mathbf{x}^{\mathrm{g}}, t) = \frac{\partial \hat{\boldsymbol{f}}(\mathbf{x}^{\mathrm{g}}, t)}{\partial t} = \left.\frac{\partial \mathbf{x}^{\mathrm{m}}}{\partial t}\right|_{\mathbf{x}^{\mathrm{g}}} \tag{2.64}$$

in which the ALE coordinate $\mathbf{x}^{\mathrm{g}}$ and material coordinate $\mathbf{X}$ in definition of material velocity are fixed. The mesh acceleration can be defined by differentiating from equation (2.64) as

$$\mathbf{a}^{\mathrm{g}} = \frac{\partial \mathbf{v}^{\mathrm{g}}(\mathbf{x}^{\mathrm{g}}, t)}{\partial t} = \frac{\partial^2 \mathbf{u}^{\mathrm{g}}(\mathbf{x}^{\mathrm{g}}, t)}{\partial t^2} \tag{2.65}$$

The mesh velocity and mesh acceleration do not show any physical meaning. In the case of Lagrangian meshes, these parameters correspond to the material velocity and acceleration.

2.7.2. Convective velocity

In ALE formulation, the convective velocity $\boldsymbol{c}$ is defined using the difference between the material and mesh velocities as (Figure 2.3)

$$c_i = v_i^{\mathrm{m}} - v_i^{\mathrm{g}} \tag{2.66}$$

where the material velocity v_i^{m} can be obtained using the chain rule expression as

$$v_i^{\mathrm{m}} = \left.\frac{\partial x_i^{\mathrm{m}}(\mathbf{x}^{\mathrm{g}},t)}{\partial t}\right|_{\mathbf{X}} = \left.\frac{\partial x_i^{\mathrm{m}}(\mathbf{x}^{\mathrm{g}},t)}{\partial t}\right|_{\mathbf{x}^{\mathrm{g}}} + \left.\frac{\partial x_i^{\mathrm{m}}(\mathbf{x}^{\mathrm{g}},t)}{\partial x_j^{\mathrm{g}}}\frac{\partial x_j^{\mathrm{g}}(\mathbf{X},t)}{\partial t}\right|_{\mathbf{X}} \tag{2.67}$$

In above equation, the material time derivative is obtained using the chain rule with respect to the ALE coordinate $\mathbf{x}^{\mathrm{g}}$ and time t. Substituting the definition of mesh velocity of equation (2.64) into (2.67) leads

$$v_i^{\mathrm{m}} = v_i^{\mathrm{g}} + \left.\frac{\partial x_i^{\mathrm{m}}}{\partial x_j^{\mathrm{g}}}\frac{\partial x_j^{\mathrm{g}}}{\partial t}\right|_{\mathbf{X}} \tag{2.68}$$

The referential velocity w_{i} can be defined as

$$w_{\mathrm{i}} = \frac{\partial g_i(\mathbf{X},t)}{\partial t} = \left.\frac{\partial x_i^{\mathrm{g}}}{\partial t}\right|_{\mathbf{X}} \tag{2.69}$$

Substituting equation (2.69) into (2.68), the material velocity can be rewritten as

$$v_i^{\mathrm{m}} = v_i^{\mathrm{g}} + \frac{\partial x_i^{\mathrm{m}}}{\partial x_j^{\mathrm{g}}} w_j \tag{2.70}$$

Using relations (2.64), (2.66) and (2.70), the convective velocity can be obtained as

$$c_i = v_i^{\mathrm{m}} - v_i^{\mathrm{g}} = \left.\frac{\partial x_i^{\mathrm{m}}}{\partial x_j^{\mathrm{g}}}\frac{\partial x_j^{\mathrm{g}}}{\partial t}\right|_{\mathbf{X}} = \frac{\partial x_i^{\mathrm{m}}}{\partial x_j^{\mathrm{g}}} w_j \tag{2.71}$$

The above relationship between the convective velocity $\mathbf{c}$, material velocity $\mathbf{v}^{\mathrm{m}}$, mesh velocity $\mathbf{v}^{\mathrm{g}}$ and referential velocity $\mathbf{w}$ is widely used in ALE formulation.

It is worth mentioning that Lagrangian and Eulerian descriptions can be derived as special cases of ALE description. In Lagrangian meshes, the material and mesh motions are identical. Thus, $\mathbf{X} = \mathbf{x}^{\mathrm{g}}$ and $\boldsymbol{g}$ is identity function, i.e. $\mathbf{X} = \boldsymbol{g}(\mathbf{X},t) = I(\mathbf{X})$. In Eulerian meshes, the spatial and ALE coordinates are coincident, i.e. $\mathbf{x}^{\mathrm{m}} = \mathbf{x}^{\mathrm{g}}$. Thus, $\hat{\boldsymbol{f}}$ is an identity function and $\mathbf{x}^{\mathrm{m}} = \hat{\boldsymbol{f}}(\mathbf{x}^{\mathrm{m}},t) = I(\mathbf{x}^{\mathrm{m}})$. In fact, in Eulerian description, $\boldsymbol{f} = \boldsymbol{g}$ and the mesh velocity is equal to zero.

2.8. ALE governing equations

The momentum equation in ALE formulation is similar to the updated Lagrangian description. The only difference is in the material time derivative terms. The momentum equation can be written as

$$\rho\, \dot{v}_i^{\mathrm{m}} = \sigma_{ji,j} + \rho b_i \tag{2.72}$$

where ρ is the density, σ the cauchy stress and b the body force. In above equation, the material time derivative of velocity $\dot{v}_i^{\mathrm{m}}$ can be defined as

$$\dot{v}_i^{\mathrm{m}} = \left.\frac{\partial v_i^{\mathrm{m}}}{\partial t}\right|_{\mathbf{x}^g} + \left.\frac{\partial v_i^{\mathrm{m}}}{\partial x_i^{\mathrm{g}}}\frac{\partial x_i^{\mathrm{g}}}{\partial t}\right|_{\mathbf{X}} = \left.\frac{\partial v_i^{\mathrm{m}}}{\partial t}\right|_{\mathbf{x}^g} + \left.\frac{\partial v_i^{\mathrm{m}}}{\partial x_j^{\mathrm{m}}}\frac{\partial x_j^{\mathrm{m}}}{\partial x_i^{\mathrm{g}}}\frac{\partial x_i^{\mathrm{g}}}{\partial t}\right|_{\mathbf{X}} \tag{2.73}$$

Substituting relations (2.69) and (2.71) into (2.73) results

$$\dot{v}_i^{\mathrm{m}} = \left.\frac{\partial v_i^{\mathrm{m}}}{\partial t}\right|_{\mathbf{x}^g} + \frac{\partial v_i^{\mathrm{m}}}{\partial x_j^{\mathrm{m}}}\frac{\partial x_j^{\mathrm{m}}}{\partial x_i^{\mathrm{g}}} w_i = \left.\frac{\partial v_i^{\mathrm{m}}}{\partial t}\right|_{\mathbf{x}^g} + \frac{\partial v_i^{\mathrm{m}}}{\partial x_j^{\mathrm{m}}} c_j \tag{2.74}$$

Substituting equation (2.74) into (2.72), the momentum equation can be therefore written as

$$\rho\left(\left.\frac{\partial v_i^{\mathrm{m}}}{\partial t}\right|_{\mathbf{x}^g} + \frac{\partial v_i^{\mathrm{m}}}{\partial x_j^{\mathrm{m}}} c_j\right) = \frac{\partial \sigma_{ji}}{\partial x_j^{\mathrm{m}}} + \rho b_i \tag{2.75}$$

In the absence of acceleration in static problems, the equilibrium equation in ALE and Lagrangian descriptions is exactly identical. Thus, we can use the Lagrangian principle of virtual power in ALE description, i.e. $\mathbf{f}^{\mathrm{int}} = \mathbf{f}^{\mathrm{ext}}$.

2.8.1. Petrov-Galerkin FE formulation

In ALE formulation, because of appearance of convective terms, i.e. $v_{i,j}$ and c_j, in momentum equation (2.75), the Galerkin finite element formulation results in numerical instabilities (Belytschko *et al.* 2000). This matter specially appears in problems with severe dynamical properties. In order to avoid this problem, the Petrov-Galerkin formulation is employed here. In this method, the shape functions applied for the discretization of velocity **v** differ from those employed for the test function $\delta\,\mathbf{v}$. The former is denoted by **N** and the latter by $\bar{\mathbf{N}}$. If $\bar{\mathbf{N}} = \mathbf{N}$, the Galerkin method can be obtained as a special case of Petrov-Galerkin method. Hence, the internal and external force vectors can be written in Petrov-Galerkin method as

$$f_{iI}^{\mathrm{int}} = \int_\Omega \frac{\partial \bar{N}_I}{\partial x_j^{\mathrm{m}}} \sigma_{ji}\, d\Omega \quad , \quad f_{iI}^{\mathrm{ext}} = \int_\Omega \bar{N}_I\, \rho\, b_i\, d\Omega + \int_\Gamma \bar{N}_I\, \bar{t}_i\, d\Gamma \tag{2.76}$$

Considering the test function in Petrov-Galerkin method by $\delta\,\tilde{v}_i$, which is defined as

$$\delta\,\tilde{v}_i = \delta\, v_{\mathrm{i}} + \delta\, v_{\mathrm{i}}^{\mathrm{PG}} \tag{2.77}$$

where $\delta\, v_{\mathrm{i}}$ is the test function of Galerkin methods, which is related to nodal values by the shape function **N**, and $\delta\, v_{\mathrm{i}}^{\mathrm{PG}}$ is the Petrov-Galerkin test function. Inserting the test function (2.77) into the weak form of momentum equation (2.72) yields

$$\int_\Omega \delta v_i \rho\, \dot{v}_i^{\mathrm{m}}\, d\Omega - \int_\Omega \delta v_{\mathrm{i}} \sigma_{ji,j} d\Omega - \int_\Omega \delta v_{\mathrm{i}} \rho\, b_i d\Omega + \int_\Omega \delta v_i^{\mathrm{PG}} (\rho\, \dot{v}_i^{\mathrm{m}} - \sigma_{ji,j} - \rho b_i)\, d\Omega = 0 \tag{2.78}$$

in which the first three terms are standard Galerkin terms and the last term serves as a stabilization term in Petrov-Galerkin method. Applying the integration by parts and the divergence theorem, the weak form of the momentum equation can be obtained as

$$\int_{\Omega} \delta v_i \rho \, \dot{v}_i^{\mathrm{m}} \, d\Omega + \int_{\Omega} \delta v_{i,j} \sigma_{ij} d\Omega - \int_{\Omega} \delta v_i \rho b_i d\Omega - \int_{\Gamma} \delta v_i \, \bar{t}_i d\Gamma + \int_{\Omega} \delta v_i^{\mathrm{PG}} (\rho \, \dot{v}_i^{\mathrm{m}} - \sigma_{ji,j} - \rho b_i) \, d\Omega = 0 \tag{2.79}$$

where the test function δv_i^{PG} can be chosen as $\delta v_i^{\mathrm{PG}} = k \, c_j \, \delta v_{i,j}$, with k denoting a constant parameter (Hughes and Mallet 1986).

The finite element discretization of the variational form (2.79) can be obtained applying the Galerkin and Petrov-Galerkin test functions into equation (2.79). The Petrov-Galerkin test function can be related to nodal values using the Galerkin test function $\delta v_i = \delta v_{iI} N_I$ as

$$\delta \tilde{v}_i = \delta v_{iI} \, N_I + k \, c_j \, N_{I,j} \, \delta v_{iI} \equiv \bar{N}_I \, \delta v_{iI} \tag{2.80}$$

where the Petrov-Galerkin shape function is defined by $\bar{N}_I = N_I + k \, c_j \, N_{I,j}$. However it must be noted that the Petrov-Galerkin approach is only necessary for problems with severe dynamical characteristics and it is not necessary for static problems.

2.8.2. ALE tangential stiffness matrix

In order to obtain the ALE tangential stiffness matrix, we need to linearize the internal nodal force. The internal nodal force can be computed in referential domain as follows

$$f_{iI}^{\mathrm{int}} = \int_{\Omega} \frac{\partial N_I}{\partial x_j^{\mathrm{m}}} \sigma_{ji} \, d\Omega \tag{2.81}$$

Applying the chain rule, the internal nodal force can be rewritten as

$$f_{iI}^{\mathrm{int}} = \int_{\hat{\Omega}} \frac{\partial N_I}{\partial x_k^{\mathrm{g}}} \frac{\partial x_k^{\mathrm{g}}}{\partial x_m^{\mathrm{m}}} \sigma_{mi} \, \hat{J} \, d\hat{\Omega} \tag{2.82}$$

where $\hat{J} = \det(\hat{\mathbf{F}})$ with $\hat{F}_{ij} = \partial x_i^{\mathrm{m}} / \partial x_j^{\mathrm{g}}$ and $d\Omega = \hat{J} \, d\hat{\Omega}$. Defining the nominal stress in referential domain as

$$\hat{P}_{ki} = \hat{J} \, \frac{\partial x_k^{\mathrm{g}}}{\partial x_m^{\mathrm{g}}} \sigma_{mi} = \hat{J} \, \hat{F}_{km}^{-1} \, \sigma_{mi} \tag{2.83}$$

Substituting equation (2.83) into (2.82) leads

$$f_{iI}^{\mathrm{int}} = \int_{\hat{\Omega}} \frac{\partial N_I}{\partial x_k^{\mathrm{g}}} \hat{P}_{ki} \, d\hat{\Omega} \tag{2.84}$$

where $\hat{\Omega}$ is fixed and the shape functions are independent of time. Thus, the time derivative of equation (2.84) can be calculated as

$$\frac{df_{iI}^{\text{int}}}{dt} = \int_{\hat{\Omega}} \frac{\partial N_I}{\partial x_k^{g}} \left.\frac{\partial \hat{P}_{ki}}{\partial t}\right|_{\mathbf{x}^{g}} d\hat{\Omega} \tag{2.85}$$

Substituting the definition of the nominal stress based on the second Piola-Kirchhoff stress into equation, we will obtain (Belytschko *et al.* 2000)

$$\frac{df_{iI}^{\text{int}}}{dt} = \int_{\Omega} N_{I,k}\left(\dot{\sigma}_{ki} - \sigma_{mi}L_{km} + v_{m,m}\sigma_{ki}\right)d\Omega + \int_{\Omega} N_{I,k}\left(\sigma_{mi}c_{k,m} - c_l\sigma_{ki,l} - c_{m,m}\sigma_{ki}\right)d\Omega \tag{2.86}$$

where $L_{km} = \hat{L}_{km} + c_{k,m}$ with $\hat{L}_{km} = \hat{F}_{km,t[\mathbf{x}^{g}]}\hat{F}_{km}^{-1}$. The first integral of equation (2.86) is associated with the Lagrangian representation and the second integral is associated with the convective effect of ALE approach. Applying the finite element discretization, df_{iI}^{int}/dt can be expressed as

$$\frac{df_{iI}^{\text{int}}}{dt} = K_{iIjJ}^{\text{lag}}\, v_{jJ}^{\text{m}} + K_{ijIJ}^{\text{ale}}\, c_{jJ} \qquad \text{or} \qquad \frac{df_{iI}^{\text{int}}}{dt} = \mathbf{K}^{\text{lag}}\,\mathbf{v}^{\text{m}} + \mathbf{K}^{\text{ale}}\,\mathbf{c} \tag{2.87}$$

where $\mathbf{K}^{\text{lag}}$ is the Lagrangian stiffness matrix which is defined by $\mathbf{K}^{\text{lag}} = \mathbf{K}^{\text{mat}} + \mathbf{K}^{\text{geo}}$, with $\mathbf{K}^{\text{mat}}$ and $\mathbf{K}^{\text{geo}}$ denoting the material and geometrical stiffness matrices, respectively. Expanding $\mathbf{K}^{\text{lag}}$ leads to

$$K_{iIlJ}^{\text{ale}} = \int_{\Omega} N_{I,k}\sigma_{km}N_{J,m}\delta_{il}\,d\Omega - \int_{\Omega} N_{I,k}\sigma_{ki,l}N_J\,d\Omega - \int_{\Omega} N_{I,k}\sigma_{ki}N_{J,l}\,d\Omega \tag{2.88}$$

where δ_{il} is the Kronecker delta. As can be seen from equation (2.88), the ALE stiffness matrix is a nonsymmetrical matrix in contrast to $\mathbf{K}^{\text{lag}}$. In addition, the gradients of the stress can be observed in $\mathbf{K}^{\text{ale}}$ while it does not appear in $\mathbf{K}^{\text{lag}}$.

2.8.3. Mesh motion in ALE formulation

The option of arbitrarily moving the mesh in ALE approach offers this possibility to move the boundary elements along with the material. In addition, the interior nodes can be moved so as to avoid excessive element distortion and entanglement. Although mesh motion in general is independent from material motion, the boundary of ALE and material domains must be coincide, i.e. at each boundary point

$$(\mathbf{v}^{\text{m}} - \mathbf{v}^{g}).\mathbf{n} = 0 \tag{2.89}$$

where $\mathbf{n}$ is the normal vector to the boundary points. There are several methods to move the grid points. In these methods, the motion of interior grid and material points is related by a linear function as

$$\mathbf{v}^{g} = \mathbf{a} + \mathbf{B}\,\mathbf{v}^{\text{m}} \tag{2.90}$$

where $\mathbf{a}$ and $\mathbf{B}$ are the vector and matrix of constant scalars, respectively. Substituting equation (2.90) into (2.87), the rate of internal nodal forces is obtained as follows

$$\frac{df_{iI}^{\text{int}}}{dt} = \mathbf{K}^{\text{lag}}\,\mathbf{v}^{\text{m}} + \mathbf{K}^{\text{ale}}\,(\,\mathbf{v}^{\text{m}} - \mathbf{a} - \mathbf{B}\,\mathbf{v}^{\text{m}}\,) \tag{2.91}$$

or

$$\frac{df_{iI}^{\text{int}}}{dt} = \left[\mathbf{K}^{\text{lag}} + \mathbf{K}^{\text{ale}}(\,\mathbf{I} - \mathbf{B}\,)\right]\,\mathbf{v}^{\text{m}} - \mathbf{K}^{\text{ale}}\,\mathbf{a} \tag{2.92}$$

in which the only unknown in equation (2.92) is $\mathbf{v}^{\text{m}}$. It must be noted that the Lagrangian motion is a special case of above scheme. Considering $\mathbf{a} = \{\mathbf{0}\}$ and $\mathbf{B} = \mathbf{I}$, with $\mathbf{I}$ denoting the identity matrix, it yields to $\mathbf{v} = \mathbf{v}^{\text{g}}$. The Eulerian motion can be also obtained by assuming $\mathbf{a} = \{\mathbf{0}\}$ and $\mathbf{B} = [\mathbf{0}]$, which yields to $\mathbf{v}^{\text{g}} = 0$.

In order to design an overall strategy for mesh motion, the Eulerian meshes may be well suited in the area where material undergoes very large deformation and consequently the elements is severely distorted. The interior nodes can be rearranged in a way that the optimized shape is acquired for the elements. However, if no severe distortion is expected for the elements in adjacent to the free boundaries, the Lagrangian meshes will be more efficient. In addition, the Eulerian meshes are more convenient in interior region whose boundaries do not move during the analysis. The nodes of free boundaries can move based on a Lagrangian motion in the direction normal to the boundary. This will guarantee the one-to-one mapping of material and mesh domains and allow for natural evaluation of free boundaries on applied loads. However, if the boundary elements are expected to undergo significant distortion, arbitrary tangential motion of exterior nodes can prevent deterioration of the elements. Otherwise, the Lagrangian motion of exterior nodes is often more efficient. Also, on boundary nodes with prescribed displacement boundary conditions, the Lagrangian motion in the direction of the applied displacement is more required.

2.8.4. Stress update procedure

The stress state in a path dependent material can be readily treated using Lagrangian meshes, because the quadrature points coincide with material points regardless of the deformation of the continuum. On the other hand, in ALE description, a Gauss point does not coincide with material points, as shown in Figure 2.4. Thus, it is necessary to update the stress field with respect to the convective velocity **c**. In nonlinear problems, the material rate of stress is usually related to the deformation history and current stress state with large variety of constitutive models. In order to obtain the constitutive equations in ALE formulation, the material rate of stress is related to the reference rate of stress by adding a convective term accounting for relative motion between material point and grid point as

$$\left.\frac{\partial \sigma_{ij}}{\partial t}\right|_{\mathbf{X}} = \left.\frac{\partial \sigma_{ij}}{\partial t}\right|_{\mathbf{x}^{\text{g}}} + \frac{\partial \sigma_{ij}}{\partial x_j^{\text{m}}}\,c_j \tag{2.93}$$

For updating the stresses, the spatial derivative of stress must be evaluated. In this procedure, the method of 'Super-convergent Patch Recovery' (SPR) is used for transformation of stresses from quadrature points to nodal points (Zienkiewicz and Zhu 1992), as demonstrated in Chapter 6. Obtaining the nodal values and using ordinary shape functions, the spatial derivative of stresses can be evaluated. It must be noted that the implementation of C^1 continuity functions for expressing element stresses does not cause the continuity of the spatial derivative of stress. Other state variables, such as: effective plastic strains, can be convected from the material coordinate to ALE coordinate in similar manner.

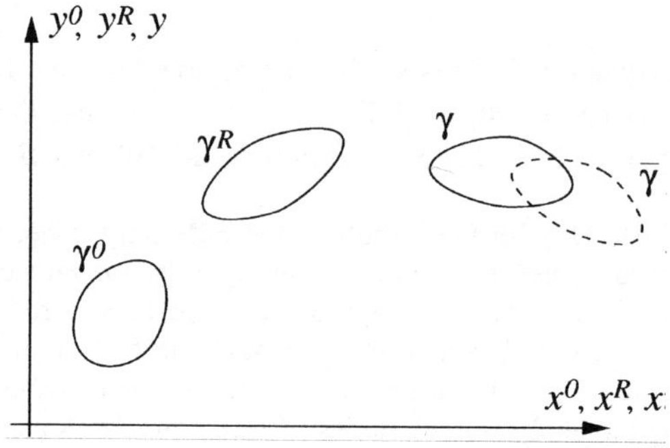

Figure 2.1. Successive configurations of the body

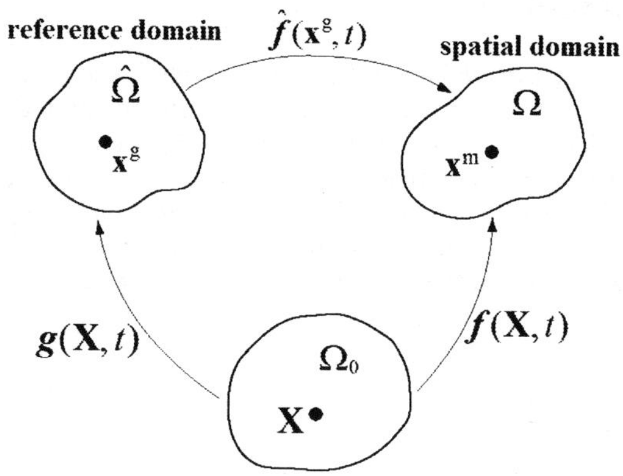

Figure 2.2. Domains and mapping between Lagrangian, Eulerian and ALE descriptions

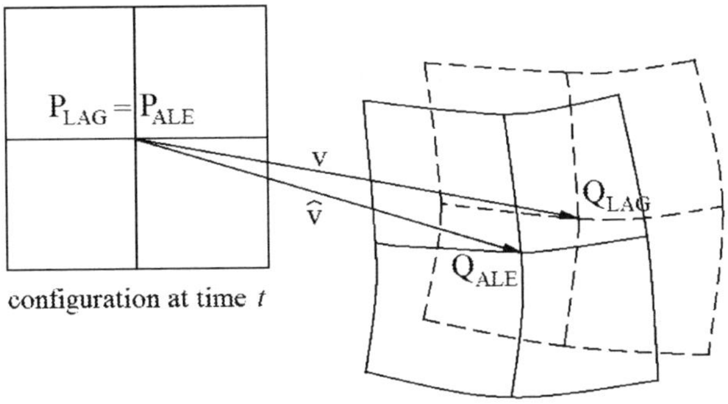

Figure 2.3. Mesh and material velocities

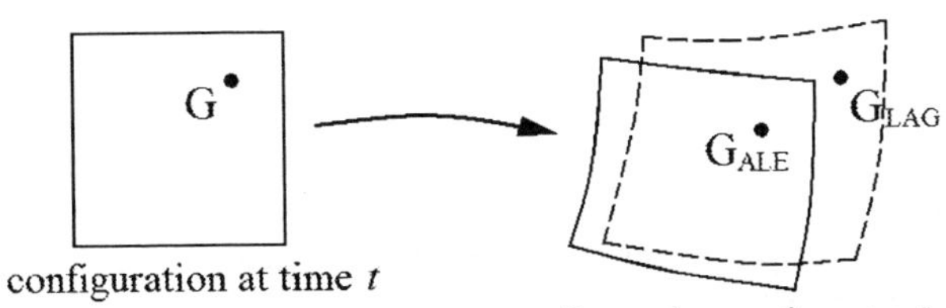

Figure 2.4. Stress update in ALE formulation

3 POWDER CONSTITUTIVE MODELS

The cold pressing process is of great importance in the manufacturing of hardmetal components from powders. If, in this process, inhomogeneities are introduced or if fracture occurs, the component will be rejected or it will perform badly in its intended use. Thus, it is of great interest to be able to accurately predict the mechanical behavior of the powder during a compaction process.

In the mechanical description of the cold pressing process, two major aspects are still at a research stage. These are the formulation of the constitutive relations for the powder materials and the formulation of the interface relations between powder and die wall. This chapter is limited to the constitutive equations of powder materials, leaving the interface problem (contact and friction) for a subsequent chapter. The aim of this investigation is to describe the mechanical behavior of metal powder using classical and modern constitutive theories. Depending on the purpose, a simple or an advanced constitutive theory may be needed. However, in any case, this choice must be based on a thorough knowledge of the mechanics involved as well as the limitations and accuracy of the constitutive theories. In this chapter, an incremental elasto-plastic material model was used to simulate the compaction process. Different plasticity theories, including: a combination of a Mohr-Coulomb and hardening cap surfaces, the double-surface plasticity model and a generalized three-invariant cap plasticity theory are developed as constitutive models to describe the nonlinear behavior of powder materials in the concept of the generalized plasticity formulation for the description of cyclic loading. A brief overview on the development of elasto-plastic constitutive modeling of powder materials and integration algorithm of stress-strain relation has been also presented.

3.1. Introduction

The constitutive modeling of powder is clearly a keystone of successful quantitative solution possibilities. Without a reasonable constitutive model, which can reproduce powder behavior under loading conditions, the computations are worthless. However, a good constitutive model, without a computational framework in which to use it, is only an academic exercise. In order to adequately reproduce powder material behavior under cyclic loading, as a sequence of loading/unloading/reloading, modifications have to be introduced to the classical constitutive models. The success for an elasto-plastic solution to a problem is dependent on

- a reasonable elasto-plastic constitutive model which can reproduce powder material behavior under complicated loading conditions;
- an accurate and stable integration algorithm for the elasto-plastic constitutive relation.

In recent years two approaches, based on 'porous material' and 'granular material' models respectively, have been developed which describe the effect of stress state on the response of the powder material.

The porous material model, generally known as a modified von-Mises, has been used for the simulation of metal forming and powder forming processes (Corapcioglu and Uz 1978, Sluzalec 1989, Hisatsune *et al.* 1991 and Mosbah *et al.* 1994). The yielding of porous material is more complicated than that for a solid material because the plastic deformation of this material will cause volumetric changes in the pores. The additional influence of hydrostatic stress must be included in the yield function and the most widely used form of yield function which has been considered by many researchers is an extension, or modification of, von-Mises's concept of yielding of fully dense materials. A number of yield criteria which include the influence of the hydrostatic stress component and also satisfies the symmetry and convexity conditions required for the development of a plasticity theory for porous materials have been developed by Kuhn and Downey (1971), Green (1972), Shima and Oyane (1976) and Doraivelu *et al.* (1984). In this model, it has always been assumed that the yield stress values under compression or tension may be treated in the same way. However, for metal powder material this is not correct since it cannot sustain significant levels of tensile stress. This model is therefore more applicable to precompacted, sintered powder or to secondary compaction processes.

The granular material model which has been used for the modeling of frictional materials such as soil or rock, is adopted to describe the behavior of metal powder (Crawford and Lindskog 1983, Haggblad 1991 and Riero and Prado 1994). The application of this model to powder forming, particularly the compaction process, has received attention by many researchers. This model reflects the yielding, frictional and densification characteristics of powder along with strain and geometrical hardening which occur during the compaction process. A number of yield criteria have been developed and tested to describe the behavior of granular material with consideration of yielding and densification characteristics (Desai and Siriwardane 1984, Chen and Baladi 1985 and Lewis and Schrefler 1987). These models generally comprise two surfaces, one to reflect shear failure and the second to capture densification. The solution yields details on the powder displacement from which it is possible to establish the stress state in the powder and the densification can be derived from consideration of the elemental volumetric strain. The most practical yield functions that have been used by authors in expressing the yielding and densification characteristics for powder materials are the critical state model, cam-clay and modified cam-clay models and cap model. Basically, these criteria have close-ended conical yield surfaces which can appropriately handle yielding under tension and compression.

Both these models, porous material and granular material models, originate from the same classical plasticity theory. They are elasto-plastic ones with isotropic or kinematic hardening. The material is considered to be elastic if the stress state is within the yield surface and plastic if the stress state exceeds the limit. The direction of the plastic straining beyond this limit will be determined by a plastic flow rule concept, which derives from a potential. Their evolution depends on one or more hardening parameters. In the field of powder compaction, the most interesting constitutive models, either for new developments, or to be integrated in numerical codes, can be classified into single- and double-surfaces plasticity, multi-surface plasticity and endochronic plasticity which is without a yield surface.

Single- and Double-surfaces plasticity

The single-surface plasticity models only have one yield function which may be used to represent a fixed or moving surface - however, in the double-surface plasticity models, two functions have been used to model the yielding behavior of powder (Figure 3.1-a, b). Two types of functions have been adopted in the double-surface plasticity models, one to represent the shear yield and the second to model hydrostatic yielding. In the first group of studies, one is fixed while the other yield function is allowed to move as in cap models. In the second type of study, both yield functions are allowed to move such as Vermeer's model (Vermeer 1978). It should be noted that the single-surface plasticity models could also be used as a series of single continuous functions, which include both the shear yield and hydrostatic yielding. Both models, single- and double-surfaces plasticity, are based on classical plasticity theory with an appended yield surface intersecting the hydrostatic axis. By using plastic flow theory, we are able to describe both loading and unloading situations. This possibility is of great importance in industrial metal powder compaction, since the process involves combined loading and unloading of the powder. With these models, it is also possible to predict nonlinear compaction and inelastic dilatancy.

Multi-surface plasticity

One way to increase the accuracy of the predictions of the plasticity models is to use more than one yield surface. Mroz (1967) and Iwan (1967) presented theories of so-called multi-surface plasticity (Figure 3.1-c). In addition to the effects which can be modelled by double-surface plasticity, the multi-surface plasticity approach more easily describes hysteresis effects in cyclic loading. Since additional parameters of the model are obtained by adding extra yield surfaces, close fits to real material behavior can be obtained. In order to describe the hydrostatic behavior of the powder with classical plasticity theory, it is necessary to have a yield surface intersecting the hydrostatic axis. One way of achieving this is to use an appended cap as in the double-surface plasticity description. It is also possible to use a continuous yield surface intersecting the hydrostatic axis. This is done in the multi-surface theory proposed by Prevost (1978), which has been extensively used in soil mechanics. Recently, a multi-surface plasticity model based on the model developed by Prevost has been used by Haggblad (1991) and Haggblad and Oldenburg (1994) for metal powder die pressing. They described the hardening behavior of powder by a set of plastic shear moduli associated with a corresponding set of nested yield surfaces in the stress space (Figure 3.1-c). In this model, individual plastic moduli are assumed to be constant on each particular surface. These surfaces have the form of ellipsoids in the stress space with the major principal axis initially coaxial with the hydrostatic axis for an initially isotropic material.

Endochronic plasticity

In the flow theory of plasticity, the transition from an elastic state to an elasto-plastic state appears more or less abruptly when the stress-point reaches a yield surface. For powder materials, it could be very difficult to define the location of a yield surface, owing to the fact that there is no distinct transition from elastic to elasto-plastic behavior. A theory of plasticity has been proposed by Valanis (1971), called the endochronic theory which has no yield surface. The fact that a yield surface is not required in the formulation of the endochronic theory is attractive. The theory can be put in a very advanced form which permits the prediction of all main phenomena. An extension of the endochronic theory which has been used mainly for concrete and rock materials has been made by Bazant *et al.* (1976). For powder materials, Haggblad (1991) used an endochronic model proposed by Bazant to describe the mechanical behavior of hard metal powders. The inelastic behavior of powder to triaxial loading was modeled by semi-empirical functions. A limitation of this model is the loss of physical significance of the powder material parameters. This increases the difficulties when fitting the parameters to the experiments.

In the present chapter, a double-surface plasticity model based on a combination of a convex yield surface consisting of a failure envelope and a hardening cap surface, and a three-invariant cap plasticity models have been developed for nonlinear behavior of powder materials in the concept of the generalized plasticity formulation for the description of cyclic loading. A brief overview of the development of elasto-plastic constitutive modeling of powder materials and an integration algorithm for the stress-strain relation will also be presented. Advanced constitutive models based on the endochronic theory and multi-surface plasticity will be dicussed in Chapter 7.

3.2. Generalized plasticity theory

The object of the mathematical theory of plasticity is to provide a theoretical description of the relationship between stress and strain or more commonly, between increments of stress and increments of strain using the assumption that the material behaves plastically only after a certain limiting value has been exceeded. In essence, plastic behavior is characterized by an irreversible straining which is not time dependent and which can only be sustained once a certain level of stress has been reached. In this section, we outline the basic assumptions and theoretical expressions for a general continuum which have been developed by Nayak and Zienkiewicz (1972), Zienkiewicz and Mroz (1984) and Zienkiewicz and Taylor (1991). In order to formulate a theory which models elasto-plastic material deformation three requirements have to be met:

- An explicit relationship between stress and strain must be formulated to describe material behavior under elastic conditions, i.e. before the onset of plastic deformation.
- A relationship between stress and strain must be developed for post-yield behavior, i.e. when the deformation is made up of both elastic and plastic components.
- A yield criterion indicating the stress level at which plastic flow commences must be postulated.

In order to calculate the tangential stiffness matrix $\mathbf{K_T}$, as defined in Section 2.4, we need to compute the consistent tangential stiffness matrix $\mathbf{D_T}$. The elasto-plastic relationship between the incremental stress $d\boldsymbol{\sigma}$ and the incremental strain $d\boldsymbol{\varepsilon}$ with respect to the incremental rotation $d\boldsymbol{\omega}$ is defined in Chapter 2, as

$$d\sigma_{ij} = D_{ijkl}(d\varepsilon_{kl} - d\varepsilon_{kl}^{0}) + \sigma_{ik} d\omega_{kj} + \sigma_{jk} d\omega_{ki} \tag{2.2}$$

In the theory of elasto-plasticity, it is assumed that the strain can be decomposed into two parts: the elastic or the recoverable strain and the plastic or the non-recoverable strain. Recoverable strain implies when the incremental stress $d\sigma_{ij}$ is reversed the incremental strain $d\varepsilon_{kl}$ can be recovered. Thus, during any increment of stress, the changes of strain are assumed to be divisible into elastic and plastic components, so that

$$d\boldsymbol{\varepsilon} = d\boldsymbol{\varepsilon}^e + d\boldsymbol{\varepsilon}^p = (\mathbf{D}^e)^{-1} d\boldsymbol{\sigma} + d\lambda \frac{dQ}{d\boldsymbol{\sigma}} \tag{3.1}$$

where $\mathbf{D}^e$ is the usual elastic matrix, Q is a plastic potential and $d\lambda$ denotes a proportionality constant termed the plastic multiplier. In equation (3.1), the elastic strain increment is related to the stress increment by the elastic matrix, which depends only on the current state of stress. The flow rule is also defined as the determination of plastic strain from the state of stress. The relationship between the plastic strain component and the stress increment is derived by utilising a further assumption regarding the material behavior. In particular, it is assumed that the plastic strain increment is proportional to the stress gradient of the plastic potential.

In the generalized plasticity formulation, the elasto-plastic constitutive matrix depends on the current state of stress and strain, its history, and importantly, the direction of the stress increment. To account for this dependence in a simple way, two different matrices $\mathbf{D}_L$ and $\mathbf{D}_U$ may be postulated for stress increments corresponding to loading and unloading situations. A loading direction vector $\mathbf{n}$ is now introduced to discriminate between loading and unloading by the sign of the projection $d\boldsymbol{\sigma}^T\mathbf{n} > 0$, i.e.

$d\boldsymbol{\sigma}^T\mathbf{n} > 0$ for loading

$d\boldsymbol{\sigma}^T\mathbf{n} < 0$ for unloading

and the situation of $d\boldsymbol{\sigma}^T\mathbf{n} = 0$ is defined as neutral loading. However, the above definition fails if strain softening occurs. Therefore, a more convenient definition was suggested by Zienkiewicz and Morz (1984) to check the loading or unloading direction by the elastic stress increment $d\boldsymbol{\sigma}^e$ as

$(d\boldsymbol{\sigma}^e)^T\,\mathbf{n} > 0$ for loading

$(d\boldsymbol{\sigma}^e)^T\,\mathbf{n} < 0$ for unloading

$(d\boldsymbol{\sigma}^e)^T\,\mathbf{n} = 0$ for neutral loading

As continuity of material behavior under neutral loading should be achieved with loading and unloading strain increments being identical, certain restrictions on the loading and unloading stiffness matrices must be enforced. Therefore, the most general forms of $\mathbf{D}_L$ and $\mathbf{D}_U$ can be written as (Zienkiewicz and Morz 1984)

$$\mathbf{D}_L = \mathbf{D}^e - \frac{\mathbf{D}^e\mathbf{n}_{gL}\mathbf{n}^T\mathbf{D}^e}{H_L + \mathbf{n}_{gL}^T\mathbf{D}^e\mathbf{n}} \tag{3.2}$$

$$\mathbf{D}_U = \mathbf{D}^e - \frac{\mathbf{D}^e\mathbf{n}_{gU}\mathbf{n}^T\mathbf{D}^e}{H_U + \mathbf{n}_{gU}^T\mathbf{D}^e\mathbf{n}} \tag{3.3}$$

where H_L and H_U are the plastic hardening/softening modulus and $\mathbf{n}_{gL}$ and $\mathbf{n}_{gU}$ are arbitrary unit stress vectors for loading and unloading directions. The last terms, in equations (3.2) and (3.3), denote the plastic matrix for loading and unloading where the plastic characteristics of the powder are incorporated. It is easily verified that continuity is satisfied for neutral loading.

Obviously, a complete elasto-plastic behavior can thus be specified by giving at each point of the stress space; loading direction vector $\mathbf{n}$, plastic flow direction vectors $\mathbf{n}_{gL}$ and $\mathbf{n}_{gU}$, plastic hardening/softening modulus H_L and H_U, and elastic matrix $\mathbf{D}^e$. It is noted that if $\mathbf{n}_{gL}$ and $\mathbf{n}_{gU}$ are different from $\mathbf{n}$, which is the case commonly referred to as non-associative plasticity, the above matrices will be non-symmetric. In the theory of plasticity the behavior of a material is specified by surfaces in the $\boldsymbol{\sigma}$ space, which enclose the elastic regions. The yield surface $F(\boldsymbol{\sigma},\kappa) = 0$ defines the loading and unloading directions, giving the unit normal vector as

$$\mathbf{n} = \frac{\partial F/\partial\boldsymbol{\sigma}}{\left[(\partial F/\partial\boldsymbol{\sigma})^T(\partial F/\partial\boldsymbol{\sigma})\right]^{1/2}} \tag{3.4}$$

Similarly, the flow direction $\mathbf{n}_g$ is defined by the plastic potential surface $Q(\boldsymbol{\sigma},\kappa) = 0$ with

$$\mathbf{n}_g = \frac{\partial Q/\partial\boldsymbol{\sigma}}{\left[(\partial Q/\partial\boldsymbol{\sigma})^T(\partial Q/\partial\boldsymbol{\sigma})\right]^{1/2}} \tag{3.5}$$

The hardening moduli H_L and H_U follow from the above definitions if we postulate that the stresses must be confined to the yield surface during 'loading', i.e. that

$$dF = \left(\frac{\partial F}{\partial \boldsymbol{\sigma}}\right)^T d\boldsymbol{\sigma} + \frac{\partial F}{\partial \kappa}\left(\frac{\partial \kappa}{\partial \boldsymbol{\varepsilon}^p}\right)^T d\boldsymbol{\varepsilon}^p \tag{3.6}$$

Substituting $d\boldsymbol{\varepsilon}^p = \mathbf{n}_g \mathbf{n}^T d\boldsymbol{\sigma}/H$ in equation (3.6), we can obtain, after some manipulation,

$$H = \frac{-(\partial F/\partial \kappa)(\partial \kappa/\partial \boldsymbol{\varepsilon})^T \mathbf{n}_g}{\left[(\partial F/\partial \boldsymbol{\sigma})^T(\partial F/\partial \boldsymbol{\sigma})\right]^{1/2}} \tag{3.7}$$

It is clear now that classical plasticity is indeed a special case of generalized plasticity. This simple and general description of generalized plasticity was introduced by Zienkiewicz and Morz (1984). It has been proved that the generalized plasticity forms are extremely useful in describing the complex behavior of powder material during compaction process as the surfaces need not be defined.

In the metal powder forming process, especially in the case of loose powder, the volume of powder reduces as compaction proceeds, so the volumetric strain is negative and under this circumstance the associated flow rule is most appropriate since the non-associated flow rule is used when the plastic strain is not in the same direction as the yield function. Thus, for metals, it has always been assumed that the yield function is identical to the potential function, $F \equiv Q$, i.e., an associated flow rule applies since it is close to experimental observation (Bleich 1972). Since the flow rule is associated, the gradient $\partial F/\partial \boldsymbol{\sigma}$ is the same as $\partial Q/\partial \boldsymbol{\sigma}$, i.e. it is normal to the yield surface at any point in its stress space.

3.2.1. Numerical computation of material property matrix

For numerical computations of plasticity problems it is convenient to present the yield function in terms of alternative stress invariants. This formulation is firstly employed by Nayak and Zienkiewicz (1972). Its main advantage is that it permits the computer coding of the yield function and the flow rule in a general form and necessitates only specification of three factors for any individual criterion. Thus, if we define the yield criterion in terms of alternative stress invariants $\left(J_1, J_{2D}^{1/2}, \theta\right)$, the yield function can be expressed by $F\left(J_1, J_{2D}^{1/2}, \theta\right) = 0$, where J_1 and J_{2D} are the first invariant of stress and second invariant of deviatoric stress, respectively, and θ is an alternative to the third invariant of deviatoric stress J_{3D}, which is defined by

$$\sin 3\theta = -\frac{3\sqrt{3}}{2}\frac{J_{3D}}{(J_{2D})^{3/2}} \tag{3.8}$$

where the invariant θ is in the range of $-30° < \theta < +30°$. The relationship between the principal stresses $(\sigma_1, \sigma_2, \sigma_3)$ and stress invariants $\left(J_1, J_{2D}^{1/2}, \theta\right)$, can be expressed by (Khan and Huang 1995)

$$\begin{Bmatrix} \sigma_1 \\ \sigma_2 \\ \sigma_3 \end{Bmatrix} = \frac{2J_{2D}^{1/2}}{\sqrt{3}} \begin{Bmatrix} \sin\left(\theta + \frac{2\pi}{3}\right) \\ \sin\theta \\ \sin\left(\theta + \frac{4\pi}{3}\right) \end{Bmatrix} + \frac{J_1}{3}\begin{Bmatrix} 1 \\ 1 \\ 1 \end{Bmatrix} \tag{3.9}$$

where $\sigma_1 > \sigma_2 > \sigma_3$. The above relation makes it easy to convert the yield surface in terms of $\left(J_1, J_{2D}^{1/2}, \theta\right)$.

In order to calculate the elasto-plastic matrix in equations (3.2) and (3.3), we need to indicate the values of loading direction vector $\mathbf{n}$, plastic flow direction vector $\mathbf{n}_g$, plastic hardening/softening modulus H and elastic matrix $\mathbf{D}^e$ at each point of stress space. The plastic flow vector defined by equation (3.4) can be written as

$$\frac{\partial F}{\partial \boldsymbol{\sigma}} = \frac{\partial F}{\partial J_1}\frac{\partial J_1}{\partial \boldsymbol{\sigma}} + \frac{\partial F}{\partial J_{2D}^{1/2}}\frac{\partial J_{2D}^{1/2}}{\partial \boldsymbol{\sigma}} + \frac{\partial F}{\partial \theta}\frac{\partial \theta}{\partial \boldsymbol{\sigma}} \tag{3.10}$$

Substituting $\partial\theta/\partial\boldsymbol{\sigma}$ from equation (3.8), the flow vector can be simplified as

$$\frac{\partial F}{\partial \boldsymbol{\sigma}} = C_1\mathbf{n}_1^T + C_2\mathbf{n}_2^T + C_3\mathbf{n}_3^T \tag{3.11}$$

where

$$\mathbf{n}_1^T = \frac{\partial J_1}{\partial \boldsymbol{\sigma}}, \quad \mathbf{n}_2^T = \frac{\partial J_{2D}^{1/2}}{\partial \boldsymbol{\sigma}}, \quad \mathbf{n}_3^T = \frac{\partial J_{3D}}{\partial \boldsymbol{\sigma}} \tag{3.12}$$

and

$$\begin{aligned} C_1 &= \frac{\partial F}{\partial J_1} \\ C_2 &= \frac{\partial F}{\partial J_{2D}^{1/2}} - \frac{\tan 3\theta}{J_{2D}^{1/2}}\frac{\partial F}{\partial \theta} \\ C_3 &= \frac{-\sqrt{3}}{2\cos 3\theta}\frac{1}{J_{2D}^{3/2}}\frac{\partial F}{\partial \theta} \end{aligned} \tag{3.13}$$

in which the constants C_1, C_2 and C_3 are then necessary to define the yield surface. Thus, the simplicity of programming can be achieved as only these three constants have to be varied between different yield surfaces.

According to the definition of stress invariants, i.e. $J_1 = \sigma_{ii}$, $J_{2D} = \frac{1}{2}\sigma'_{ij}\sigma'_{ij}$ and $J_{3D} = \det(\sigma'_{ij})$ in which the deviatoric stress tensor defined as $\sigma'_{ij} = \sigma_{ij} - \frac{1}{3}\delta_{ij}\sigma_{mm}$, the vectors $\mathbf{n}_1$, $\mathbf{n}_2$ and $\mathbf{n}_3$ can be calculated as

$$\begin{aligned} \mathbf{n}_1^T &= \frac{\partial J_1}{\partial \boldsymbol{\sigma}} = \{1,\ 1,\ 1,\ 0,\ 0,\ 0\} \\ \mathbf{n}_2^T &= \frac{\partial (J_{2D})^{1/2}}{\partial \boldsymbol{\sigma}} = \frac{1}{2(J_{2D})^{1/2}}\{\sigma'_x,\ \sigma'_y,\ \sigma'_z,\ 2\tau_{yz},\ 2\tau_{zx},\ 2\tau_{xy}\} \\ \mathbf{n}_3^T &= \frac{\partial J_{3D}}{\partial \boldsymbol{\sigma}} = \{(\sigma'_y\sigma'_z - \tau_{yz}^2 + J_{2D}/3),\ (\sigma'_x\sigma'_z - \tau_{xz}^2 + J_{2D}/3),\ (\sigma'_y\sigma'_x - \tau_{yx}^2 + J_{2D}/3), \\ &\qquad 2(\tau_{xz}\tau_{xy} - \sigma_x\tau_{yz}),\ 2(\tau_{xy}\tau_{yz} - \sigma_{yy}\tau_{xz}),\ 2(\tau_{yz}\tau_{xz} - \sigma_{zz}\tau_{xy})\} \end{aligned} \tag{3.14}$$

where $(\sigma_1, \sigma_2, \sigma_3)$ are the principal deviatoric stresses. A similar approach can be employed to calculate the plastic flow direction $\mathbf{n}_g$ to the plastic potential surface Q.

3.3. Plasticity models for porous material

Although, for the analysis of metal powder forming problems, various theories and methods were developed, they were not capable of being applied to the deformation of porous materials. In conventional plasticity theory, on which these theories and methods are based, zero volume changes are assumed for the material undergoing deformation, and this assumption applies also to pore-free metals. In the deformation of porous metals, however, the volume does not remain constant. Kuhn and Downey (1971) and Green (1972) independently proposed yield criteria and stress-strain relations for these materials by taking the effect of hydrostatic stress on yielding into consideration. In this theory, simple models are used to generate an explicit form for a second-order yield surface parameterized by the void ratio and relative density which leads to a stress-strain relation involving the yield criterion and a differential equation between strain, stress and relative density.

Shima and Oyane (1976) and Oyane *et al.* (1973, 1978) established a theory of plasticity for porous metals by extending the yield criteria to a general stress state. Oyane's theory has been applied to the slab method by Oh and Lee (1985), the slip-line field method by Oyane and Tabata (1974) and the upper-bound method by Shima *et al.* (1976) for analysis of forming of sintered powders. Gurson (1977) developed approximate yield criteria and flow rules for porous materials to be applied to the process of ductile fracture. A plasticity theory for porous materials with application to sintered metal powders, along with a criterion for the minimum stress level called 'limiting condition' to provide the determination of the beginning of the plastic deformation, was presented by Corapcioglu and Uz (1978).

In the development of a plasticity theory for porous materials, it is necessary to establish a yield criterion and a flow rule from which the stress-strain relations can be derived. The yielding of porous materials is more complicated than that of fully dense materials because the onset of yielding is influenced not only by the deviatoric stress component but also by the hydrostatic stress. It is for this reason that a von-Mises yield function cannot be used for the development of a plasticity theory for porous materials. Therefore, a yield function for porous materials, which can be considered as an extension of von-Mises's concept of yielding of fully dense materials, has been considered by many researchers as

$$A\,J_{2D} + B\,J_1^2 = \delta\,Y_0^2 = Y_\rho^2 \tag{3.15}$$

where Y_0 and Y_ρ are the yield stress of the solid material and yield stress of aggregate or partially dense material having relative density ρ, respectively. The parameters A, B and δ are functions of relative density. A typical yield surface for a porous material with a relative density ρ, expressed by the above equation, has the form of an ellipsoid whose major axis coincides with the σ_m axis, and is shown in Figure 3.2. Equation (3.15) represents a prolate spheroid in principal stress space which is a smooth, convex, bounded surface of a very simple form. The analogy with the comparable arguments for the use of the von-Mises yield criterion for non-porous materials is very compelling. In this model, if $A = 3$, $B = 0$ and $\delta = 1$ then we are left with the well known yield criterion of von-Mises.

This yield function includes the influence of the hydrostatic stress component and also satisfies the symmetry and convexity conditions required for the development of a plasticity theory for porous materials. Many different expressions have been proposed for A, B and δ in terms of relative density. Most of the expressions which have been derived are based upon experimental results and heuristic arguments. The expressions proposed by a number of investigators for A, B and δ are tabulated in Table 3.1. Also, the functions which make use of these constants are graphically represented in $\sqrt{3J_{2D}}/Y_0$ and J_1/Y_0 space for different relative densities, 0.75, 0.85 and 1.0, as shown in Figure 3.3(a)–(h) (Doraivelu *et al.* 1984). As can be observed, most of the case studies reporting compaction start at a relative density of about 0.7 - however, for the powder compaction process, the relative density of the loose powder fill is about 0.25 to 0.40. Furthermore, this type of material model neglects the hardening factor associated with the densification process where this is one of the main features in the process. Thus, the use of a yield criterion for a porous material is not suitable for loose metal powder and is more applicable to precompacted, sintered powder or secondary compaction processes and therefore, the adoption of a model which represents a frictional granular material is more applicable.

3.4. Plasticity models for granular materials

Considerable attention has been given since the late 1960's to the development of constitutive equations for soil and granular materials. Although many different models have been proposed, there is not yet firm agreement among researchers. Further, many of the constitutive models presented seem unnecessarily arbitrary, and at this stage it is important to emphasize that, to be satisfactory, a material model idealization should possess the following *necessary* properties:

1. The model should be *complete*, i.e. able to make statements about the material behavior for all stress and strain paths, and not merely restricted to a single class of paths (e.g., axial symmetry or pure shear);
2. It should be possible to identify the model parameters by means of *a small* number of standard, or simple material tests;
3. The model should be founded on some *physical interpretation* of the ways in which the material is responding to changes in applied stress or strain (e.g., the material should not be modeled as elastic if permanent deformations are observed upon unloading).

Among the earlier models which were basically developed to consider the plastic behavior of soil and granular material are Mohr-Coulomb (Coulomb 1776) and Drucker-Prager (1952). Nayak and Zienkiewicz (1972) converted these models into convenient forms to make the computational aspect simple and presented a generalized approach to elasto-plastic finite element analysis. Several types of constitutive models have been developed for granular material based on an extension and modification of Mohr-Coulomb's and Drucker-Prager's concepts. An extension of Drucker-Prager yield criterion was developed by Zienkiewicz *et al.* (1975), Zienkiewicz and Pande (1977) and Naylor (1978). Lade and Duncan (1975) developed a single parameter yield surface for cohesionless materials. Roscoe *et al.* (1963) formulated a cam-clay model for over-consolidated clays using an isotropic hardening plasticity formulation. A modification and extention of this model to a three-dimensional stress state was then developed by Roscoe and Burland (1968) as a modified cam-clay model. The critical state model was proposed by Naylor (1978) and Desai *et al.* (1981) to introduce a distinction between yielding and ultimate collapse.

In this section, we first present a review of various yield criteria which have been developed by researchers for soil and granular materials and then, the plasticity models for powders will be followed in the next section.

3.4.1. Mohr-Coulomb and Drucker-Prager yield criteria

The Mohr-Coulomb yield criterion is a generalization of the Coulomb failure law defined by

$$\tau + \sigma_n \tan\phi - c = 0 \tag{3.16}$$

where c and ϕ are the cohesion and internal friction, and τ and σ_n are the shear and normal stresses, respectively. From the conventional Mohr circle diagram, we have

$$-\frac{1}{2}(\sigma_1 - \sigma_3)\cos\phi = c - \left(\frac{\sigma_1 + \sigma_3}{2} + \frac{\sigma_1 - \sigma_3}{2}\sin\phi\right)\tan\phi \tag{3.17}$$

in which on rearranging gives

$$(\sigma_1 - \sigma_3) = 2c\cos\phi - (\sigma_1 + \sigma_3)\sin\phi \tag{3.18}$$

If $\phi = 0$, equation (3.18) results in $(\sigma_1 - \sigma_3)/2 - c = 0$, which represents the Tresca yield criterion of maximum shear stress. The Mohr-Coulomb criterion neglects the influence of the intermediate principal stress. In addition, numerical difficulties will be encountered for the Mohr-Coulomb yield surface around the corners, which lead to singularities and cause numerical ill-conditioning.

An approximation to the Mohr-Coulomb law was presented by Drucker and Prager (1952) as a modification of the von-Mises yield criterion. The influence of the hydrostatic stress component on yielding was introduced by inclusion of an additional term in the von-Mises expression as

$$\alpha J_1 + (J_{2D})^{1/2} = k \tag{3.19}$$

where α and k are the material constants, which may be related to Coulomb's material parameters c and ϕ. In order to make the Drucker-Prager circle coincide with the outer apices of the Mohr-Coulomb hexagon at any section, it can be shown that

$$\alpha = \frac{2\sin\phi}{\sqrt{3}(3 - \sin\phi)}, \quad k = \frac{6c\cos\phi}{\sqrt{3}(3 - \sin\phi)} \tag{3.20}$$

This is referred to the Drucker-Prager yield criterion I. The coincidence with the inner apices of the Mohr-Coulomb hexagon, referred to the Drucker-Prager yield criterion II, is provided by

$$\alpha = \frac{2\sin\phi}{\sqrt{3}(3 + \sin\phi)}, \quad k = \frac{6c\cos\phi}{\sqrt{3}(3 + \sin\phi)} \tag{3.21}$$

As the values of c and ϕ, which can be determined by using the conventional triaxial compression tests, are different from those obtained by a plane strain condition, the parameters α and k for a plane strain condition can be expressed as (Drucker-Prager yield criterion III)

$$\alpha = \frac{\tan\phi}{(4 + 12\tan^2\phi)^{1/2}}, \quad k = \frac{3c}{(9 + 12\tan^2\phi)^{1/2}} \tag{3.22}$$

The two material parameters α and k for the Drucker-Prager model can be determined from the slope and intercept of the failure envelope plotted in the space of J_1–$(J_{2D})^{1/2}$. If $\alpha = 0$ (i.e., $\phi = 0$), the yield surface reduces to a von-Mises criterion.

3.4.2. Naylor-Zienkiewicz model

A generalized yield surface similar to that of Drucker-Prager criterion was developed by Zienkiewicz *et al.* (1975), Zienkiewicz and Pande (1977) and Naylor (1978). This yield surface has a right circular family of conical yield surfaces defined by

$$F(p,q) = q - Mp - N = 0 \tag{3.23}$$

where $q = \sqrt{3}(J_{2D})^{1/2}$ and $p = J_1/3$ which is the mean stress. Based on different values of parameters M and N, the following yield surfaces have been derived;

i) Compromise cone, called the Nayor-Zienkiewicz model

$$F(p,q) = q - 2\sin\phi\, p - 2c\cos\phi = 0 \tag{3.24}$$

ii) Extended von-Mises model

$$F(p,q) = q - \frac{6\sin\phi}{3-\sin\phi}p - \frac{6c\cos\phi}{3-\sin\phi} = 0 \tag{3.25}$$

iii) Axial extension cone

$$F(p,q) = q - \frac{6\sin\phi}{3+\sin\phi}p + \frac{6c\cos\phi}{3+\sin\phi} = 0 \tag{3.26}$$

The Drucker-Prager criterion together with above associated models, i.e. compromise cone, extended von-Mises and axial-extension cone, are considered as a three dimensional approximation of the Mohr-Coulomb failure criterion with a simple smooth surface. These criteria are expressed as a linear combination of the first invariant of stress J_1 and square root of the second invariant of deviatoric stress J_{2D} along with two material parameters α and k. However, these models with associated flow rules can not predict the volumetric strain observed in experiments and cause excessive dilatancy (Naylor 1978).

3.4.3. Lade's model

Lade and Duncan (1975) developed a single parameter yield surface based on the data obtained from triaxial experimental tests for cohesionless materials. This model accounts different characteristics of the material, such as: the effect of principal stress, hydrostatic pressure and noncircular trace on the deviatoric plane. The model is expressed in terms of the first and third invariants of stress J_1 and J_3 as

$$F(J_1, J_3) = J_1^3 - k_1 J_3 = 0 \tag{3.27}$$

where k_l is the material parameter that depends on the density (i.e., the initial void ratio). For numerical computation, the third invariant of stress, i.e. $J_3 = \det(\sigma_{ij})$, can be expressed in terms of alternative stress invariants $(J_1, J_{2D}^{1/2}, \theta)$ as $J_3 = -\frac{2}{3\sqrt{3}} \sin 3\theta (J_{2D})^{3/2} - \frac{1}{2\sqrt{3}} J_1 J_{2D} + \frac{1}{27} J_1^3$.

An extension of Lade-Duncan model was derived by Lade (1977) in order to take into account the curvature of the failure envelope. This model which is also expressed in terms of the first and third invariants of stress is defined by

$$F(J_1, J_3) = (J_1^3 / J_3 - 27)(J_1 / P_a)^m - k_2 = 0 \tag{3.28}$$

where k_2 and m are the material parameters and P_a is the atmospheric pressure. In equation (3.28), the material parameters k_2 and m can be obtained by using the results of triaxial compression tests.

An isotropic yield surface which is more versatile than Lade-Duncan model was proposed by Van-Eekelen (1980). The model is expressed in terms of three invariants of stress as

$$F(J_1, J_{2D}, \theta) = (J_{2D})^{1/2} - \left(a - \frac{J_1}{3}\right) \alpha (1 - \beta \sin 3\theta)^n = 0 \tag{3.29}$$

where $\alpha = \sin \phi$ and β is a constant chosen in the range of $-1 < \beta < +1$. a is the constant parameter in which for cohesionless material $a = 0$ and n is a parameter which indicates the type of surface. In the case of $n = -0.229$ the model is optimal and always convex in nature (Van-Eekelen 1980).

In Table 3.2, the effective stress levels along with the equivalent yield stresses are given for different yield criteria employed for soil and granular material. Also presented in this table are the values of constants C_i, defined in equation (3.13), which are necessary in the calculation of plastic flow vector in elasto-plastic analysis.

3.5. Plasticity models for powder materials

In the analysis of powder compaction problems, the Mohr-Coulomb and Drucker-Prager criteria suffer from two deficiencies. Firstly, the assumption of an associated flow rule leads to excessive dilatency and secondly it has been seen that both models imply that the material can support an unlimited hydrostatic compression. These deficiencies can be removed by use of the close-ended conical yield surfaces, which assume that the yield surface comprises two distinct parts. The most practical yield functions which have been employed for powder materials are the Critical State model, Cam-clay model, modified Cam-clay model and cap model. These models reflect the yielding, frictional and densification characteristics of powder along with strain and geometrical hardening which occurs during the compaction process.

Apparently, Drucker *et al.* (1957) were the first to suggest an elasto-plastic work-hardening material model. They proposed that successive yield surfaces might resemble extended Drucker-Prager cones with convex end spherical caps. As the material strain hardens, both the cone and the end cap expand, the position and size of which depend upon the hydrostatic strain. In this model, there are two important innovations. The first is the introduction of the idea of a spherical cap fitted to the cone. The second is the use of current material density (specific volume or void ratio) as the strain-hardening parameter to determine the successive loading surfaces. This model demonstrates such phenomena as an increase in volume due to a decrease in hydrostatic stress and hence a decrease in yield strength and an increase or decrease in plastic volume due to yielding, the rate of which is a function of the state of yielding.

The application of this model to describe the behavior of frictional granular materials has been well established by many authors. An isotropic-kinematic hardening plasticity model which includes distortion, translation and rotation of the yield surfaces in the stress space was extended by Baltov and Sawczuk (1965) and developed for simulating experimental behavior of various geologic materials by Desai (1980) and Desai and Siriwardane (1980). The experimental results of Watson and Wert (1993) and Brown and Abou-Chedid (1994) demonstrated that the constitutive modeling of geological and frictional materials can be utilized to construct the suitable phenomenological constitutive models which capture the major features of the response of initially loose powders to the complex deformation processing histories encountered in the manufacture of engineering components by powder metallurgy techniques. In particular, they suggested that a 'two-mechanism-model', such as: Drucker-Prager or Mohr-Coulomb and elliptical cap models, which are widely used for geological materials and exhibit pressure dependent behavior can be useful for modeling the response of powder materials. These models consist of two yield surfaces; a 'distortion surface' and a 'consolidation' or 'cap' surface, which has an elliptical shape. The distortion surface controls the ultimate shear strength of material and the cap surface captures the hardening behavior of material under compression. A generalized cap model was also proposed by DiMaggio and Sandler (1971) for granular materials and then adopted for powder materials by Crawford and Lindskog (1983), Trassoras *et al.* (1989) and Haggblad and Oldenburg (1994).

In this section, we intend to present a detailed review of the most practical yield functions, the Cam-clay model, Desai's kinematic and isotropic hardening model and DiMaggio-Sandler's cap model which have been widely utilised by researchers for granular material and subsequently developed for powder forming analysis. Also, a combination of the Mohr-Coulomb and elliptical cap (MCEC), a double-surface plasticity and, a three-invariant single plasticity models, which have been employed in this work, will be explained in the next sections.

3.5.1. Cam-clay model

The cam-clay model is one of the most interesting models used for granular materials. This model was originally formulated by Roscoe *et al.* (1963) and Schofield and Wroth (1963) for normally and lightly over-consolidated clays, and then developed to a general three-dimensional stress state by Roscoe and Burland (1968) as the modified cam-clay model. They have shown that elliptical yield surfaces fit experimental data more closely than previous solutions (Drucker 1957 and Roscoe *et al.* 1963) and that the phenomena of strain hardening and softening are automatically accounted for in the problem formulation. The model introduces a distinction between yielding and ultimate collapse by using the concept of a critical state line in conjunction with a strain dependent yield surface. The material is in a critical state if during deformation, there is no change in the void ratio and effective stress component. In this model, the material undergoing shear deformation can pass through a yield point without collapse and continue to deform until eventually the critical state line is reached.

A yield surface of this form, which is only a function of the first two stress invariants, is shown in Figures 3.4(a) and (b). These figures are drawn so that the 'critical-state line', or ultimate-failure surface, corresponds to a generalization of the Mohr-Coulomb representation given by Drucker and Prager. The normality principle assuming associated plastic behavior applies to the elliptical surface, thus defining the direction of plastic straining. Since at the intersection of the critical-state line and the ellipse the normal to the surface is vertical, there is at this point no component of plastic volumetric strain. Consequently, any intersection to the right of this causes a plastic volumetric decrease and hence strain hardening, whereas any intersection to the left results in a plastic volumetric increase and strain softening.

The yield surface equation of the Cam-clay model (Figure 3.4-a), which is based on the critical state principle where the volumetric strain is zero, is given by Schofield and Wroth (1968) as

$$F(p, q, p_0) = \frac{q}{M_{cs}\, p} + \ln\left(\frac{p}{p_0}\right) = 0 \tag{3.30}$$

where M_{cs} is the slope of the failure line in the (p, q) plane with $p = J_1/3$ and $q = \sqrt{3J_2'}$ and p_0 is the hardening parameter.

A modification of the cam-clay model to a generalized three-dimensional stress and strain condition was extended by Roscoe and Burland (1968) and called the 'modified' Cam-clay model. The modified Cam-clay model, which fits experimental data quite satisfactorily, was the first comprehensive and entirely self-consistent mathematical model for the prediction of the mechanical behavior of ideal 'wet' clay. The model was described in terms of the effective stress quantities p and q (Figure 3.4-b) which are relevant to the discussion of soil response in conventional triaxial tests. This model comprised a series of elliptical yield surfaces with all the ellipses passing through the same point at the origin. The ellipse yield surface in the (p, q) plane is defined by the equation (Lewis and Schrefler 1987)

$$F(p, q, p_c) = \frac{q^2}{M_{cs}^2} - 2pp_c(\varepsilon_v^p) + p^2 = 0 \tag{3.31}$$

where p_c is the current semi-diameter of the ellipse in the p–direction, being a function of the plastic volumetric strain ε_v^p, defined by

$$\varepsilon_v^p = \frac{\lambda - K}{1 + e_o} \log_{10} \frac{p_c'}{p_c} \tag{3.32}$$

where λ and K are the compression and swelling indices, respectively, e_o is the initial void ratio and p_c' is the pre-consolidation pressure. As the ellipse changes in magnitude, the locus of the critical state points is a pyramid with its apex at the origin shown by the critical state line in the (p, q) plane. The modified Cam-clay model is an isotropic, nonlinear elasto-plastic strain-hardening model. Only the volumetric strain is assumed to be partially recoverable; i.e. the elastic shearing strain is assumed to be identically zero. The elastic volumetric strain is nonlinearly dependent on the hydrostatic stress and is independent of the deviatoric (shear) stresses.

A generalization of the modified Cam-clay model was proposed by Zienkiewicz *et al.* (1975) by introducing the effect of the stress invariant in term of θ. This model has the form of an ellipsoid whose section in the π – plane is similar to that of Mohr-Coulomb criterion. Replacing the critical state line by the Mohr-Coulomb envelope, it leads to the following yield surface

$$F(p,q,\theta,p_{co}) = \frac{3q^2}{(p_{co}\tan\phi)^2} + \frac{(p - p_{co} + a)^2}{p_{co}^2} - 1 = 0 \tag{3.33}$$

where p_{co} is the semi-diameter of the ellipse in the $p-$direction and indicates the initial pre-consolidation pressure. In equation (3.33), $a = \overline{C}/\tan\phi$ where

$$\overline{C} = \frac{3c\cos\phi}{\sqrt{3}\cos\theta - \sin\theta\sin\phi} \tag{3.34}$$

and

$$\tan\phi = \frac{3\sin\phi}{\sqrt{3}\cos\theta - \sin\theta\sin\phi} \tag{3.35}$$

The hardening rule is expressed based on the volumetric stress change by means of a consolidation test as

$$\Delta p_{co} = p_{co}\exp(-\chi\,\varepsilon_v^p) \tag{3.36}$$

where χ is a constant defined by $\chi = (1+e_o)/(\lambda - K)$, with e_o denoting the initial voids ratio, λ the compression index and K the swelling index.

The application of the Cam-clay model to powder compaction processes was proposed by Sun and Kim (1997). However, it must be noted that the most successful applications of the Cam-clay, or modified Cam-clay model, have been to problems involving the loading of geotechnical constructions on clay which exhibit high cohesion stresses and therefore the model is not appropriate for loose powder materials.

3.5.2. Desai's kinematic-isotropic hardening model

The kinematic and isotropic hardening model which includes distortion, translation and rotation of the yield surfaces in the stress space, was extended by Baltov and Sawczuk (1965) based on a generalization form of the von-Mises criterion. A special case of this model was presented by Desai (1980) and Desai and Siriwardane (1980) under various stress paths in a true triaxial device for various geologic materials. They assumed that the influence of all the three invariants of stress states, J_1, J_2 and J_3, should be included in the yield, failure and plastic potential functions for geological media. The idea of expressing behavior of a material by using functional representations was considered by a function f expressed as a (complete) polynomial in terms of these three invariants. A large number of combinations as truncated forms of the function f can be chosen by retaining terms and many of the resulting expressions in yield, failure and plastic potential functions. One of these yield surfaces, expressed in terms of J_1, J_{2D} and J_3 and shown in Figure 3.4(c), is (Desai and Siriwardane 1984)

$$f(\boldsymbol{\sigma},\beta) = J_{2D} + \alpha J_1^2 - \beta J_1 (J_3)^{1/3} - \gamma J_1 - k^2 = 0 \tag{3.37}$$

where α, γ and k are material parameters which can be expressed as functions of quantities such as strains and density. β is treated as the hardening or growth function. The function in equation (3.37) is convex and continuous in the stress space (Figure 3.4-c). It is interesting and useful to note that this model plots as a series of single continuous functions, which include both the yielding and ultimate, yield (or failure) behavior. This is a significant simplification when compared to models such as the 'cap' and 'critical state' which involve two separate functions for yielding and failure.

Recently, a plasticity model involving a non-associated flow rule with respect to the response of cohesionless materials was developed by Desai and Hashmi (1989), as shown in Figure 3.4(d). In this approach, a basic associative model was established to describe the behavior of an initially isotropic material hardening isotropically and following an associative rule. Corrections were then introduced to obtain different representations such as non-associative, anisotropic hardening and damage with strain softening. The yield function of Desai-Hashmi's model is given by

$$f(\boldsymbol{\sigma},\alpha)=(J_{2D})^2-(-\alpha J_1^n+\gamma J_1^2)(1-\beta J_r)^m=0 \tag{3.38}$$

and the plastic potential function is expressed by $Q=f+h$, in which h is the correction function. In equation (3.38), β, γ and m are response functions, associated with the ultimate yield surface, n is a constant associated with the transition point from contractive to dilative response, and α denotes the hardening parameter. J_r is the stress ratio defined as $J_r=(J_{2D})^{1/3}/(J_{2D})^{1/2}$.

For metal powder materials, Sava (1997) assumed that in equation (3.38), J_r is constant (no phase change) and the yield function can be given by

$$f(\boldsymbol{\sigma},\alpha)=(J_{2D})^2-\gamma J_1^2+\alpha J_1^n=0 \tag{3.39}$$

3.5.3. DiMaggio-Sandler's cap model

A generalized cap model, based on classical plasticity theory, originally proposed by DiMaggio and Sandler (1971) and Sandler *et al.* (1976), allows the control of dilatancy by means of a hardening cap that intersects a fixed failure envelope in a nonsmooth fashion. A commonly accepted interpretation of this model, given by Desai and Siriwardane (1984), Chen and Baladi (1985), Simo *et al.* (1988) and Hofstetter *et al.* (1993), assumes a one-to-one correspondence between the hardening of the cap and the plastic volume change. However, if a one-to-one hardening law is postulated, softening response may occur when the stress point is located at the compressive corner region. Motivated by this softening behavior, Sandler and Rubin (1979) proposed a modified hardening law which prevents softening response. The generalized cap models have also been expanded by Nelson and Baladi (1977), Nelson (1978) and Sandler and Baron (1979) to include rate effects, and anisotropic behavior within the yield surface and visco-plastic behavior during yielding.

The cap models are based on the concept of continuous yielding of materials, but they are expressed in terms of a three-dimensional state of stress and are formulated on the basis of consistent mechanics principles. The yield surface of this elasto-plastic model has a moving cap, intersecting the hydrostatic loading line, whose position is a function of plastic volumetric strain (Figure 3.4-e). The main features of the cap model include a failure surface and an elliptical yield cap which closes the open space between the failure surface and the hydrostatic axis. The yield cap expands in the stress space according to a specified hardening rule. The functional forms for these surfaces are (Sandler *et al.* 1976)

$$f_1(\boldsymbol{\sigma})=(J_{2D})^{1/2}+DJ_1+Ce^{-BJ_1}-A=0 \tag{3.40}$$

$$f_2(\boldsymbol{\sigma},\kappa)=\frac{1}{R}\left([X(\kappa)-L(\kappa)]^2-[J_1-L(\kappa)]^2\right)^{1/2}=0 \tag{3.41}$$

in which

$$X(\kappa)=\kappa-RF_1(\kappa), \qquad L(\kappa)=\begin{cases}\kappa & if\ \kappa<0\\ 0 & if\ \kappa\geq 0\end{cases} \tag{3.42}$$

where κ is the hardening parameter, which defined as a functional of the plastic volumetric strain, and the quantities *A*, *B*, *C*, *D* and *R* in the above equations are material parameters. This model is shown to control excessive dilatancy, allows hysteresis and an apparent Bauschinger effect, and satisfies continuity and uniqueness conditions when used with an associated flow rule and linear elastic behavior.

There are certain differences between the cap and Cam-clay models. In the cap models, the portion of the cap above the fixed yield, or failure surface, (Figure 3.4-e) is omitted and only the yield surfaces $f_1(\boldsymbol{\sigma})$ and $f_2(\boldsymbol{\sigma},\kappa)$ are considered. In the Cam-clay models, the moving cap plays the main role in defining yielding, and the fixed yield surface is used essentially to define the critical state. On the other hand, in the cap model, both the fixed and moving surfaces are used to define the yielding process. For this reason, the cap model is referred to as a Cam-clay model and has been used in the analysis of nonlinear behavior of powder forming processes (Crawford and Lindskog 1983, Trassoras *et al.* 1989 and Haggblad and Oldenburg 1994).

3.6. MCEC plasticity model

Many studies of powder compaction have been made, most of them concerned with the relationship between compacting pressure and density, basic equations, such as yield criteria were proposed by means of which it is possible to predict the response of pressed powders. Most of these theories concern elasto-plastic laws with isotropic or kinematic hardening, and associated or non-associated flow rule, which indicate the direction of the plastic strain. In this section, the nonlinear behavior of powder materials is described using a combination of the Mohr-Coulomb and Elliptical Cap (MCEC) yield model. The yield function consists of a perfectly-plastic (failure) portion fitted to a strain-hardening elliptical cap. An associated flow rule is employed for the failure and cap functions. The functional forms for both the perfectly-plastic and the strain-hardening portions may be quite general and would allow for the fitting of a wide range of material properties. The movement of the cap is controlled by the increase, or decrease, of the plastic volumetric strain. Thus, strain-hardening can be reversed and it is this mechanism that leads to an effective control of dilatancy, which can be kept quite small (effectively zero) as required for many powders.

The cap model is a plasticity model defined by a non-softening convex yield surface and a plastic strain rate vector that is normal to the yield surface in stress space. The yield surface is defined by means of a failure envelope and a hardening cap. In Figure 3.5, the two-invariant-rate-independent cap model is defined by a convex yield surface, which consists of a failure envelope, such as Mohr-Coulomb yield surface, $F_1(\boldsymbol{\sigma})$ and an elliptical yield cap surface $F_2(\boldsymbol{\sigma},\sigma_c)$, where σ_c is the hardening parameter being a function of plastic volumetric strain. The functional form for the failure envelope is

$$F_1(\boldsymbol{\sigma})=\frac{1}{3}J_1\sin\varphi+\sqrt{J_{2D}}\cos\theta-\sqrt{\frac{J_{2D}}{3}}\sin\theta\cos\varphi-c\cos\varphi=0, \quad \text{for} \quad \sigma_m>L(\sigma_c) \tag{3.43}$$

where σ_m is the mean stress, defined as $\sigma_m = \frac{1}{3}J_1$ and θ is the dilatancy angle. Note that σ_m is chosen as being positive in tension. The parameters c and φ are the cohesion and the internal friction angle of the metal powder. These parameters can be easily determined by a series of standard tests such as a direct shear or triaxial test.

It is implicitly understood that the yield cap function satisfies the relation

$$0 \le F_2(\boldsymbol{\sigma}, \sigma_c) \le F_1(\boldsymbol{\sigma}) \qquad \text{for} \qquad L(\sigma_c) \ge \sigma_m \ge \chi(\sigma_c) \tag{3.44}$$

where $L(\sigma_c)$ and $\chi(\sigma_c)$ define the first stress invariant range of the cap (Figure 3.5). Certain limitations must be placed on the functional form that define the model. In order to avoid work-softening in the model, the functions $L(\sigma_c)$ and $\chi(\sigma_c)$ must be continuous and monotonically increasing functions and

$$\frac{\partial F_2}{\partial \sigma_m} \ge 0 \qquad \text{and} \qquad \frac{\partial F_2}{\partial \sigma_c} < 0 \tag{3.45}$$

The cap must extend from the σ_m–axis to a point on the failure envelope. The functional form for the cap, which is elliptically shaped, is given by

$$F_2(\boldsymbol{\sigma}, \sigma_c) = J_2' + M^2\left[(\sigma_m + \sigma_c)^2 - \left(\sigma_c + \frac{c}{\tan\varphi}\right)^2\right] = 0, \qquad \text{for} \quad L(\sigma_c) \ge \sigma_m \ge \chi(\sigma_c) \tag{3.46}$$

where M is the slope of the Mohr-Coulomb failure surface in the space $\left(\sqrt{J_{2D}}, \sigma_m\right)$ and σ_c is the critical stress governed by the function of plastic volumetric strain, which is based on a 'hardening rule'. A hardening rule is used to define the dependence of the yield surface on the degree of plastic straining. The evolution of the yield surface must respond to either 'hardening' or 'softening'. The 'hardening' is caused by the elimination of voids with an increase in particle resistance to plastic flow and consequently the powder material itself becomes harder. Conversely, 'softening' is caused due to void expansion, the compacted powder may crack especially at corners or at other points of high stress concentration, which will cause a reduction of the ultimate strength of the finished component.

The hardening behavior of the material is reflected in a movement of the densification line and its intersection with the shear envelop (Naylor *et al.* 1981) and was applied by Gethin *et al.* (1995) for powder material. The hardening parameter $L(\sigma_c)$ is chosen so that

$$L(\sigma_c) = \langle \sigma_c \rangle = \begin{cases} \sigma_c & \text{if } \sigma_c < 0 \\ 0 & \text{if } \sigma_c \ge 0 \end{cases} \tag{3.47}$$

and the point of intersection of the cap with the σ_m–axis is defined as

$$\chi(\sigma_c) = 2\sigma_c - \frac{c}{\tan\varphi} \tag{3.48}$$

where σ_c is defined in term of ε_v^p as

$$\sigma_c = f(\varepsilon_v^p) = \sigma_{c0} \exp\left(-\frac{\bar{\varepsilon}_v^p}{\bar{\chi}}\right) \tag{3.49}$$

where σ_{c0} is the initial hydrostatic stress of compaction that is derived from a consideration of die filling and assumed to be uniform, and $\bar{\chi}$ is the plastic hardening coefficient which is derived from an uniaxial compression/relaxation test. ε_v^p denotes the plastic volumetric strain which, in the present case of an associative flow rule, takes the form

$$\dot{\varepsilon}_v^p = \dot{\lambda} \frac{\partial F}{\partial \sigma_{ij}} \delta_{ij} \tag{3.50}$$

where λ denotes a proportionality constant termed the plastic multiplier. The effective plastic strain $\bar{\varepsilon}^p$ may be defined in a rate form as

$$\dot{\bar{\varepsilon}}^p = \sqrt{\frac{2}{3} \dot{e}_{ij}^p \dot{e}_{ij}^p} \tag{3.51}$$

where e_{ij} are the deviatoric strains, e.g. $e_{ij} = \varepsilon_{ij} - \frac{1}{3}\delta_{ij}\varepsilon_{kk}$. In equation (3.51), $\bar{\varepsilon}_v^p$ is the effective volumetric plastic strain, given by the differential relation

$$\dot{\bar{\varepsilon}}_v^p = \begin{cases} \dot{\varepsilon}_v^p & \text{if } \dot{\varepsilon}_v^p \leq 0 \text{ or } \sigma_c < \sigma_m \text{ and } \sigma_c < 0 \\ 0 & \text{otherwise} \end{cases} \tag{3.52}$$

The mathematical condition $\sigma_c < \sigma_m$ when $\dot{\varepsilon}_v^p > 0$ in the first line of equation (3.52) is a consequence of the stability requirement that the cap be prevented from acting as a softening yield surface and the model behaves as in perfect plasticity. If dilatancy occurs when $\sigma_m > 0$, equation (3.52) limits the shrinking of the cap to $\sigma_c = L(\sigma_c) = 0$ and this condition ensures that the cap remains finite.

The models can handle a wider range of powder material characteristics from loose powders, with $\bar{\rho} = 0.2$, to dense material, with $\bar{\rho} = 1.0$, where $\bar{\rho}$ denotes the relative density. It should be mentioned that the parameters c and φ change as the powder is compressed, approaching the dense (solid) material where $\varphi = 0$ and the Mohr-Coulomb function approaches the von-Mises form.

In this model, the difficulty of yielding at stresses below the failure envelope can be avoided if it is recognized that yielding may occur before the material reaches a failure envelope. Figure 3.6 shows a family of yield criteria forming 'caps' to the open-ended Mohr-Coulomb envelope. As the stress exceeds the yield stress value at point A and densification occurs, the plastic strain vector, normal to the original ellipse, produces a plastic volumetric decrease which causes the material to harden and the ellipse expands until eventually the point B is reached. The elliptical yield surface is strain dependent and expands when the material hardens, which is associated with a decrease in volume. Conversely, this contracts when the material softens and is associated with an increase in volume. As the loading increases to point C, a crack, or failure, occurs since softening takes place due to the increase in volumetric strain. The 3D representation of MCEC plasticity is shown in Figure 3.7.

3.7. Double-surface plasticity model

The double surface cap plasticity model is one of the most practical yield surfaces employed for powder materials. The model originally proposed by Sandler *et al.* (1976, 1979), are formulated in three-dimensional stress states. It comprises three distinct parts; a tension cut-off, a failure envelope and a moving cap, which expands in the stress space, as shown in Figure 3.8. The functional forms of these surfaces are as follows

$$f_1 = (J_{2D})^{1/2} - \theta J_1 + \gamma e^{-\beta J_1} - \alpha = 0 \tag{3.53}$$

$$f_2 = R^2 J_{2D} + (J_1 - L)^2 - R^2 b^2 = 0 \tag{3.54}$$

$$f_3 = J_1 - T = 0 \tag{3.55}$$

where α, β, γ and θ are the parameters of fixed yield surface f_1, which controls the deviatoric stress limits. In above equations, X denotes the value of J_1 at the intersection point of function f_2 with J_1 −axis and, L and b are the points of intersection of f_1 and f_2 in J_1 and $\sqrt{J_{2D}}$ directions, respectively. The fixed yield surface f_1 is defined by an exponential function which consists of two different Drucker-Prager yield surfaces. The cap yield surface f_2 is an elliptical function, with R denoting the ratio of two elliptical cap's diameters. The function f_3 indicates the tension cutoff zone, with T denoting the material's tension limit.

The hardening rule for moving cap is related to the volumetric plastic strain ε_v^p as

$$X(\kappa) = X(\varepsilon_v^p) = \frac{-1}{D} \ln\left(\frac{1-\varepsilon_v^p}{W}\right) + X_0 \tag{3.56}$$

where D and W are material parameters and X_0 refers to the position of initial cap surface. The plastic hardening/softening modulus H is zero for f_1 and f_3. Figure 3.9 shows how the cap yield surface grows with densification due to increase of the volumetric plastic strain. The 3D representation of the above double-surface plasticity is shown in Figure 3.10 for different values of volumetric plastic strain.

3.7.1. Material property matrix for double surface plasticity

In order to compute the elasto-plastic matrix, the values of loading direction vector $\mathbf{n}$, plastic flow direction vector $\mathbf{n}_g$, plastic hardening/softening modulus H and elastic matrix $\mathbf{D}^e$ must be determined at each stress Gauss point. Considering the yield criterion in terms of first invariant of stress J_1 and the second and third invariants of deviatoric stresses J_{2D} and J_{3D}, the flow direction vector introduced by (3.4) can be written as

$$\frac{\partial F}{\partial \boldsymbol{\sigma}} = C_1 \frac{\partial J_1}{\partial \boldsymbol{\sigma}} + C_2 \frac{\partial (J_{2D})^{1/2}}{\partial \boldsymbol{\sigma}} + C_3 \frac{\partial J_{3D}}{\partial \boldsymbol{\sigma}} \tag{3.57}$$

where

$$C_1 = \frac{\partial F}{\partial J_1}, \quad C_2 = \frac{\partial F}{\partial (J_{2D})^{1/2}}, \quad C_3 = \frac{\partial F}{\partial J_{3D}} \tag{3.58}$$

in which the constants C_1, C_2 and C_3 are necessary to define the fixed, moving cap and tension cut-off yield surfaces.

3.7.1.1. The fixed yield surface

Consider the associated flow rule, the values of constants C_1, C_2 and C_3 for the fixed yield surface are

$$C_1 = -\theta - \beta\gamma e^{-\beta J_1}$$
$$C_2 = 1 \tag{3.59}$$
$$C_3 = 0$$

In above relations, the value of C_1 is less than zero. It means that the normal vector to the plastic potential surface is in opposite direction of J_1 −axis. On the other hand, the associated flow rule results in dilation of material (increase of volume) and the stress state lays on the fixed yield surface at the ultimate shear stress. As the ultimate shear stress shows no volume changes in material, the non-associated flow rule can be therefore utilized, in which the normal vector to the plastic potential surface is perpendicular to J_1 −axis. Thus, the values of constants C_1, C_2 and C_3 for plastic flow vector $\mathbf{n}_g$ are as follows

$$C_1 = 0$$
$$C_2 = 1 \tag{3.60}$$
$$C_3 = 0$$

3.7.1.2. The moving cap surface

In order to calculate the values of constants C_1, C_2 and C_3 and parameter of hardening/softening modulus H for moving cap surface, function f_2 in equation (3.54) is rewritten as

$$\sqrt{R^2 J_{2D} + (J_1 - L)^2} - Rb = 0 \tag{3.61}$$

Thus, the values of constants C_i in flow direction vector are

$$C_1 = \frac{J_1 - L}{\sqrt{R^2 J_{2D} + (J_1 - L)^2}}$$
$$C_2 = \frac{R^2 \sqrt{J_{2D}}}{\sqrt{R^2 J_{2D} + (J_1 - L)^2}} \tag{3.62}$$
$$C_3 = 0$$

The plastic hardening/softening modulus can be determined using equation (3.7). Considering the hardening parameter κ equal to ε_v^p, it leads to

$$H = -\left(\frac{\partial f_2}{\partial \varepsilon_v^p}\right)\left(\frac{\partial \varepsilon_v^p}{\partial \boldsymbol{\varepsilon}}\right)^T \mathbf{n}_g = -\left(\frac{\partial f_2}{\partial \varepsilon_v^p}\right)(n_{g\,ii}) \tag{3.63}$$

where

$$\frac{\partial f_2}{\partial \varepsilon_v^p} = \frac{\partial f_2}{\partial L}\frac{\partial L}{\partial \varepsilon_v^p} + \frac{\partial f_2}{\partial b}\frac{\partial b}{\partial \varepsilon_v^p} \tag{3.64}$$

To determine the plastic hardening/softening modulus, we need to evaluate the values of derivatives in right side of equation (3.64).

$$\frac{\partial f_2}{\partial L} = 2(J_1 - L) \tag{3.65}$$

$$\frac{\partial f_2}{\partial b} = -2R^2 b \tag{3.66}$$

From the definition of moving cap surface, parameters L and b are defined as

$$L = X - Rb \tag{3.67}$$

$$b = \theta L - \gamma e^{-\beta L} + \alpha \tag{3.68}$$

Substituting equation (3.68) into (3.67) lead to

$$L(1 + R\theta) - R\gamma e^{-\beta L} + R\alpha = X \tag{3.69}$$

By taking derivative from the above equation, we will arrive to

$$\frac{\partial L}{\partial \varepsilon_v^p} = \frac{\partial X}{\partial \varepsilon_v^p}\,\frac{1}{1 + R\theta + R\beta\gamma e^{-\beta L}} \tag{3.70}$$

where

$$\frac{\partial X}{\partial \varepsilon_v^p} = \frac{1}{D(W - \varepsilon_v^p)} \tag{3.71}$$

Finally, the definition of $\partial b / \partial \varepsilon_v^p$ used in equation (3.64) can be calculated using equation (3.67) as

$$\frac{\partial b}{\partial \varepsilon_v^p} = \frac{1}{R}\left(\frac{\partial X}{\partial \varepsilon_v^p} - \frac{\partial L}{\partial \varepsilon_v^p}\right) \tag{3.72}$$

3.7.1.3. The tension cut-off yield surface

The values of constants C_1, C_2 and C_3 in plastic flow vector can be simply calculated according to the definition of the tension cut-off yield surface given in equation (3.55) as

$$C_1 = -1.0$$
$$C_2 = 0 \qquad (3.73)$$
$$C_3 = 0$$

and the plastic hardening/softening modulus for tension cut-off surface is assumed to be zero.

3.7.2. Parameter determination of double-surface plasticity

In order to evaluate the parameters of double-surface plasticity, it is necessary to obtain these values for the fixed, moving cap and tension cut-off yield surfaces. The fixed yield surface has an exponential form which is composed of an initial portion of the Drucker-Prager envelope joined smoothly to the subsequent Drucker-Prager line, as demonstrated in Figure 3.11(a). The slope of the first Drucker-Prager line, i.e. line I, is grater than the slope of the second Drucker-Prager line, i.e. line II. The logic for adopting the second Drucker-Prager surface in higher stresses is based on the observation that at higher stresses the material behave like a liquid. This was adopted particularly to simulate behavior of cohesionless materials subjected to high stresses.

In order to determine the parameters of fixed yield surface f_1, the results of true triaxial tests are necessary at the ultimate shear stresses. These ultimate points are presented in Figure 3.11(b) in $J_1 - \sqrt{J_{2D}}$ plane. Considering the state of stress when J_1 is equal to zero and substituting $J_1 = 0$ in equation (3.53) lead to

$$f_1 = \sqrt{J_{2D}} + \gamma - \alpha = 0 \qquad (3.74)$$

It means that the intersection of the fixed yield surface, or the Drucker-Prager line I, with $\sqrt{J_{2D}}$ −axis is $\alpha - \gamma$. As β is assumed to be a positive quantity and compression is taken as positive, the quantity of $e^{-\beta J_1}$ will be very small for large value of J_1. Thus, it leads to

$$f_1 = \sqrt{J_{2D}} - \theta J_1 - \alpha = 0 \qquad (3.75)$$

The above equation represents the Drucker-Prager yield criterion, which is shown by line II in Figure 3.11. As can be observed the slope of this line is θ and its intersection with $\sqrt{J_{2D}}$ −axis is α. The parameter β in fixed yield surface can be evaluated using an arbitrary point on transition curve, as shown in Figure 3.11.

In order to determine the parameters of moving cap surface f_2, we need to investigate the plastic deformations under compression. The increase of compression causes the increase of plastic volumetric strain and hence, the expansion of cap surface in J_1 −axis. The values of D, W and X_0 can be computed by using equation (3.56). The value of X_0 is zero when there is no significant initial yielding cap. For granular and porous materials, there is no significant initial stress and $X_0 = 0$. The values of D and W can be obtained from the confining pressure, or hydrostatic pressure test. From these tests, the value of X obtained by equation (3.56) is equal to J_1. The elastic volumetric strain ε_v^e can be evaluated from the variation of J_1 with ε_v on unloading portion curve obtained from the hydrostatic pressure test. The plastic volumetric strains ε_v^p can be expressed in terms of total and elastic components of strain as

$$\varepsilon_v^p = \varepsilon_v - \varepsilon_v^e \tag{3.76}$$

The values of D and W can be estimated using the variation of J_1 and X with ε_v^p. The parameter R can be determined using the confining pressure and a set of triaxial tests. For the cap yield surface with a given value of X, the shape of surface can be specified in $J_1 - \sqrt{J_{2D}}$ plane for the same values of plastic volumetric strain from different hydrostatic and triaxial tests.

3.8. The three-invariant single plasticity model

The single cone-cap plasticity can be written using the invariants of stress and deviatoric stress tensors in the following manner

$$F(\boldsymbol{\sigma},\eta) = \psi J_{3D}^{2/3} + J_{2D} + \left(\frac{f_d}{f_h}\right)^2 J_1^2 - f_d^2 = 0 \tag{3.77}$$

where f_h is a positive increasing function of relative density, which controls the intersection of J_1 axis with yield surface function at the maximum value of compression, and f_d is a function of the first invariant of stress tensor and relative density. This parameter controls the size of yield surface in the perpendicular direction to J_1 axis. Different combinations of f_h and f_d lead to different shapes of the yield surface.

The following polynomial functions are proposed for the dependence of f_h and f_d on relative density

$$f_h = \left(\frac{1}{1-\eta^b}\right)\left(a_0 + a_1\eta + a_2\eta^2 + a_3\eta^3\right) \tag{3.78}$$

$$f_d = (\alpha - \beta J_1) + \left(c_1\eta + c_2\eta^2 + c_3\eta^3 + c_4\eta^4\right) \tag{3.79}$$

where η is the relative density defined as

$$\eta = \frac{\rho}{\rho_s} \tag{3.80}$$

where ρ is the current total density, ρ_s denotes the density of solid particles, in which η_0 indicates the position of first yield surface according to the initial value of relative density, and ψ, b, a_0, a_1, a_2, a_3, α, β, c_1, c_2, c_3, c_4 are parameters of the powder.

In order to demonstrate different aspects of the yield surface of equation (3.77), we investigate the effects of f_h and f_d on its behavior. Consider that the first two terms of equation (3.77) are zero, it can be written as

$$\left(\frac{f_d}{f_h}\right)^2 J_1^2 = f_d^2 \tag{3.81}$$

The above equation generally yields to three roots, the points of intersection of yield surface with J_1–axis, i.e. $J_1 = \pm f_h$ and one more from $f_d = 0.0$, which has been defined in equation (3.79) as a linear function of J_1.

If the third condition (i.e. $f_d = 0.0$) does not lead to any value for J_1, for example $\beta = 0.0$, the yield surface of equation (3.77) yields to two roots for J_1, the positive and negative values of f_h, which results in an elliptical shape in meridian plane, as illustrated in Figure 3.12(a). This yield surface is very similar to the elliptical yield functions developed by authors for porous metal and sintered powder based on an extension of von-Mises's concept (Green 1972, Shima and Oyane 1976 and Doraivelu *et al.* 1984). In the special case of $f_d = cte$, where the parameters c_1, c_2, c_3, c_4 are zero, the yield surface (3.77) results in special forms of elliptical shape, as shown in Figure 3.12(b). If $f_d = 0.0$ leads to a positive value of J_1 more than f_h, the yield surface of equation (3.77) results in the curved elliptical shape in meridian plane, as demonstrated in Figure 3.12(c). In this case, the points of intersection of yield surface with J_1–axis are $-f_h$ and $+f_h$, and the other value obtained by $f_d = 0.0$ is not acceptable. By choosing the zero values for parameters c_1, c_2, c_3, c_4, special forms of yield surface can be produced, as presented in Figure 3.12(d).

If $f_d = 0.0$ leads to the value of J_1 between $-f_h$ and $+f_h$, the cone-cap yield surface will be produced from equation (3.77) based on the intersection points of $J_1 = -f_h$ and the value obtained from $f_d = 0.0$, as demonstrated in Figure 3.12(e). In the case of $f_d = cte$ (i.e. c_1, c_2, c_3, $c_4 = 0.0$), it results in special forms of cone-cap plasticity, as illustrated in Figure 3.12(f). This yield surface is very similar to the double-surface cap models, i.e. a combination of Mohr-Coulomb or Drucker-Prager and elliptical surfaces, which has been extensively used by authors to demonstrate the behavior of powder and granular materials (Aydin *et al.* 1996, Brandt and Nilsson 1999, Gu *et al.* 2001 and Lewis and Khoei 2001). It is worth mentioning that as the double-surface plasticity consists of two different yield functions, special treatment should be made to avoid numerical difficulties in the intersection of these two surfaces (Gu *et al.* 2001), however, the single yield surface (3.77) does not have such a drawback. Finally, if $f_d = 0.0$ leads to a value for J_1 less than $-f_h$, it does not result to any yield surface. A summary of above explanation is illustrated in Table 3.3.

The three-dimensional yield surface (3.77) is generated in the principal stress space, as shown in Figure 3.13, for different values of relative density, i.e. 0.5, 0.7 and 1.0, where the intersection point of yield surface with J_1–axis are $-f_h$ and $+f_h$. This representation clearly shows how the yield surface grows with densification, eventually becoming independent of the hydrostatic stress J_1 at full density, where the von-Mises yield surface is generated. In Figure 3.14, the 3D representation of yield surface (3.77) is shown for different values of relative density, where the points of intersection of yield surface with J_1–axis are $-f_h$ and the value obtained from $f_d = 0.0$. As can be observed from the figure the yield surface grows with densification and reduces to the Drucker-Prager yield function for full dense bodies.

In Figure 3.15, the effects of parameter ψ on the shape of yield surface (3.77) are shown in principal stress space for the value of $\psi = 0.1$. As can be observed from the figure, the value of ψ greater than zero causes triangularity of deviatoric trace along the hydrostatic axis. This yield surface is similar to the irregular hexagonal pyramid of the Mohr-Coulomb and cone-cap yield surface employed by researchers for description of soil and geomaterial behavior (Lade and Kim-Moon 1995).

3.8.1. Powder property matrix for single-cap plasticity

In order to derive the constitutive elasto-plastic matrix and its components, we need to calculate $\mathbf{D}^e$, $\mathbf{n}$, $\mathbf{n}_g$ and H, defined in equations (3.2) and (3.3), as functions of powder relative density. For the proposed cap plasticity, the nonlinear elastic behavior is considered similar to the definition of f_h given in equation (3.78). The Young modulus is calculated by $E = 3(1-2\upsilon)K$, where K is the bulk modulus of powder defined as

$$K = K_0\left(\frac{1}{1-\eta^b}\right)\left(a_0 + a_1\eta + a_2\eta^2 + a_3\eta^3\right) \tag{3.82}$$

where K_0 is the powder parameter.

The normal vector to the yield surface is determined by

$$\mathbf{n} = \frac{\partial F}{\partial \boldsymbol{\sigma}} = \frac{\partial F}{\partial J_1}\frac{\partial J_1}{\partial \boldsymbol{\sigma}} + \frac{\partial F}{\partial J_{2D}}\frac{\partial J_{2D}}{\partial \boldsymbol{\sigma}} + \frac{\partial F}{\partial J_{3D}^{2/3}}\frac{\partial J_{3D}^{2/3}}{\partial \boldsymbol{\sigma}} \tag{3.83}$$

where

$$\begin{aligned} \frac{\partial F}{\partial J_1} &= 2J_1\frac{f_d^2}{f_h^2} - 2\beta f_d \frac{J_1^2}{f_h^2} - 2\beta f_d \\ \frac{\partial F}{\partial J_{2D}} &= 1.0 \\ \frac{\partial F}{\partial J_{3D}^{2/3}} &= \psi \end{aligned} \tag{3.84}$$

The hardening plastic modulus H can be determined from $H = -(\partial F/\partial \kappa)(d\kappa/d\lambda)$ as

$$H = -\frac{\partial F}{\partial \eta}\frac{\partial \eta}{\partial \boldsymbol{\varepsilon}}\frac{\partial \boldsymbol{\varepsilon}}{\partial \boldsymbol{\varepsilon}_p}\mathbf{n}_g \tag{3.85}$$

Substituting the yield surface (3.77) into equation (3.85), the hardening plastic modulus can be computed as

$$H = -\left(\frac{\partial F}{\partial f_h}\frac{\partial f_h}{\partial \eta} + \frac{\partial F}{\partial f_d}\frac{\partial f_d}{\partial \eta}\right)\frac{\eta^2}{\eta_0} n_{g\,ii} \tag{3.86}$$

where

$$\frac{\partial F}{\partial f_h} = f_d^2 J_1^2\left(\frac{-2}{f_h^3}\right)$$

$$\frac{\partial F}{\partial f_d} = f_d J_1^2 \left(\frac{2}{f_h^2} \right)$$
$$\frac{\partial f_h}{\partial \eta} = \frac{b\eta^{b-1}}{\left(1-\eta^b\right)^b}\left(a_0 + a_1\eta + a_2\eta^2 + a_3\eta^3\right) + \frac{1}{1-\eta^b}\left(a_1 + 2a_2\eta + 3a_3\eta^2\right) \tag{3.87}$$
$$\frac{\partial f_d}{\partial \eta} = c_1 + 2c_2\eta + 3c_3\eta^2 + 4c_4\eta^3$$

Figure 3.16 presents the variation of the hardening plastic modulus with hydrostatic pressure in meridian plane for a typical yield surface. As can be observed, the maximum value of hardening plastic modulus occurs in the cap part of cone-cap yield surface, in which for $H = 0.0$ the normal vector to yield surface $\mathbf{n}$ is at the maximum value of second deviatoric stress J_{2D}.

3.8.2. Parameter determination of single-cap plasticity

In order to determine the material parameters of single-cap plasticity in prediction of powder behavior, a series of experimental tests performed by Doremus *et al.* (1995) is analyzed numerically. Both uniaxial and triaxial compression tests are driven to determine and calibrate the parameters of the model. The parameters of function f_h are evaluated using the confining pressure test, where the values of J_{2D} and J_{3D} in equation (3.77) are zero. Considering an initial value for parameter b in equation (3.78), parameters a_0 , a_1, a_2 and a_3 can be estimated using a least square method. The parameters of function f_d are estimated performing the least square method on the data obtained by the true-triaxial tests. The triaxial tests consist of an initial isostatic compaction step up to pressure value of 400 MPa, followed by a subsequent uniaxial compaction step. This step is carried out by keeping pressure constant and increasing the axial stress up to the maximum value of 1250 MPa. Table 3.4 presents the evolution of the density versus the hydrostatic pressure performed by Doremus *et al.* (1995). The relationship between the axial stress and axial strain and between the axial stress and density are given in Table 3.5 at the confining pressure P = 150 MPa.

A computer program is developed for the determination of the material parameters. Figures 3.17-19 present the shape of yield surfaces obtained by fitting the stress path to triaxial tests at initial confining pressure of 150 MPa. Also presented in these figures are the material parameters of the proposed cap plasticity. These material parameters are calibrated by comparing the numerical simulations and experimental results given in Table 3.4. In Figure 3.20, the shape of yield surfaces obtained by fitting the stress path to both triaxial and confining pressure tests is presented. The material parameters calibrated for these yield surfaces are given in Table 3.6. The applicability of the single cap plasticity model to handle the triaxial test is evident in these figures.

3.9. Integration of the constitutive relation

Plasticity theory has gained widespread acceptance in large-scale numerical simulations of practical material engineering problems, due to its extreme versatility and accuracy in modeling real engineering materials behavior. Building upon pioneering works of Drucker and Prager (1952), the modern trend has been toward the development of more and more elaborate and complicated elastoplastic constitutive models which resemble the behavior of real engineering materials more closely. Most constitutive equations describe the non-linear and irreversible behavior of engineering materials in terms of increments or rates of strain and stress. These mathematical models are generally complicated and cannot provide an analytical simulation of the stress-strain response of the samples that are uniformly strained and stressed in the laboratory. In most instances, they must be integrated numerically to generate simulated stress-strain curves comparable to experimental results. The simulation of material responses to generalized loadings has several purposes. In research on material behavior, it assesses how constitutive models perform along cyclic and multi-dimensional loadings. In engineering practice, it transforms the constitutive equations into simulated stress-strain curves that can be calibrated with experimental results. In computational mechanics, it gives bench-mark solutions to evaluate the efficient but approximate stress calculation routines of finite element programs.

In computational plasticity, the accuracy and stability of the global solution is strongly affected by the accuracy and stability of the computation of the stress increments, i.e. the integration of the stress-strain relation and the convergence of nonlinear iteration. The numerical solution of elastic-plastic boundary value problems is based on an iterative solution of the discretized momentum balance equations. Typically, for every load/time step, solution involves the following steps which gives a converged configuration at step n;

1. The discretized momentum equations are used to compute a new configuration for step (n+1) via an incremental motion which is used to compute the incremental strains $\Delta\boldsymbol{\varepsilon}$ at each Gauss point;
2. At each stress Gauss point, for the given incremental strains $\Delta\boldsymbol{\varepsilon}$, the new values of state variables $(\boldsymbol{\sigma}_{n+1}, \boldsymbol{q}_{n+1})$ and $\boldsymbol{\varepsilon}_{n+1}^{p}$ are obtained by integration of the local constitutive equations;
3. From the new computed stresses, the balance of momentum is checked and if violated the iterations are performed by returning to step 1.

In this section, attention is focused on step 2 which may be regarded as the central problem of computational plasticity since it is the main role played by the constitutive equations in the computations. In finite element programming, the elasto-plastic constitutive equations are usually incorporated through a separate set of constitutive subroutines. The purpose of these subroutines is the integration of the elastic-plastic constitutive equations that is given a deformation history at each stress Gauss point. The role of the constitutive-equation subroutine is to return the corresponding stress history. Exact analytical solutions for the elastic-plastic evolution problem are available only for the simplest elastic-plastic models. The first exact solution was obtained by Krieg and Krieg (1977) for the case of the isotropic elastic-perfectly plastic von-Mises model. Yoder and Whirley (1984) extended the solution by applying to the von-Mises model with arbitrary combination of kinematic and isotropic hardening. Loret and Prevost (1986) developed an exact solution for the isotropic Drucker-Prager model with linear hardening and arbitrary degree of non-associativity. Although their solution is error-free, it is computationally slow to be used in actual calculations. As the exact analytical solutions are not available for more complex models, the elastic-plastic models are therefore implemented in numerical analysis using an integration algorithm called the stress-point algorithm. Evidently, the accuracy and stability of the global solutions is to be strongly affected by the accuracy and stability of the stress-point algorithm. Also, the cost of the analysis is most strongly affected by the efficiency of the stress-point algorithm. The best algorithm is therefore the one which combines computational efficiency with accuracy.

The first stress-point algorithm was developed by Wilkins (1964) for the elastic-perfectly plastic von-Mises model using the radial return algorithm. The algorithm was then extended by Krieg and Key (1976) to accommodate isotropic and kinematic hardening laws. It was also implemented for different computational aspects by Krieg and Krieg (1977), Schreyer et al. (1979), Yoder and Wirley (1984) and Ortiz and Popov (1985). A subincremental strategy with successive radial stress corrections was proposed by Nayak and Zienkiewicz (1972), Chen (1975), Vermeer (1980) and Loret and Prevost (1986) to correct the inherent stress drift away from the yield surface. The integration algorithm for cap plasticity model was developed by Sandler and Rubin (1979). However, there were no general frameworks for developing the consistent, accurate and stable algorithm until recently, when the concept of an elastic predictor with a plastic return mapping was developed by Ortiz and Popov (1985). Since the elastic-plastic evolution problem is of a strain driven nature, the integration process is split into an elastic predictor and a plastic return mapping to restore plastic consistency. The return mapping algorithm is achieved by integrating the nonlinear plastic evolution equations by Simo and Ortiz (1985), Simo and Taylor (1986), Ortiz and Simo (1986) and Simo and Hughes (1987). In Table 3.7, the elastic predictor-plastic corrector algorithm for rate independent isotropic hardening elasto-plasticity is presented.

3.9.1. Explicit algorithm

A simple and practical integration algorithm for the incremental constitutive relation is commonly based on an explicit integration scheme within each increment. The simplest scheme for such an explicit solution is obtained essentially by employing the elasto-plastic matrix evaluated at the beginning of the incremental step, i.e.

$$\boldsymbol{\sigma}_{n+1} = \boldsymbol{\sigma}_n + \Delta\boldsymbol{\sigma}_n \tag{3.88}$$

with

$$\Delta\boldsymbol{\sigma}_n = \mathbf{D}(\boldsymbol{\sigma}_n, \kappa_n)\Delta\boldsymbol{\varepsilon}_n \tag{3.89}$$

in which $\mathbf{D}(\boldsymbol{\sigma}_n, \kappa_n)$ is the elasto-plastic constitutive matrix evaluated at the beginning of step n with stress $\boldsymbol{\sigma}_n$ and parameter κ_n, which is the internal parameter for history dependency. The explicit integration can be also performed by using more accurate schemes, e.g. the modified Euler and Runge-Kutta schemes.

It should be noted that in each time step, several iterations are generally needed to obtain a converged solution for the nonlinear equation system. Starting from the last accepted stress-strain state $\boldsymbol{\sigma}_n$ and $\boldsymbol{\varepsilon}_n$ at time station t_n, the incremental displacement $d\mathbf{u}_n^i$ in i^{th} iteration is first obtained. Then, by using the strain-displacement relation, the strain increment $d\boldsymbol{\varepsilon}_n^i$ in this iteration is determined. The total incremental strain in the time duration (t_n, t_{n+1}) can be written as

$$\Delta\boldsymbol{\varepsilon}_n^i = \sum_{k=1}^{i} d\boldsymbol{\varepsilon}_n^i \tag{3.90}$$

However, in a large deformation analysis the relationship between strain and displacement is no longer linear and therefore equation (3.90) is invalid. In this case, the total incremental strain in one time step has to been evaluated by using total increment displacements. Corresponding to the strain increment $\Delta\boldsymbol{\varepsilon}_n^i$, the stress increment $\Delta\boldsymbol{\sigma}_n^i$ is determined by an integration of the form

$$\Delta\boldsymbol{\sigma}_n^i = \int_0^{\Delta\boldsymbol{\varepsilon}_n^i} \mathbf{D}(\boldsymbol{\sigma}_n, \kappa_n)\, d\boldsymbol{\varepsilon} \tag{3.91}$$

and the predicted stress for the next iteration is thus given by

$$\boldsymbol{\sigma}_{n+1}^{i+1} = \boldsymbol{\sigma}_n + \Delta\boldsymbol{\sigma}_n^i \tag{3.92}$$

It should be pointed out that incrementing the stress during the iterations should always be referred to the last accepted stress-strain state as represented in equation (3.92). Otherwise, the error arising from irreversible plastic strains will accumulate. The simplest scheme for the integration of equation (3.91) is the Euler scheme, which is written as

$$\boldsymbol{\sigma}_{n+1}^{i+1} = \boldsymbol{\sigma}_n + \Delta\boldsymbol{\sigma}^a \tag{3.93}$$

where

$$\Delta\boldsymbol{\sigma}^a = \mathbf{D}(\boldsymbol{\sigma}_n, \kappa_n)\Delta\boldsymbol{\varepsilon}_n^i \tag{3.94}$$

where $\mathbf{D}(\boldsymbol{\sigma}_n, \kappa_n)$ is the constitutive matrix evaluated at the beginning of the nth time step with stress $\boldsymbol{\sigma}_n$ and parameter κ_n. A more accurate estimate of $\boldsymbol{\sigma}_{n+1}^{i+1}$ may be obtained from the modified Euler scheme, which gives

$$\hat{\boldsymbol{\sigma}}_{n+1}^{i+1} = \boldsymbol{\sigma}_n + \frac{1}{2}\left(\Delta\boldsymbol{\sigma}^a + \Delta\boldsymbol{\sigma}^b\right) \tag{3.95}$$

where $\Delta\boldsymbol{\sigma}^a$ is given by equation (3.94) and

$$\Delta\boldsymbol{\sigma}^b = \mathbf{D}(\boldsymbol{\sigma}_{n+1}^{i+1}, \kappa_{n+1}^{i+1})\Delta\boldsymbol{\varepsilon}_n^i \tag{3.96}$$

In which $\boldsymbol{\sigma}_{n+1}^{i+1}$ is given by equation (3.93) and κ_{n+1}^{i+1} is the updated internal parameter at the end of the Euler integration (equations (3.93) and (3.94)).

For a given strain increment, the local truncation error in the Euler scheme is of second order, whereas the local truncation error in the modified Euler scheme is of third order. A comparison of the results from these two schemes can thus give an indication of the local error in the Euler integration (equations (3.93) and (3.94)). Subtracting equation (3.93) from equation (3.95), we have

$$\mathbf{e} = \frac{1}{2}\left(\Delta\boldsymbol{\sigma}^a - \Delta\boldsymbol{\sigma}^b\right) \tag{3.97}$$

In order to obtain an accurate stress integration, a modified Euler scheme with strain subincrementation can be performed for a large strain increment. The strain increment $\Delta\boldsymbol{\varepsilon}_n^i$ is firstly divided into M subincrementations of equal size and then the stress is calculated as

$$\boldsymbol{\sigma}_{n+1}^{i+1} = \boldsymbol{\sigma}_n + \sum_{j=1}^{M}\Delta\boldsymbol{\sigma}_{nj} \tag{3.98}$$

where,

$$\Delta\boldsymbol{\sigma}_{nj} = \mathbf{D}(\boldsymbol{\sigma}_{nj}, \kappa_{nj})\Delta\boldsymbol{\varepsilon}_n^i / M \tag{3.99}$$

in which

$$\boldsymbol{\sigma}_{nj} = \boldsymbol{\sigma}_n + \sum_{r=1}^{j-1} \Delta\boldsymbol{\sigma}_{nr} \tag{3.100}$$

and κ_{nj} is the updated internal parameter at stress $\boldsymbol{\sigma}_{nj}$. The error estimator (3.97) may be used as a guide for selecting the number of subdivisions *M*. From equation (3.97), a normalized relative error can be defined as

$$\eta = \frac{\| \mathbf{e} \|}{\left\| \frac{1}{2}(\Delta\boldsymbol{\sigma}^a + \Delta\boldsymbol{\sigma}^b) \right\|} = \frac{\left\| (\Delta\boldsymbol{\sigma}^b - \Delta\boldsymbol{\sigma}^a) \right\|}{\left\| (\Delta\boldsymbol{\sigma}^a + \Delta\boldsymbol{\sigma}^b) \right\|} \tag{3.101}$$

In the above, the L_2 norm is conveniently used, which is defined by $\| \mathbf{e} \| = (\mathbf{e}^T \mathbf{e})^{1/2}$. The stress integration error may be controlled by ensuring that the relative error given in equation (3.101) at each time step is less than some specified tolerance $\bar{\eta}$. Therefore, the number of subdivisions *M* can be estimated as

$$M = \frac{\eta}{\bar{\eta}} \tag{3.102}$$

There are even more sophisticated schemes for the stress integration, though the computational cost needed may increase considerably. For instance, instead of the uniform strain subdivisions, we may control the error at each iteration by predicting the size of each strain subdivision, or instead of using the Euler scheme; we may use other schemes of higher order such as a Runge-Kutta scheme. Its accuracy increases, of course with the number of subincrements used. In general it is difficult a priori to decide on this number and accuracy of prediction is not easy to determine. The subincrementation strategy does not guarantee the stability of integration algorithm.

3.9.2. Implicit algorithm

In this section, the general closest-point-projection algorithm developed by Simo *et al.* (1985, 1986, 1988), Borja (1990, 1991) and Matzenmiller and Taylor (1994), is introduced. Following the standard notation, the elasto-plastic equations may be formulated as

$$\dot{\boldsymbol{\sigma}} = \mathbf{D}^e \left(\dot{\boldsymbol{\varepsilon}} - \dot{\boldsymbol{\varepsilon}}^p \right) \tag{3.103}$$

and

$$\dot{\boldsymbol{\varepsilon}}^p = \dot{\phi} \frac{\partial Q}{\partial \boldsymbol{\sigma}} \tag{3.104}$$

where $Q(\boldsymbol{\sigma}, \kappa)$ is the plastic potential function, κ a hardening parameter, ϕ a plastic multiplier, $\dot{\boldsymbol{\varepsilon}}$ the total strain and $\dot{\boldsymbol{\varepsilon}}^p$ the plastic strain. The loading and unloading can be expressed by the following condition which relate the plastic multiplier to a suitably chosen yield function $F(\boldsymbol{\sigma}, \kappa) = 0$. These conditions which is called Kuhn-Tucher conditions, are

$$F \le 0, \quad \dot{\phi} \ge 0 \quad \text{and} \quad \dot{\phi} F = 0 \tag{3.105}$$

The plastic consistency condition can be derived from the above conditions. This requires that the stress state must be on the yield surface during plastic loading, i.e. $F(\boldsymbol{\sigma}, \kappa) = 0$.

After time discretization of equation (3.103) over the time step (t_n, t_{n+1}), we have

$$\boldsymbol{\sigma}_{n+1} - \boldsymbol{\sigma}_n = \mathbf{D}^e(\boldsymbol{\varepsilon}_{n+1} - \boldsymbol{\varepsilon}_n) - \mathbf{D}^e(\boldsymbol{\varepsilon}_{n+1}^p - \boldsymbol{\varepsilon}_n^p) \tag{3.106}$$

or

$$\boldsymbol{\sigma}_{n+1} = \boldsymbol{\sigma}_{n+1}^{tr} - \mathbf{D}^e(\boldsymbol{\varepsilon}_{n+1}^p - \boldsymbol{\varepsilon}_n^p) \tag{3.107}$$

where $\boldsymbol{\sigma}_{n+1}^{tr} = \boldsymbol{\sigma}_n + \mathbf{D}^e(\boldsymbol{\varepsilon}_{n+1} - \boldsymbol{\varepsilon}_n)$ is the trial stress. From equation (3.104), we can write

$$\boldsymbol{\varepsilon}_{n+1}^p - \boldsymbol{\varepsilon}_n^p = \lambda\left[(1-\alpha)\left(\frac{\partial Q}{\partial \boldsymbol{\sigma}}\right)_n + \alpha\left(\frac{\partial Q}{\partial \boldsymbol{\sigma}}\right)_{n+1}\right] \tag{3.108}$$

where $F(\boldsymbol{\sigma}_{n+1}, \kappa) = 0$ and $\lambda = \Delta\phi$. The approximation is particularly simple and stable for $\alpha = 1$ (backward difference). If $\alpha = 1$, the above nonlinear system of equations (3.107) and (3.108) can be rewritten as

$$\psi_1 = (\mathbf{D}^e)^{-1}(\boldsymbol{\sigma}_{n+1} - \boldsymbol{\sigma}_{n+1}^{tr}) + \lambda\left(\frac{\partial Q}{\partial \boldsymbol{\sigma}}\right)_{n+1} = 0 \tag{3.109}$$

and

$$\psi_2 = F(\boldsymbol{\sigma}_{n+1}, \kappa) = 0 \tag{3.110}$$

Applying the Newton-Raphson process leads to

$$\psi_1^i + (\mathbf{D}^e)^{-1}\delta\boldsymbol{\sigma}_{n+1}^i + \lambda^i\left(\frac{\partial^2 Q}{\partial \boldsymbol{\sigma}^2}\right)_{n+1}^i + \delta\lambda^i\left(\frac{\partial Q}{\partial \boldsymbol{\sigma}}\right)_{n+1}^i = 0 \tag{3.111}$$

and

$$\psi_2^i + \left[\left(\frac{\partial F}{\partial \boldsymbol{\sigma}}\right)_{n+1}^i\right]^T \delta\boldsymbol{\sigma}_{n+1}^i - H^i\delta\lambda^i = 0 \tag{3.112}$$

where $H = -(\partial F/\partial\kappa)(\partial\kappa/\partial\lambda)$ is the hardening/softening plastic modulus. Now we obtain the following linearized equation

$$\mathbf{A}^i\begin{Bmatrix}\delta\boldsymbol{\sigma}_{n+1}^i \\ \delta\lambda_{n+1}^i\end{Bmatrix} = -\begin{Bmatrix}\psi_1^i \\ \psi_2^i\end{Bmatrix} = \boldsymbol{\Psi}^i \tag{3.113}$$

where

$$\mathbf{A}^i = \begin{bmatrix} (\mathbf{D}^e)^{-1} + \lambda^i \left(\dfrac{\partial^2 Q}{\partial \boldsymbol{\sigma}^2}\right)_{n+1}^i & \left(\dfrac{\partial Q}{\partial \boldsymbol{\sigma}}\right)_{n+1}^i \\ \left[\left(\dfrac{\partial F}{\partial \boldsymbol{\sigma}}\right)_{n+1}^i\right]^T & -H^i \end{bmatrix} \tag{3.114}$$

The so-called closest-point-projection algorithm can be therefore proceeded, as given in Table 3.8. In general this implicit algorithm can not be easily performed for complicated elasto-plastic constitutive models which resemble the behavior of materials closely. For this reason, the explicit algorithms and its variations are still widely adopted.

Another important issue in computational plasticity is the derivation of tangent elasto-plastic matrix. A general elasto-plastic matrix was first derived by Yamada *et al.* (1968) and Zienkiewicz *et al.* (1969) and has been widely used in plastic analysis. This elasto-plastic matrix is referred as the continuum elastoplastic tangent matrix in order to make the distinction. However it is now well recognized that such elasto-plastic tangent matrix is known to destroy the important property of asymptotic quadratic convergence in Newton-Raphson method particularly when the step size is large. In what follows, we shall derive a so-called consistent elasto-plastic tangent matrix developed by Simo and Taylor 1985 to improve the convergence of Newton-Raphson iteration.

Differentiating equations (3.106) and (3.108) in time at step n+1, we obtain the following relation for $\alpha = 1.0$

$$d\boldsymbol{\sigma} = \mathbf{D}^e\left(d\boldsymbol{\varepsilon}_{n+1} - d\boldsymbol{\varepsilon}_{n+1}^p\right) \tag{3.115}$$

and

$$d\boldsymbol{\varepsilon}_{n+1}^p = d\lambda\left(\frac{\partial Q}{\partial \boldsymbol{\sigma}}\right)_{n+1} - \lambda\left(\frac{\partial^2 Q}{\partial \boldsymbol{\sigma}^2}\right)_{n+1} d\boldsymbol{\sigma}_{n+1} \tag{3.116}$$

Combining equations (3.115) and (3.116), we obtain the algorithm relation

$$d\boldsymbol{\sigma}_{n+1} = \mathbf{D}_{n+1}^*\left[d\varepsilon_{n+1} - d\lambda\left(\frac{\partial Q}{\partial \boldsymbol{\sigma}}\right)_{n+1}\right] \tag{3.117}$$

where

$$\mathbf{D}_{n+1}^* = \left[\left(\mathbf{D}^e\right)^{-1} + \lambda\left(\frac{\partial^2 Q}{\partial \boldsymbol{\sigma}^2}\right)_{n+1}\right]^{-1} \tag{3.118}$$

or

$$\mathbf{D}_{n+1}^* = \left[\mathbf{I} + \lambda\,\mathbf{D}^e\left(\frac{\partial^2 Q}{\partial \boldsymbol{\sigma}^2}\right)_{n+1}\right]^{-1}\mathbf{D}^e \tag{3.119}$$

Differentiation of the consistency condition $F = 0$ yields

$$\left(\frac{\partial F}{\partial \boldsymbol{\sigma}}\right)_{n+1}^{T} d\boldsymbol{\sigma}_{n+1} - H d\lambda = 0 \tag{3.120}$$

Now, from equations (3.117) and (3.120) we obtain

$$d\lambda = \frac{\left(\frac{\partial F}{\partial \boldsymbol{\sigma}}\right)_{n+1}^{T} \mathbf{D}_{n+1}^{*} \, d\boldsymbol{\varepsilon}_{n+1}}{H + \left(\frac{\partial F}{\partial \boldsymbol{\sigma}}\right)_{n+1}^{T} \mathbf{D}_{n+1}^{*} \left(\frac{\partial Q}{\partial \boldsymbol{\sigma}}\right)_{n+1}} \tag{3.121}$$

Finally, substitution of (3.121) into (3.117) yields the expression for the algorithmic consistent elasto-plastic matrix

$$\frac{d\boldsymbol{\sigma}_{n+1}}{d\boldsymbol{\varepsilon}_{n+1}} = \mathbf{D}_{n+1}^{*} - \frac{\mathbf{D}_{n+1}^{*} \left(\frac{\partial Q}{\partial \boldsymbol{\sigma}}\right)_{n+1} \left(\frac{\partial F}{\partial \boldsymbol{\sigma}}\right)_{n+1}^{T} \mathbf{D}_{n+1}^{*}}{H + \left(\frac{\partial F}{\partial \boldsymbol{\sigma}}\right)_{n+1}^{T} \mathbf{D}_{n+1}^{*} \left(\frac{\partial Q}{\partial \boldsymbol{\sigma}}\right)_{n+1}} \tag{3.122}$$

The above matrix is often very complicated though it can remain the property of asymptotic quadratic convergence in Newton-Raphson iteration. The simplification of the closet point projection algorithm and consistent tangent elasto-plastic matrix can be obtained for elasto-plastic models with yield function $F(p, q, \kappa)$ or $F(q, \kappa)$ and plastic potential function $Q(p, q, \kappa)$ or $Q(q, \kappa)$; e.g. von-Mises, Drucker-Prager and Cam-clay models; where p and q are defined as $p = \frac{1}{3}\boldsymbol{m}^T\boldsymbol{\sigma}$ and $q = \sqrt{\frac{3}{2}}\|\boldsymbol{\tau}\|$ with $\boldsymbol{\tau}$ denoting the deviatoric stress as $\boldsymbol{\tau} = \boldsymbol{\sigma} - \boldsymbol{m}p$.

3.9.2.1. A Simplified Procedure

In this section, a simplified procedure is presented for the models which are independent of the third invariant of the stresses. Considering a plastic potential function $Q(p, q, \kappa)$, we have

$$\Delta\boldsymbol{\varepsilon}^{p} = \lambda\left(\frac{\partial Q}{\partial p}\frac{1}{3}\mathbf{m} + \frac{\partial Q}{\partial q}\sqrt{\frac{3}{2}}\hat{\mathbf{n}}\right) \tag{3.123}$$

or

$$\Delta\boldsymbol{\varepsilon}^{p} = \frac{1}{3}\mathbf{m}\Delta\varepsilon_{v}^{p} + \Delta\boldsymbol{e}^{p} \tag{3.124}$$

where $\Delta\varepsilon_{v}^{p} = \mathbf{m}^{T}\Delta\boldsymbol{\varepsilon}^{p} = \lambda(\partial Q/\partial p)$, $\Delta\boldsymbol{e}^{p} = \Delta\boldsymbol{\varepsilon}^{p} - \frac{1}{3}\mathbf{m}\Delta\varepsilon_{v}^{p} = \lambda(\partial Q/\partial q)\sqrt{\frac{3}{2}}\hat{\mathbf{n}}$ and $\hat{\mathbf{n}} = \frac{\boldsymbol{\tau}_{n+1}}{\|\boldsymbol{\tau}_{n+1}\|}$. By introducing the elastic matrix in the following form

$$\mathbf{D}^{e} = K\mathbf{m}\otimes\mathbf{m} + 2\mu\left(\mathbf{I} - \frac{1}{3}\mathbf{m}\otimes\mathbf{m}\right) \tag{3.125}$$

where K is the elastic bulk modulus and μ is the elastic shear modulus. From equations (3.107) and (3.125), we obtain the following volumetric part of $\boldsymbol{\sigma}_{n+1}$

$$p_{n+1} = p_{n+1}^{tr} - K\Delta\varepsilon_v^p = p_{n+1}^{tr} - K\lambda\frac{\partial Q}{\partial p} \tag{3.126}$$

and its deviatoric part

$$\boldsymbol{\tau}_{n+1} = \boldsymbol{\tau}_{n+1}^{tr} - 2\mu\,\Delta\mathbf{e}^p = \boldsymbol{\tau}_{n+1}^{tr} - 2\mu\lambda\frac{\partial Q}{\partial q}\sqrt{\frac{3}{2}}\hat{\mathbf{n}} \tag{3.127}$$

$$\left\|\boldsymbol{\tau}_{n+1}\right\| = \left\|\boldsymbol{\tau}_{n+1}^{tr}\right\| - 2\mu\left\|\Delta\mathbf{e}^p\right\| = \left\|\boldsymbol{\tau}_{n+1}^{tr}\right\| - 2\mu\lambda\frac{\partial Q}{\partial q}\sqrt{\frac{3}{2}} \tag{3.128}$$

$$q_{n+1} = q_{n+1}^{tr} - 3\mu\,\Delta\varepsilon_s^p = q_{n+1}^{tr} - 3\mu\lambda\frac{\partial Q}{\partial q} \tag{3.129}$$

where $p_{n+1}^{tr} = \frac{1}{3}\mathbf{m}^T\boldsymbol{\sigma}_{n+1}^{tr}$, $\boldsymbol{\tau}_{n+1}^{tr} = \boldsymbol{\sigma}_{n+1}^{tr} - \mathbf{m}\,p_{n+1}^{tr}$, $q_{n+1}^{tr} = \sqrt{\frac{3}{2}}\left\|\boldsymbol{\tau}_{n+1}^{tr}\right\|$ and $\Delta\varepsilon_s^p = \lambda\frac{\partial Q}{\partial q} = \sqrt{\frac{2}{3}}\left\|\Delta\mathbf{e}^p\right\|$. Considering the yield function $F(p, q, \kappa)$, we can write the elasto-plastic equations as follows

$$\frac{1}{K}\left(p_{n+1} - p_{n+1}^{tr}\right) + \lambda\left(\frac{\partial Q}{\partial p}\right)_{n+1} = 0 \tag{3.130}$$

$$\frac{1}{3\mu}\left(q_{n+1} - q_{n+1}^{tr}\right) + \lambda\left(\frac{\partial Q}{\partial q}\right)_{n+1} = 0 \tag{3.131}$$

and

$$F\left(p_{n+1}, q_{n+1}, \kappa\right) = 0 \tag{3.132}$$

In order to solve the above nonlinear equation system, the standard Newton-Raphson procedure is employed. The resulting equation system for i^{th} iteration can be expressed as

$$\boldsymbol{\Psi}^i + \left.\frac{\partial\boldsymbol{\Psi}}{\partial\mathbf{x}}\right|_{\mathbf{x}=\mathbf{x}^i}\delta\mathbf{x}^i = 0 \tag{3.133}$$

where

$$\delta\mathbf{x}^i = \begin{Bmatrix}\delta p_{n+1}^i \\ \delta q_{n+1}^i \\ \delta\lambda^i\end{Bmatrix}, \qquad \mathbf{x}^i = \begin{Bmatrix}p_{n+1}^i \\ q_{n+1}^i \\ \lambda^i\end{Bmatrix} \tag{3.134}$$

and

$$\boldsymbol{\Psi}^i = \begin{Bmatrix}\frac{1}{K}\left(p_{n+1}^i - p_{n+1}^{tr}\right) + \lambda(\partial Q/\partial p)_{n+1}^i \\ \frac{1}{3\mu}\left(q_{n+1}^i - q_{n+1}^{tr}\right) + \lambda(\partial Q/\partial q)_{n+1}^i \\ F\left(p_{n+1}^i, q_{n+1}^i, \kappa\right)\end{Bmatrix} \tag{3.135}$$

Equation (3.133) can be solved to obtain the incremental values of δp_{n+1}^{i} and δq_{n+1}^{i}. The updated value of the stress invariants can be calculated as

$$p_{n+1}^{i+1} = p_{n+1}^{i} + \delta p_{n+1}^{i} \tag{3.136}$$

$$q_{n+1}^{i+1} = q_{n+1}^{i} + \delta q_{n+1}^{i} \tag{3.137}$$

and

$$\boldsymbol{\sigma}_{n+1}^{i} = p_{n+1}^{i}\mathbf{m} + \sqrt{\frac{2}{3}}q_{n+1}^{i}\hat{\mathbf{n}} \tag{3.138}$$

As $\boldsymbol{\tau}_{n+1} = \|\boldsymbol{\tau}_{n+1}\|\hat{\mathbf{n}}$, it results from equation (3.127) that the unit normal $\hat{\mathbf{n}}$ can be determined in terms of the trial elastic stress $\boldsymbol{\sigma}_{n+1}^{tr}$ according to

$$\hat{\mathbf{n}} = \frac{\boldsymbol{\tau}_{n+1}^{tr}}{\|\boldsymbol{\tau}_{n+1}^{tr}\|} \tag{3.139}$$

The variation of p_{n+1}^{tr}, q_{n+1}^{tr} and $\hat{\mathbf{n}}$ with respect to $\boldsymbol{\varepsilon}_{n+1}$ can be expressed as (Simo and Taylor 1985)

$$\frac{\partial p_{n+1}^{tr}}{\partial \boldsymbol{\varepsilon}_{n+1}} = K\mathbf{m} \tag{3.140}$$

$$\frac{\partial q_{n+1}^{tr}}{\partial \boldsymbol{\varepsilon}_{n+1}} = 2\mu\sqrt{\frac{3}{2}}\hat{\mathbf{n}} \tag{3.141}$$

and

$$\frac{\partial \hat{\mathbf{n}}}{\partial \boldsymbol{\varepsilon}_{n+1}} = \frac{2\mu}{\sqrt{\frac{3}{2}}q_{n+1}^{tr}}\left(\mathbf{I} - \frac{1}{3}\mathbf{m}\otimes\mathbf{m} - \hat{\mathbf{n}}\otimes\hat{\mathbf{n}}\right) \tag{3.142}$$

By considering $\boldsymbol{\sigma}_{n+1} = p_{n+1}\mathbf{m} + \sqrt{\frac{2}{3}}q_{n+1}\hat{\mathbf{n}}$, the consistent tangent elasto-plastic matrix can be then obtained by directly evaluating the variation

$$\frac{\partial \boldsymbol{\sigma}_{n+1}}{\partial \boldsymbol{\varepsilon}_{n+1}} = \mathbf{m}\otimes\frac{\partial p_{n+1}}{\partial \boldsymbol{\varepsilon}_{n+1}} + \sqrt{\frac{2}{3}}q_{n+1}\frac{\partial \hat{\mathbf{n}}}{\partial \boldsymbol{\varepsilon}_{n+1}} + \sqrt{\frac{2}{3}}\hat{\mathbf{n}}\otimes\frac{\partial q_{n+1}}{\partial \boldsymbol{\varepsilon}_{n+1}} \tag{3.143}$$

The derivatives of $\partial p_{n+1}/\partial \boldsymbol{\varepsilon}_{n+1}$ and $\partial q_{n+1}/\partial \boldsymbol{\varepsilon}_{n+1}$ can be determined by using the consistency condition $F(p, q, \kappa) = 0$ and equations (3.126), (3.129), (3.140) and (3.141).

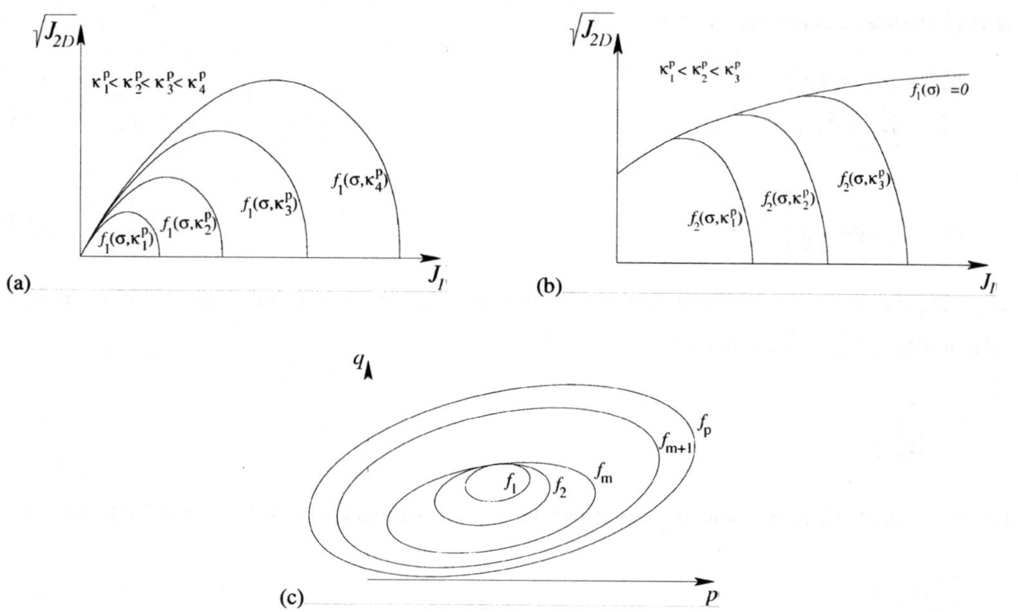

Figure 3.1. Various constitutive models in powder forming analysis; a) Single-surface plasticity, b) Double-surface plasticity, c) Multi-surface plasticity

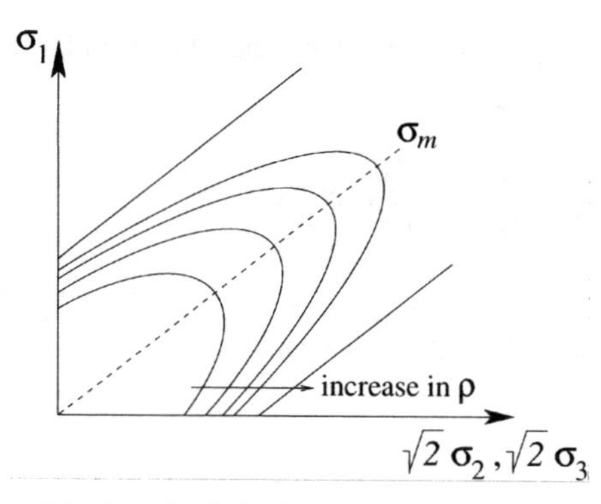

Figure 3.2. A typical yield surface for porous material

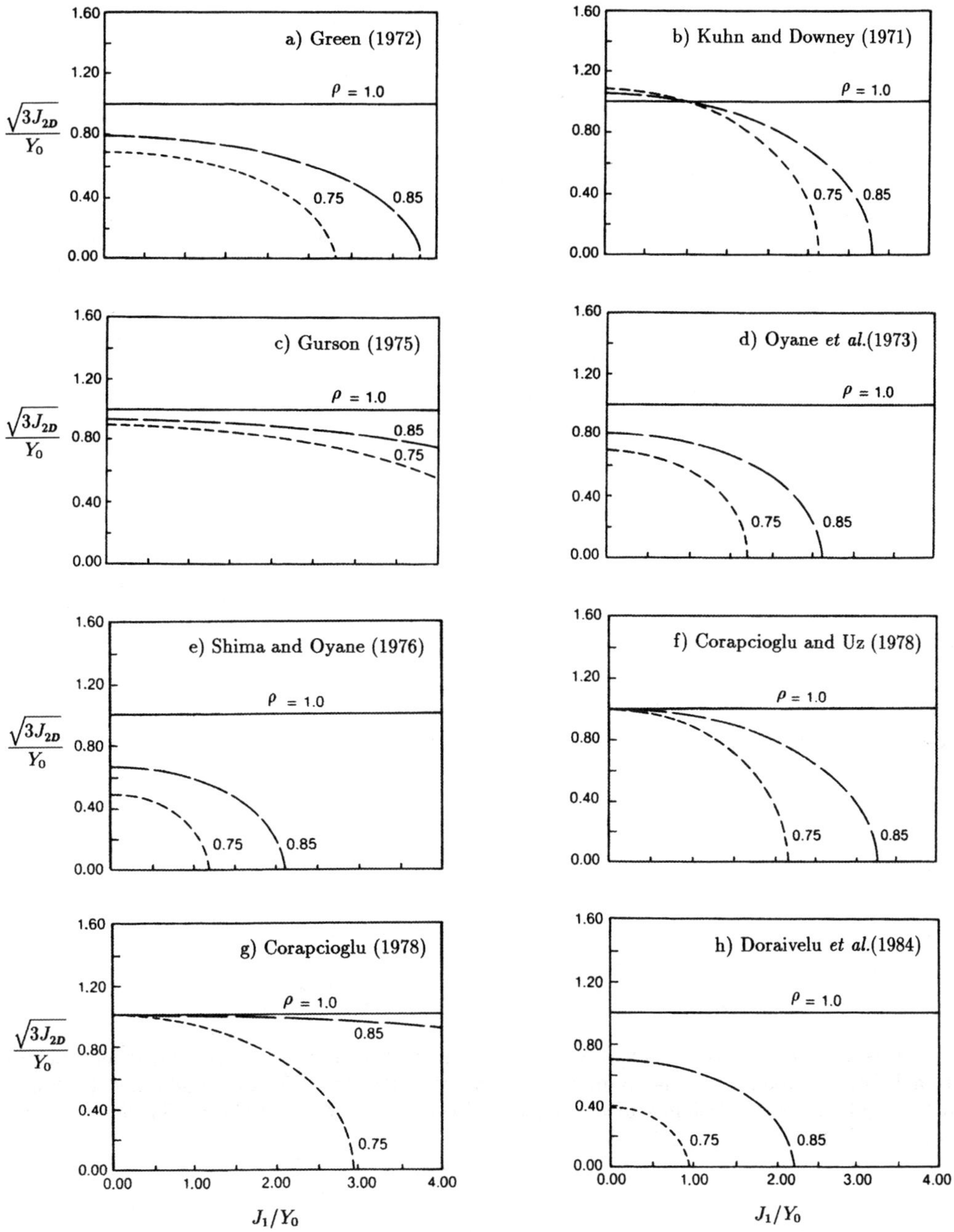

Figure 3.3. Plasticity models proposed by researchers for porous materials

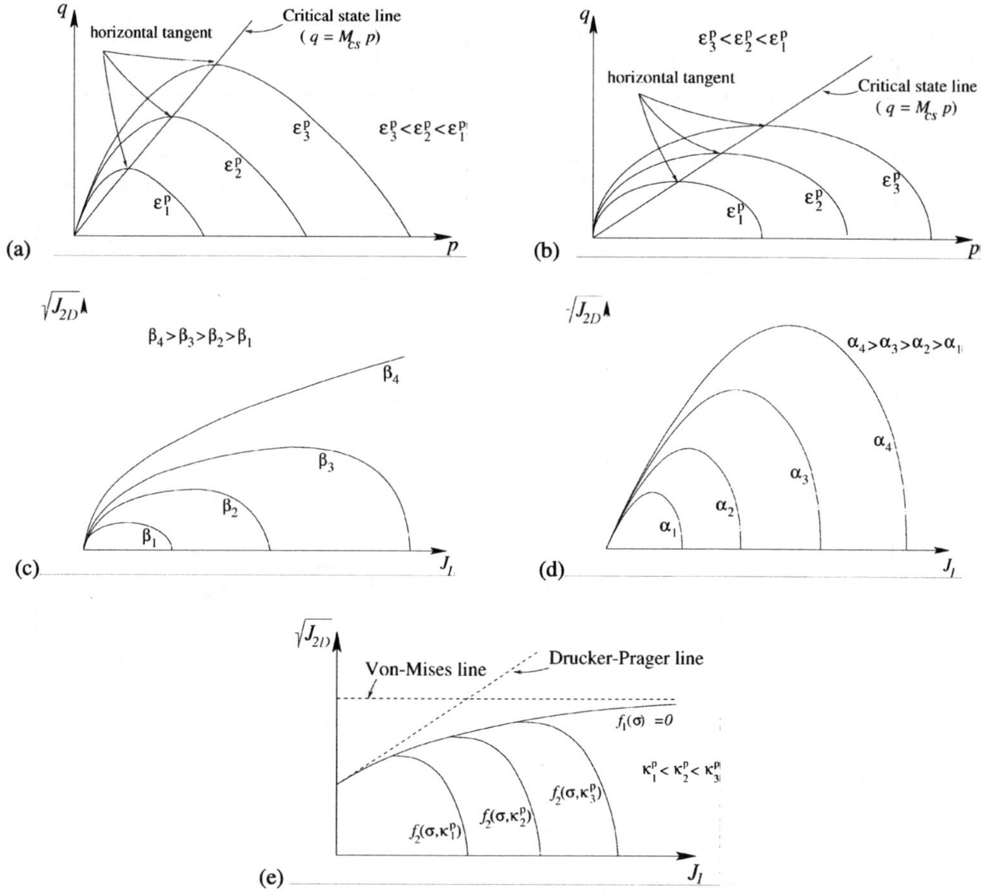

Figure 3.4. Plasticity models proposed by researchers for granular materials; a) Cam-clay model (Schofield and Wroth 1968), b) Modified Cam-clay model (Lewis and Schrefler 1987), c) Desai's model (Desai and Siriwardane 1984), d) Desai-Hashmi's model (Desai and Hashmi 1989), e) DiMaggio-Sandler's cap model (Sandler *et al.* 1976)

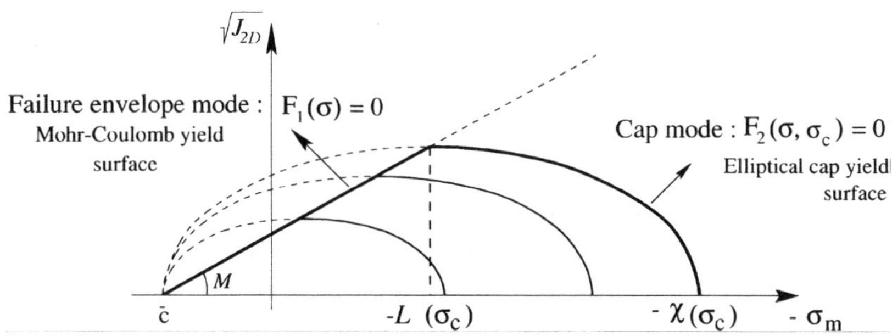

Figure 3.5. A combination of the Mohr-Coulomb and Elliptical Cap models (MCEC)

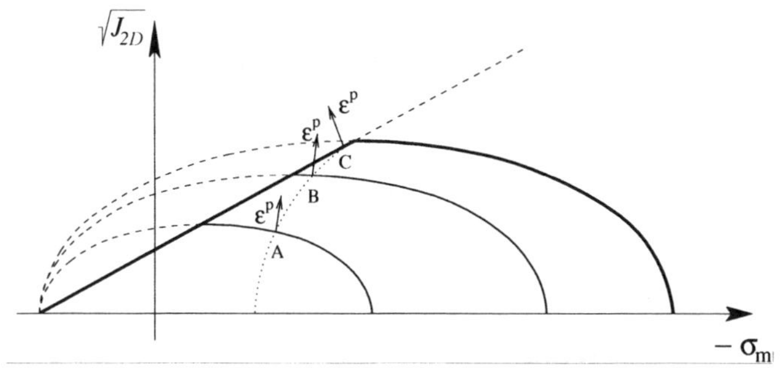

Figure 3.6. Stress paths in strain hardening capped yield criterion

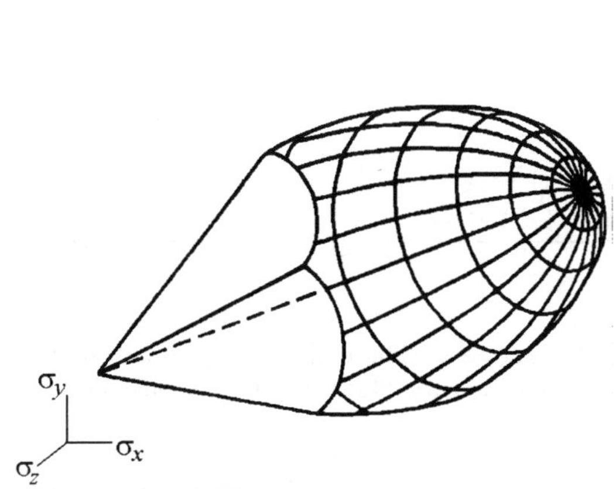

Figure 3.7. 3D representation of the Mohr-Coulomb yield surface and elliptical cap in principal stress space

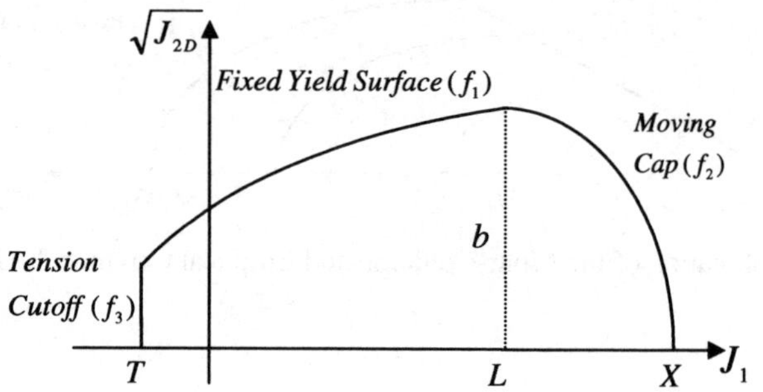

Figure 3.8. The double-surface cap plasticity model

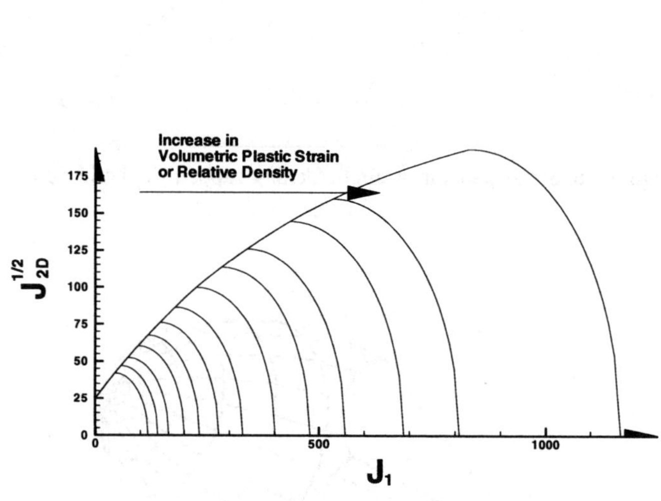

Figure 3.9. The expansion of moving cap surface with increasing the volumetric plastic strain, or relative density

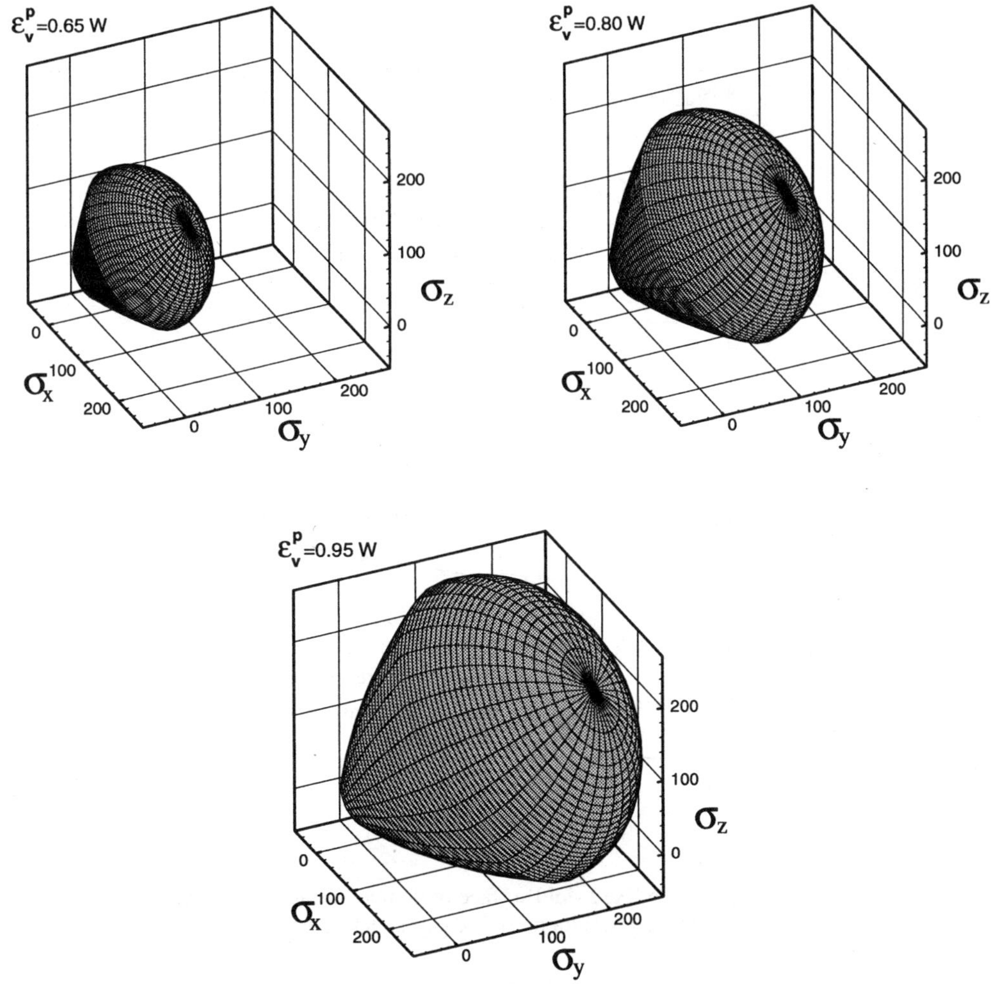

Figure 3.10. 3D representation of the double-surface plasticity at different relative densities

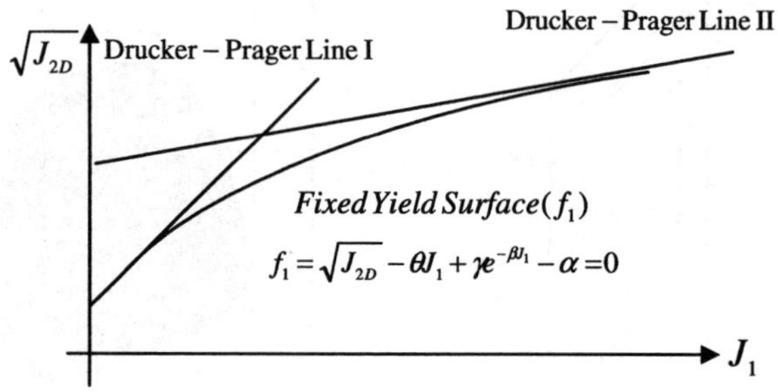

(a)

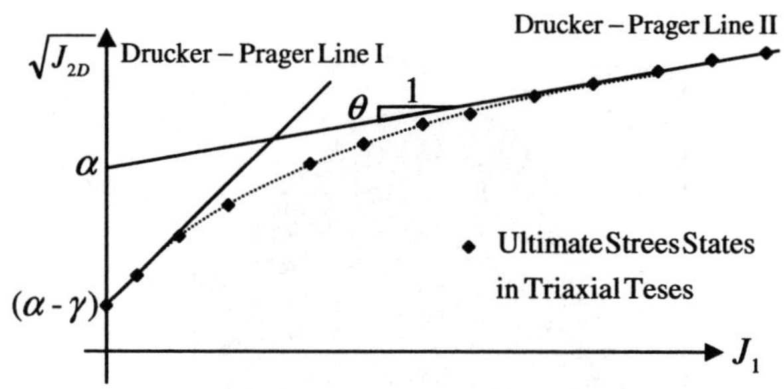

(b)

Figure 3.11. The fixed yield surface of double-surface plasticity;
a) The model description, b) The parameter determination

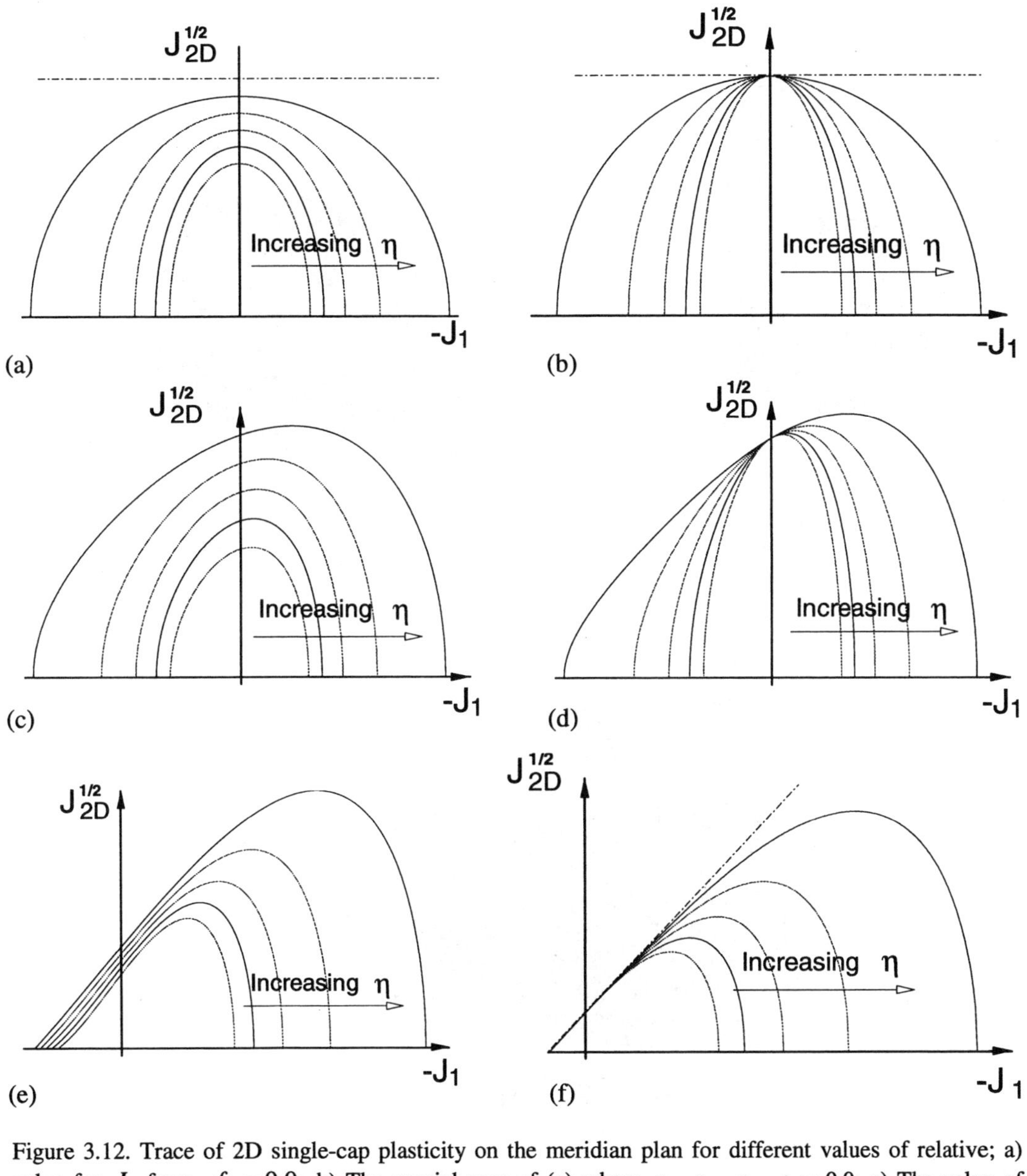

Figure 3.12. Trace of 2D single-cap plasticity on the meridian plan for different values of relative; a) No value for J_1 from $f_d = 0.0$; b) The special case of (a) where c_1, c_2, c_3, c_4= 0.0; c) The value of J_1 obtained by $f_d = 0.0$ greater than $+f_h$; d) The special case of (c) where c_1, c_2, c_3, c_4= 0.0; e) The value of J_1 obtained by $f_d = 0.0$ between $-f_h$ and $+f_h$; f) The special case of (e) where c_1, c_2, c_3, c_4= 0.0

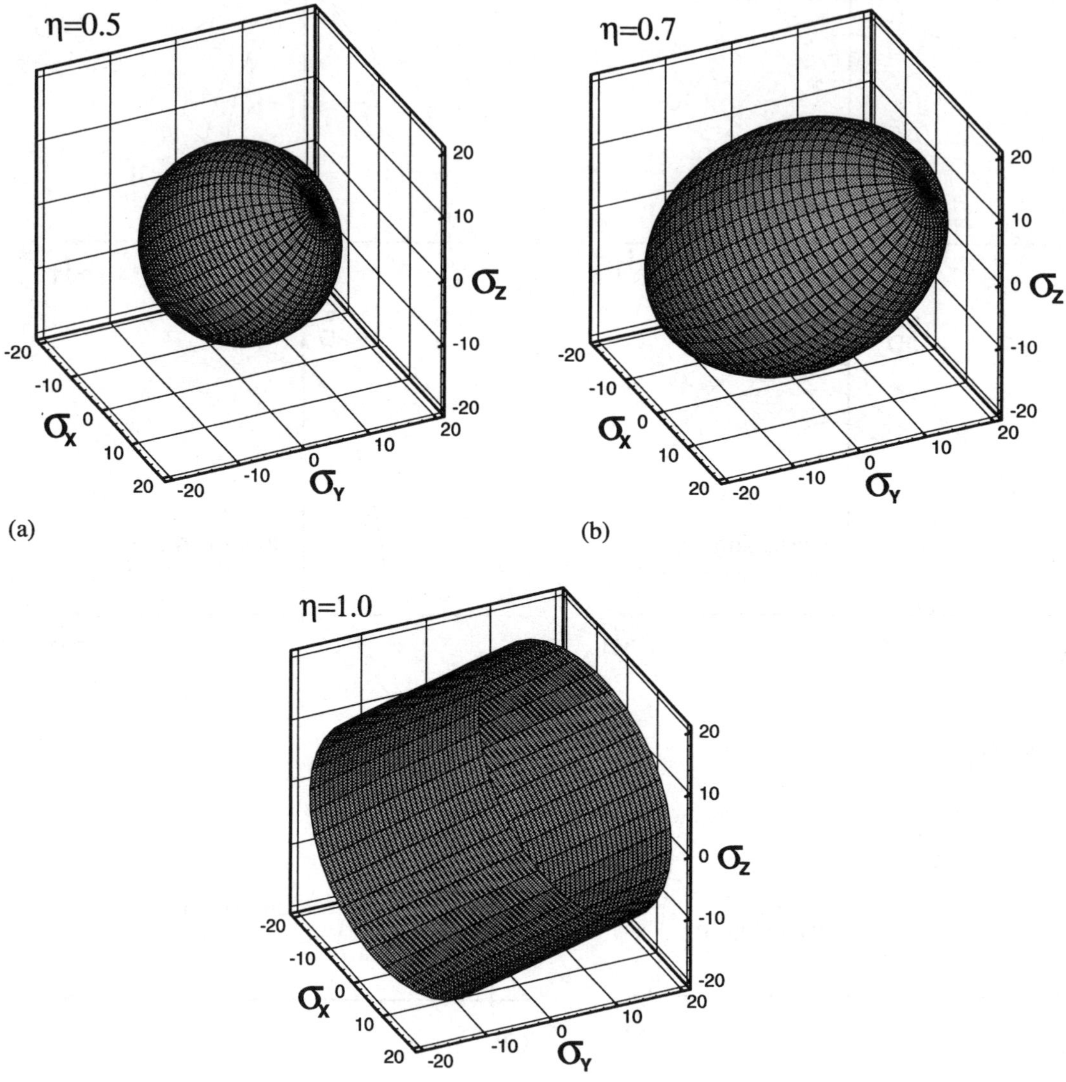

Figure 3.13. Trace of 3D elliptical yield function in principal stress space for different values of relative density; No value for J_1 from $f_d = 0.0$

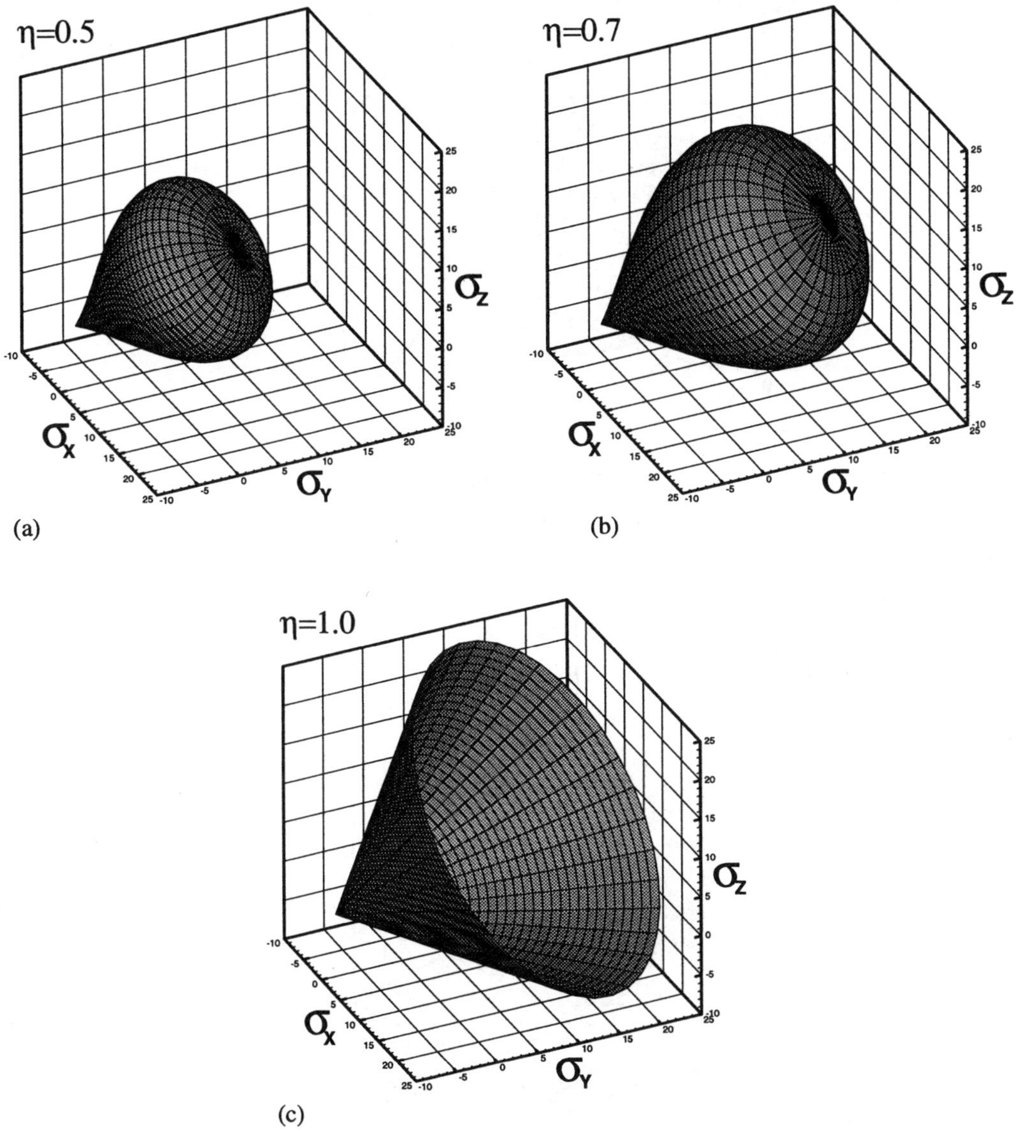

Figure 3.14. Trace of 3D cone-cap yield function in principal stress space for different values of relative density; The value of J_1 obtained by $f_d = 0.0$ between $-f_h$ and $+f_h$

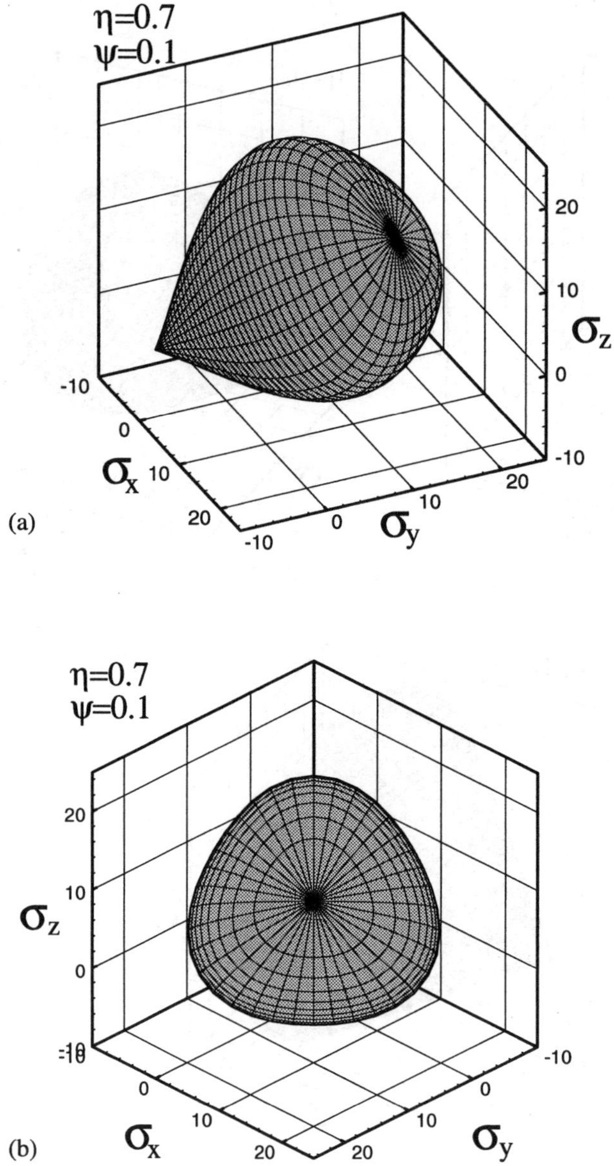

Figure 3.15. 3D irregular hexagonal pyramid of the Mohr-Coulomb and cone-cap yield function in principal stress space for $\psi = 0.1$

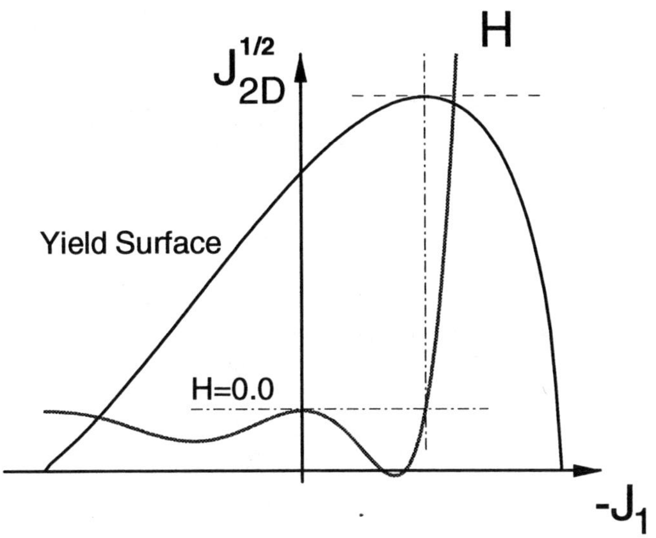

Figure 3.16. The variation of the hardening plastic modulus with hydrostatic pressure in meridian plane for a typical yield surface

The model parameters

f_h		f_d	
b	0.2615	α (MPa)	1702.834
a_0 (MPa)	- 45.746	β	3.792 e-2
a_1 (MPa)	- 368.869	c_1 (MPa)	- 1816.213
a_2 (MPa)	997.772	c_2 (MPa)	- 1943.106
a_3 (MPa)	- 571.576	c_3 (MPa)	1529.148
		c_4 (MPa)	795.287

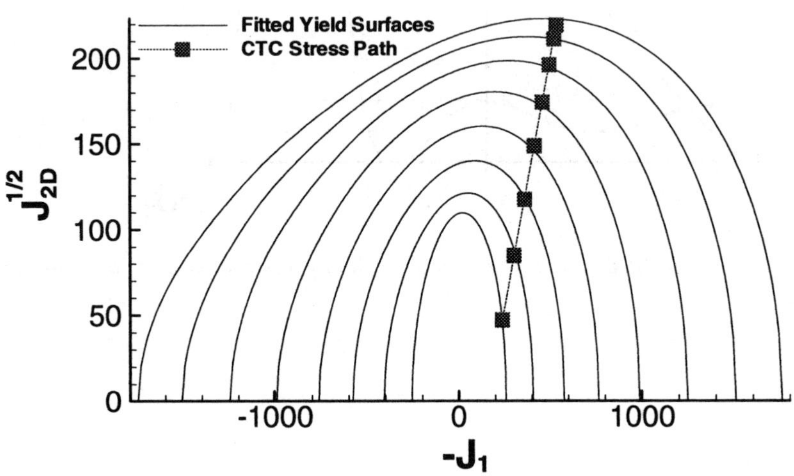

Figure 3.17. Model 1; The shape of yield surfaces obtained by fitting the stress path to triaxial tests performed by Doremus *et. al.* (1995) at initial confining pressure of 150 MPa

The model parameters

f_h		f_d	
b	0.8135	α (MPa)	1798.993
a_0 (MPa)	610.307	β	8.382 e-2
a_1 (MPa)	- 4010.446	c_1 (MPa)	- 2020.630
a_2 (MPa)	6748.890	c_2 (MPa)	- 1941.796
a_3 (MPa)	- 3328.278	c_3 (MPa)	1819.745
		c_4 (MPa)	593.164

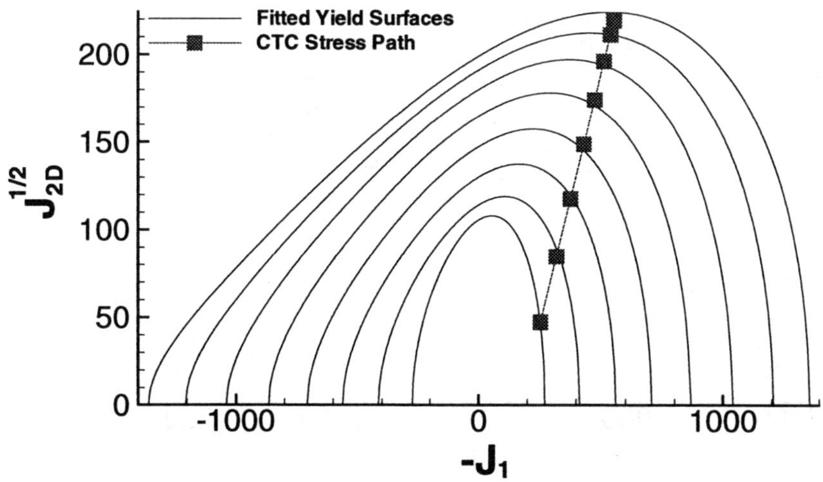

Figure 3.18. Model 2; The shape of yield surfaces obtained by fitting the stress path to triaxial tests performed by Doremus *et. al.* (1995) at initial confining pressure of 150 MPa

The model parameters

f_h		f_d	
b	0.9449	α (MPa)	2230.538
a_0 (MPa)	1463.873	β	7.116 e-2
a_1 (MPa)	- 5731.414	c_1 (MPa)	- 2859.303
a_2 (MPa)	8353.101	c_2 (MPa)	- 2053.762
a_3 (MPa)	- 4061.744	c_3 (MPa)	3306.836
		c_4 (MPa)	- 393.539

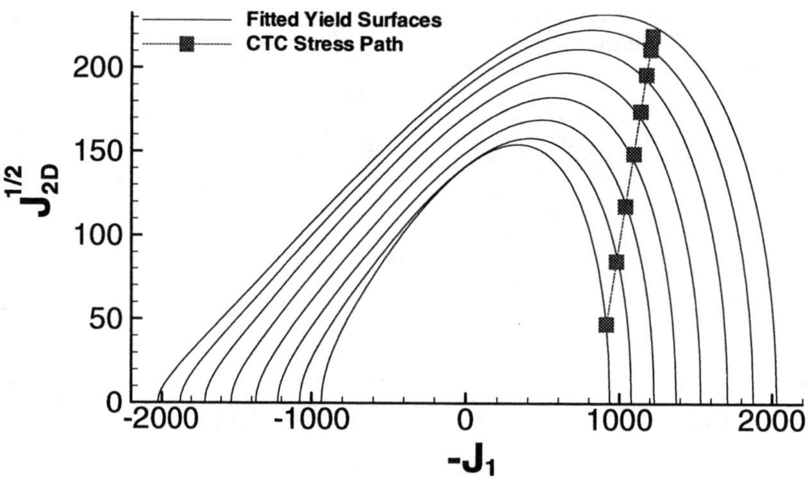

Figure 3.19. Model 3; The shape of yield surfaces obtained by fitting the stress path to triaxial tests performed by Doremus *et. al.* (1995) at initial confining pressure of 150 MPa

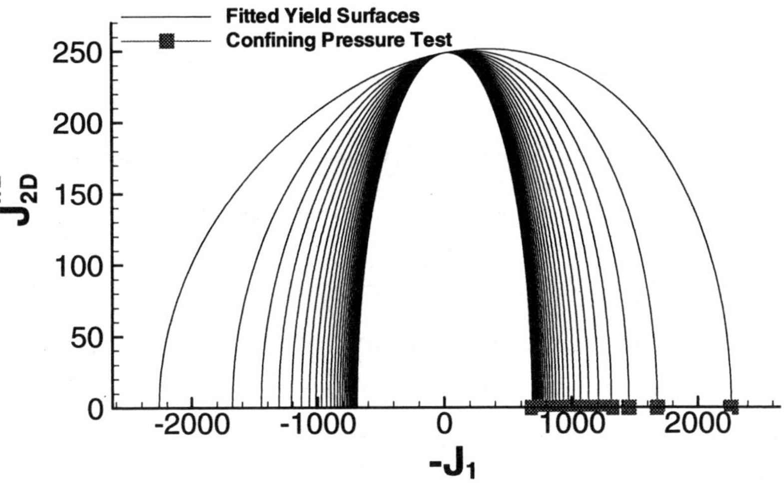

(a)

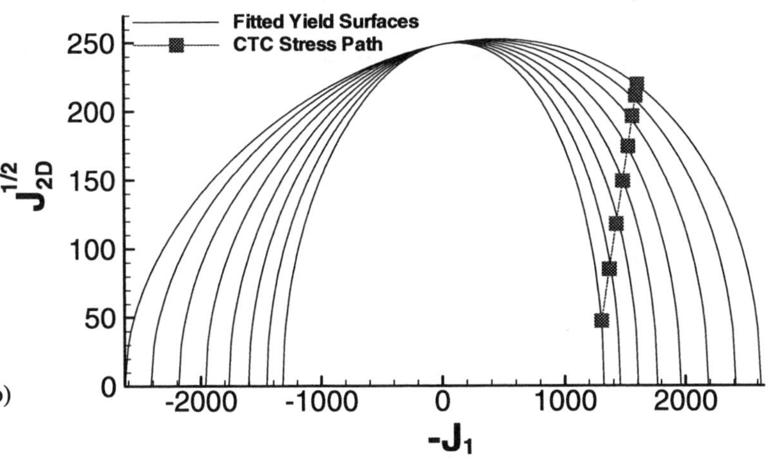

(b)

Figure 3.20. Model 4; The shape of yield surfaces obtained by fitting the stress path to both triaxial and confining pressure tests performed by Doremus *et. al.* (1995) at initial confining pressure of 150 MPa

Table 3.1. Yield functions proposed by researchers for porous materials

Reference	A, B and δ as a function of relative density ρ
Kuhn and Downey (1971)	$A=2+\rho^{1.92},\ B=\dfrac{(1-\rho^{1.92})}{3},\ \delta=1$
Green (1972)	$A=3,\ B=\dfrac{1}{4}\left[\dfrac{3(1-(1-\rho)^{1/3})}{(3-2(1-\rho)^{1/4})\ln(1-\rho)}\right]^2,\ \delta=\left[\dfrac{3(1-(1-\rho)^{1/3})}{(3-2(1-\rho)^{1/4})}\right]^2$
Oyane *et al.* (1973)	$A=3,\ B=\dfrac{1}{(1+\sqrt{\rho/(1-\rho)})^2},\ \delta=\rho^2$
Gurson (1975)	$A=3,\ B=\dfrac{(1-\rho)^2}{8},\ \delta=\rho^2-\rho+1$
Shima and Oyane (1976)	$A=3,\ B=\dfrac{(2.49(1-\rho)^{0.514})^2}{9},\ \delta=\rho^5$
Corapcioglu (1978)	$A=3,\ B=\dfrac{2.09(1-\rho)^2-0.26(1-\rho)}{\rho^2},\ \delta=1$
Corapcioglu and Uz (1978)	$A=3,\ B=\dfrac{0.361(1-\rho)^2+0.398(1-\rho)}{\rho^2},\ \delta=1$
Doraivelu *et al.* (1984)	$A=2+\rho^2,\ B=\dfrac{(1-\rho^2)}{3},\ \delta=2\rho^2-1$

Table 3.2. The effective stress, equivalent yield stress and various constants of the plastic flow vector for different yield criteria

Yield criterion	Stress level (effective stress)	Equivalent yield stress	Constants of plastic flow vector		
			C_1	C_2	C_3
Tresca	$2(J_{2D})^{1/2}\cos\theta$	σ_y	0	$2\cos\theta(1+\tan\theta\tan 3\theta)$	$\frac{3\sin\theta}{J_{2D}\cos 3\theta}$
von-Mises	$\sqrt{3}(J_{2D})^{1/2}$	σ_y	0	$\sqrt{3}$	0
Mohr-Coulomb	$\frac{1}{3}J_1\sin\phi+(J_{2D})^{1/2}\times\left(\cos\theta-\sin\theta\sin\phi/\sqrt{3}\right)$	$c\cos\phi$	$\frac{1}{3}\sin\phi$	$\cos\theta[(1+\tan\theta\tan 3\theta)+\sin\phi(\tan 3\theta-\tan\theta)/\sqrt{3}]$	$\frac{\sqrt{3}\sin\theta+\cos\theta\sin\phi}{2J_{2D}\cos 3\theta}$
Drucker-Prager	$\alpha J_1+(J_{2D})^{1/2}$	k	α	1.0	0
Compromise cone	$\frac{2J_1}{3}\sin\phi+\sqrt{3}(J_{2D})^{1/2}$	$2c\cos\phi$	$\frac{2}{3}\sin\phi$	$\sqrt{3}$	0
Axial-extension cone	$2J_1\sin\phi+\sqrt{3}(3+\sin\phi)(J_{2D})^{1/2}$	$6c\cos\phi$	$2\sin\phi$	$\sqrt{3}(3+\sin\phi)$	0
Extended von-Mises	$2J_1\sin\phi+\sqrt{3}(3-\sin\phi)(J_{2D})^{1/2}$	$6c\cos\phi$	$2\sin\phi$	$\sqrt{3}(3-\sin\phi)$	0
Critical state model I	$-\frac{J_1^2}{27J_{2D}}-\frac{2p_cJ_1}{9J_{2D}}$	$\frac{1}{M_{cs}^2}$	$-\frac{2}{9J_{2D}}\left(\frac{2}{3}J_1+p_c\right)$	$\frac{2J_1}{9(J_{2D})^{3/2}}\left(\frac{J_1}{3}+2p_c\right)$	0
Critical state model II	$J_{2D}\left[\frac{3\cos\theta-\sin\theta\sin\phi}{\sin\phi}\right]^2+\left[-\frac{J_1}{3}-p_{co}+\frac{c\left(\sqrt{3}\cos\theta-\sin\theta\sin\phi\right)}{3\sin\phi}\right]^2$	p_{co}^2	$\frac{2}{3}\left[\frac{J_1}{3}+p_{co}-\frac{c}{3\sin\phi}\times\left(\sqrt{3}\cos\theta-\sin\theta\sin\phi\right)\right]$	$\frac{2(J_{2D})^{1/2}}{\sin^2\phi}\left(\sqrt{3}\cos\theta-\sin\theta\sin\phi\right)^2+\frac{2\tan 3\theta}{J_{2D}}\left(\sqrt{3}\sin\theta+\cos\theta\sin\phi\right)\times\left[\frac{\sqrt{3}\cos\theta-\sin\theta\sin\phi}{\sin^2\phi}\left(J_{2D}+\frac{c^2}{9}\right)-\frac{1}{3}\sin\phi\left(\frac{J_1}{3}+p_{co}\right)\right]$	$\frac{\sqrt{3}}{(J_{2D})^{3/2}}\left(\sqrt{3}\sin\theta+\cos\theta\sin\phi\right)\times\left[\frac{\left(\sqrt{3}\cos\theta-\sin\theta\sin\phi\right)}{\sin^2\phi}\left(J_{2D}+\frac{c^2}{9}\right)-\frac{c}{3}\sin\phi\left(\frac{J_1}{3}+P_{co}\right)\right]$

Table 3.2. The effective stress, equivalent yield stress and various constants of the plastic flow vector for different yield criteria (continued)

Yield criterion	Stress level (effective stress)	Equivalent yield stress	Constants of plastic flow vector		
			C_1	C_2	C_3
Lade-Duncan model	$J_1^3(J_3)^{-1}$	k_1	$\frac{J_1^2}{J_3}\left[3+\frac{1}{J_3}\left(\frac{J_1 J_{2D}}{2\sqrt{3}}-\frac{J_1^3}{9}\right)\right]$	$\frac{1}{\sqrt{3}}J_1^4(J_{2D})^{1/2}J_3^{-2}$	$-J_1^3(J_3)^{-2}$
Lade model	$\left(J_1^3/J_3-27\right)\left(\frac{J_1}{P_a}\right)^m$	k_2	$\frac{1}{P_a^m}\left(\frac{J_1^{m+3}J_{2D}}{2\sqrt{3}}-\frac{J_1^{m+5}}{9}\right)(J_3)^{-2}$ $+\frac{1}{P_a^m}(m+3)\frac{J_1^{m+2}}{J_3}$ $-27m\frac{J_1^{m-1}}{P_a^m}$	$\frac{1}{\sqrt{3}P_a^m}J_1^{m+4}(J_{2D})^{1/2}(J_3)^{-2}$	$-\frac{1}{P_a^m}J_1^{m+3}(J_3)^{-2}$
Eekelen model	$\frac{\left(a-\frac{J_1}{3}\right)(1-\beta\sin 3\theta)^n}{(J_{2D})^{1/2}}$	$\frac{1}{\alpha}$	$-\frac{1}{3}\frac{(1-\beta\sin 3\theta)^n}{(J_{2D})^{1/2}}$	$\frac{1}{J_{2D}}\left(a-\frac{J_1}{3}\right)(1-\beta\sin 3\theta)^n$ $\times\left(-1+\frac{3\beta n\sin 3\theta}{1-\beta\sin 3\theta}\right)$	$\frac{3\sqrt{3}}{2}\frac{\beta n}{(J_{2D})^2}\left(a-\frac{J_1}{3}\right)$ $\times(1-\beta\sin 3\theta)^{n-1}$
Sandler's cap model	$f_1:\ (J_{2D})^{1/2}-\theta J_1+\gamma e^{-\beta J_1}-\alpha$	α	$C_1=-\theta-\beta\gamma e^{-\beta J_1}$	1.0	0
	$f_2:\ \sqrt{R^2 J_{2D}+(J_1-L)^2}$	Rb	$\frac{J_1-L}{\sqrt{R^2 J_{2D}+(J_1-L)^2}}$	$\frac{R^2\sqrt{J_{2D}}}{\sqrt{R^2 J_{2D}+(J_1-L)^2}}$	0
Desai's model	$J_{2D}+\alpha J_1^2-\beta J_1(J_3)^{1/3}-\gamma J_1$	k^2	$2\alpha J_1-\gamma-\beta(J_3)^{1/3}$ $+\beta\left(\frac{J_1 J_{2D}}{6\sqrt{3}}-\frac{J_1^3}{27}\right)(J_3)^{-2/3}$	$(J_{2D})^{1/2}\left(2+\frac{1}{3\sqrt{3}}\frac{\beta J_1^2}{(J_3)^{2/3}}\right)$	$\beta\frac{J_1}{3}(J_3)^{-2/3}$

Table 3.3. The shape of three-invariant single yield surface in meridian plane

The value of J_1 obtained by $f_d = 0.0$	The intersection points of yield surface with J_1 axis	The shape of yield surface
no value for J_1	$J_1 = \pm f_h$	Figure 3.12(a), (b)
$J_1 > +f_h$	$J_1 = \pm f_h$	Figure 3.12(c), (d)
$-f_h < J_1 < +f_h$	$J_1 = -f_h$ and J_1 obtained by $f_d = 0.0$	Figure 3.12(e), (f)
$J_1 < -f_h$	no available roots	no yield surface

Table 3.4. The variation of density with hydrostatic pressure

Confining pressure (MPa)	Relative density	Density (g/cm^3)
0	0.486737401	3.67
50	0.679045093	5.12
100	0.762599469	5.75
150	0.815649867	6.15
200	0.848806366	6.40
250	0.881962865	6.65
300	0.901856764	6.80
350	0.921750663	6.95
400	0.938992042	7.08

Table 3.5. The variation of axial stress with axial strain and density at $P = 150$ MPa

Axial stress (MPa)	Axial strain	Relative density	Density (g/cm^3)
150	0.00	0.810345	6.110
232	0.05	0.850796	6.415
297	0.10	0.883289	6.660
354	0.15	0.909814	6.860
408	0.20	0.929708	7.010
452	0.25	0.945623	7.130
490	0.30	0.957560	7.220
516	0.35	0.965517	7.280
530	0.40	0.970822	7.320

Table 3.6. The model parameters of single-cap plasticity

f_h		f_d	
b	1.2908	α (MPa)	248.620
a_0 (MPa)	1087.135	β	1.6990 e-2
a_1 (MPa)	- 2875.383	c_1 (MPa)	0.0
a_2 (MPa)	4406.547	c_2 (MPa)	0.0
a_3 (MPa)	- 2568.542	c_3 (MPa)	0.0
		c_4 (MPa)	0.0

Table 3.7. The elastic predictor-plastic corrector algorithm for rate-independent isotropic hardening elasto-plasticity

i) Evaluate elastic predictor

$$\boldsymbol{\sigma}_{n+1}^{\text{trial}} = \mathbf{D}^e\left(\boldsymbol{\varepsilon}_{n+1} - \boldsymbol{\varepsilon}_n^p\right)$$

ii) Check yield condition

$$f_{n+1}^{\text{trial}} := f\left(\boldsymbol{\sigma}_{n+1}^{\text{trial}}, \boldsymbol{q}_n\right)$$

IF $f_{n+1}^{\text{trial}} \leq 0$ THEN /Elastic State/

Set $\Xi_{n+1} = \Xi_{n+1}^{\text{trial}} \leq 0$ and EXIT

ELSE /Plastic Corrector/

iii) Evaluate the plastic corrector

$$\left(\mathbf{D}^e\right)^{-1}\left(\boldsymbol{\sigma}_{n+1} - \boldsymbol{\sigma}_{n+1}^{\text{trial}}\right) + \lambda_{n+1}\left.\frac{\partial f}{\partial \boldsymbol{\sigma}}\right|_{(\boldsymbol{\sigma}_{n+1}, \boldsymbol{q}_{n+1})} = 0$$

$$\boldsymbol{q}_{n+1} - \boldsymbol{q}_n + \lambda_{n+1}\left.\frac{\partial f}{\partial \boldsymbol{q}}\right|_{(\boldsymbol{\sigma}_{n+1}, \boldsymbol{q}_{n+1})} = 0$$

$$f(\boldsymbol{\sigma}_{n+1}, \boldsymbol{q}_{n+1}) \leq 0, \quad \lambda_{n+1} \geq 0, \quad \lambda_{n+1} f(\boldsymbol{\sigma}_{n+1}, \boldsymbol{q}_{n+1}) = 0$$

Table 3.8. The closest-point-projection algorithm

i) Define the residual $\boldsymbol{\psi}_1^i$ and ψ_2^i at iteration i

ii) Check whether convergence is attained, i.e.

IF $\|\boldsymbol{\psi}\|^i <$ TOLERANCE THEN EXIT

ELSE CONTINUE

iii) Compute the tangent matrix $\boldsymbol{A}^i$

iv) Solve the linearized system $\boldsymbol{A}^i\{\delta\sigma_{n+1}^i \quad \delta\lambda_{n+1}^i\}^T = \boldsymbol{\Psi}^i$

v) Update the solution

$$\lambda_{n+1}^{i+1} = \lambda_{n+1}^i + \delta\lambda_{n+1}^i \quad \text{and} \quad \sigma_{n+1}^{i+1} = \sigma_{n+1}^i + \delta\sigma_{n+1}^i$$

vi) GO TO (i)

4 BOUNDARY FRICTON MODEL

The influence of powder-tool friction on the mechanical properties of the final product is significant in pressing metal powders. Friction between the powder and tool affects the density distribution in the compact. A non-homogeneous density distribution induces cracks and residual stresses during compaction and sintering which is detrimental to the strength of the component. Friction also affects the final density, pressing and ejection forces and die wear. As the shape of a component gets more complex the consequences of friction get more difficult to foresee. Thus, it is important to predict the behavior of the powder-tool friction.

Contact problems in powder forming processes are often associated with very complex geometries and nonlinear constitutive behavior. For such problems only numerical methods, like the finite element method, yield solutions of the associated mathematical model. In the present chapter, we discuss the general problem of formulating continuum models for frictional phenomena and of developing computational methods for analyzing these phenomena. It is also concerned with the numerical modeling of frictional contact between a rigid tool and a deformable material. The finite element approach adopted is characterized by the use of 'interface' elements in which a plasticity theory of friction is incorporated to simulate sliding resistance at the powder-tool interface. The constitutive relations for friction are derived from a Coulomb friction law.

4.1. Introduction

Numerical analysis techniques have so improved in the last decades that such methods have been adapted to many problems of metal forming processes. However, to perform such analyses, knowledge of the friction phenomenon is extremely important. The friction phenomenon occurs during the metal forming process and is one of the most important factors, which contributes to material deformation. When two solid bodies in contact slide against each other a resistance to this sliding is generated along their interface. The existence of wedge-shaped protrusions on the tool surface may produce asperities on the surface of the workpiece, when pressing these protrusions onto the workpiece and sliding the tool along the workpiece. These asperities move in a wave-like motion along the surface layer and cause plastic deformation through a specified depth under the surface. This plastic deformation combines with local friction between the tool and the workpiece along the asperity interface to produce resistance to sliding.

In powder compaction forming, it is well known that the friction between the powder and the tools limits the performances of the process and the mechanical characteristics of the parts. Friction can also result in poor density distributions, which leads to differential retreat during compaction and sintering, and the heterogeneity of the final properties of the part. After compaction, when the tooling retreats on a multi-levels press, or during ejection, the forces due to friction may result in fatal cracks of different size. Many aspects of the process are affected by friction e.g. density distribution, pressing forces, final shape, tool wear, residual stresses and cracks.

One way to overcome these problems is to mix a solid phase lubricant into the powder. Even if the friction force cannot be ultimately reduced to zero, trying to reduce it appears to be a reasonable approach. The use of a well-appropriated lubricant, a good coating or surface treatment, or the employment of a deformable die (Holownia 1996) is other possible alternative solutions to this problem. Decreasing the friction is of great industrial importance. The resulting increase in density could alter the mechanical characteristics of the parts obtained by the process of powder compaction to be similar to those obtained by other processes such as forging.

Powder friction can be separated into two categories; internal friction caused by friction forces between the powder particles and contact friction between powder particles and the die wall. The material model handles the internal friction while a separate friction law is used to describe the wall friction effects. The internal friction between the powder particles, along with the constitutive equations of powder materials, have been described in Chapter 3. The aim of this chapter is to describe the mechanical behavior of contact friction between powder particles and die wall using classical and modern constitutive theories. Modeling of the interaction between the powder and the die is very dependent on both the friction model and the constitutive law for the powder itself.

During the contact between the powder and the tool the friction forces prevent complete compaction. Since the surfaces are rough, the true area of contact is much smaller than the apparent area of contact (Figure 4.1). As a result, the true contact surfaces may often support higher localised pressures that can produce plastic flow. With increasing load, the extent of these asperity contacts will increase and fresh contacts will be made. The influence of friction increases with the complexity of the powder material part being analysed. The non-linear nature of the numerical contact problem arise from the fact that the size of the powder-tool contact surface changes during pressing. Because of its microscopic nature, friction cannot be modelled adequately in the numerical analysis of macroscopic metal powder forming processes. The frictional resistance to sliding is therefore simulated by assuming mathematical relationships between the stresses and other state variables, which describe the contact conditions. These relationships can be derived from the micromechanical models of friction and then incorporated into an appropriate constitutive theory. In this study, the constitutive modeling of the frictional behavior of the metal powder is modelled by Coulomb's friction law and the plasticity theory of friction in the context of an interface element formulation.

4.2. Physical aspects of friction

Contact problems have been subjected to extensive research work due to their importance and complexity. Contact problems always involve friction problems, which constitutes a broad scientific field in itself. Frictional effects may, however, be neglected in many cases if the frictional forces on the contacting boundaries are sufficiently small. Therefore, one can have a frictionless contact problem, or a frictional contact problem, with the former much simpler than the latter. The nature of the friction forces developed between bodies in contact is extremely complex and is affected by a long list of factors (Oden and Martins 1985); the constitution of the interface, the time scales and frequency of the contact, the response of the interface to normal forces, inertia and thermal effects, roughness of the contacting surfaces, history of loadings, wear and general failure of the interface materials, the presence or absence of lubricants *etc*. Thus, friction is not a single phenomenon but is a collection of many complex mechanical and chemical phenomena entwined in a mosaic whose features cannot be grasped simply by means of isolated simple experiments.

Successful computational methods in nonlinear mechanics are generally those based on a thorough familiarity with the natural phenomena being simulated and on a good understanding of the mathematical models available to describe it. A major issue in the computer simulation of many natural events is the absence of a physically and mathematically sound model of the phenomena. The recognition that it may not be reasonable to attempt to describe all friction effects by a single theory seems to be a major step forward. It is convenient to divide the various types of frictional effects into three general categories (Oden and Martins 1985); quasi-static friction, dynamic sliding friction and wear and ploughing phenomena.

Type I. Quasi-static friction

The quasi-static frictional mechanisms occur when two metallic surfaces are slowly pressed together and are in a state of static equilibrium or are slowly being displaced relative to one another, and for which normal loads are sufficiently small such that no severe penetration of the contact interface occurs. The most conspicuous mechanisms contributing to this type of friction are the plastic deformation of asperities, the formation of the elasto-plastic junctions on the contact surface, the strong coupling between normal and tangential plastic deformations, and the dissipative mechanisms afforded by the history-dependence of junction deformation and possible fracture of these junctions. The frictional forces may depend upon histories of micro-tangential displacements of particles on the contact surface. Theories describing such frictional effects are basically static in nature; inertia effects are generally not important and the modeling should capture the elasto-plastic deformation of the interface in some way. Models and computational procedures for studying these classes of problems have been advanced by Fredriksson (1976), Michalowski and Morz (1978), Campos *et al.* (1982) and Oden and Pires (1983).

Type II. Dynamic sliding friction

The dynamic sliding friction refers to a large and important class of truly dynamic problems, which include such effects as frictional damping, dynamic sliding, stick-slip motion, chattering, *etc.* The constitution of the material interface is essentially stable; there is no marked penetration, or normal plastic deformation, of the interface and, at least from a global point of view, the frictional forces developed on the contact surface appear to depend on the sliding velocity of one surface relative to another. To obtain reproducible experimental data for these types of frictional effects, it is generally necessary to rub the contacting surfaces together for a period of time until a steady-state condition of the interface is reached. Such pre-conditioning may fracture and/or work-harden the surface asperities and result in a smoother contact surface for which, unlike category I, plastic deformations normal to the interface are not significant or do not occur at all. A significant distinguishing feature between Type I and II friction is that of scale. In Type I, the local frictional forces on the interface depend upon surface micro-displacements, which may be of an order of magnitude of ten microns; in Type II, much larger rigid relative motions can occur and frictional forces vary with sliding velocity.

Type III. Wear and ploughing phenomena

The wear and plowing phenomena encompasses cases in which substantial damage and penetration of the metallic interfaces occurs; there may be phase changes in the interface materials, and portions of the material may be removed or relocated during contact. These types of contact situations may involve the transfer of large normal forces across a small contact area. Wear may be experienced in all three of the categories mentioned, but in categories I and II it is either ignored or assumed to be negligible. We shall not consider these types of frictional phenomena in the present study. We also do not consider here problems of collision and impact, which differ from those mentioned earlier due to the occurrence of significant jumps in particle velocities at the contact surface, and instantaneous transfer of momentum from one body to another, and the propagation and reflection of stress waves. Moreover, while all contact problems with friction involve thermal effects to varying degrees, we shall ignore such effects in the present investigation.

The present chapter is devoted to the study of models and computational procedures for analyzing contact and friction problems employing in metal powder forming processes that fall into category II: sliding friction and stick-slip motions. Indeed, the friction category of type II can be characterized by 'large rigid relative motions' which is similar to the elasto-plastic theory by decomposition of the displacement into an elastic deformation and sliding distance component. The first and perhaps most difficult step in the analysis of dynamic friction is the development of an acceptable model for the contact interface. It is well known that the interface through which bodies contact is a mechanically-complicated media, the constitution of which is different from that of the parent material of either contacting body. Therefore, it is natural in developing phenomenological models of friction, to assign to the interface a separate structure characterized by constitutive equations independent of those characterizing the parent metals.

4.3. Plasticity theory of friction

In spite of the resemblance, which exists between plastic and frictional phenomena, the degrees of development reached by the theories of plasticity and friction are rather disparate. Indeed, although plasticity has long received a rigorous and general framework in the name of the classical theory of plasticity and encompassing a wide variety of material behavior, the theory of friction has had very sporadic attention dominated almost exclusively limited by Coulomb's law of perfect friction, which only covers a restricted range of tribological situations. This lag of the development of friction theory is even more flagrant when considering the progresses accomplished during the last decades with regards to the formulation of contact problems by means of variational inequalities (De Pater and Kalker 1975 and Oden and Pires 1983), calling for more sophisticated laws of friction. In this study, we have employed one of the developments of Curnier (1984), Rodic and Owen (1989) and Peric and Owen (1992) in the plasticity theory of friction for the analysis of powder compaction processes.

The objective of the mathematical theory of friction is to provide a theoretical description of motion at the interface of bodies in contact. The plasticity theory of friction can be achieved by an analogy between plastic and frictional phenomena. In order to formulate such a theory of friction several requirements have to be considered (Fredriksson 1976, Michalowski and Morz 1978, Kravchuk 1981, Curnier 1984 and Rodic and Owen 1989). These requirements, which are similar to the requirements, which have to be considered in the theory of elasto-plasticity, are

- *Stick (or adhesion) law;* a mathematical description of the stress state under sticking (elastic) conditions,
- *Stick-slip law;* a theoretical description of the relationship between stress and stick-slip (elasto-plastic) conditions,
- *Wear and tear rule;* a hardening and softening rule during sliding,
- *Slip criterion;* a yield criterion indicating the stress level at which relative slip motion occurs,
- *Slip rule;* a flow rule indicating the relationship between stress and slip motion.

To formulate a plasticity theory of friction, we consider two deformable bodies, one called the slave and the other the master. In principle the roles of the slave and master are interchangeable, however, in practice it is preferable to take the slave to be the more convex of the two bodies (and therefore the master as the flatest and stiffest one) in the presumed area of surface contact if at all possible. In this section, we present a summary account of the theories, which have been proposed based on the above requirements to explain the mathematical theory of friction.

4.3.1. Stick (or adhesion) law

Many of the early theories of friction, developed during the eighteenth and nineteenth centuries, attempted to explain the frictional behavior in terms of the surface roughness and the interlocking of the surface asperities. According to Moore (1975), in the classic laws of friction, as evolved from early studies in the past centuries, when two metallic bodies in contact are subjected to applied forces which tend to produce relative sliding motion, friction stresses develop on the interface that tend to oppose that motion. The friction force at the onset of sliding and during sliding is proportional to the normal contact force, i.e.

$$|F| = \mu N \tag{4.1}$$

where N is the compressive normal force, F is the friction force and μ denotes the coefficient of friction which is independent of the apparent area of contact. Often two values of μ are quoted; the coefficient of static friction, μ_s, which applies to the onset of sliding and the coefficient of kinetic friction, μ_k, which applies during sliding motion and is independent of the sliding velocity. The coefficient of static friction is greater than the kinetic coefficient.

The above equation is generally observed to hold for gross motions of effectively rigid bodies. However, it should be noticed that deviations from this equation have been reported at various circumstances; an increase of the coefficient of friction for light loads (Bowden and Tabor 1964), or a decrease of the friction coefficient for very high loads when the true area of contact approaches the magnitude of the apparent area of contact and bulk plastic deformation of the bodies in contact occurs (Bay and Wanheim 1976). Another important characteristic of the friction force is when tangential motion occurs, the friction force acts in the same direction of the tangential sliding velocity, $\dot{u}_T$, but in an opposite sense

$$F = -\mu N \, \dot{u}_T / |\dot{u}_T| \tag{4.2}$$

In this theory, it was assumed that the sliding of two contacting bodies involved the riding of rigid asperities of one surface over the other. If the average asperity angle is θ the coefficient of friction would be equal to $\tan\theta$ and consequently it would be independent of the load or the size of the contacting surfaces. The assumption that the asperities on one surface could traverse the gap between asperities on the other provided an explanation for the fact that often the kinetic friction is smaller than the static one. The main weakness of this early work on friction is that basically the models proposed are non-dissipative and friction is certainly a dissipative process (Tabor 1981).

4.3.2. Stick-slip law

A substantially different explanation for the friction phenomena, in the context of an adhesion-ploughing theory, was firstly established by Bowden and Tabor (1964). This theory has been the most widely accepted in recent decades among researchers of solid contact phenomena. The interfacial friction between metallic bodies is attributed essentially to two causes; the formation and shearing of metallic junctions between the surface asperities and the plastic deformation of the softer surface by hard asperities. As a consequence, the friction coefficient can be given as the sum of two components resulting from each of the above effects, i.e.

$$\mu = \mu_a + \mu_p \tag{4.3}$$

where μ_a results from the adhesion (welding) and can be obtained using a yield criterion of the combined normal and tangential loading of the asperities. μ_p results from the plastic deformation (ploughing) and can be estimated using the simplified model of a hard conical asperity grooving on a softer surface. These friction coefficients are given by

$$\mu_a = \frac{F_a}{N} = \frac{1}{\alpha^{1/2}\left(K^{-2}-1\right)^{1/2}}, \qquad \mu_p = \frac{F_p}{N} = \frac{2}{\pi}\tan\theta \tag{4.4}$$

where α is a constant of order 10 and K is a variable in the range of 0–1. For $K \to 1$, $\mu_a \to \infty$ which corresponds to perfectly clean surfaces of very ductile metals, where junction growth goes on indefinitely. For a very weak interface, such as $K < 0.2$, the junction growth is very small which corresponds to the presence of a good boundary lubricant or a thin film of a softer metal. In equation (4.4), θ is the slope of the asperity. Due to the small slopes of the asperities, μ_p leads to small values of the order of 0.07 to 0.14 and the contribution of the plowing component of friction will be negligible unless the adhesion is small.

Despite the wide acceptance of the theory described above and its effectiveness in explaining the basic laws of friction, several serious criticisms have been offered and some alternative theories have been proposed. According to Bikerman (1976), most of the solids are covered with various films, which prevent atomic contact between the bulk materials of two contacting bodies. Thus, the force required to pull apart two contacting bodies is zero. Furthermore, from several experimental observations on adhesion and friction, it has been concluded that the importance of the adhesion in frictional phenomena is negligible, as long as the term friction is used in the sense of a force, which, for a given path, is reproducible and is a definite function of the normal load.

In a similar manner to this microscopic model of friction, i.e. the 'adhesion-ploughing theory', Curnier (1984) developed a 'stick-slip' theory. The theory is based upon the decomposition of the relative displacement at the contact surface into the sum of two parts; one reversible, rather unusual, called 'stick' (or 'adherence'), can be attributed to the elastic deformation of asperities and denoted by $\mathbf{u}^A$ and the other irreversible, more familiar, called 'slip', may be attributed partly to the plastic deformation of these asperities and denoted by $\mathbf{u}^S$. The relative displacement $\mathbf{u}$ can be resolved into normal and tangential components as follows,

$$\begin{aligned} \mathbf{u} &= \mathbf{u}^A + \mathbf{u}^S = \left(\mathbf{u}_N^A + \mathbf{u}_N^S\right) + \left(\mathbf{u}_T^A + \mathbf{u}_T^S\right) \\ &= \mathbf{u}_N + \mathbf{u}_T \end{aligned} \tag{4.5}$$

This double decomposition is implicitly based upon two assumptions regarding the relative displacement $\mathbf{u}$ and normal variation N of contact, respectively (Figure 4.2). Firstly, for the decomposition of the relative displacement into 'stick' and 'slip' to remain meaningful, it is essential to conserve the same origin M (defined as the initial contact point of the slave on the master surface) for all subsequent measurements of the relative displacement of contact attached to the slave point S by definition $\mathbf{u} = S - M$. Secondly, for the resolution of the relative displacement into normal and tangential components to remain meaningful, it is imperative that the direction of the outward normal N to the master surface remains nearly constant throughout the sliding process.

The decomposition of the relative displacement in equation (4.5) may be compared to the strain decompositions into elastic, $\boldsymbol{\varepsilon}^e$, and plastic, $\boldsymbol{\varepsilon}^p$, parts on one hand and into bulk, $\bar{\boldsymbol{\varepsilon}}$, and deviatoric, $\boldsymbol{\varepsilon}'$, components on the other hand as used in plasticity theory. These decompositions are subject to finite strain and sometimes it is preferable to deal with strain rates instead, are (Curnier 1984)

$$\begin{aligned} \boldsymbol{\varepsilon} &= \boldsymbol{\varepsilon}^e + \boldsymbol{\varepsilon}^p = \left(\bar{\boldsymbol{\varepsilon}}^e + \bar{\boldsymbol{\varepsilon}}^p\right) + \left(\boldsymbol{\varepsilon}'^e + \boldsymbol{\varepsilon}'^p\right) \\ &= \bar{\boldsymbol{\varepsilon}} + \boldsymbol{\varepsilon}' \end{aligned} \tag{4.6}$$

As mentioned above, the relative displacements $\mathbf{u}^A$ and $\mathbf{u}^S$, called 'stick' and 'slip' can be decomposed to the elastic and plastic deformations of the asperities of two bodies in contact. The terms 'stick' and 'slip' must be taken in an enlarged acceptance of these words, since they include not only the usual tangential relative displacements but also the normal ones. This double connotation prepares the simultaneous treatment of both the normal impenetrability condition and the tangential slip condition, which is convenient for the numerical formulation. Since the normal displacement is assumed to be zero, a tangential displacement is only considered, which consists of 'stick' and 'slip' decompositions and is in principle the same as the decomposition of elastic and plastic behavior.

4.3.3. Wear and tear rule

The kinematic variables of stick, $\mathbf{u}^A$, and slip, $\mathbf{u}^S$, are associated with the friction force and also wear and tear forces. The forces of wear and tear characterize two different forms of a single and similar phenomenon (Curnier 1984); i.e. the running-in (grinding) of contact surfaces in relative sliding motion. The first form occurs when two bodies rub against one another in arbitrary directions, resulting in isotropic wear. The second form occurs when the sliding motion of two bodies is oriented along a preferential direction, resulting in an anisotropic tear of the contact surfaces. Both processes produce a reduction in the force of friction and are therefore analogous to softening processes, which occur in plasticity theories.

The forces of friction, tear and wear are the analogues of the stress, kinematic stress and isotropic (or equivalent) stress, respectively, in plasticity. Since existing experimental data do not provide information accurate enough for the precise modeling of frictional behavior, it is difficult to obtain appropriate friction factors and thus, the modeling of the influence of wear and tear forces to sliding resistance remains only a dream in the numerical simulations of metal powder forming. The modeling of friction considered in this study is therefore analogous to the treatment of perfectly plastic materials in plasticity.

4.3.4. Slip criterion

In order to perform the decomposition of the relative displacement into stick and slip, a slip criterion should be introduced. A slip criterion, also referred to as a friction criterion, is a theoretical description of the relationship between stress and stick-slip motion at the interface of bodies in contact. It is interrogated to decide which one of the two modes, sticking or slipping, occurs. If stick reaches (i.e. drops below) a certain threshold, called the slip or stick limit, then the relative motion contributes exclusively to slip (stick) respectively. As mentioned in Section 4.3.3, the modeling of friction considered in the present work is analogous to the treatment of perfectly plastic materials in plasticity. The law of perfect friction states that the force of friction is proportional to the load and is independent of the apparent area of contact and the other state variables.

The onset of irreversible slip is governed by a slip criterion and can be presented in the space of normal and tangential stresses as shown in Figure 4.3. Analogous to the theory of plasticity, the stress states within the surface are defined by a slip criterion, which produces only reversible relative motions between the bodies in contact, while the stress states, which lie on the slip surface, produce irreversible relative motions. One of the most common slip criteria, which have been used to describe the plasticity behavior of friction contact, is Coulomb's friction law.

Although specific details may differ, most current theories of the origin of friction single out plastic deformation on the contact neighborhood as a major phenomenon associated with the frictional sliding of two bodies. On the other hand, the model of a particle sliding with friction on a surface is often presented as a prototype of plastic behavior. A question, which then arises, is why classical plasticity theory has not been applied successfully to friction problems. Drucker (1954) and Michalowski and Mroz (1978) pointed out that the sliding rule dictated by Coulomb's friction theory is not an associated flow rule as defined in the classical theory of plasticity. In particular, a pointwise version of Coulomb friction is encompassed in the relations

$$|\tau| \le \mu |\sigma_n|, \qquad \sigma_n \le 0 \tag{4.7}$$

and

$$\begin{aligned} \dot{u}_T &= 0 && \textit{if } |\tau| < \mu |\sigma_n| \\ -\dot{u}_T &= \lambda \tau && \textit{if } |\tau| = \mu |\sigma_n|, \quad \lambda \ge 0 \end{aligned} \tag{4.8}$$

where $\dot{u}_T$ is the tangential sliding velocity at a point on the contact surface, σ_n is the normal stress (the 'contact pressure'), τ is the tangential stress (the frictional stress) and λ is a real number. Thus, we can define a slip surface, such as that shown in Figure 4.3, in the spirit of a yield surface in plasticity, such that contact stress states on the surface correspond to states at which sliding occurs. Since $-\dot{u}_T = \lambda\tau$ on this surface, classical friction does not satisfy the classical normality rule and, therefore, most of the standard arguments of classical plasticity do not have a straightforward extension to frictional phenomena.

In the present study, a type of slip criterion based on Mohr-Coulomb's law with a tension cut-off model, as illustrated in Figure 4.4, has been considered. The slip criterion can be written as

$$\begin{aligned} f_{1f} &= \sigma_n = 0 \\ f_{2f} &= |\tau| - \sigma_n \tan\varphi_f - c_f = 0 \end{aligned} \tag{4.9}$$

where c_f is the cohesion between the powder and tool, φ_f is the tool friction angle and $\mu_f = \tan\varphi_f$ denotes the Coulomb friction coefficient. The parameters c_f and φ_f can be obtained from simple laboratory tests such as compaction experiments (Gethin *et al.* 1994). In this model, when $f_{2f} < 0$ a 'stick' situation occurs, i.e. there is no relative movement between the powder and tool. Conversely, if $f_{2f} \ge 0$ a 'slip' situation occurs and the shear stress at the interface is constrained to the value $f_{2f} = 0$. It must be noted that if a positive value of σ_n is registered, then the 'tension cut-off' condition is applied, i.e. $f_{1f} = 0$, and consequently $\tau = 0$.

Avitzur and Nakamura (1986) developed a micromechanical friction model with a transition region between Coulomb's law and the constant shear rule (Figure 4.5). They demonstrated a mathematical modeling and simulation of friction during steady state sliding of metals, based on the upper-bound approach. The relation between the normal pressure (stress) and the sliding (tangential) stress is established for the entire range of pressure levels from zero to infinity. This slip criterion is given by

$$\begin{aligned} f_{1f} &= |\tau| - \mu |\sigma_n| = 0 \qquad && if\ |\sigma_n| \le p_c \\ f_{2f} &= |\tau| - m\sigma_0/\sqrt{3} = 0 \qquad && if\ |\sigma_n| > p_c \end{aligned} \tag{4.10}$$

where m is the friction factor and $\sigma_0/\sqrt{3}$ denotes the maximum shear stress that the deformed material can withstand according to the von-Mises yield criterion. The critical value of p_c depends on the interface conditions and the properties of the materials in contact. Some values of p_c are suggested by Avitzur and Nakamura (1986). In the numerical simulations considered by Rodic and Owen (1989), p_c is assumed to be proportional to the σ_0 value of the material, i.e. $p_c = \alpha\sigma_0$ where typical values of α are in the range between 1 and 1.5.

4.3.5. Slip rule

The direction of slip is governed by an appropriate slip rule, which can be derived from the gradient of a convex potential Z as

$$\dot{\gamma} = \dot{\lambda}\frac{\partial Z}{\partial \tau} \quad , \qquad \dot{\varepsilon}_n = \dot{\lambda}\frac{\partial Z}{\partial \sigma_n} \tag{4.11}$$

where λ is a, yet arbitrary, constant expressing the collinearity of the slip increment with the outward normal to the potential Z. The fact that the two increments derive from the same potential characterizes a model of 'standard generalized friction'.

If the potential Z is replaced by the slip criterion f_f, the slip rule becomes associated. Although in plasticity the flow rules associated with the standard criteria prove realistic for relatively large classes of materials, the slip rule associated to the usual friction criterion (4.9) is not acceptable. Indeed, the relative movement at the interface derived from the associated potential $Z = f_f$ by using the normality rule (4.11), results in the creation of gaps (separation) as compaction proceeds. Thus, in order to avoid the slave body separation from the contacting surface, a non-associated flow rule is typically adopted.

Michalowski and Mroz (1978) suggested that in such problems the slip rule needs to be specified only in a tangential direction, where the slip direction can be derived from a convex potential Z (Figure 4.6). Hence, the slip potential Z can be given as a cylinder with radius $|\tau|$ for isotropic frictional contact, i.e.

$$Z = |\tau| \tag{4.12}$$

and the slip direction is defined as the outward normal to the slip potential as

$$\dot{\gamma} = \dot{\lambda}\frac{\tau}{|\tau|} \tag{4.13}$$

4.4. Modeling of friction

In the powder compaction process, the required shape changes are obtained by pressing the material between tools, which are much stiffer than the shaped material. Since there is a large difference between the deformability of the tool and the material, relative movement occurs in the contact area, the magnitude of which will depend on the geometry and frictional conditions at the interface. The implementation of these frictional conditions into the compaction finite element model plays a crucial role in the estimation of the density distribution. In Chapter 2, it has been postulated that on the boundaries of the forming domain, either displacements (when contact with the forming tool occurs), or tractions (as in the case of a free surface or surface loaded with prescribed forces) are specified. In most practical problems another boundary condition occurs, i.e., one in which relative slippage occurs between the forming tool and the material. The treatment of such phenomena is difficult, especially as the physics involved is still under debate.

From the finite element point of view the modeling of interface friction can be categorised into three routes:

- *Traction boundary condition;* In this method, frictional forces can be appended to the external force vector as a traction boundary force.
- *Contact node algorithm;* This model is an iterative implementation of the boundary conditions on velocity or displacement depending on the nodal forces at the interface friction boundary.
- *Interface element;* As an alternative to contact node algorithm, friction can be modelled by considering the utilization of thin elements, having very high aspect (length/breadth) ratios, along the friction boundary.

Early studies on contact problems were largely related to a linear geometry and often involved node-to-node contacts when two boundaries come into contact (Chan and Tuba 1971, Fracavilla and Zienkiewicz 1975, Hughes *et al.* 1976 and Beer 1985). Such a node-to-node model can only be applied to problems in which relative sliding displacements of the two contact boundaries are sufficiently small. Once significant non-linear contact deformations were introduced, methods tended to switch to node-on-segment or node-on surface contact (Wriggers and Simo 1985, Ju *et al.* 1987, Parisch 1989 and Hallquist *et al.* 1985, 1992). Contacts between general contact boundaries can then be represented by contacts between contact nodes and contact segments. The contact node is also called a slave node, a projectile, a hitting node, a contactor *etc.* The contact segment is called a master segment or a target segment.

There are important links between the finite element contact problem and mathematical programming techniques. Indeed, because the contact problems usually involve inequality constraints (varying contact areas), the mathematics can be related to the method of variational inequalities. Basically, two main constraint methods of solution have been employed in the finite element solution of contact problems; the method of Lagrangian multipliers and the penalty approach.

The method of Lagrangian multipliers has been applied to friction often with the aid of a plasticity technique which follows from the observation of the close links between friction and plasticity made by Michalowski and Mroz (1978) (Chaudaray and Bathe 1986 and Gallego and Anza 1989). In this method, contact forces are taken as primary unknowns and the non-penetration condition is enforced. The algorithm takes a non-associative flow rule so that the tangent stiffness matrix is non-symmetric. The Lagrange multiplier method does expand the original global stiffness matrix but does not affect the original global tangent stiffness matrix if small displacements are assumed, or if a quasi-Newton method is used under large displacements and large rotation. Within a general non-linear finite element context, important issues are associated with contact detection. The Lagrange multiplier method is simpler with lower order elements and, partially for this reason these lower-order elements are still very popular.

The penalty approach is closely related to techniques, which attempt to introduce a genuine (although high and possibly non-linear) stiffness for the contact region (Wriggers *et al.* 1985, Curnier and Alart 1988 and Peric and Owen 1992). In this method, penetration between two contacting boundaries is introduced and the normal contact force is related to the penetration by a penalty parameter. One alternative procedure involves the elimination of degrees of freedom and can be considered as a form of non-linear master slave approach (Jelenic and Saje 1995). Although the penalty method does not introduce any new unknown it is not necessarily more efficient than the Lagrange multiplier method, which does introduce new unknowns. This is because the penalty method will change the profile and skyline of the global tangent stiffness matrix, which will greatly influence the solution time.

Based on the two basic constraint methods, other constraint techniques have been proposed and applied. In order to eliminate the zero diagonal terms introduced in the Lagrange multiplier method, Simo *et al.* (1985) proposed a perturbed Lagrangian formulation by introducing a certain amount of penetration as a function of the normal contact load. Many concepts from the mathematical programing literature have been incorporated within finite element algorithms - in particular the augmented Lagrangian technique. Wriggers *et al.* (1985) and Simo and Laursen (1992) introduced an augmented Lagrange method by the combination of the two basic constraint methods. Within a finite element context, one of the most important recent developments involved the 'consistent linearisation' of the changing geometrical contact relationships (Wriggers and Simo 1985 and Parisch 1989).

In relation to these techniques, important issues are associated with the use of a 'one-pass' or 'two-pass' algorithm. For two bodies, the master contact surface and the slave contact surface slide against each other. We can assume that only the slave contact surface is subjected to contact constraints and is thus not allowed to penetrate through the master surface. The master surface can, however, penetrate through the slave surface. This is the so called 'one-pass' treatment of the contact interface. In a 'two-pass' treatment, neither the slave surface nor the master surfaces are allowed to penetrate which is obviously preferable from an accuracy point of view. In particular, Taylor and Papadopoulos (1991) have shown that a two-pass formulation is essential if the 'contact path test' is to be passed.

In this section, we first introduce a model of contact problems with friction based on the penalty method, which has been proposed by Peric and Owen (1992) in the unilateral contact of flexible three-dimensional structures. An interface element model will be then described in the simulation of powder forming process in which a plasticity theory of friction is incorporated to simulate sliding resistance at the powder-tool interface.

4.5. Continuum model of friction

Consider two bodies, a master (target) and a slave body, as shown in Figure 4.2, with initial configurations denoted by Ω^S and Ω^T. The relative displacement from the point $\mathbf{x}$ on the contact surface of the slave body to the point $\mathbf{y}$ on the contact surface of the target may be defined as

$$g_N(\mathbf{x}) = \{\chi(\mathbf{x}) - \chi(\mathbf{y})\}\cdot\mathbf{N} \qquad \text{on} \qquad \partial\chi(\Omega_C) = \partial\chi(\Omega^S)\cap\partial\chi(\Omega^T) \tag{4.14}$$

where $g_N(\mathbf{x})$ is the gap between the two bodies, $\chi(\mathbf{x})$ and $\chi(\mathbf{y})$ are the configuration mapping of the slave and target bodies, $\partial\chi(\Omega^S)$ and $\partial\chi(\Omega^T)$ are the slave and target body surfaces respectively, $\partial\chi(\Omega_C)$ denotes a surface where contact between the two bodies occurs and $\mathbf{N}$ is the unit outward normal vector on the target surface. During the contact and sliding of the bodies (Figure 4.2), we define p_N and p_T as the normal and tangential load acting on the point $\mathbf{x}$, respectively. The contact conditions may be expressed in the standard Kuhn-Tucker form as

$$g_N \geq 0 \qquad p_N \leq 0 \qquad p_N \cdot g_N = 0 \tag{4.15}$$

which is best suited for a variational formulation. Since there is no gap between the two bodies in the sliding contact problems of metal powder forming processes, the normal displacement is assumed to be zero and a tangential displacement is only considered, which consists of stick and slip decompositions and is in principle the same as the decomposition of elastic and plastic behavior. Thus, replacing Figure 4.2. by Figure 4.7, the decomposition of the tangential displacement at the contact surface can be given from equation (4.5) as

$$\mathbf{u}_T = \mathbf{u}_T^A + \mathbf{u}_T^S = \mathbf{u}_T^e + \mathbf{u}_T^p \tag{4.16}$$

where $\mathbf{u}_T$ is the tangential part of the displacement described by $\mathbf{u}_T = (\mathbf{I} - \mathbf{N} \otimes \mathbf{N})\mathbf{u}$.

In relation (4.15), the kinetic constraint of impenetrability of two bodies can be satisfied as well as the static condition of compressive normal load. To resolve the resulting unilateral contact problem the Lagrange multiplier method, or the penalty method, are typically used. In the case of the Lagrange multiplier method, however, a large number of additional unknown variables need to be included owing to incorporation of $\mathbf{p}_N$ as new variables. On the contrary, the penalty method needs no additional variable, because the impenetrability condition is approximately satisfied (by constraint via embedding very stiff springs on the contact surface). Consequently, the normal load $\mathbf{p}_N$ can be obtained from multiplication of the penalty factor k_N and the displacement in the normal direction $\mathbf{u}_N$. Similarly, the stick (or elastic) component of the tangential load may be obtained by multiplying the penalty factor k_T and elastic part of the displacement in the tangential direction $\mathbf{u}_T^A$. The penalty factors k_N and k_T can be considered as being the normal stiffness constant and shear stiffness constant, respectively. Constitutive laws for the contact loads can now be summarized as

$$\mathbf{p}_N = \left(\mathbf{D}_f^e\right)_N \mathbf{u}^e \tag{4.17}$$

$$\mathbf{p}_T = \left(\mathbf{D}_f^e\right)_T \mathbf{u}^e \tag{4.18}$$

where $\left(\mathbf{D}_f^e\right)_N$ and $\left(\mathbf{D}_f^e\right)_T$ are the normal and tangential parts of the elastic modulus tensor for friction defined as

$$\left(\mathbf{D}_f^e\right)_N = -k_N(\mathbf{N} \otimes \mathbf{N}) \tag{4.19}$$

$$\left(\mathbf{D}_f^e\right)_T = -k_T(\mathbf{I} - \mathbf{N} \otimes \mathbf{N}) \tag{4.20}$$

In order to perform the additive decomposition of displacement into adherence and slip, a slip criterion must be introduced. To this end the slip surface F_f is postulated in the contact stress space on which slip will occur. In addition, the coefficient of friction is assumed to depend on the adopted state variables, i.e. the frictional work parameter w and the normal force p_N. This results in the following slip criterion

$$F_f(\mathbf{p}, w) = \|\mathbf{p}_T\| - \mu_f(p_N, w)\|\mathbf{p}_N\| \quad \begin{cases} = 0 & \text{slip (or grap)} \\ < 0 & \text{adherence} \end{cases} \tag{4.21}$$

the above slip criterion is depicted as a generalized cone in Figure 4.8(b) in the space $(\mathbf{p}_N, \mathbf{p}_T)$ due to dependence on the state variable $\mathbf{p}_N$, in contrast to the standard Amontons and Coulomb laws of friction depicted in Figure 4.8(a). As frictional work hardening, or softening, is assumed, this slip criterion will expand, or shrink, in the radial direction with the amount of frictional work (Figure 4.8c).

As mentioned before, the associated flow rule of slip criterion (4.21) would generate a normal force. In order to avoid the slave body separation from the contacting surface, a non-associated flow rule is typically adopted (Curnier 1984). Hence, the slip potential Z is introduced as a cylinder with radius $\|\mathbf{P}_T\|$ for isotropic frictional contact and the slip direction is defined as the outward normal to the slip potential Z. Consequently, the plastic part of the tangential displacement $\mathbf{u}_T^p$, in equation (4.16), can be defined by the definition of the slip rule as

$$\dot{\mathbf{u}}_T^p = \dot{\lambda}\frac{\partial Z}{\partial \mathbf{p}_T} = \dot{\lambda}\mathbf{T} \tag{4.22}$$

where $\mathbf{T} = \frac{\mathbf{p}_T}{\|\mathbf{p}_T\|}$ is the unit tangential vector.

4.5.1. Continuum tangent matrix

The constitutive model for the contact problem with a non-linear frictional evolution law is conveniently summarized in Table 4.1. Following the standard arguments of elasto-plasticity, the continuum tangent tensor for the contact problem with non-linear frictional evolution is derived using the consistency condition

$$\left(\frac{\partial F_f(\mathbf{p}, w)}{\partial \mathbf{p}_T}\right)\dot{\mathbf{p}}_T + \left(\frac{\partial F_f(\mathbf{p}, w)}{\partial \mathbf{p}_N}\right)\dot{\mathbf{p}}_N + \left(\frac{\partial F_f(\mathbf{p}, w)}{\partial w}\right)\dot{w} = 0 \tag{4.23}$$

and substituting $\dot{\lambda}$ into the constitutive law, a linearized equation is obtained in the rate form as (Peric and Owen 1992)

$$\dot{\mathbf{p}} = \mathbf{D}_f^{ep}\,\dot{\mathbf{u}} \tag{4.24}$$

where the continuum tangent tensor $\mathbf{D}_f^{ep}$ is

$$\begin{aligned} \mathbf{D}_f^{ep} = & -k_N(\mathbf{N}\otimes\mathbf{N}) - k_T(\mathbf{I} - \mathbf{N}\otimes\mathbf{N} - \mathbf{T}\otimes\mathbf{T}) \\ & -k_T\frac{1}{\beta}\left(\frac{\mu_f(p_N, w)}{k_T}\|\mathbf{p}_N\|^2\left(\frac{\partial \mu_f}{\partial w}\right)\right)(\mathbf{T}\otimes\mathbf{T}) \\ & -k_N\frac{1}{\beta}\left(\mu_f(p_N, w) + \|\mathbf{p}_N\|\left(\frac{\partial \mu_f}{\partial \|\mathbf{p}_N\|}\right)\right)(\mathbf{T}\otimes\mathbf{N}) \end{aligned} \tag{4.25}$$

where

$$\beta = 1 + \frac{\mu_f(p_N, w)}{k_T} \|\mathbf{p}_N\|^2 \left(\frac{\partial \mu_f}{\partial w} \right) \tag{4.26}$$

In equation (4.25), the first term of the left hand side continuum tangent tensor $\mathbf{D}_f^{ep}$ indicates the stiffness in the normal direction to the contact surface. The second term denotes the adhesion stiffness perpendicular to the sliding direction on the contact surface. The third and forth terms indicate the adhesion and slip stiffness with hardening, or softening, in the sliding direction, respectively. In the case of frictional slip without hardening, or softening, equation (4.25) can be simplified as

$$\mathbf{D}_f^{ep} = -k_N(\mathbf{N} \otimes \mathbf{N}) - k_T(\mathbf{I} - \mathbf{N} \otimes \mathbf{N} - \mathbf{T} \otimes \mathbf{T}) - \mu_f k_N(\mathbf{T} \otimes \mathbf{N}) \tag{4.27}$$

Evidently, the non-associative slip rule (4.22) results in non-symmetry of the slip modulus tensor, which is defined by (4.25) under the conditions of frictional slip.

4.5.2. Consistent tangent matrix

The resemblance between computational contact models with friction and the classical elasto-plasticity models allows us to utilize a similar numerical integration scheme to that used in elasto-plasticity. Among several numerical integration schemes (e.g. Ortiz and Simo 1986, Mitchell and Owen 1988 and Crisfield 1991), the implicit backward Euler method, which gives good convergency and accuracy, is introduced here. The numerical integration algorithm is summarized in Table 4.2. Even thought the local friction consistency condition is satisfied, the *global* equilibrium condition may be violated. In order to obtain the solution which satisfies both the *consistency condition* and the *equilibrium condition*, we have to carry on some iterations until satisfying both conditions. In addition, we must employ the consistent tangent modulus at each iteration to obtain quadratic convergence.

From Table 4.2, the response function can be calculated as

$$\begin{aligned} \hat{\mathbf{p}}(\mathbf{p}_n, w_n, \Delta\mathbf{u}_{n+1}) &= \mathbf{p}_n + (\Delta\mathbf{p}_N)_{n+1} + (\Delta\mathbf{p}_T)_{n+1} \\ &= \mathbf{p}_n + \left(\mathbf{D}_f^e\right)_N \Delta\mathbf{u}_{n+1} + \left(\mathbf{D}_f^e\right)_T \Delta\mathbf{u}_{n+1} - k_T \Delta\gamma \mathbf{T}_{n+1} \end{aligned} \tag{4.28}$$

The total differential of equation (4.28) can be written as

$$\begin{aligned} d\mathbf{p}_{n+1} &= \frac{\partial \hat{\mathbf{p}}(\mathbf{p}_n, w_n, \Delta\mathbf{u}_{n+1})}{\partial \mathbf{u}} \\ &= \left(\mathbf{D}_f^e\right)_N d(\Delta\mathbf{u}_{n+1}) + \left(\mathbf{D}_f^e\right)_T d(\Delta\mathbf{u}_{n+1}) - k_T \mathbf{T}_{n+1} d(\Delta\gamma) - k_T \Delta\gamma \left(\frac{\partial \mathbf{T}}{\partial \mathbf{p}_T} \right) d(\Delta\mathbf{p}_T)_{n+1} \end{aligned} \tag{4.29}$$

After solving equation (4.29) for $d(\Delta\gamma)$ by using the consistency condition

$$(F_f)_{n+1} = \left\| (\mathbf{p}_T)_{n+1}^{\text{trial}} \right\| - k_T \Delta\gamma - (\mu_f)_{n+1} \left\| (\mathbf{p}_N)_{n+1} \right\| = 0 \tag{4.30}$$

and frictional work parameter

$$w_{n+1} = w_n + (\mu_f)_{n+1} \left\| (\mathbf{p}_N)_{n+1} \right\| \Delta\gamma \tag{4.31}$$

we obtain the linearized function, as

$$d(\mathbf{p}_N)_{n+1} = \widehat{\mathbf{D}}_f^{ep} \, d\mathbf{u}_{n+1} \tag{4.32}$$

where the consistent tangent tensor $\widehat{\mathbf{D}}_f^{ep}$ is expressed as

$$\begin{aligned}\widehat{\mathbf{D}}_f^{ep} = &-k_N (\mathbf{N} \otimes \mathbf{N}) - k_T \frac{\left\| (\mathbf{p}_T)_{n+1} \right\|}{\left\| (\mathbf{p}_T)_{n+1}^{\text{trial}} \right\|} (\mathbf{I} - \mathbf{N} \otimes \mathbf{N} - \mathbf{T} \otimes \mathbf{T}) \\ &- k_T \frac{1}{\gamma} \left(\frac{(\mu_f)_{n+1}}{k_T} \left\| \mathbf{p}_N \right\|^2 \left(\frac{\partial \mu_f}{\partial w} \right)_{n+1} \right) (\mathbf{T} \otimes \mathbf{T}) \\ &- k_N \frac{1}{\gamma} \left((\mu_f)_{n+1} + \left\| \mathbf{p}_N \right\| \left(\frac{\partial \mu_f}{\partial \left\| \mathbf{p}_N \right\|} \right) \right) (\mathbf{T} \otimes \mathbf{N})\end{aligned} \tag{4.33}$$

where

$$\gamma = 1 + \left(\frac{(\mu_f)_{n+1}}{k_T} \left\| (\mathbf{p}_N)_{n+1} \right\| - \Delta\gamma \right) \left(\frac{\partial \mu_f}{\partial w} \right)_{n+1} \left\| (\mathbf{p}_N)_{n+1} \right\| \tag{4.34}$$

It can be observed from equation (4.33) that the accumulation of siding distance may increase, or decrease, the stiffness matrix.

4.6. Interface element formulation

Friction occurs in a very thin layer at the tool-powder interface. Modeling of this phenomenon by standard finite element techniques would lead to numerical ill conditioning due to the high aspect ratios present. The idea of overcoming this problem by introducing special elements is not new (Ghaboussi *et al.* 1973, Beer 1985 and Schellekens and de Borst 1993, 1994). In this case, the interface behavior between the boundary of the tool and the deforming material is modelled by using interface elements, which have one side representing the die and the other side representing the powder. The top nodes of the interface elements are used to monitor contact with the tool. When contact is registered, the appropriate integration points within the interface element become active and the frictional resistance is simulated.

Evaluation of the stiffness matrices for the interface elements requires linearization of the governing equations for the frictional contact problem. Strict mathematical linearization results in non-symmetric constitutive matrices $\mathbf{D}_f$, due to the non-associated slip rules employed. To preserve the symmetry of the numerical formulation the off-diagonal term in $\mathbf{D}_f$, which represents the coupling between the normal and tangential stresses at the interface, is neglected, so that its effect is brought into the formulation via residual 'pseudo loads'. In this way the problem is artificially decomposed into a pure contact in the normal direction and frictional resistance in the tangential direction, which are linearized separately. In the incremental manner, the stress components are related to the strains through the material property matrix $\mathbf{D}_f$, as follows

$$\Delta\boldsymbol{\sigma} = \mathbf{D}_f \, \Delta\boldsymbol{\varepsilon} \tag{4.35}$$

in which the stress vector of two-dimensional interface element is assumed to have only two components; the normal stress σ_n and shear stress τ, where $\Delta\boldsymbol{\sigma} = \{\Delta\tau,\ \Delta\sigma_n\}^T$ and $\Delta\boldsymbol{\varepsilon} = \{\Delta\gamma,\ \Delta\varepsilon_n\}^T$. In 3D interface elements, the stress vector includes three components; the normal stress σ_n and shear stresses τ_1 and τ_2, where $\Delta\boldsymbol{\sigma} = \{\Delta\tau_1,\ \Delta\tau_2,\ \Delta\sigma_n\}^T$ and $\Delta\boldsymbol{\varepsilon} = \{\Delta\gamma_1,\ \Delta\gamma_2,\ \Delta\varepsilon_n\}^T$. In equation (4.35), the material matrix $\mathbf{D}_f$ needs to capture the details of the physical processes taking place such as asperity contact, adhesion and the consequent 'stick-slip' behavior. These requirements may be considered by using the constitutive frictional law in the matrix $\mathbf{D}_f$. Since there is no volume change due to shearing strains, the shear and normal components of deformation are therefore uncoupled.

The width w of the interface element is considered to be very small, i.e. $w \to 0$ so that the deformation of the element in the normal direction does not have a significant influence on the compaction process. The normal strains are measured only to monitor the normal stress by means of the linear equation, i.e. $\sigma_n = E_f \, \varepsilon_n$, where E_f is chosen as an arbitrary large number for numerical convenience. Note that only compressive normal stresses are allowed, i.e. $\sigma_n \le 0$.

Since the tangential micro-shifts due to adherence are negligible in comparison to the micro-slips due to sliding, the adherence does not require a very precise numerical treatment. The incremental form of the shear stress-strain relationship under adherence is defined as $\Delta\tau = G_A \Delta\gamma$, where G_A is proportional to the stick shear modulus of the powder and tool. The frictional shear force, however, is limited by the slip criterion. In two dimensional cases, the shear stress-strain relationship is defined, as shown in Figure 4.9, where $\mathcal{T} = \mu_f |\sigma_n|$ is considered to be a known state variable.

Based on the above description, the constitutive material matrix $\mathbf{D}_f$ in equation (4.35) can be defined as

$$\mathbf{D}_f = \begin{bmatrix} G_f & 0 \\ 0 & E_f \end{bmatrix} \tag{4.36}$$

for 2D interface elements, whereas for 3D interface elements

$$\mathbf{D}_f = \begin{bmatrix} G_{f_1} & 0 & 0 \\ 0 & G_{f_2} & 0 \\ 0 & 0 & E_f \end{bmatrix} \tag{4.37}$$

The frictional non-linearity is modelled by an appropriate variation of G_f which can be obtained from the shear stress-strain $(\tau - \gamma)$ relationship for the stick and slip region, as shown in Figure 4.9. It must be noted that the shear modulus G_f is divided into two components, the first where the compact has not moved but there is a rapid buildup in load. Under this condition, a stick shear modulus $G_f = G_A$ is derived directly from the slope of the $\tau - \gamma$ curve. On commencement of movement, the slip shear modulus condition $G_f = G_B$ is now appropriate and this is derived similarly.

4.6.1. 2D interface elements

In this section, the numerical modeling of frictional contact between a rigid tool and a deformable material is characterized by the use of 'interface' elements, which is a good approximation for the contact conditions between the tool and workpiece during powder forming operations. The workpiece is represented by standard finite elements, while the surface where the contact with the tool is expected, is covered with interface elements (Figure 4.10). The working surfaces of the tool are defined at discrete points and their position is derived from the movement and rotation of a reference point. The top nodes of the interface elements are used to monitor contact with the tool. When the contact is registered, the appropriate integration points within the interface element become active and the frictional resistance is simulated.

The relative displacements at any points along the interface element, shown in Figure 4.10, are given by

$$\boldsymbol{\delta} = \begin{Bmatrix} \delta_s \\ \delta_n \end{Bmatrix} = \begin{Bmatrix} u_s \\ u_n \end{Bmatrix}_{\text{top}} - \begin{Bmatrix} u_s \\ u_n \end{Bmatrix}_{\text{bot}} \tag{4.38}$$

where δ_s and δ_n are the relative displacements in the local s and n axes, $(u_s)_{\text{top}}$ and $(u_n)_{\text{top}}$ are the displacements in the local s and n axes direction of the top side of the element, and $(u_s)_{\text{bot}}$ and $(u_n)_{\text{bot}}$ are the displacements in the local s and n axes direction of the bottom side of the element, respectively.

The shear and normal strains can be defined from the relative displacements as

$$\gamma = \frac{1}{w}\delta_s \,, \qquad \varepsilon_n = \frac{1}{w}\delta_n \tag{4.39}$$

Displacements at any point on the bottom and top sides of the element can be related to the nodal values by

$$\begin{aligned} (u_s)_{\text{bot}} &= \sum_{i=1}^{3} N_i(\xi)\,\bar{u}_i \,, \qquad (u_n)_{\text{bot}} = \sum_{i=1}^{3} N_i(\xi)\,\bar{v}_i \\ (u_s)_{\text{top}} &= \sum_{i=4}^{6} N_i(\xi)\,\bar{u}_i \,, \qquad (u_n)_{\text{top}} = \sum_{i=4}^{6} N_i(\xi)\,\bar{v}_i \end{aligned} \tag{4.40}$$

where $\bar{u}_i$ and $\bar{v}_i$ are displacements of node i in the local s and n axes, respectively. The shape functions $N_i(\xi)$ associated with node i are expressed in terms of the local coordinate ξ by $N_i = \xi\xi_i(1+\xi\xi_i)/2$ for $i = 1, 3, 4, 6$ and $N_i = 1-\xi^2$ for $i = 2, 5$. For the finite element formulation it is useful to express the relative displacement vector $\boldsymbol{\delta}$ defined by (4.38) in terms of the displacements in global coordinates as

$$\boldsymbol{\delta} = \{-\mathbf{N}_{\text{bot}} \quad \mathbf{N}_{\text{top}}\} \begin{Bmatrix} \bar{\mathbf{u}}_{\text{bot}} \\ \bar{\mathbf{u}}_{\text{top}} \end{Bmatrix} = \mathbf{N}\,\bar{\mathbf{u}} \tag{4.41}$$

where $\mathbf{N}_{\text{bot}} = \{\mathbf{N}_1, \mathbf{N}_2, \mathbf{N}_3\}$, $\mathbf{N}_{\text{top}} = \{\mathbf{N}_4, \mathbf{N}_5, \mathbf{N}_6\}$ and the displacements in global coordinates are $\bar{\mathbf{u}}_i = \{\bar{u}_i, \bar{v}_i\}^T$. The shape functions $\mathbf{N}_i$ is defined as

$$\mathbf{N}_i = \begin{bmatrix} N_i & 0 \\ 0 & N_i \end{bmatrix} \begin{bmatrix} \cos\theta & \sin\theta \\ -\sin\theta & \cos\theta \end{bmatrix} \tag{4.42}$$

where

$$\theta = \arctan\left(\frac{\partial y/\partial\xi}{\partial x/\partial\xi}\right) \tag{4.43}$$

The stiffness matrix of an interface element is defined in the standard finite element manner as

$$\mathbf{K}_f = \int_\Omega \mathbf{B}_f^T \mathbf{D}_f \mathbf{B}_f \, d\Omega \tag{4.44}$$

where the $\mathbf{B}_f$ matrix is equal to $\frac{1}{w}\mathbf{N}$. The evaluation of the stiffness matrix $\mathbf{K}_f$ in equation (4.44), can be performed by a line integral as

$$\mathbf{K}_f = \int_{-1}^{+1} \mathbf{B}_f^T \mathbf{D}_f \mathbf{B}_f \left[\left(\frac{\partial x}{\partial\xi}\right)^2 + \left(\frac{\partial y}{\partial\xi}\right)^2\right]^{1/2} d\xi \tag{4.45}$$

It must be noted that τ and σ_n are decoupled in the stiffness matrix $\mathbf{K}_f$. However, in the stress calculation σ_n is evaluated first so that τ can then be obtained from the $(\tau-\gamma)$ relationship. Also, if a positive value of σ_n is registered, then the tension cut-off condition is applied, i.e. $f_{1f} = 0$, and consequently $\tau = 0$. Table 4.3 presents the computational procedure based on the Newton-Raphson procedure implemented to calculate the stiffness matrix $\mathbf{K}_f$.

4.6.2. 3D interface elements

The vector of relative displacements between two homologous points can be obtained from the displacement fields associated to the element faces (top and bottom), as shown in Figure 4.11, by

$$\boldsymbol{\delta} = \begin{Bmatrix} \delta_s \\ \delta_t \\ \delta_n \end{Bmatrix} = \begin{Bmatrix} u_s \\ u_t \\ u_n \end{Bmatrix}_{\text{top}} - \begin{Bmatrix} u_s \\ u_t \\ u_n \end{Bmatrix}_{\text{bot}} \tag{4.46}$$

where s and t represent the tangential directions and n the normal direction. A local co-ordinate system is established at a point on the contact by obtaining the vector normal to the contact surface, as shown in Figure 4.12. This is obtained as a cross-product of two vectors

$$\mathbf{n} = \frac{1}{A} \begin{Bmatrix} \dfrac{\partial x}{\partial\xi} \\ \dfrac{\partial y}{\partial\xi} \\ \dfrac{\partial z}{\partial\xi} \end{Bmatrix} \times \begin{Bmatrix} \dfrac{\partial x}{\partial\eta} \\ \dfrac{\partial y}{\partial\eta} \\ \dfrac{\partial z}{\partial\eta} \end{Bmatrix} \tag{4.47}$$

where A is the length of vector normal to the contact surface and represents the unit mapped area of the contact plane. The derivatives in equation (4.47) are coefficients of the Jacobian matrix of the co-ordinate transformation. The two tangent vectors can be formed as

$$\mathbf{s}=\begin{Bmatrix}1\\0\\0\end{Bmatrix}\times\mathbf{n}\,,\qquad \mathbf{t}=\mathbf{s}\times\mathbf{n} \tag{4.48}$$

If the direction of $\mathbf{n}$ is exactly in the $x-$direction then $\mathbf{s}$ is obtained by

$$\mathbf{s}=\begin{Bmatrix}0\\1\\0\end{Bmatrix}\times\mathbf{n} \tag{4.49}$$

The relative displacements of equation (4.46) can be therefore written as

$$\boldsymbol{\delta}=\begin{Bmatrix}\delta_s\\\delta_t\\\delta_n\end{Bmatrix}=\boldsymbol{\theta}^T\left(\mathbf{N}_{\text{top}}\overline{\mathbf{u}}_{\text{top}}-\mathbf{N}_{\text{bot}}\overline{\mathbf{u}}_{\text{bot}}\right) \tag{4.50}$$

or

$$\boldsymbol{\delta}=\boldsymbol{\theta}^T\left\{-\mathbf{N}_{\text{bot}}\quad \mathbf{N}_{\text{top}}\right\}\begin{Bmatrix}\overline{\mathbf{u}}_{\text{bot}}\\\overline{\mathbf{u}}_{\text{top}}\end{Bmatrix} \tag{4.51}$$

where $\boldsymbol{\theta}=\{\mathbf{s},\ \mathbf{t},\ \mathbf{n}\}$, $\mathbf{N}_{\text{bot}}=\mathbf{N}_{\text{top}}=\{N_1\mathbf{I},\ N_2\mathbf{I},\ \cdots,\ N_n\mathbf{I}\}$ and $\overline{\mathbf{u}}_i=\{\overline{u}_i, \overline{v}_i, \overline{w}_i\}^T$.

The stiffness matrix of 3D interface element can be defined by equation (4.44), in which for the numerical integration of interface elements, the integral in equation (4.44) is replaced by an integration over the iso-parametric co-ordinates ξ and η

$$\mathbf{K}_f=\int_{\xi=-1}^{\xi=+1}\int_{\eta=-1}^{\eta=+1}\mathbf{B}_f^T\mathbf{D}_f\mathbf{B}_f\det\mathbf{J}\,d\xi\,d\eta \tag{4.52}$$

where $\det\mathbf{J}$ denotes the determinant of the Jacobian matrix.

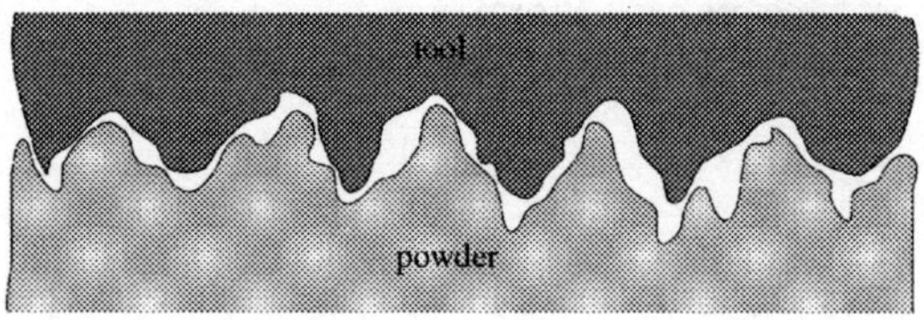

Figure 4.1. Microscopic geometrical conditions at the powder-tool interface

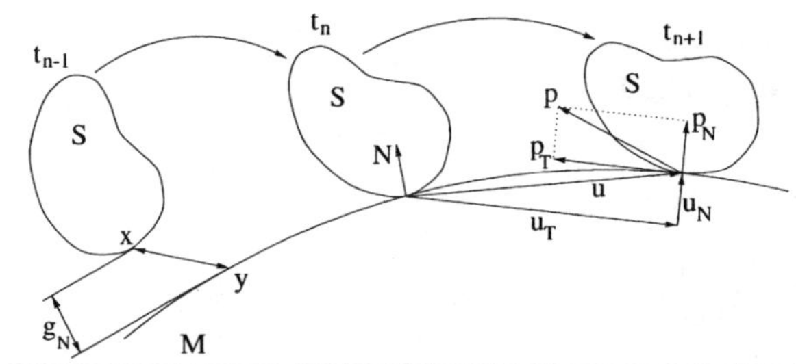

Figure 4.2. Definition of contact and sliding between two bodies

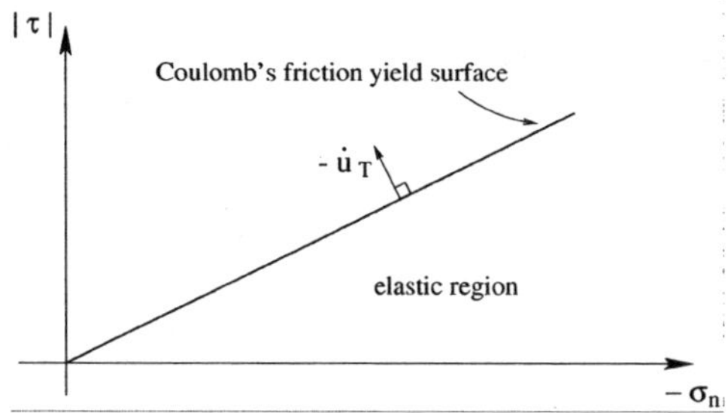

Figure 4.3. A slip criterion corresponding to Coulomb's law for which the normality rule is clearly violated

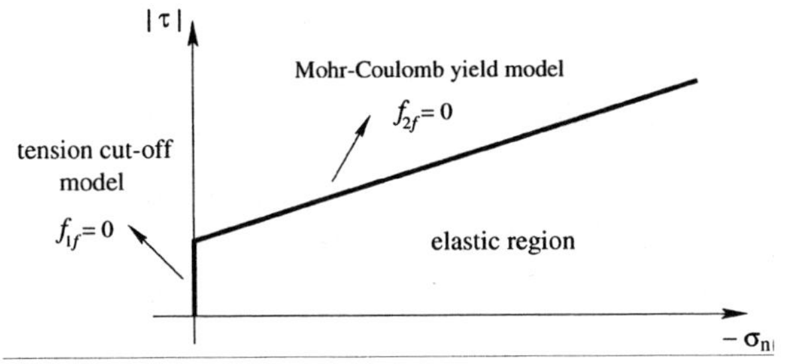

Figure 4.4. Mohr-Coulomb's slip criterion with tension cut-off model

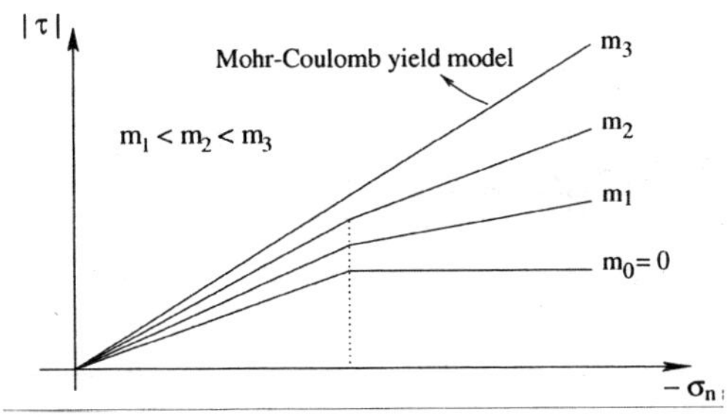

Figure 4.5. Avitzur's slip criterion

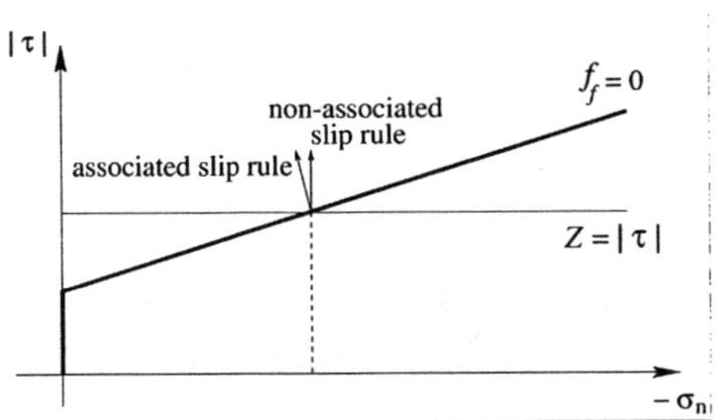

Figure 4.6. Associated and non-associated slip rules for Mohr-Coulomb's slip criterion

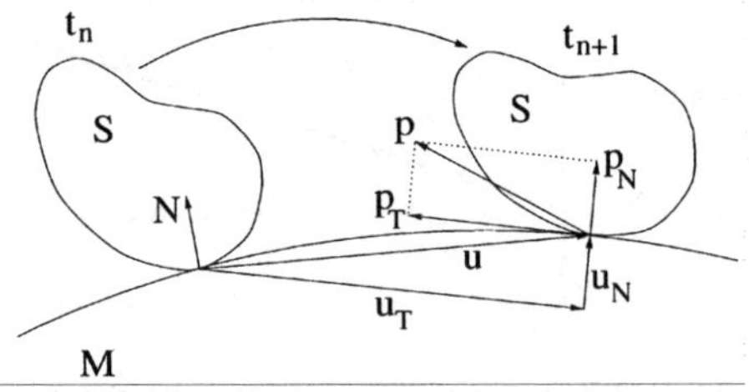

Figure 4.7. Definition of sliding contact between two bodies

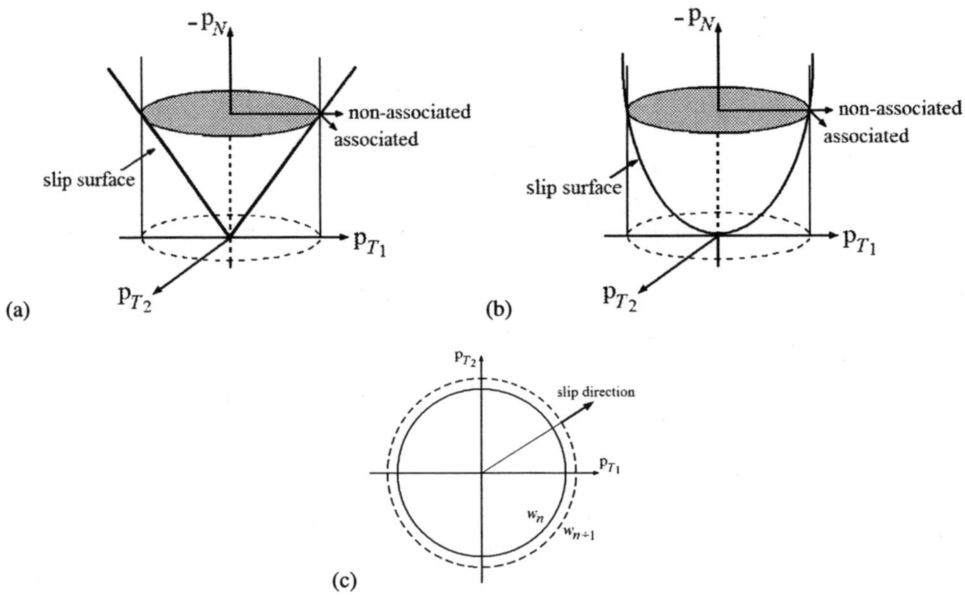

Figure 4.8. Slip criteria and slip rule; a) Linear friction law, b) Nonlinear friction law, c) Non-associated slip rule

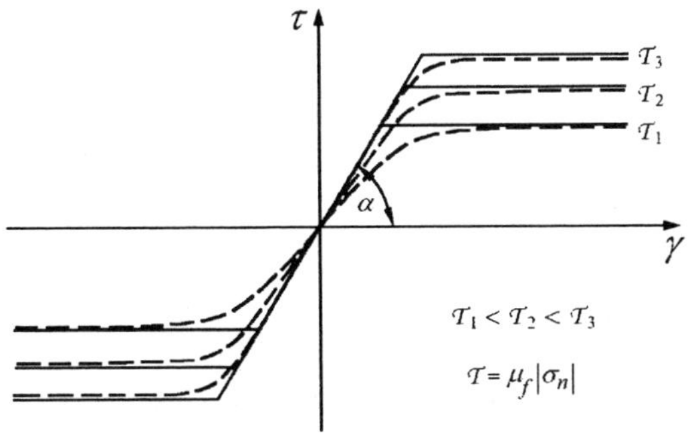

Figure 4.9. Frictional shear stress-strain relationship

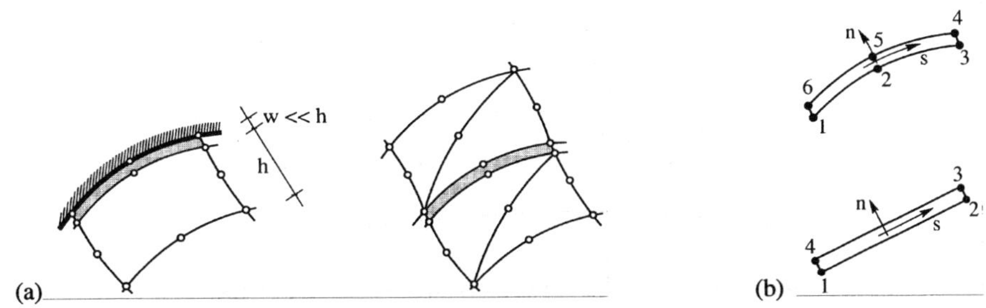

Figure 4.10. 2D interface elements; a) Definition, b) Quadratic and linear elements

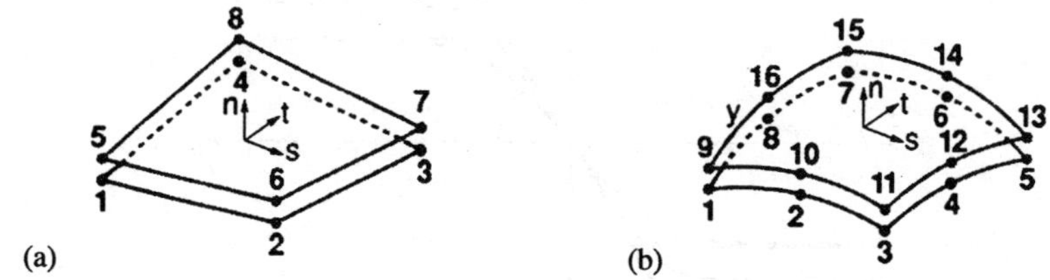

Figure 4.11. 3D interface elements; a) Linear element, b) Quadratic element

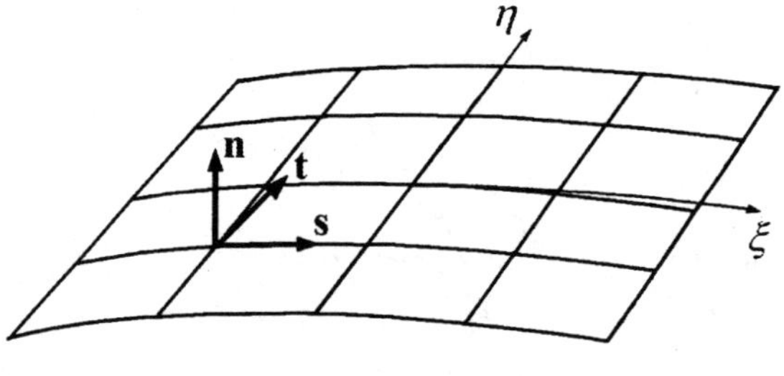

Figure 4.12. Definition of vectors in 3D interface elements

Table 4.1. Constitutive model for the plasticity theory of friction with non-linear evolution

i) Decomposition of displacement rate

$$\mathbf{u}_T = \mathbf{u}_T^A + \mathbf{u}_T^S \quad \text{where} \quad \mathbf{u}_T = (\mathbf{I} - \mathbf{N} \otimes \mathbf{N})\mathbf{u}$$

ii) Constitutive laws for contact forces

$$\mathbf{p}_N = \left(\mathbf{D}_f^e\right)_N \mathbf{u}^e \quad \text{and} \quad \mathbf{p}_T = \left(\mathbf{D}_f^e\right)_T \mathbf{u}^e$$

iii) Slip criterion

$$F_f(\mathbf{p}, w) = \|\mathbf{p}_T\| - \mu_f(p_N, w)\|\mathbf{p}_N\|$$

iv) Slip potential and slip rule

$$Z(\mathbf{p}) = \|\mathbf{p}_T\|, \quad \dot{\mathbf{u}}_T^p = \dot{\lambda}\frac{\partial Z}{\partial \mathbf{p}_T} = \dot{\lambda}\frac{\mathbf{p}_T}{\|\mathbf{p}_T\|}$$

v) Evolution law for the state variable w

$$\dot{w} = \mathbf{p}_T \dot{\mathbf{u}}_T^p = \dot{\lambda}\|\mathbf{p}_T\|$$

vi) Kuhn Tucker loading /unloading conditions

$$F_f \leq 0 \qquad \dot{\lambda} \geq 0 \qquad \dot{\lambda} \cdot F_f = 0$$

Table 4.2. Numerical integration algorithm for frictional problem with non-linear friction evolution

i) Update configuration, normal force and slip surface *

$$\mathbf{X}_{n+1}^{i} = \mathbf{X}_{n+1}^{i-1} + \Delta\mathbf{u}_{n+1}^{i} , \quad (\mathbf{p}_N)_{n+1} = (\mathbf{p}_N)_n + (\mathbf{D}_f^e)_N \Delta\mathbf{u}_{n+1} ,$$

$$\mu_f\left((p_N)_{n+1}^i, w_n\right), \quad \frac{\partial\mu_f\left((p_N)_{n+1}^i, w_n\right)}{\partial w}, \quad \frac{\partial\mu_f\left((p_N)_{n+1}^i, w_n\right)}{\partial\|\mathbf{p}_N\|}$$

ii) Elastic predictor phase

1) Calculate elastic trial tangential load

$$(\mathbf{p}_T)_{n+1}^{\text{trial}} = (\mathbf{p}_T)_n + (\mathbf{D}_f^e)_T \Delta\mathbf{u}_{n+1}$$

2) Check the current slip/stick condition

$$\text{IF} \quad (F_f)_{n+1}^{\text{trial}} = \left\|(\mathbf{p}_T)_{n+1}^{\text{trial}}\right\| - \mu_f\left((p_N)_{n+1}^i, w_n\right)\left\|(\mathbf{p}_N)_{n+1}\right\| \le 0 \quad \text{THEN}$$

3) Update variables and exist

$$\text{SET} \quad \left\{\mathbf{p}_T, w, \mu_f, \frac{\partial\mu_f}{\partial\|\mathbf{p}_N\|}, \frac{\partial\mu_f}{\partial w}\right\}_{n+1} = \left\{\mathbf{p}_T, w, \mu_f, \frac{\partial\mu_f}{\partial\|\mathbf{p}_N\|}, \frac{\partial\mu_f}{\partial w}\right\}_{n+1}^{\text{trial}} \quad \text{EXIT}$$

ELSE

iii) Slide corrector phase

1) Solve the local equations in terms of $\Delta\gamma$

$$(F_f)_{n+1}^{\text{trial}} + \left(\frac{\partial F_f}{\partial\mathbf{p}_T}\right)_{n+1}^{\text{trial}} \Delta\mathbf{p}_T + \left(\frac{\partial F_f}{\partial w}\right)_{n+1}^{\text{trial}} \Delta w = 0$$

where $\Delta\mathbf{p}_T = \Delta\gamma(\mathbf{D}_f^e)_T \mathbf{T}_{n+1}$, $\Delta w = (\mu_f)_{n+1}^{\text{trial}} \left\|(\mathbf{p}_N)_{n+1}\right\| \Delta\gamma$, and

$$\Delta\gamma = \frac{\left\|(\mathbf{p}_T)_{n+1}^{\text{trial}}\right\| - (\mu_f)_{n+1}^{\text{trial}}\left\|(\mathbf{p}_N)_{n+1}\right\|}{k_T + (\mu_f)_{n+1}^{\text{trial}}\left(\partial\mu_f/\partial w\right)_{n+1}^{\text{trial}}\left\|(\mathbf{p}_N)_{n+1}\right\|^2}$$

2) Evaluate the new state of stress and slip surface

$$\left\|(\mathbf{p}_T)_{n+1}\right\| = \left\|(\mathbf{p}_T)_{n+1}^{\text{trial}}\right\| + \Delta\gamma\, k_T \mathbf{T} , \quad w_{n+1} = w_n + \left\|(\mathbf{p}_T)_{n+1}\right\| \Delta\gamma_{n+1} ,$$

$$\mu_f\left((p_N)_{n+1}^i, w_{n+1}\right), \quad \partial\mu_f\left((p_N)_{n+1}^i, w_{n+1}\right)/\partial w$$

3) Check the subsequent slip/stick condition

$$\text{IF} \quad (F_f)_{n+1} = \left\|(\mathbf{p}_T)_{n+1}\right\| - \mu_f\left((p_N)_{n+1}^i, w_{n+1}\right)\left\|(\mathbf{p}_N)_{n+1}\right\| \le 0 \quad \text{THEN EXIT}$$

ELSE

$$\text{SET} \quad \left\{\mathbf{p}_T, w, \mu_f, \frac{\partial\mu_f}{\partial\|\mathbf{p}_N\|}, \frac{\partial\mu_f}{\partial w}\right\}_{n+1}^{\text{trial}} = \left\{\mathbf{p}_T, w, \mu_f, \frac{\partial\mu_f}{\partial\|\mathbf{p}_N\|}, \frac{\partial\mu_f}{\partial w}\right\}_{n+1}$$

GO TO (**iii** – 1)

* Note that the slip surface at $(\mathbf{p}_N)_n$ must be shifted to the new one at $(\mathbf{p}_N)_{n+1}$ when the new normal load $(\mathbf{p}_N)_{n+1}$ is calculated.

Table 4.3. The computational procedure of interface element using the Newton-Raphson procedure

For each time step $\Delta t_{n+1} = t_{n+1} - t_n$

i) Read the incremental displacements of the nodal points

ii) Set the top nodes of interface elements perpendicular to the bottom nodes

iii) Search the top nodes of interface elements which are in contact with the tool and prescribe their displacements at the current time step

iv) Define the active interface elements and the corresponding integration (Gauss) points

For each iteration i within the time step (t_n, t_{n+1}) at each Gauss point

v) Evaluate the stiffness matrix $\mathbf{K}^i_f$ using the appropriate G_f modulus (In the first iteration of first time step, set $G_f = G_A$. In subsequent iterations, G_f is calculated from stage **x**)

vi) Solve the global system of equations $\mathbf{K}^i_f \, \delta\overline{\mathbf{u}}^i = \delta\mathbf{f}^i$

vii) Compute the incremental nodal displacements at the next iteration $\Delta\overline{\mathbf{u}}^i = \Delta\overline{\mathbf{u}}^{i-1} + \delta\overline{\mathbf{u}}^i$ with $\delta\overline{\mathbf{u}}^0 = 0$

viii) Evaluate the incremental strain $\Delta\boldsymbol{\varepsilon}^i_{n+1}$ and the total strain $\boldsymbol{\varepsilon}^i_{n+1} = \boldsymbol{\varepsilon}_n + \Delta\boldsymbol{\varepsilon}^i_{n+1}$

ix) Calculate the stress at the normal direction according to $\sigma_n = E_f \, \varepsilon_n$ and evaluate $\mathcal{T} = \mu_f |\sigma_n|$ at t_{n+1} (Note that only non-positive values of normal stress are allowed)

x) Evaluate the frictional shear stress τ and G_f for the next iteration, i.e. $\tau^i_{n+1} = \tau_n + G_A \Delta\gamma^i_{n+1}$ and $G_f = G_A$. If $|\tau| > \mathcal{T}$ at t_{n+1} then correct the current values of τ and G_f according to

$$\tau = \mathcal{T}\frac{\Delta\gamma}{|\Delta\gamma|}, \quad G_f = \frac{\mathcal{T}}{|\Delta\gamma|}$$

xi) Evaluate the out of balance, or residual forces. Computational steps (**v**) to (**xi**) are repeated until the norm of residual forces and maximum residual are both less than prescribed tolerances

5 APPLICATION OF CAP PLASTICITY MODELS

Numerical simulation is increasingly used for the prediction, or the adjustment of the industrial processes of metal powder forming. The derivation and the formulation of the numerical algorithm for a large displacement finite element solution has been given in the preceding chapters. In the present chapter, the finite element computer code is used to calculate various type of problems including the static and transient dynamic behavior of metal powder forming processes. A combined experimental and numerical investigation is described for powder compaction processes. It utilizes material properties derived from experimental work into the nonlinear analysis of finite element model. The powder material parameters, such as: cohesion, internal friction angle and Young modulus of powders are experimentally established as functions of relative density. These nonlinear functions are incorporated into the nonlinear analysis of finite element model in order to establish model input parameters during different stages of compaction.

In order to demonstrate the applicability of the present formulation, the powder behavior during the compaction of simple and complicated die geometries are analyzed numerically. The predicted compaction forces at different displacements, the variation with time of the displacement, relative density and stress contours are obtained. A good agreement was achieved between the predicted and measured density profiles. It is shown that the proposed large displacement elasto-plastic finite element approach is capable of simulating metal powder during compaction. The results clearly show that the updated Lagrangian formulation together with the cap plasticity model, such as: a combination of the Mohr-Coulomb and elliptical cap, or a generalized single-cap plasticity model, can be effective in simulating metal powder compaction. This chapter adequately predicts the compaction forces and density distribution within the complicated die geometries. In the next chapter we will show how mesh adaptivity can be effectively used for simulating large deformations in metal powder forming.

5.1. Introduction

Powder compaction is becoming an increasingly important industrial manufacturing process where components of rather complicated geometry are being manufactured. Today, complex parts can be produced by multi-level tools that were inconceivable just a few years ago. The often extremely high tolerance requirements of the parts and the cost for hard machining of a sintered component are a challenge for die pressing. The physical and mechanical properties of powder metallurgy (PM) components are closely related to their final density. There is wide agreement in the industry that the major source of any physical flaw in a product arises from an inhomogeneous density distribution within the powder compact and internal cracks initiated on an internal shear plane. It is important to understand densification behavior of mixed powders, since powder manufactured has been considered as a major route for processing composite materials. Except for the very simplest part geometries it is not possible to achieve a homogeneous green density distribution by die compaction. Minimizing the density gradients is an important consideration when high and consistent mechanical performance is required. The final density is determined by the press-and-sinter process parameters and by the material characteristics and filling conditions. The density distribution of the material in the as-poured condition has effects that are propagated throughout the subsequent PM processes. Understanding and quantifying the main factors that influence the fill density could be a platform for 'tailoring' the initial density in the die.

The density distribution is dependent on the combination of many factors, such as geometrical shape, mechanical properties of the powder, powder-tool frictional behavior and the pressing cycle. To date, the design and manufacture of powder based components is usually founded on empirical methods and experience, together with a tooling development programme. Thus, an analysis of this process is necessary to improve and to develop a deeper understanding of the effects of process variables on the product with the main objective of producing a powder compact which is free from internal cracks and has near uniform density. An analysis of powder compaction is important because the compact powder density influences the further densification of the samples during sintering and hence the performance of the final part. These variations in density will also cause shape distortion after sintering along with regions of elevated stress and stress concentrations. For these reasons, the control of compact density is very important for the manufacture of reliable products, particularly where they are subjected to a significant level of stress in their duty cycle.

A faster alternative to the conventional method of trial-and-error is to optimize the whole process by computer simulation. Finite element simulations of the compaction process in combination with appropriate material laws for the powder allow quantitative predictions of the tool loading, of the green density distribution and of the sinter distortions. The global behavior of the cold pressing process of iron powder is dependent on the process kinematics, material response in the powder as well as the friction effects in the contact between the powder and the tool walls. The analysis of powder pressing requires accurate material models of the various powder mixes that are used. In order to perform such analyses, the complex mechanisms in the compaction process must be expressed in a mathematical formulation with the knowledge of material and frictional behavior. Following the Lagrangian finite element formulation for large deformation, a hardening cap model for the nonlinear behavior of powder material and an interface element formulation for the frictional behavior of the contact surface, described in preceding chapters, an elasto-plastic incremental solution strategy is used which integrates the constitutive equation and iterative scheme to solve the equilibrium equation. A full Newton-Raphson method is employed with an explicit algorithm for the integration of the stress-strain relation. A direct iterative method is then utilized which is more economical and also sufficient since the stress computation is incorporated with a direct integration procedure to improve the accuracy.

5.2. Experimental investigation

In powder compaction simulation, the numerical model must capture the physics of friction internal to the powder and at the die wall and the yielding of the powder bulk. Also as densification takes place, the material is expected to work harden and the yield is dependent consequently on both hydrostatic and deviatoric stress states. In powder compaction, the behavior of stress and strain during the process is characterized by a non-linear relationship started from the beginning of the compaction stage. Upon unloading, the powder will not return to its initial configuration but clearly remains at some stage of compaction. In this section, a series of experiments, which needs to be conducted in understanding the powder behavior during the compaction process, will be illustrated.

5.2.1. Compaction test

A compaction test equipment was designed and manufactured by Gethin *et al.* (1994) to enable the production of a simple cylindrical bush of compacted powder in the laboratory. This was a single-ended uniaxial closed-die apparatus where during compaction the top punch moved downward to the desired level, while the bottom punch remained fixed. A sectional view of this system is shown in Figure 5.1. It consists of a top and bottom punches, a sleeve and core rod. The equipment is designed to measure a complete force balance on the compact. A typical result for an iron powder mix of 20 mm fill depth is presented in Figure 5.2. For convenience, all the forces are plotted as positive values, however it should be noted that the top punch force is opposite in sign to the remaining three forces.

5.2.2. Shear test

In order to characterize the yield behavior of the powder at different density levels, a shear box is employed by Gethin *et al.* (1994) to enable shear of cylindrical compacts under different normal load. It is adopted to relate shear stress at failure directly to normal stress to define the Mohr-Coulomb failure envelope. Figure 5.3 shows diagrammatically the shear box during shearing the powder. Four measurements are taken; the normal load, normal displacement, shear load and shear displacement.

In shear box test, the powder considered is the same as those in the compaction of the plain bush and the steps in the experiment comprise:

a) The top and bottom boxes are aligned and powder is filled into the cylindrical cavity to the top surface of the die,

b) The horizontal load cell and displacement transducer are placed into position,

c) The normal load is applied to the prescribed level and sustained at this value under load control,

d) The shear load is applied and force and displacement monitored.

The same procedures from step (a) to (d) are repeated for different normal applied loads up to a maximum allowable for the equipment or to practical industrial limits.

The results from the shear box test are presented in Figure 5.4 in the form of shear force against shear displacement for a series of normal loads for a mixed iron powder. Clearly, the ultimate shearing force increases when the normal load applied increases and this is because the powder becomes harder and more load is needed to fail and shear the particle interaction. It should be mentioned that this test is not as accurate as the triaxial test in examining the shear-strength of particulate material due to the fact that the shear plane predefined by the split box. There are two important points need to be considered. Firstly, the shearing load level must be taken to establish the yield criteria to represent the powder behavior, and secondly, the variation of shear stress should be considered through the sample section.

There are two possible choices in establishing the shearing load level. One is based on the initial shear yield point and the other on the ultimate shear stress (Gethin *et al.* 1994). Figure 5.5 shows the shear yield point and the ultimate stress. The true shear stress curve (A) was calculated by considering the reduction of cross section area during shear displacement. The other curve (B) is the calculation of shear load based on constant cross section area. In each case, the divergence point is considered to be the shear yield point. This choice is important in the determination of the internal friction angle, which represents the shear failure behavior of the powder. Obviously, the ultimate shear stress gives a higher internal friction angle than the shear yield point value.

5.3. Parameter determination

In order to perform a nonlinear finite element analysis, the experimental results given by Gethin *et al.* (1994) are incorporated into the numerical simulation of powder compaction process. In the numerical modeling of powder compaction, there are few material parameters, such as: cohesion, internal friction angle and Young modules of powder, which need to be established experimentally as functions of relative density. These nonlinear functions will be incorporated into the nonlinear FE analysis in order to establish model input parameters during different stages of compaction.

5.3.1. Cohesion and internal friction angle

In powder compaction process, there are two criteria which define the boundary of elastic-plastic limits, notably, shear and hydrostatic. However, the physical behavior after the stress state reaches the yield point will be different for these criteria. In die powder compaction, the material undergoes continuous yielding (i.e. decrease in volume) until it reaches a state where particle fracture occurs. By taking the initial shear yield point in establishing the Mohr-Coulomb envelop this may not truly represent the continuous densification unless incorporating a non-associated flow rule. Since the densification criterion can be calculated by incorporating the cap model, the ultimate shear stress is appropriate for representing the shear failure criteria. In order to construct the Mohr-Coulomb shear failure envelop, the ultimate shear stress is taken for every normal load given in Figure 5.4 (Ransing *et al.* 1994). These are depicted in Figure 5.6 where the two parameters associated with the Mohr-Coulomb failure criterion, i.e. the internal friction angle and cohesion, are established.

5.3.2. Young modulus

One of the most important material parameters, which effects on powder yielding under hydrostatic and shear stresses is the powder Young modulus. This can be simply determined by means of a powder compaction experiment and establishing the stress-strain curves at different levels of density. An experimental investigation was carried out by Gethin *et al.* (1994), in order to measure the complete force balance for the compaction phase and the punch load during the ejection phase. Figure 5.7 illustrates the variation of the Young modulus with relative density for the iron powder. This curve was found by experiment utilizing unloading and reloading of the powder. During unloading the elastic spring-back was measured and the same compacted powder was used in reloading to establish a relationship between the Young modulus and axial load at different density levels. The relationship between this modulus and relative density was then established for data input to the numerical simulation.

5.3.3. Friction coefficient

In order to estimate the Coulomb friction coefficient between the powder and die, a relationship between the axial and radial stresses need to be established using the compaction phase experiments. A complete set of compaction phase results is shown in Figure 5.8 for the iron powder (Ransing *et al.* 1994). In the compaction phase, all the forces start from zero and build up in a non-linear manner as the compaction proceeds. From the curves, it can be observed that there is a reduction in stress during pressure transmission from the top to the bottom punch due to friction in the die and core rod surfaces. In the determination of the friction coefficient two assumptions are made. The first is that the axial stress is uniform over any horizontal section and the second is that the same value of radial stress acts on the die and core rod surface. These are reasonable assumptions for a thin walled bush.

Considering the stresses acting on a small element *dz* of the bush, as shown in Figure 5.9, the force balance on the element gives,

$$d\sigma_z = \frac{4\mu\sigma_r}{D_0 - D_i} dz \tag{5.1}$$

The solution of equation (5.1) requires a relationship between the axial and radial stresses. Two approaches are presented by Kuhn and Downey (1971) and Tabata *et al.* (1980). In the first method, the axial and radial stresses are assumed to be constant and their relationship is based on an elastic model, as

$$\sigma_r = \frac{\nu}{1-\nu}\sigma_z \tag{5.2}$$

As the relationship in equation (5.2) is based on an elastic model of the powder, its application to the compaction process is therefore an approximation of real problem. In the compaction stage, it is likely to be most applicable when high density levels are achieved and the compact assumes the properties of the solid material. Substituting equation (5.2) into (5.1), the relationship between the stresses at the top and bottom of the compact can be written as

$$\sigma_{zb} = \sigma_{za} \exp\left(-\frac{4\alpha\mu L}{D_0 - D_i}\right) \tag{5.3}$$

where $\alpha = \nu/(1-\nu)$.

The second approach embodies an empirical relationship between the axial and radial stress values derived from experimental. It was derived by Tabata *et al.* (1980) for iron powder as

$$\sigma_r = 0.087\sigma_z^{1.38} \tag{5.4}$$

Thus, equation (5.1) becomes

$$\sigma_{zb} = \left[\sigma_{za}^{-0.38} + \frac{0.1322\mu L}{D_0 - D_i}\right]^{-2.63} \tag{5.5}$$

Equations (5.3) and (5.5) were applied to the compact having a fill depth of 20 mm. This was chosen since the radial stress will be most nearly uniform over the compact height. The results of applying equations (5.3) and (5.5) are shown in Figure 5.10 for lubricated and unlubricated powder where a nearly constant value is derived (Tabata *et al.* 1980).

5.4. Application of MCEC plasticity model

In order to illustrate the applicability of the MCEC plasticity model, the powder behavior during the compaction of a plane bush, a cutting tool, a rotational flanged and an axisymmetric shaped tip component are analysed numerically. The problems have been solved with displacement control by increasing the punch movement, *d*, and predicting the compaction forces at different displacements. An incremental manner is also used to simulate the dynamic analysis of a compaction process subjected to a surface step loading. The iron powder properties chosen in the analysis are typical for industrial application (Gethin *et al.* 1995). These material properties are as follows for all examples: $E = 40.0\ \text{N/mm}^2$, $\upsilon = 0.35$, $c = 5.0\ \text{N/mm}^2$ and $\varphi = 30^\circ$. These values are used for the Mohr-Coulomb yield criterion with a cap model. The die wall friction is simulated using a set of interface elements with the Coulomb friction coefficient $\mu = 0.08$. The initial relative density is $\overline{\rho}_0 = 0.4$. The plastic hardening coefficient $\chi = 0.16$ and the initial hydrostatic stress of compaction $\sigma_{c0} = 0.2\ \text{N/mm}^2$ are assumed.

In this study, we refer frequently to the 'relative density' which is obtained using either the initial (fill) density in the total Lagrangian description (TLD) or the density field of the previous increment in an updated Lagrangian description (ULD) as

$$\overline{\rho}_i = \frac{\overline{\rho}_0}{1+\varepsilon_\upsilon^{0i}} \qquad \text{for TLD method} \tag{5.6}$$

$$\overline{\rho}_i = \frac{\overline{\rho}_{i-1}}{1+\varepsilon_\upsilon^{i-1}} \qquad \text{for ULD method} \tag{5.7}$$

where $\overline{\rho}_0$ is the initial (fill) relative density and i and i–1 denote the ith and $(i-1)^{th}$ increment. $\varepsilon_\upsilon^{0i}$ is the volumetric strain at the ith increment with respect to the original configuration and $\varepsilon_\upsilon^{i-1}$ is the volumetric strain at the ith increment with respect to the previous increment configuration.

5.4.1. A plain bush component

The first example is chosen to show that the updated Lagrangian description (ULD), using a combination of the Mohr-Coulomb and elliptical cap (MCEC) model, can give good answers in simulating the metal powder during compaction. The finite element modeling of a plain bush component is performed employing an axisymmetric representation which is illustrated in Figure 5.11(a). The geometry and boundary conditions of this component are shown in Figure 5.11(b). Since a displacement based formulation is used, the implementation of loading in the finite element code is achieved by the use of prescribed nodal displacements. The direction of this displacement is always in a vertical plane which represents the axial punch load. Fixed nodal values are employed to represent the fixed bottom punch. These infer an assumption that there is no relative movement between the punch and the powder. Where there is a relative movement, the interface element is used which in this case is along the sleeve (die) and core rod surface (Figure 5.11).

In order to compare the performance of the total and updated Lagrangian formulation for both the Mohr-Coulomb (MC) and combination of Mohr-Coulomb and Elliptical Cap (MCEC) model, the predictive compaction forces at different displacements are shown in Figure 5.12(a). Also plotted in this figure are the experimental results and numerical solutions corresponding to a plain bush component given by Tran *et al.* (1993) and Gethin *et al.* (1994). Obviously the Mohr-Coulomb constitutive model fails to predict the compaction forces - however, the results can be improved significantly when a combination of the Mohr-Coulomb and elliptical cap model is adopted. Results were obtained by applying the ULD method with the MCEC model using various elements and are shown in Figure 5.12(b), i.e.

i) the triangular linear continuous displacement elements (T3C) with the four-noded linear interface elements (I4C);

ii) the triangular quadratic continuous displacement elements (T6C) with the six-noded quadratic interface elements (I6C); and

iii) the eight-noded quadrilateral elements (Q8C) with the six-noded quadratic interface elements (I6C).

It can be observed that the Q8C elements give a remarkable improvement in accuracy for the predicted compaction forces.

In Figures 5.13 and 5.14, the numerical results of the dynamic analysis of a plain bush component subjected to a surface step loading are shown. The variation with time of the displacements at the surface of the top punch for different values of ramp time interval, i.e. $t_0 = 10, 20, 50$ and $100\,\text{sec}$, are presented in Figure 5.13. Also plotted in Figure 5.14 is the variation with time of the relative density at different nodes of a plain bush component indicated in Figure 5.11(b).

The displacement and relative density contours at different powder movements are shown in Figures 5.15 and 5.16. Clearly, from an analysis of the displacement contours the flow pattern is a consequence of two distinct effects. First, the single punch motion gives maximum powder movement at the top punch and reduces to zero movement at the bottom punch. Secondly, the effect of friction causes different powder movement through the radius of the bush. The powder can flow more freely further from the tooling surface due to the reduced effect of friction at the powder-tooling interface. The density variation which occurs over the compact length is the result of powder flow and wall friction. At the end of the compaction phase, the relative density contour shows the highest density values at the top corners and reduces gradually to the bottom punch surface. The results indicate that the density distribution at the end of the compaction phase, as obtained by the present formulation, can be compared with other work as proposed by Brown and Weber (1988) for a cylindrical component. It can be observed that the proposed finite element modeling of large deformation with a cap model is capable of simulating the metal powder during compaction.

5.4.2. A cutting tool component

The second example is chosen to demonstrate the performance of the MCEC plasticity model for the simulation of plane strain problems. A cutting tool component, with top punch motion and a fixed bottom punch as shown in Figure 5.17(a), was numerically analysed by various meshes using the eight-noded quadrilateral elements (Q8C) and six-noded quadratic interface elements (I6C). The geometry and boundary conditions of this component are shown in Figure 5.17(b).

In order to compare the effect of various meshes in predicting compaction forces at different displacements, four type of meshes, as shown in Figures 5.17(b) and 5.18, are considered. It can be observed from the punch compaction forces in Figure 5.19 that meshes 3 and 4, using Q8C elements, give a remarkable improvement in accuracy of the predicted compaction forces. Although meshes 2, 3 and 4 give similar accuracy, it can be found that mesh 4, with 521 nodes, is the most accurate and mesh 2, with 107 nodes, is the most economic. In Figure 5.20, the numerical results of the dynamic analysis of a cutting tool component subjected to a surface step loading are shown. The variation with time of the displacements at the surface of the top punch are presented in Figure 5.20(a) for different values of ramp time interval, i.e. $t_0 = 10, 20, 50$ and $100\,\text{sec}$. Also plotted in Figure 5.20(b) is the variation with time of the relative density at different nodes of a cutting tool indicated in Figure 5.17(b).

The powder displacement contours at the half and final movements of the compaction process are shown in Figure 5.21. It can be observed from the top moving punch that the powder is displaced so that the isolines indicate an almost uniform level where the maximum powder movement occurs at the top and reduces to zero at the bottom punch. In Figure 5.22, the relative density contours are plotted at two different stages of movement. The results of the density distribution during the process of compaction, show the highest values of density at the left gradually reducing to the right, where the width of the cutting tool is considered to be wider. Finally, the normal stress contours σ_x and σ_y at the half and final stages of compaction are plotted in Figures 5.23 and 5.24. From the contours, it can be seen that the stresses at the left of the cutting tool are greater than on the right handside.

5.4.3. A rotational flanged component

The third compaction example is of a multi-level component which is modelled by an axisymmetric representation as illustrated in Figure 5.25(a). The geometry and boundary conditions are shown in Figure 5.25(b) for this rotational flanged component. Since the crucial numerical problem in this more practical geometry is associated with the distorted mesh, the simulation has been performed employing the Lagrangian and arbitrary Lagrangian-Eulerian techniques to demonstrate the performance of two different approaches. In Lagrangian analysis, the study has been conducted using a fixed mesh analysis to identify an optimum tooling motion. Three case studies have been solved to test the modeling procedures necessary for a multi-level part and the effect of using different tooling movement histories are established. Because of severe distortion of elements in single top, or bottom punch motion, the analysis is carried out using the ALE method to keep mesh distortion and element entanglement under control.

5.4.3.1. Numerical simulation of Lagrangian analysis

Case 1: Single top punch motion with a fixed bottom punch

This case refers to a single action of the top punch moving as shown in Figure 5.26(a). A displacement control was employed by increasing the top punch movement whilst fixing the bottom punch and the exerted forces at the moving top are then computed. The variation of the predicted top punch forces at different displacements are also plotted in this figure. The deformed mesh due to the single top punch is plotted in Figure 5.26(b). In Figure 5.27, a comparison between the predicted density profile, obtained by the ULD finite element formulation, and those given by Gethin *et al.* (1994), taken along the section *a-a* at the final stage of a single-ended compaction, is illustrated.

The powder displacement and relative density contours at half and final stages of the top punch motion, $d_1 = 3.03$ and 6.06 mm, are shown in Figures 5.28 and 5.29. The legends indicated are relative to the initial position of the powder. In the case of the top punch motion, it is evident that the powder displacement is greatest at the top and lowest at the bottom surface of the flange and bottom punch. It is clear from the isoline that the powder is free to move around the corner region.

Case 2: Single bottom punch motion with a fixed top punch

This case refers to a single movement of the bottom punch as shown in Figure 5.30(a). The prescribed nodal displacements are applied at the bottom punch and fixed nodal values are employed to represent the fixed top punch. The variation of the predicted bottom punch forces at different displacements are also plotted in this figure. In this case, some elements have become severely distorted and the solution breaks down, thus prohibiting further advance of the punch. It has been computed that the maximum bottom punch movement is $d_2 = 5.10$ mm when mesh distortion occurs. The deformed mesh due to the single bottom punch is plotted in Figure 5.30(b).

In Figure 5.31, a comparison between the predicted density profile, obtained by the ULD finite element formulation and those given by Gethin *et al.* (1994), taken along the section *b-b* at the final stage of a single-ended compaction, $d_2 = 5.10$ mm, is presented. In Figures 5.32 and 5.33, the powder displacement and density contours at two stages of the bottom punch motion, $d_2 = 3.40$ and 5.10 mm, are plotted. Clearly the compaction occurs principally in the bush region where the powder is forced to move near the corner junction. This has a significant effect on the mesh distortion in this area when using a fixed mesh analysis.

Case 3: Double punch motion

This case refers to the double action when both the top and bottom punches are moving as shown in Figure 5.34. It has been shown, in cases 1 and 2, that for single-ended compaction, the density variation within the compact is quite significant and often crack failure can accur at the corner. In practice, a double-ended compaction process should be used for this type of stepped section. In this case, when simultaneous top and bottom punch motions are applied, mesh distortion does not occur, due to the powder being compacted by both moving punches. The loading characteristics are achieved by the use of prescribed nodal displacements for the top and bottom punches.

The variation of the top and bottom punch forces at different displacements is plotted in Figure 5.34(a). These results can be compared with the predicted compaction forces for the single-ended compaction process as given in Figures 5.26 and 5.30. The deformed mesh at final powder movement is also shown in Figure 5.34(b) for a double punch motion. A comparison between the measured and predicted density profile, taken along sections identical to those in the experimental measurements (Lewis and Tran. 1994), is presented in Figure 5.35. The density profiles show good agreement with the experimental data. In Figures 5.36 and 5.37, the numerical results for the dynamic analysis of a flanged component, simultaneously subjected to a surface step loading at both the top and bottom punches, are shown. The variation with time of the displacements at the surfaces of both punches for different values of ramp time interval, $t_0 = 10, 20, 50$ and $100\,\text{sec}$, are presented in Figure 5.36. Also plotted in Figure 5.37 are the variations with time of the relative densities at different nodes of the flanged component, depicted in Figure 5.25.

The powder displacement contours at the half and final movements of the compaction process are shown in Figure 5.38. It can be observed that the powder is displaced so that the isolines indicate an almost uniform level. In Figure 5.39, the relative density contours at two stages of movement are presented. It is evident that the powder movement in this type of component shape results in density variations throughout the section. Finally, the normal stress contours σ_x and σ_y at the half and final stages of compaction are plotted in

Figures 5.40 and 5.41. From the contours, it can be seen that the stresses at the flange and bush region are almost uniform. However, at the corner region there are larger changes and the highest stress value is at the corner itself.

5.4.3.2. Numerical simulation of ALE analysis

In the case studies 1 and 2, the Lagrangian meshes lead to highly distorted elements around the corner, which in turn result in spurious oscillations of the relative density distribution. The mesh distortion is therefore controlled in an effective way by means of an ALE strategy. In this case, the original rectangular shape of the elements is maintained throughout the whole compaction process, and as a consequence, a smoother field of relative densities can be obtained, with no spurious oscillations. In the simulation presented here the friction effects are neglected. Therefore, the results are a qualitative approximation to the real compaction process, where the friction effects have to be taken into account. In Figure 5.42, the relative density distribution together with deformed meshes are presented at three different movements of top and bottom punches, i.e. d_t = 6.06 mm and d_b = 7.70 mm. As can be seen in Figure 5.42(c), the distribution of relative density is almost homogeneous with difference less than 20% in final stage of compaction. The higher value of relative density can be seen in top corner of component and the lower value in right bottom corner of flange. The results are qualitatively similar to those presented in double punch motion.

5.4.4. An axisymmetric shaped tip

The next compaction example is of an engineering tip component in hard metal powder which includes several complications not present in the plain bush, cutting tool and flanged compaction simulated in previous sections. Some problems are caused by discontinuities in geometry, for example flow around corners and multipunch set up, i.e. an upper and lower punch. In the production of a shaped tip component the compaction starts from loose powder by first moving the upper punch downward to its final position (~11 mm), followed by moving the lower punch upward by 5.5 mm (Haggblad and Oldenburg, 1994). In order to illustrate a comparison between the density distribution obtained by the MCEC plasticity model and those given by Haggblad and Oldenburg (1994), the simulation has been performed using the remaining pressing distance of 3.0 mm from above and an under-pressing of 1.45 mm.

The finite element modeling of a shaped tip component using the eight-noded quadrilateral elements (Q8C) and six-noded quadratic interface elements (I6C) is performed employing an axisymmetric representation which is illustrated in Figure 5.43(a). Also plotted in this figure are the geometry and boundary conditions of this component. Since the crucial numerical problem in this complicated die geometry is associated with the distorted mesh, a double-ended compaction process has been used employing a fixed mesh analysis. In this case, when simultaneous top and bottom punch motions are applied, mesh distortion does not occur, due to the powder being compacted by both moving punches. The loading characteristics are achieved by the use of prescribed nodal displacements for the top and bottom punches. In Figure 5.43(b), the deformed mesh at final powder movement is plotted for a double punch motion.

The numerical results of the dynamic analysis of a shaped tip component, simultaneously subjected to a surface step loading at both the top and bottom punches are shown in Figures 5.44 and 5.45. The variation with time of the displacements at the surfaces of both punches for different values of ramp time interval, $t_0 = 10, 20, 50$ and 100 sec are presented in Figure 5.44. Also plotted in Figure 5.45 are the variations with time of the relative densities at different nodes of the shaped tip component. A comparison between the measured density distribution given by Haggblad and Oldenburg (1994) and proposed ULD finite element formulation is illustrated in Figure 5.46. The relative density distributions are presented in 3D plots where the third direction represents the relative density. It shows a good agreement between the predicted and measured density distribution. Finally, the normal stress contours σ_x and σ_y at different stages of compaction are plotted in Figures 5.47 and 5.48. From the contours, it can be seen that the stresses at the corner and right region of the axisymmetric tip are greater than other sides. This tip component example shows the applicability of the MCEC model in simulating metal powder compaction.

5.5. Application of three-invariant single plasticity model

In this section the capability of three-invariant single plasticity model, presented in Section 3.8, is illustrated throughout the compaction of a set of complex forming processes. The constitutive equations of single-cap plasticity, derived in Section 3.8, along with the Lagrangian finite element formulation, given in Chapter 2, have been implemented in a nonlinear finite element code to evaluate the capability of the model in simulating powder compaction process. The simulation is presented for efficiency and accuracy in the modeling of an automotive component and a conical shaped-charge liner. The problems have been solved with displacement control by increasing the punch movement and predicting the compaction forces at different displacements. The distribution of stress and relative density of powder are presented at the final stage of compaction. In the FE simulations, the tools are modeled as rigid bodies, because the elastic deformation of the tools has only an insignificant influence on the density distribution in the green component.

The raw material is composed of iron, copper, wax and zinc stearate, in which the last two components are admixed as internal lubricants. The density of the solid phase is about 7.54 g/cm^3 and the tap powder density is about 3.67 g/cm^3. The particles have irregular shapes and their sizes are between 10 and 100 μm. The compacted specimen has an initial height H_0 of 42 cm and diameter D_0 of 20 cm, as shown in Figure 5.49. The triaxial tests consist of an initial isostatic compaction step up to pressure value of 400 MPa, followed by keeping pressure constant and increasing the axial stress up to the maximum value of 1250 MPa (Doremus *et al.* 1995). The material parameters are obtained by fitting the stress path to confining pressure and triaxial tests, which are given in Table 3.6. The variations of the density with axial stress are shown in Figure 5.50 for different values of hydrostatic pressure. As can be observed, the proposed model captures the behavior of powder in complete triaxial experiment. Figure 5.51 presents the evolution of the density versus the hydrostatic pressure. This evolution is the characteristic of metal powders. It shows a good agreement between the experimental and numerical results for the isostatic compression step. Figure 5.52 corresponds to the complete triaxial compression tests. The variations of the axial stress with axial strain are shown in Figure 5.52(a) for different values of density. Also plotted in Figure 5.52(b) is the variation of the density with axial strain for different values of the hydrostatic pressure. Remarkable agreements can be observed between experimental and numerical results.

5.5.1. An automotive component

In the first example, an axisymmetric automotive part which is compacted from iron powder with a mechanical press and a multi-platen die set is simulated numerically using the generalized single plasticity. The simulated and measured density distributions are obtained for this component by Shen *et al.* (2001). The simulation is performed with the iron powder material parameters presented in Table 3.6. The geometry and FE mesh of the uncompacted powder in their position before compaction are presented in Figure 5.53(a). The component is modeled by 2D axisymmetric structured mesh of four-node quadrilateral elements. The compaction is employed by means of the action of two bottom punches; a lower inner punch and a lower outer punch labeled 'a' and 'b', respectively, in Figure 5.53(a). The simulation has been performed using the remaining pressing distance of lower inner punch of 9.75 mm and lower outer punch of 31.10 mm. In the simulation presented here the friction effects are neglected. Therefore, the results are a qualitative approximation to the real compaction process, where the friction effects have to be taken into account (Shen *et al.* 2001). Figure 5.53(b) presents the final FE mesh and geometry of the component at the final stage of compaction.

In Figure 5.54, the predicted relative density and normal stress σ_y distributions are shown at the final stage of compaction. The result of density distribution is qualitatively similar to that presented by Shen *et al.* (2001). The variation of the top punch force with displacement is plotted in Figure 5.55. In Figure 5.56, the stress paths with their associated yield surfaces are presented in meridian plane at different Gauss points of an automotive component indicated in Figure 5.53(a). As the densification occurs and stress exceeds, the plastic strain vector, normal to the current yield surface, produces a plastic volumetric decrease which causes the material to harden and the surface expands until the next yield surface is reached. This representation clearly shows how the yield surface grows with densification and expands when the material becomes harder. The applicability of the single cap plasticity model to handle the compaction simulation of powder is clearly evident in this figure.

5.5.2. A conical shaped-charge liner

The next example is of a shaped-charge liner, which is extensively used for civilian oil and steel sectors in geophysical prospecting, mining, and quarrying (Gu *et al.* 2001). Most liners used in the civilian sector have a conical shape, and are often made from a mixture of different metallic powders. In the present simulation, a conical liner is pressed from the iron powder with the material parameters given in Table 3.6. The schematic of the process to form a conical shaped charge liner from iron powder along with the geometry and FE mesh before compaction are presented in Figure 5.57(a). The loading characteristics are achieved by the use of prescribed nodal displacements for the top punch movement. The die and upper punch were modeled as rigid surfaces. The simulation has been performed using the remaining pressing distance of 11.78 mm from above. This component has been simulated by Gu *et al.* (2001) using different punch set up. Figure 5.57(b) shows the geometry and FE mesh of the pressed conical liner at the final stage of compaction.

The distribution of relative density and normal stress σ_y are presented in Figure 5.58 at final stage of compaction. In the simulation presented here the friction effects are neglected. Therefore, the results are a qualitative approximation to the real compaction process, where the friction effects have to be taken into account. Figure 5.59 shows the measured axial load on the upper punch versus vertical displacement. The stress paths with associated yield surfaces are presented in Figure 5.60 in meridian plane at four Gauss points depicted in Figure 5.57(a).

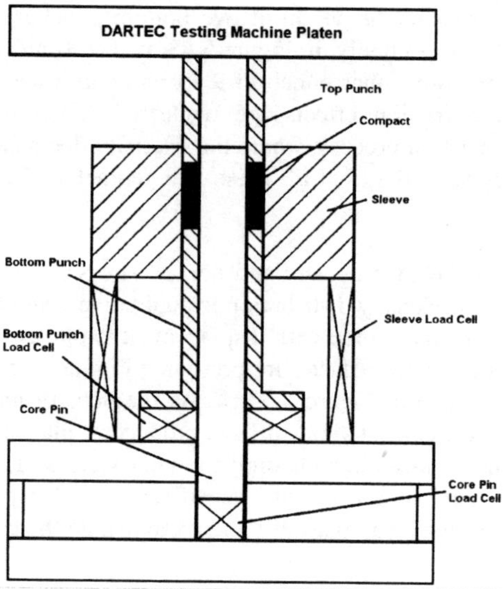

Figure 5.1. A sectional view of the compaction test

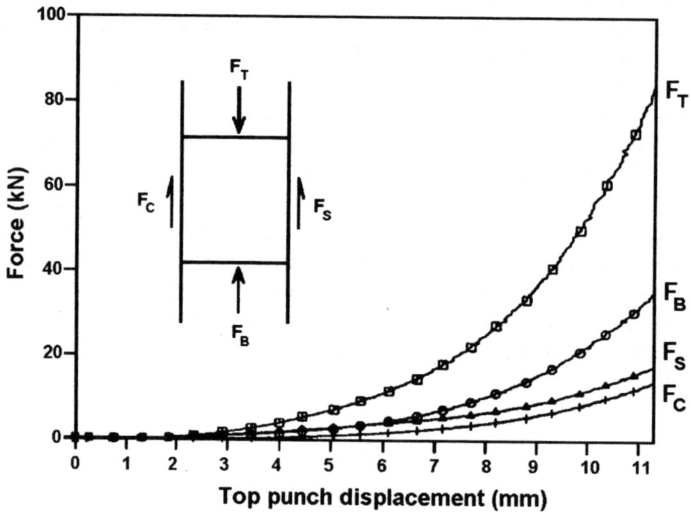

Figure 5.2. Typical complete cycle of iron powder compaction process for 20 mm length

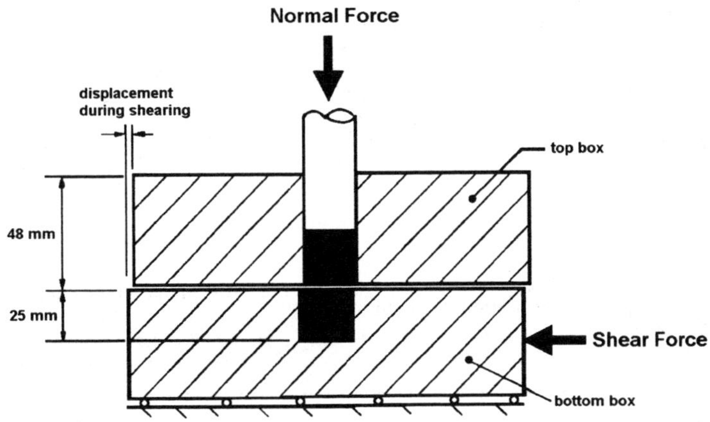

Figure 5.3. Shear box during shearing the powder

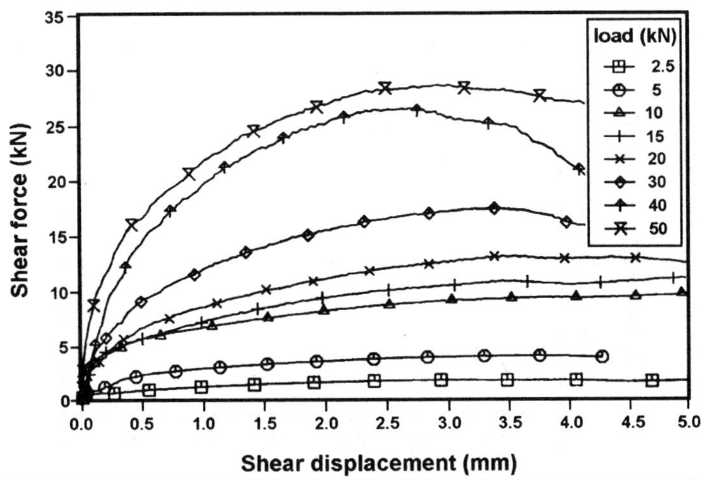

Figure 5.4. Variation of shear force with shear displacement for different normal load

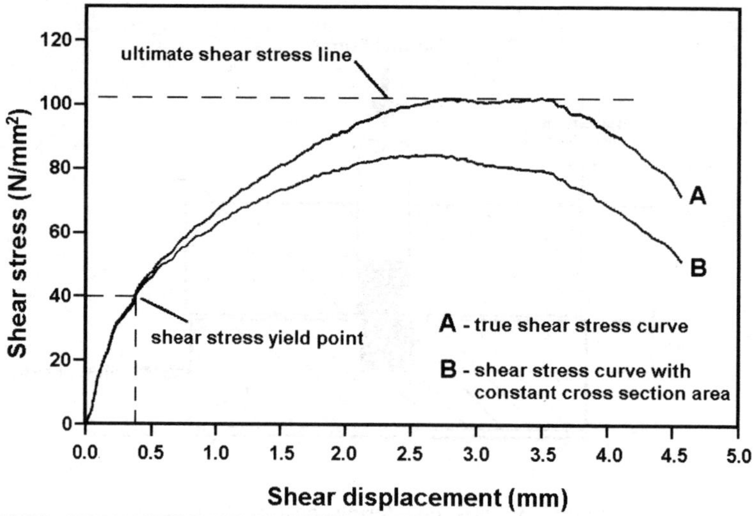

Figure 5.5. Typical shear stress - shear displacement curves at normal stress, 127 N/mm^2

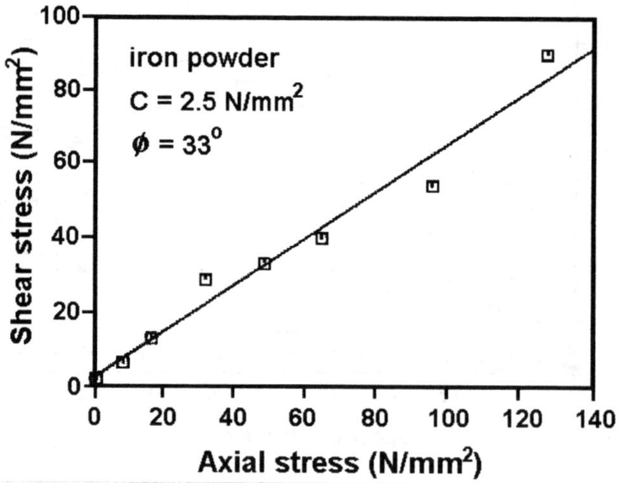

Figure 5.6. Cohesion and internal friction angle for iron powder

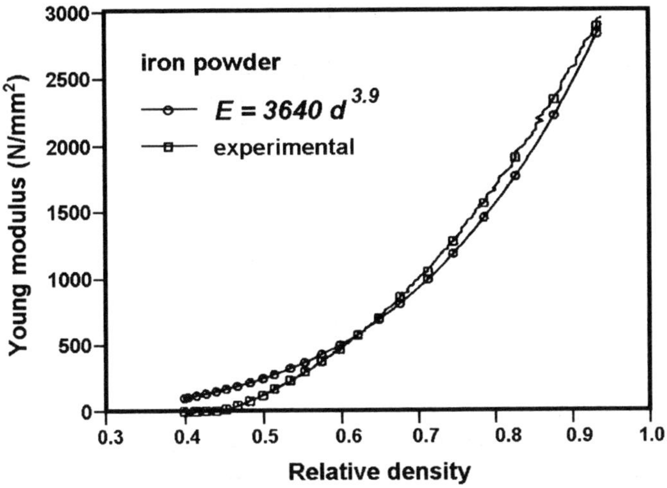

Figure 5.7. Variation of Young modulus with relative density for iron powder

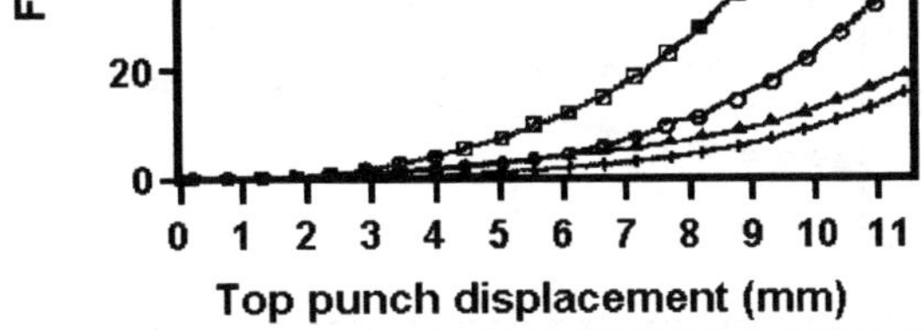

Figure 5.8. Compaction force balance for iron powder

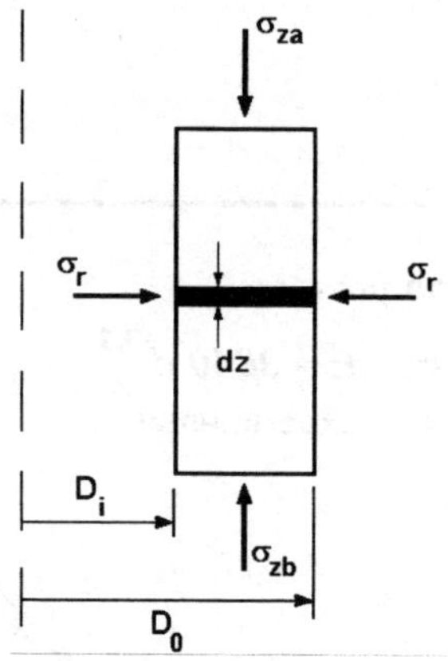

Figure 5.9. A typical element in compact

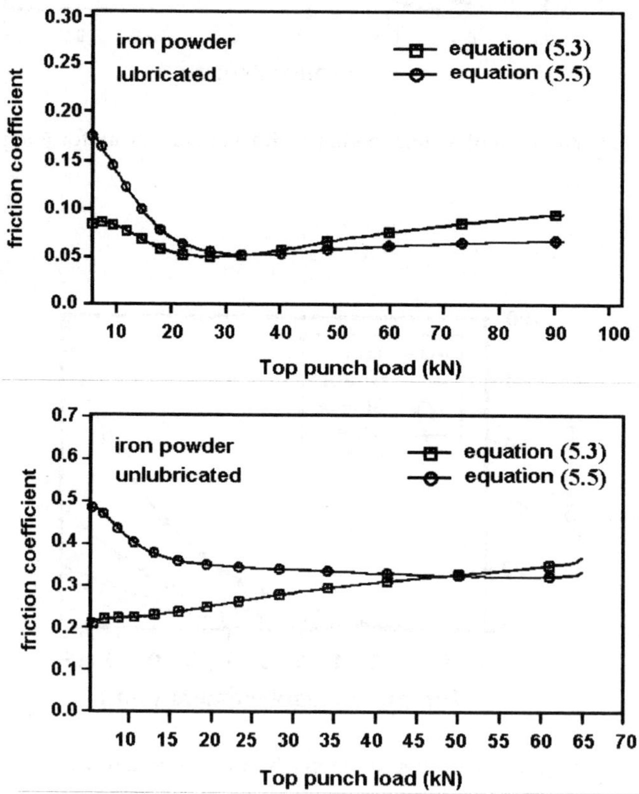

Figure 5.10. Friction coefficient over complete compaction for lubricated and unlubricated powder

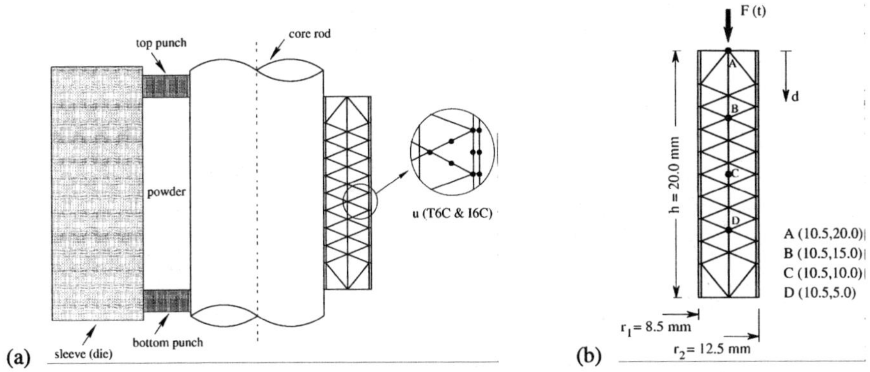

Figure 5.11. A plain bush component; a) Finite element modeling, b) Geometry and boundary conditions

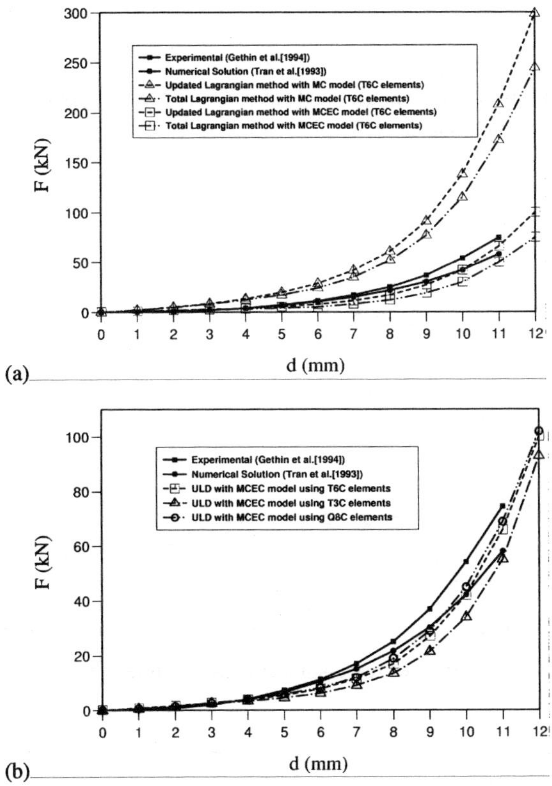

Figure 5.12. Predicted compaction forces at different displacements for a plain bush component; a) Comparison of total and updated Lagrangian formulations using the MC and MCEC models with the experimental and numerical solutions, b) Comparison of predicted compaction forces using various elements

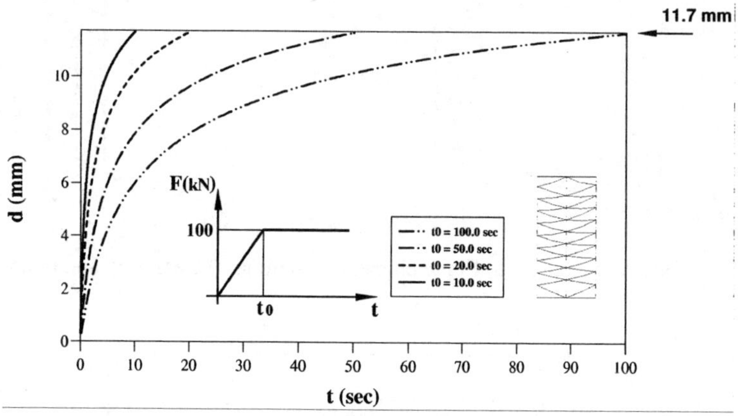

Figure 5.13. Variation with time of the displacements of a plain bush component subjected to a surface step loading for various ramp time intervals t_0

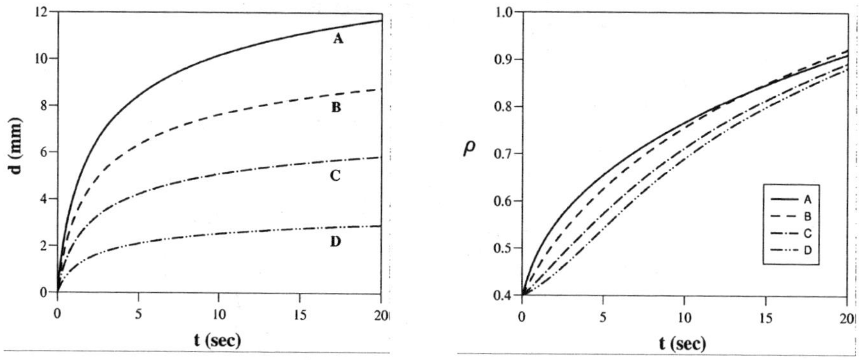

Figure 5.14. Variation with time of the displacements and relative densities at different nodes

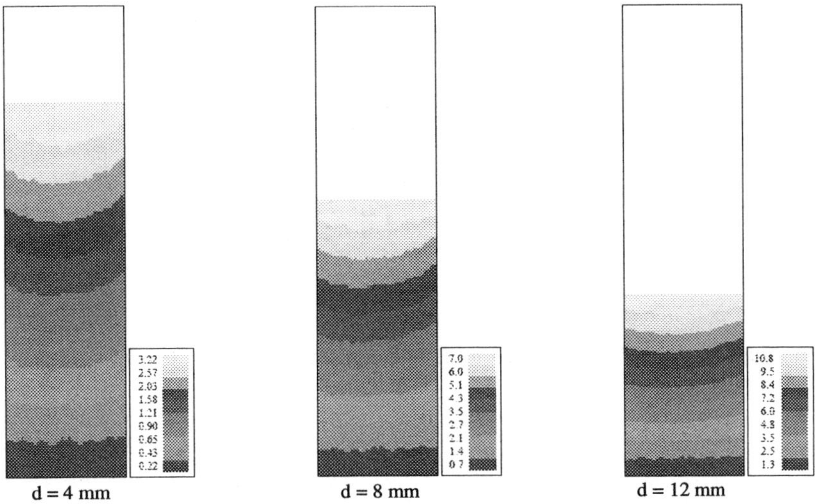

Figure 5.15. Displacement contours at different powder movements

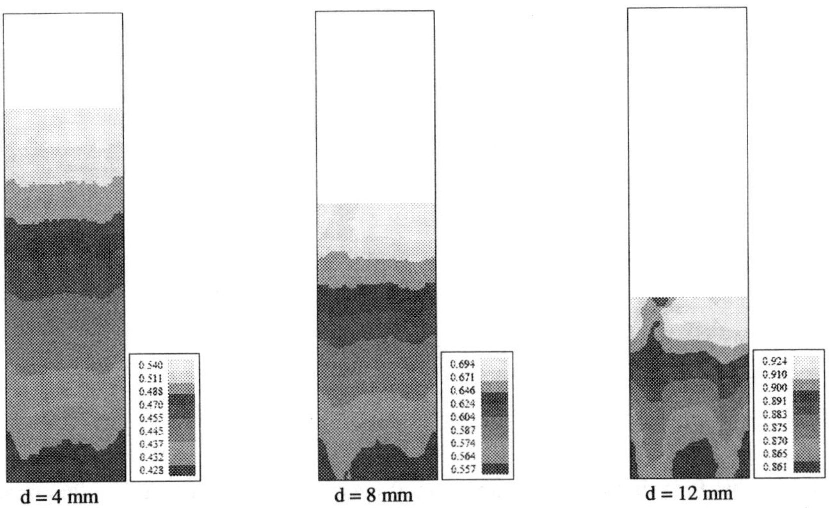

Figure 5.16. Relative density contours at different powder movements

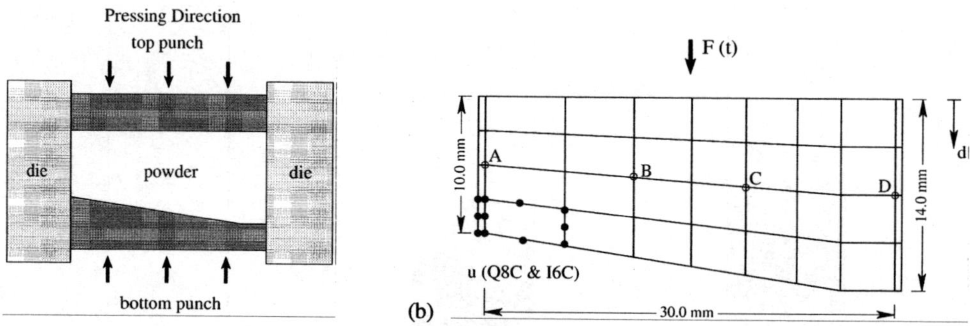

Figure 5.17. a) Shape of the modeled cutting tool, b) Geometry and finite element modeling with a typical mesh (mesh 2: 107 nodes)

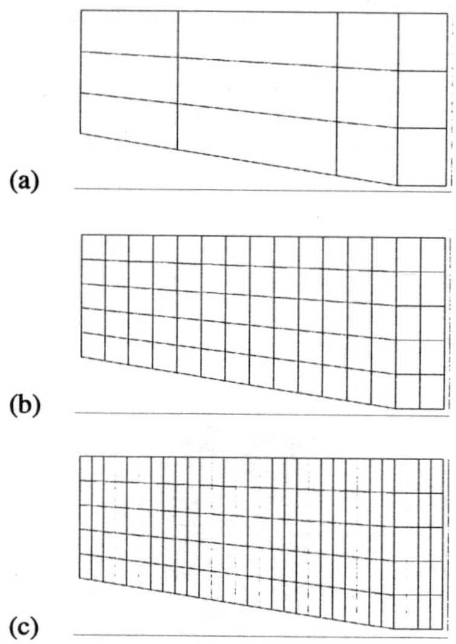

Figure 5.18. FE meshes for a cutting tool component; a) mesh 1: 62 nodes, b) mesh 3: 266 nodes, c) mesh 4: 521 nodes

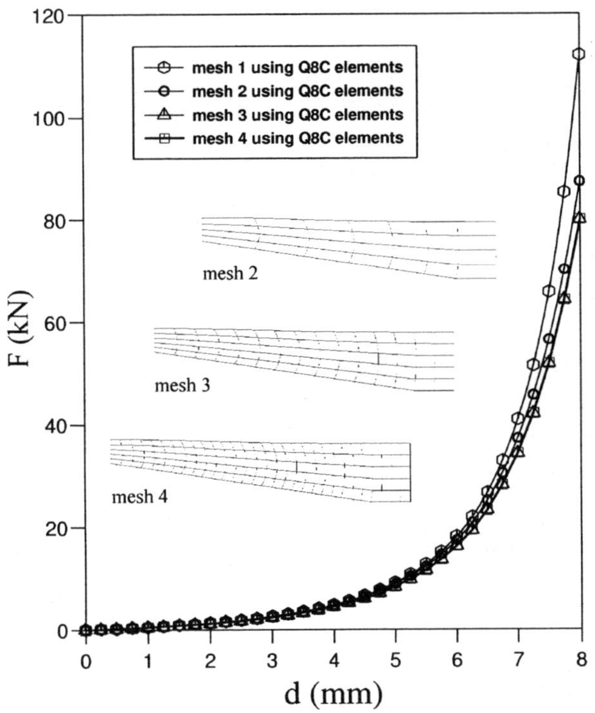

Figure 5.19. Predicted compaction forces at different displacements for a cutting tool using various elements

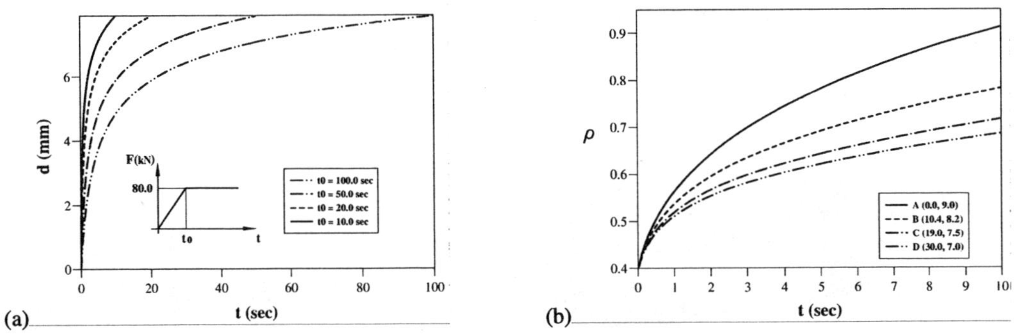

Figure 5.20. a) Variation with time of the displacements of a cutting tool subjected to a surface step loading for various ramp time intervals t_0, b) Variation with time of the relative densities at different nodes

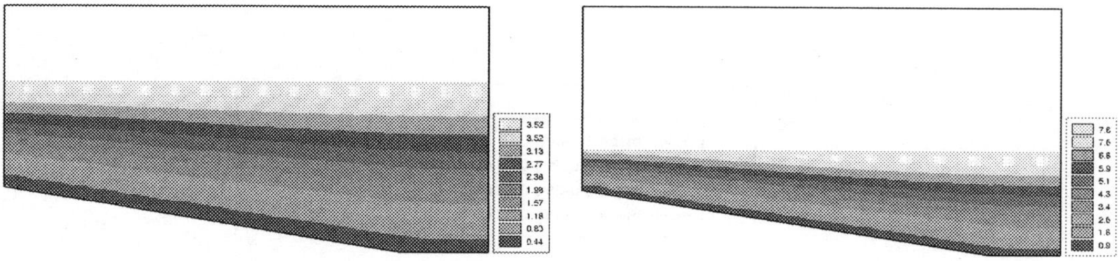

Figure 5.21. Displacement contours at the half and final stages of compaction

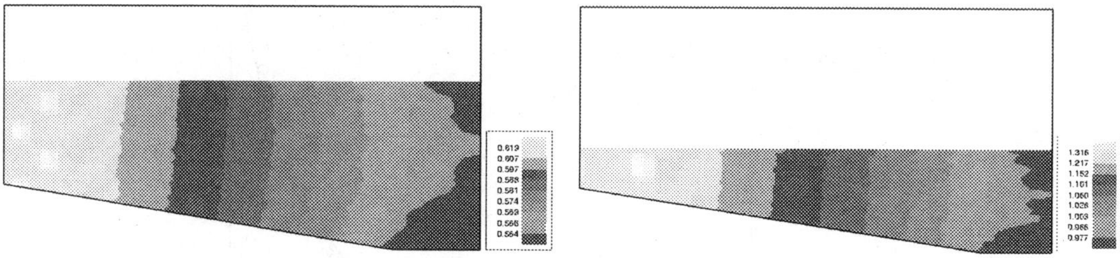

Figure 5.22. Relative density contours at the half and final stages of compaction

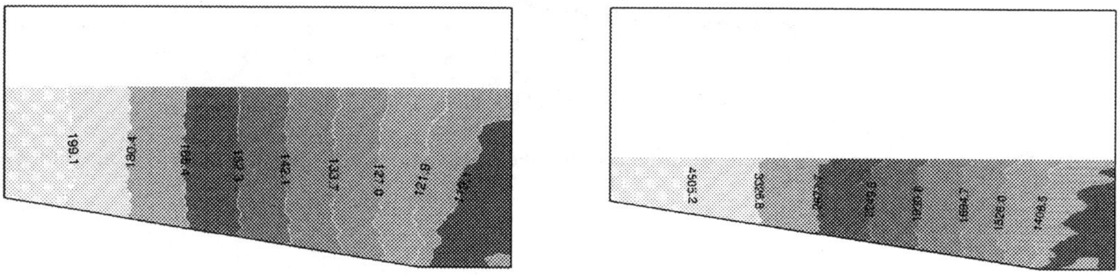

Figure 5.23. Normal stress σ_x contours at the half and final stages of compaction

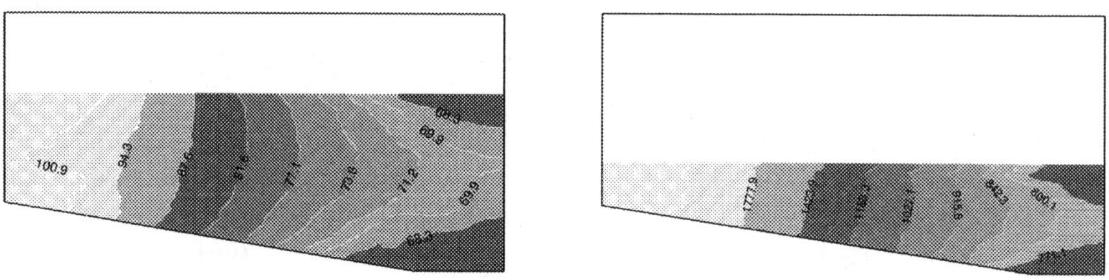

Figure 5.24. Normal stress σ_y contours at the half and final stages of compaction

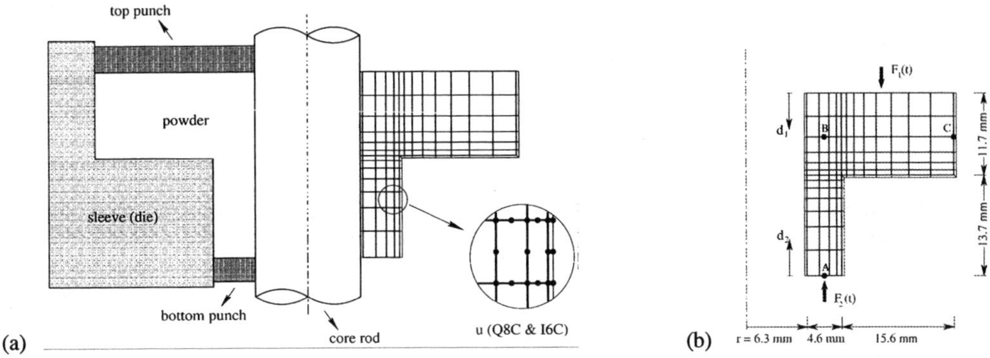

Figure 5.25. A rotational flanged component; a) Finite element modeling, b) Geometry and boundary conditions

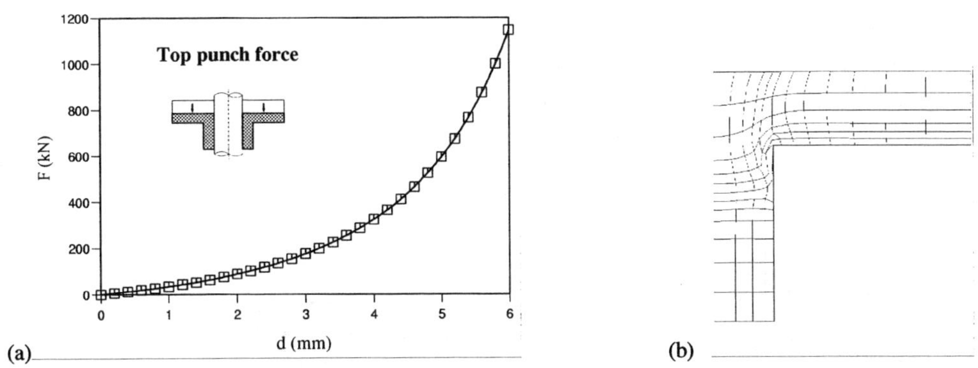

Figure 5.26. A rotational flanged component due to the single top punch; a) Predicted top punch compaction forces at different displacements, b) Deformed mesh at $d_1 = 6.06$ mm

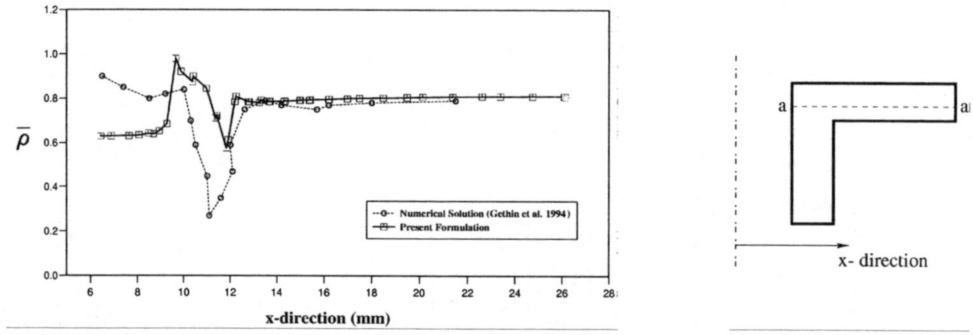

Figure 5.27. Comparison of the relative density profile by the MCEC plasticity model and predicted results given by Gethin *et al.* (1994) at section *a-a* for a flanged component

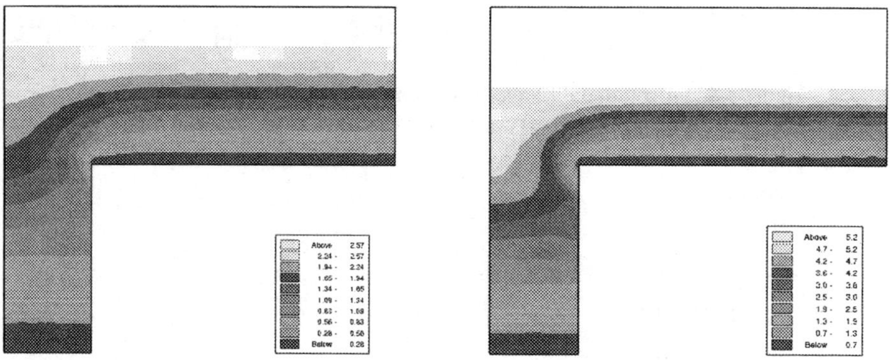

Figure 5.28. Displacement contours at the half and final stages of the top punch movement, $d_1 = 3.03$ and 6.06 mm

Figure 5.29. Relative density contours at the half and final stages of the top punch movement, $d_1 = 3.03$ and 6.06 mm

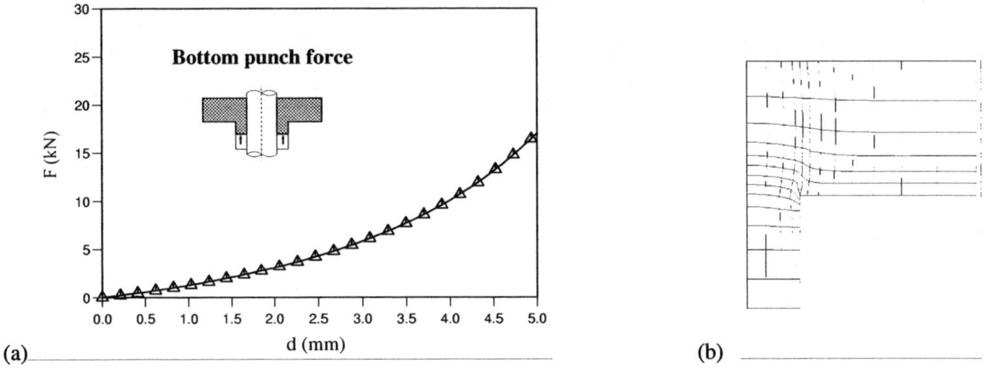

Figure 5.30. A rotational flanged component due to the single bottom punch; a) Predicted bottom punch compaction forces at different displacements, b) Deformed mesh at $d_2 = 5.10$ mm

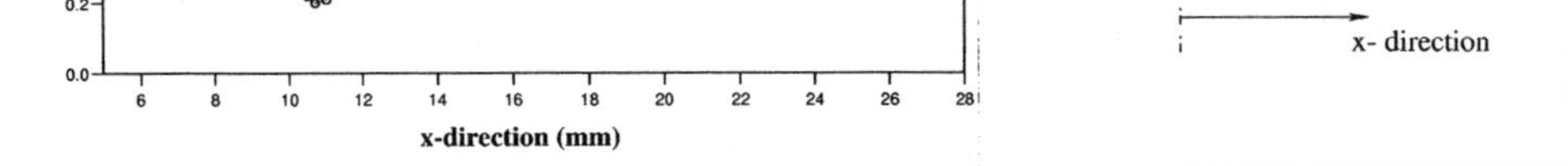

Figure 5.31. Comparison of the relative density profile by the ULD finite element formulation and predicted results given by Gethin *et al.* (1994) at section *b-b* for a flanged component

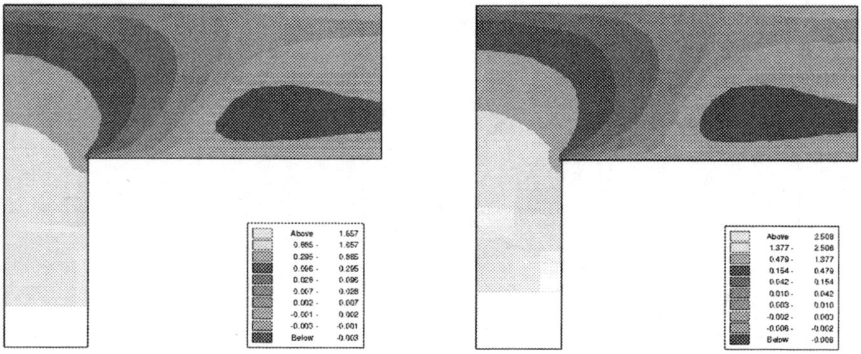

Figure 5.32. Displacement contours at the half and final stages of bottom punch movement, $d_2 = 3.40$ and 5.10 mm

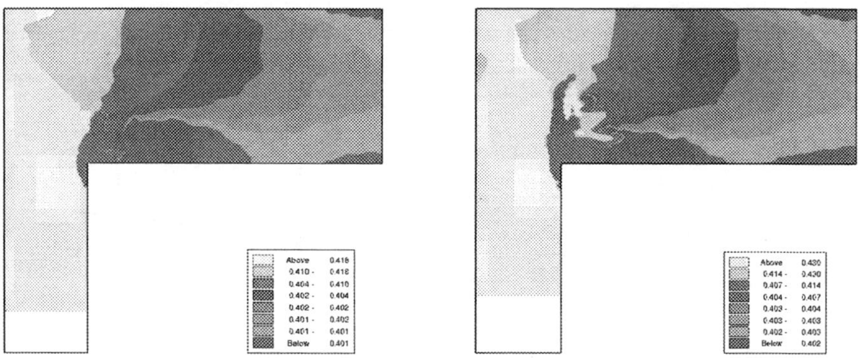

Figure 5.33. Relative density contours at the half and final stages of bottom punch movement, $d_2 = 3.40$ and 5.10 mm

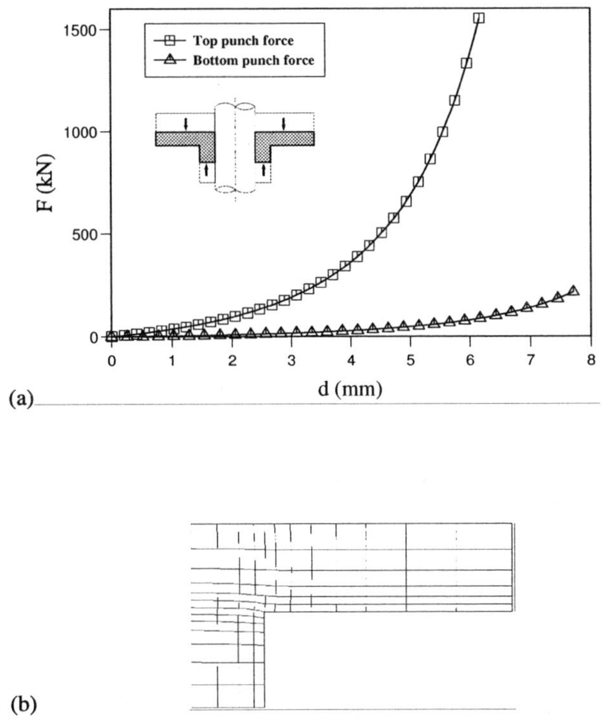

Figure 5.34. A rotational flanged component due to the top and bottom punches; a) Predicted top and bottom punches compaction forces at different displacements, b) Deformed mesh at $d_1 = 6.06$ mm and $d_2 = 7.70$ mm

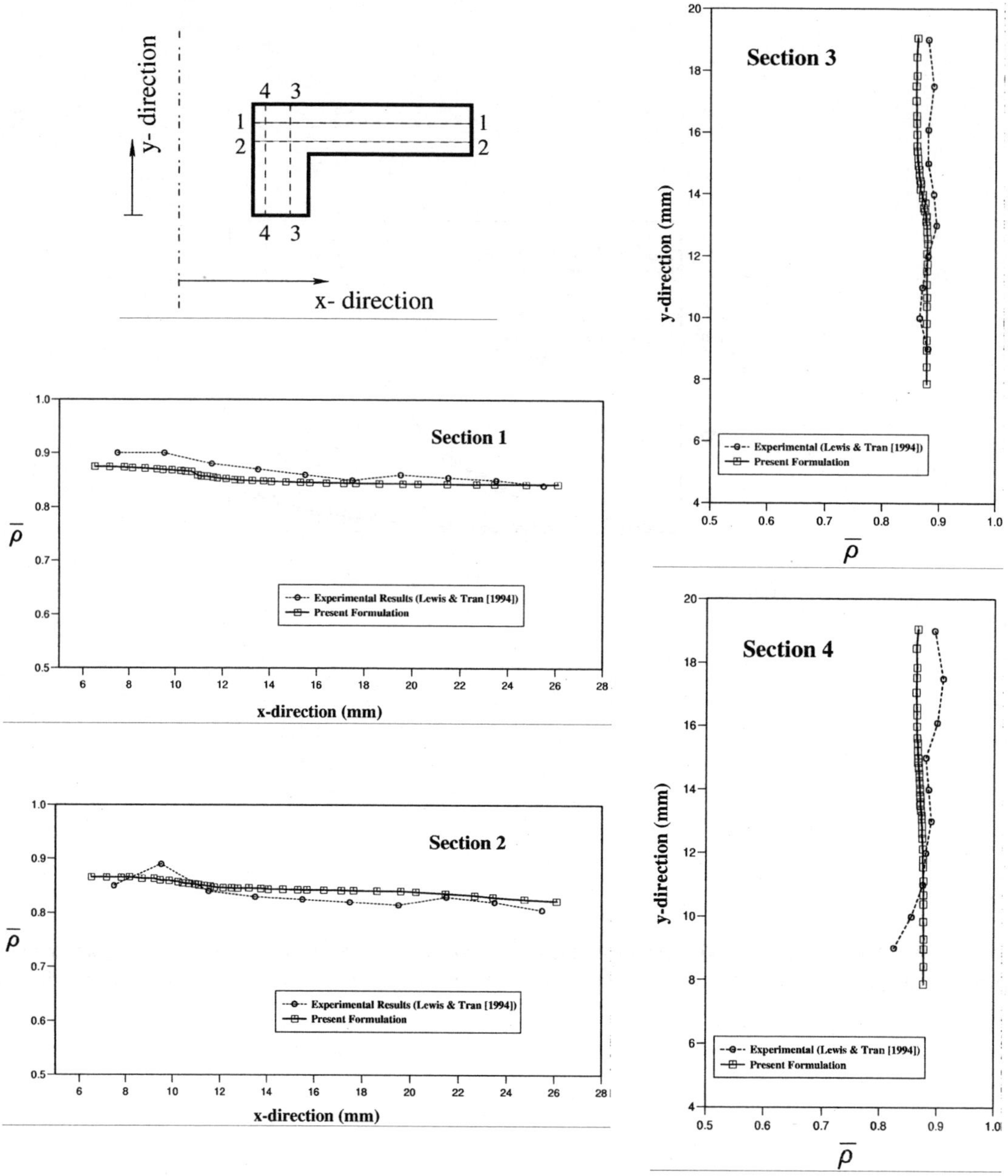

Figure 5.35. Comparison of the relative density profile by the MCEC plasticity model and experimental results given by Lewis and Tran (1994) for a flanged component

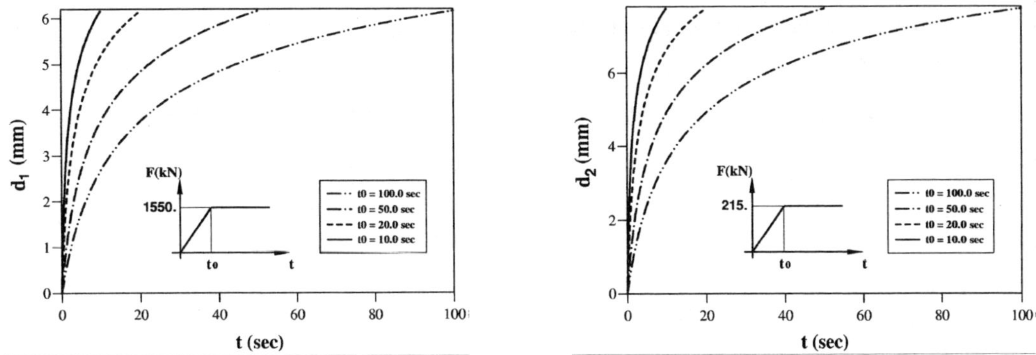

Figure 5.36. Variation with time of the displacements of a flanged component subjected to a surface step loading for various ramp time intervals t_0

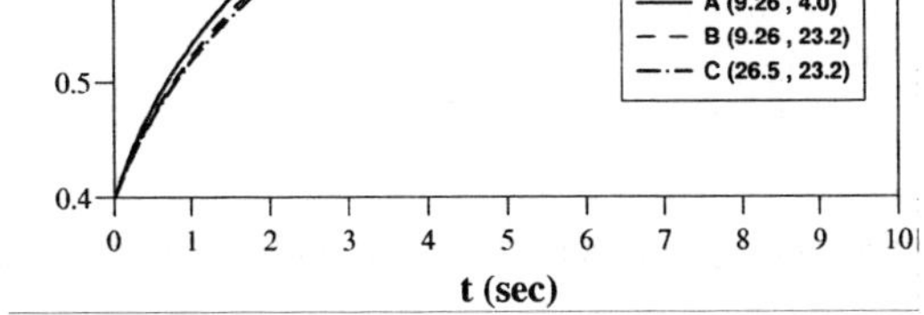

Figure 5.37. Variation with time of the relative densities at different nodes

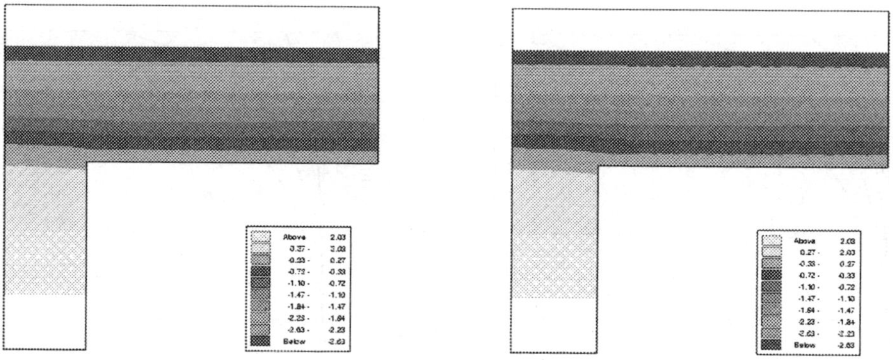

Figure 5.38. Displacement contours at the half and final stages of powder compaction

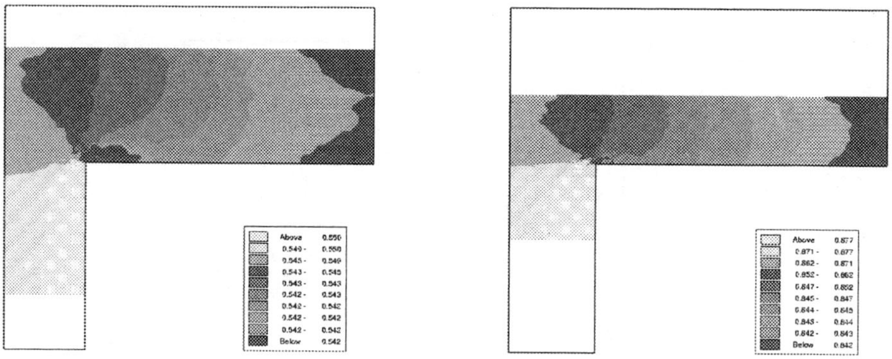

Figure 5.39. Relative density contours at the half and final stages of powder compaction

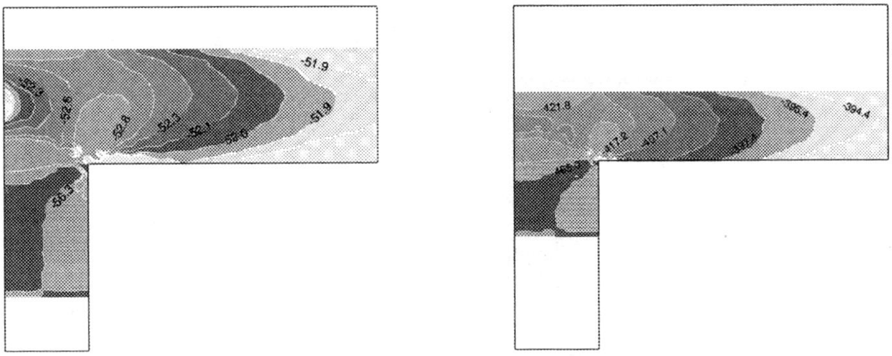

Figure 5.40. Normal stress σ_x contours at the half and final stages of powder compaction

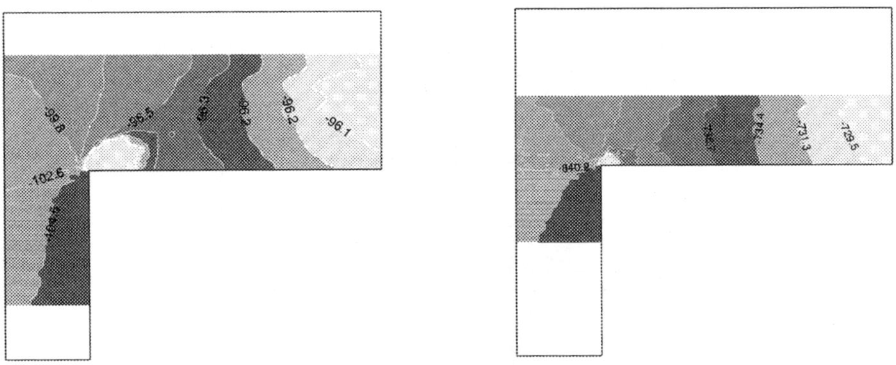

Figure 5.41. Normal stress σ_y contours at the half and final stages of powder compaction

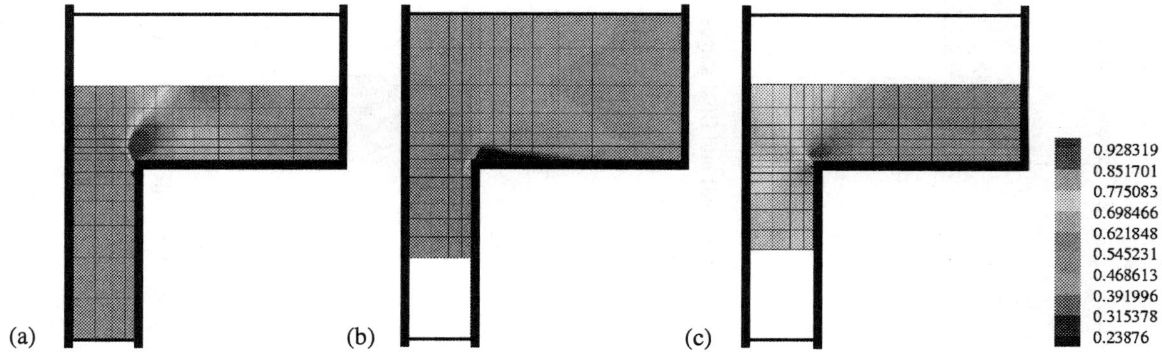

Figure 5.42. Relative density distributions at different movements of punches using ALE formulation;
a) $d_t = 6.06$ mm, b) $d_b = 7.70$ mm, c) $d_t = 6.06$ mm and $d_b = 7.70$ mm.
(see colour plate section at beginning of book)

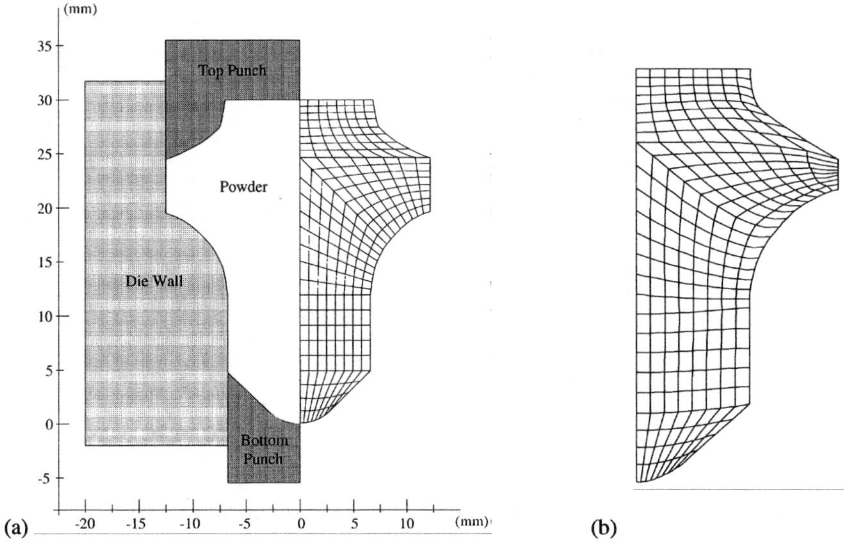

Figure 5.43. An axisymmetric shaped tip; a) Finite element modeling, geometry and boundary conditions, b) Deformed mesh at $d_1 = 3.00$ mm and $d_2 = 1.60$ mm

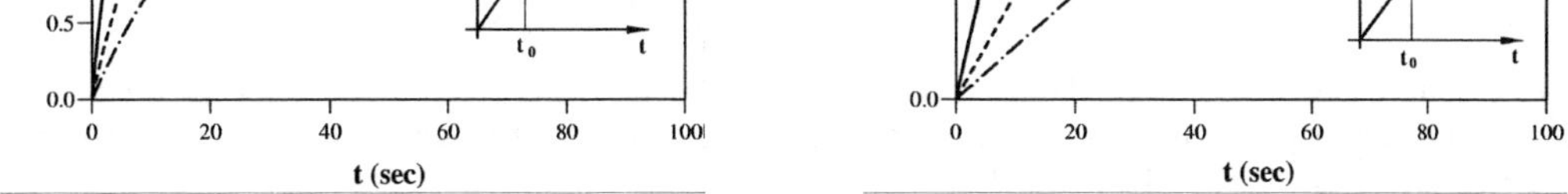

Figure 5.44. Variation with time of the displacements of an axisymmetric tip subjected to a surface step loading for various ramp time intervals t_0

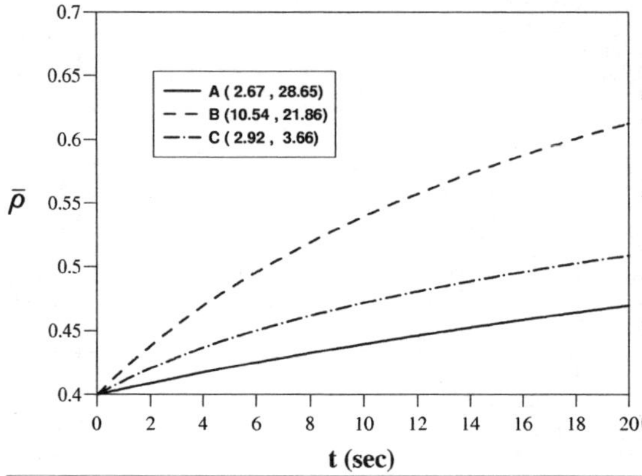

Figure 5.45. Variation with time of the relative densities at different nodes

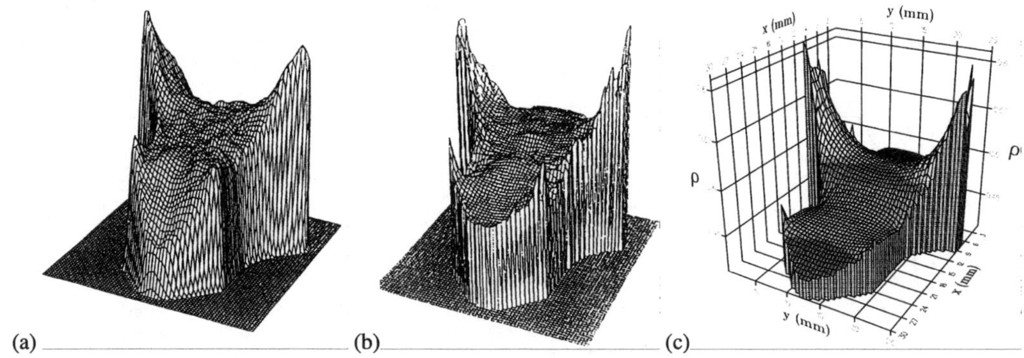

Figure 5.46. Comparison of the density distribution by the MCEC plasticity model and predicted results for an axisymmetric tip; a) The measured density distribution and, b) The numerical solution by Haggblad and Oldenburg (1994), c) The MCEC model

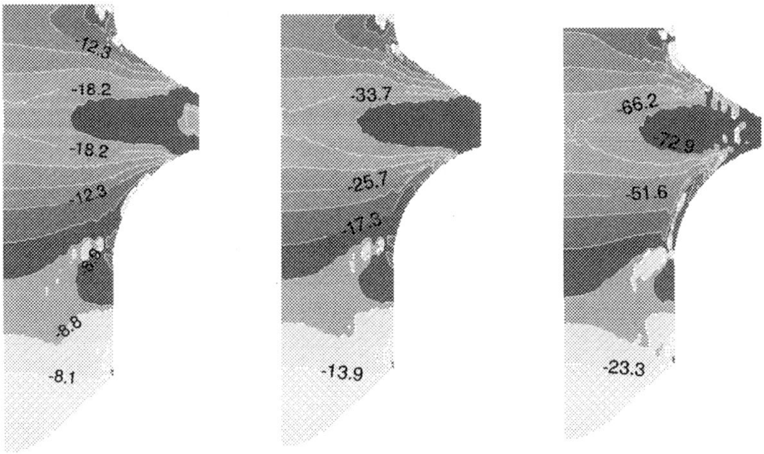

Figure 5.47. Normal stress σ_x contours at different powder movements

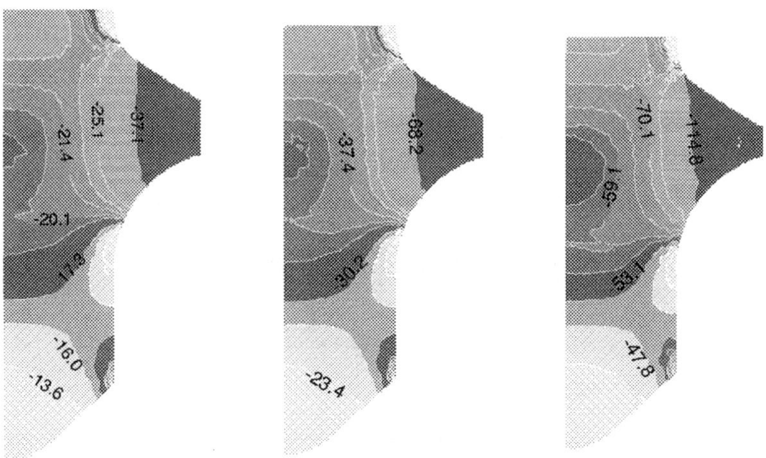

Figure 5.48. Normal stress σ_y contours at different powder movements

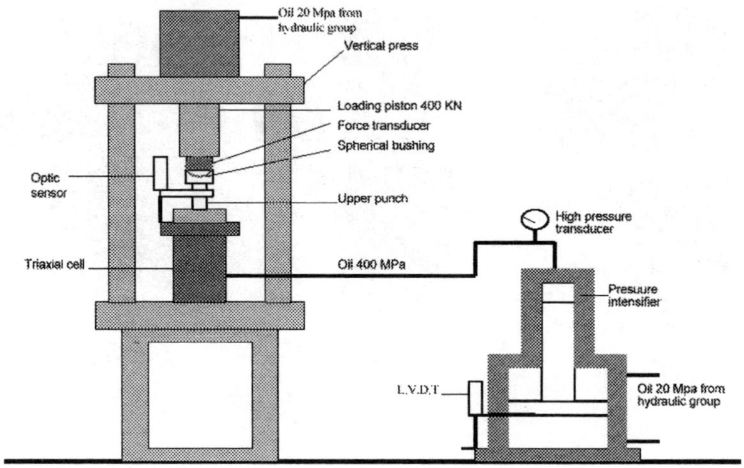

(a)

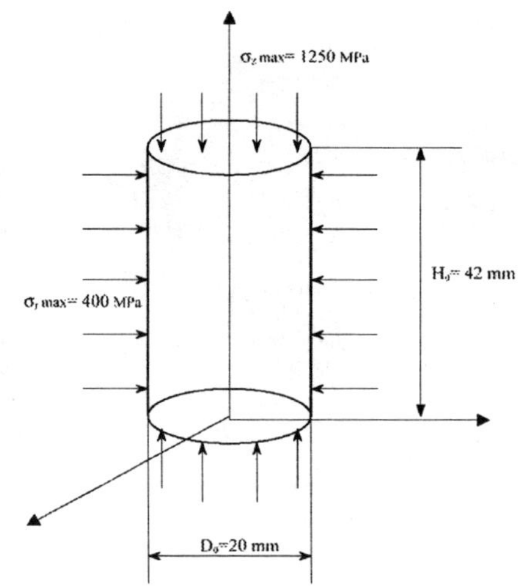

(b)

Figure 5.49. The triaxial test; a) Schematic diagram of apparatus, b) Initial geometry of specimen and maximum triaxial stress applied (Doremus *et. al.*, 1995)

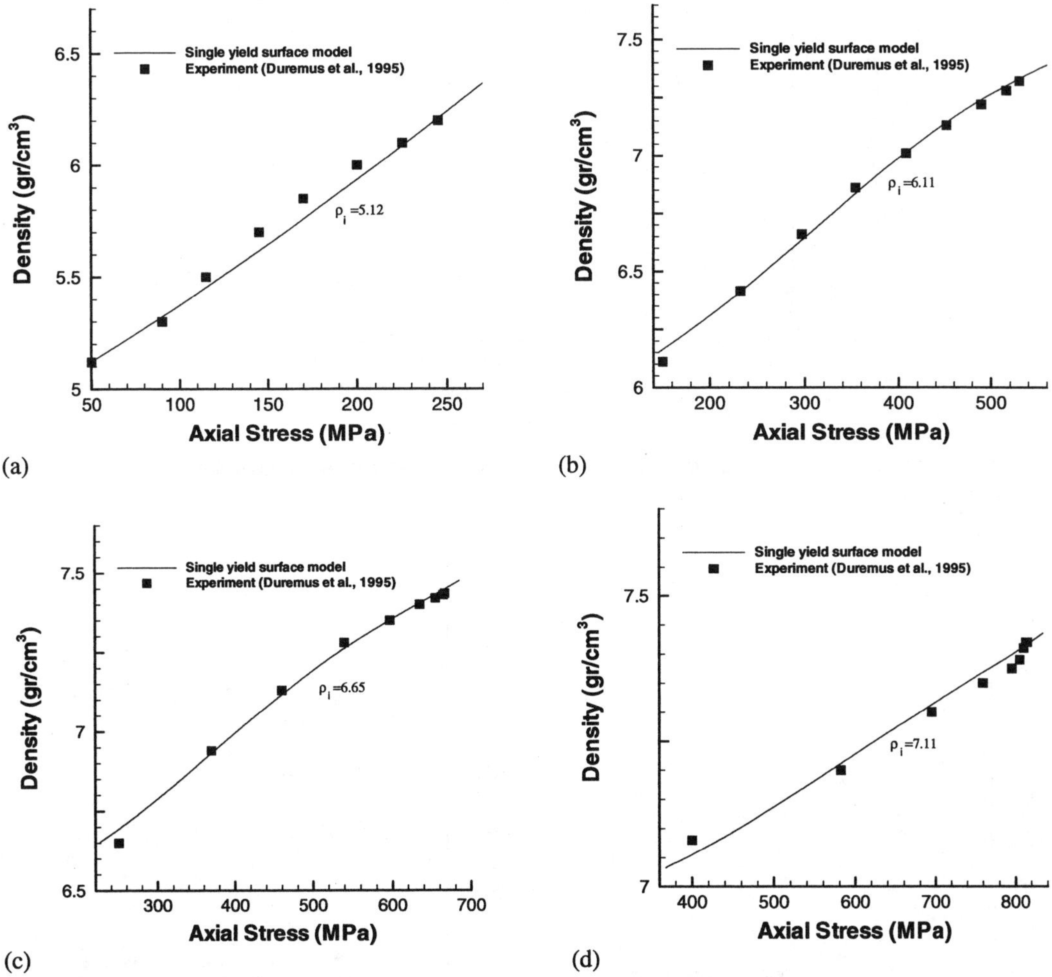

Figure 5.50. The variations of the density with axial stress for different values of hydrostatic pressure; A comparison of numerical and experimental results

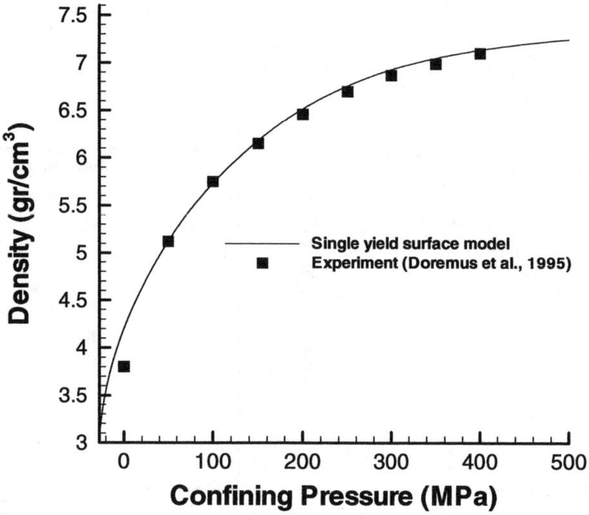

Figure 5.51. The variation of the density with hydrostatic pressure; A comparison of numerical and experimental results

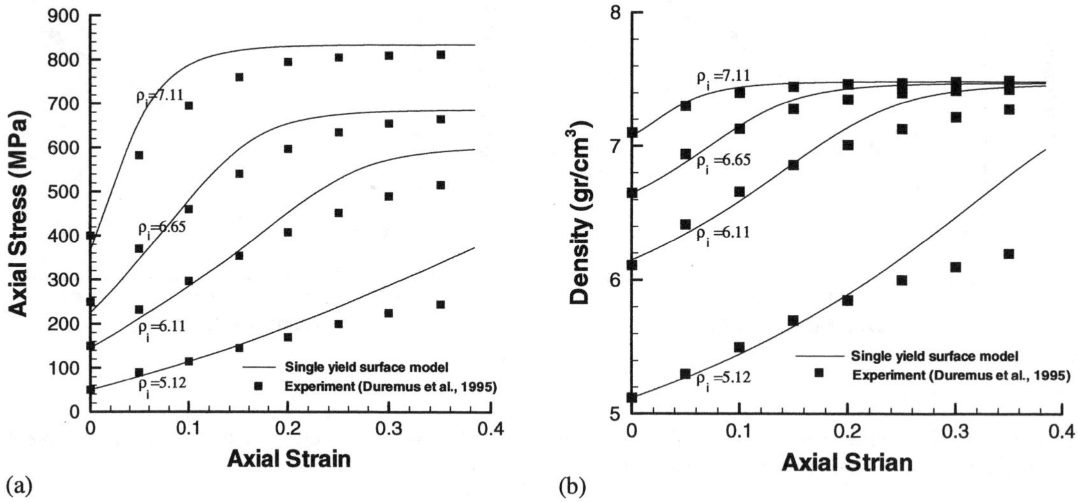

Figure 5.52. A comparison of numerical and experimental results in triaxial tests; a) The axial stress versus axial strain, b) The density versus axial strain

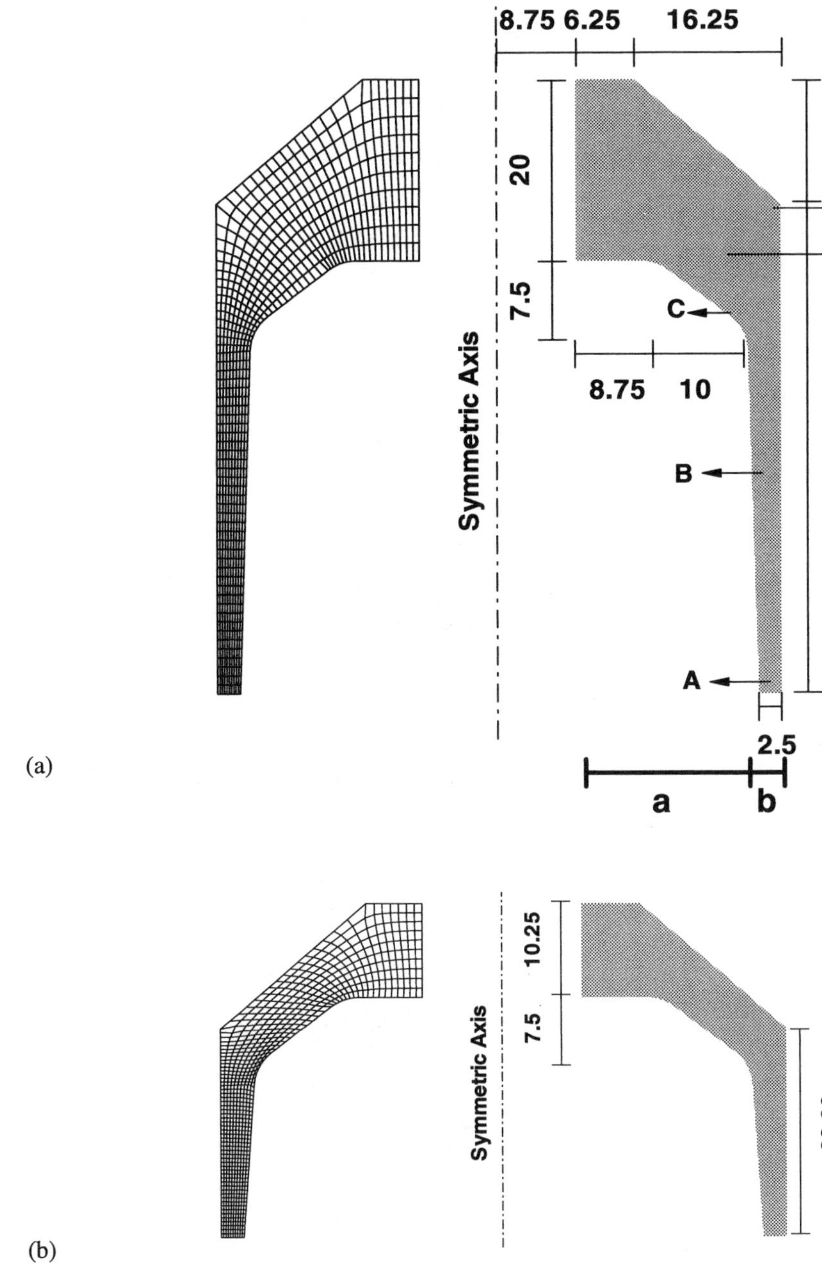

Figure 5.53. An automotive component; a) The geometry, boundary conditions and FE mesh of the uncompacted powder; b) The final geometry and FE mesh of the component at final stage of compaction

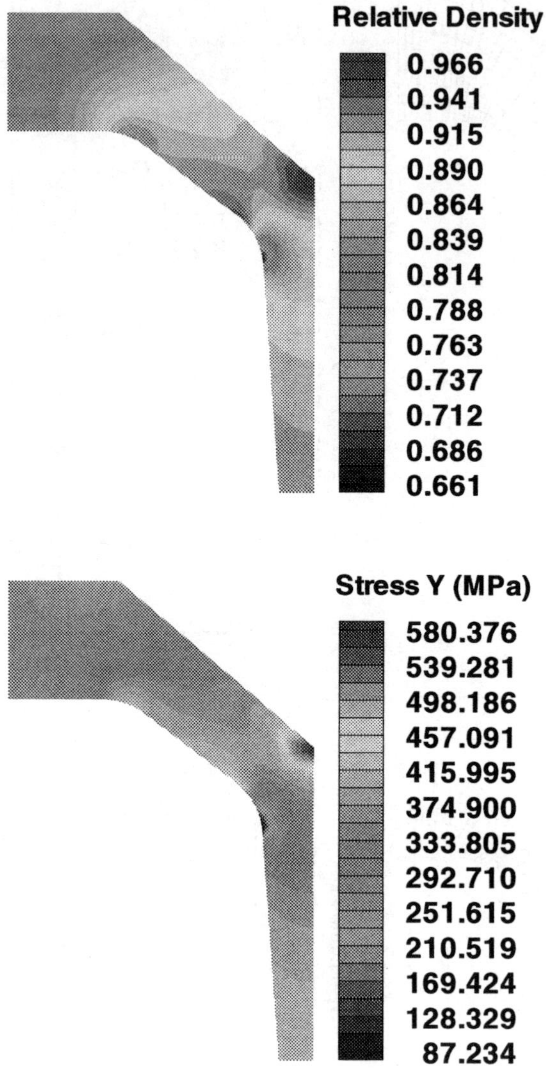

(a)

(b)

Figure 5.54. An automotive component; a) The predicted relative density contour and; b) The normal stress σ_y distribution at the final stage of compaction

(see colour plate section at beginning of book)

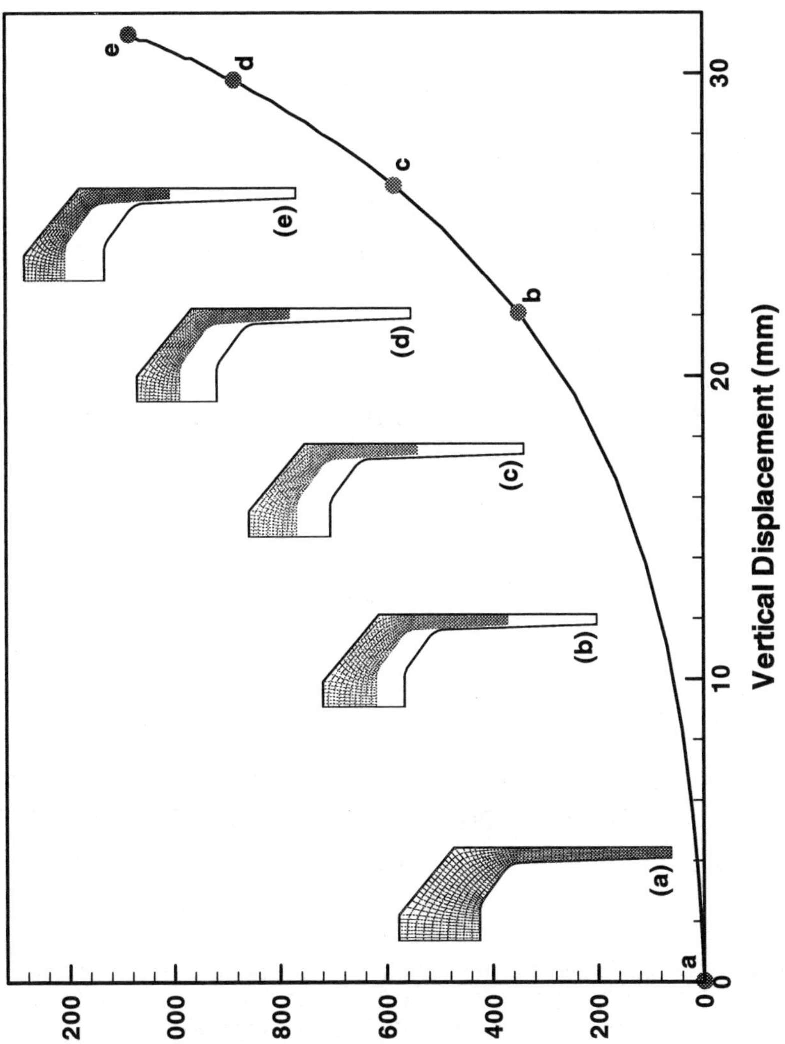

Figure 5.55. An automotive component; The variation of the top punch force with displacement

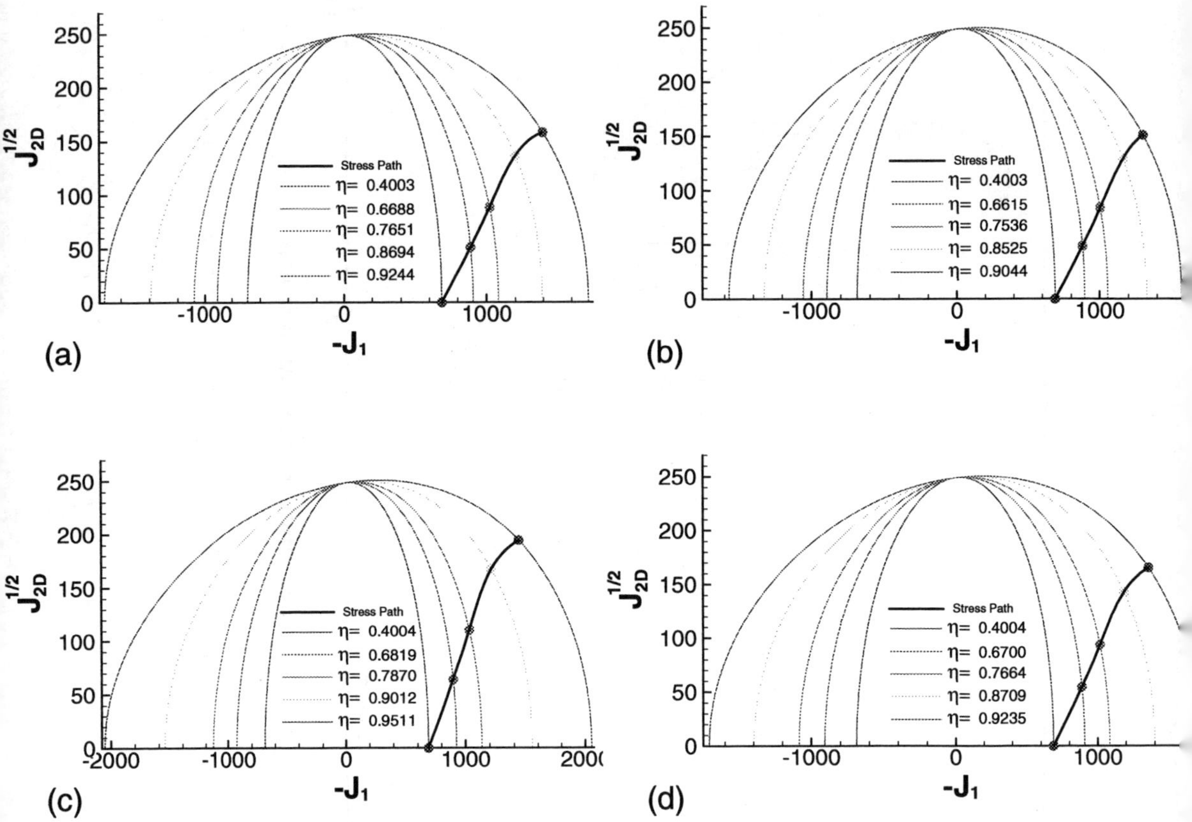

Figure 5.56. An automotive component; The stress paths with their associated yield surfaces at different Gauss points; a) A(30.2, -23.0), b) B(30.0, 0.0), c) C(25.6, 17.7), d) E(28.4, 24.4)
(see colour plate section at beginning of book)

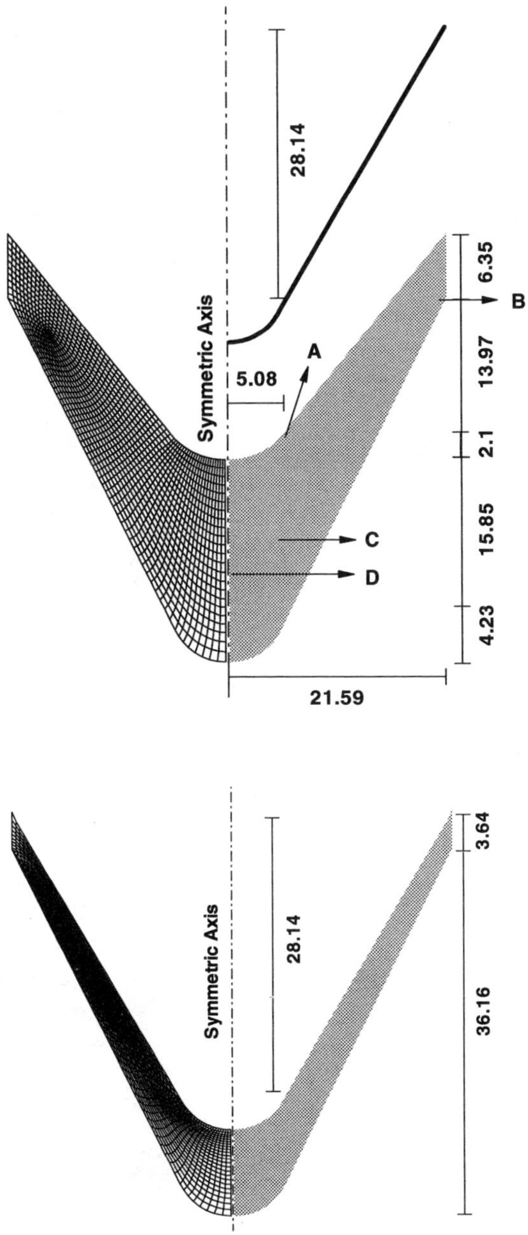

(a)

(b)

Figure 5.57. A conical shaped-charge liner; a) The schematic of the process, geometry, boundary conditions and FE mesh of the uncompacted powder; b) The final geometry and FE mesh of the component at final stage of compaction

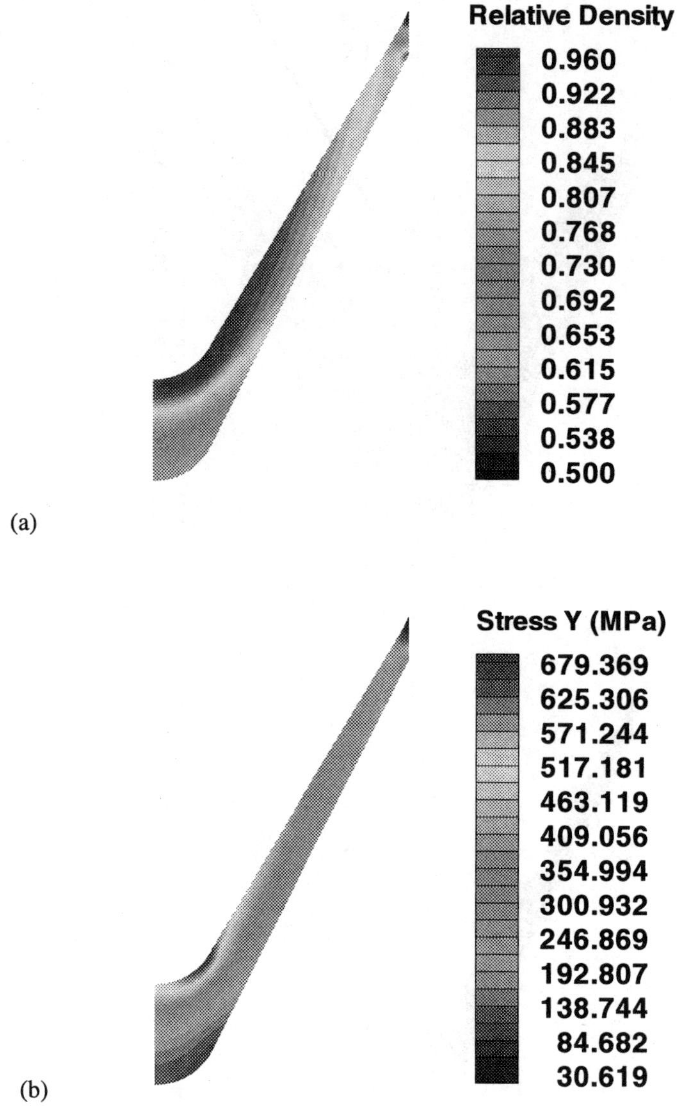

Figure 5.58. A conical shaped-charge liner; a) The predicted relative density contour and; b) The normal stress σ_y distribution at the final stage of compaction

(see colour plate section at beginning of book)

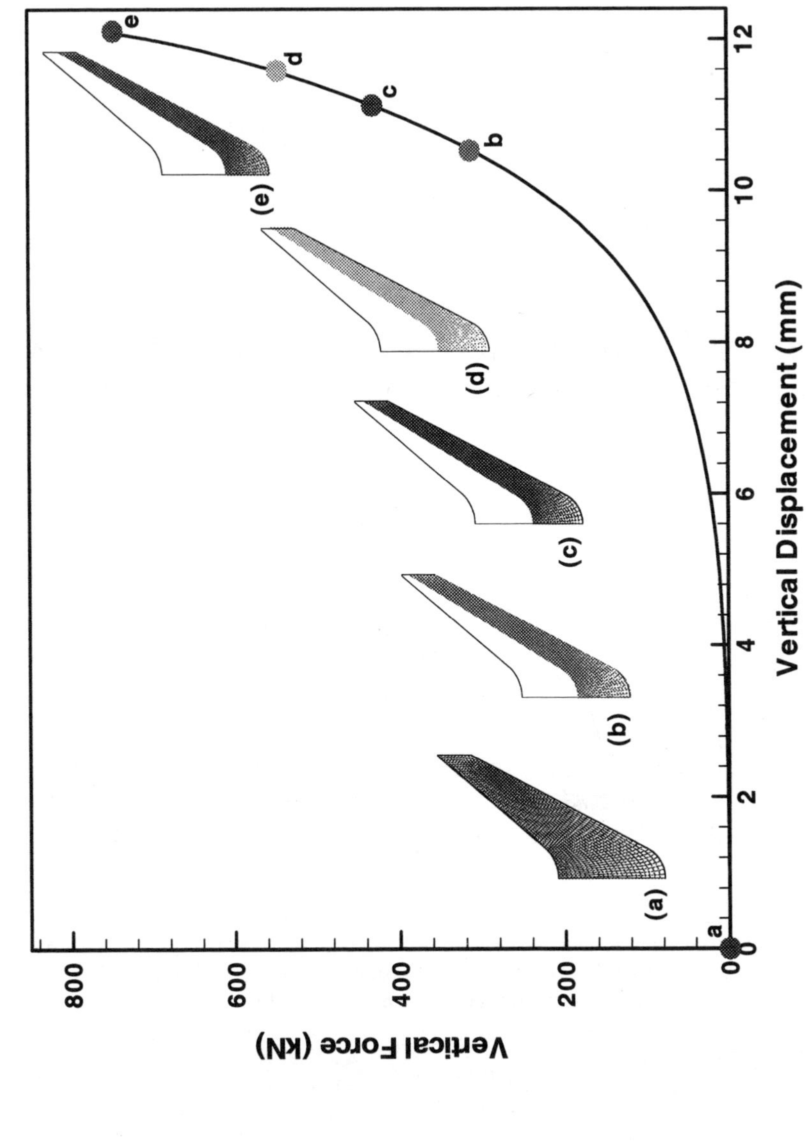

Figure 5.59. A conical shaped-charge liner; The measured axial load on the upper punch versus vertical displacement

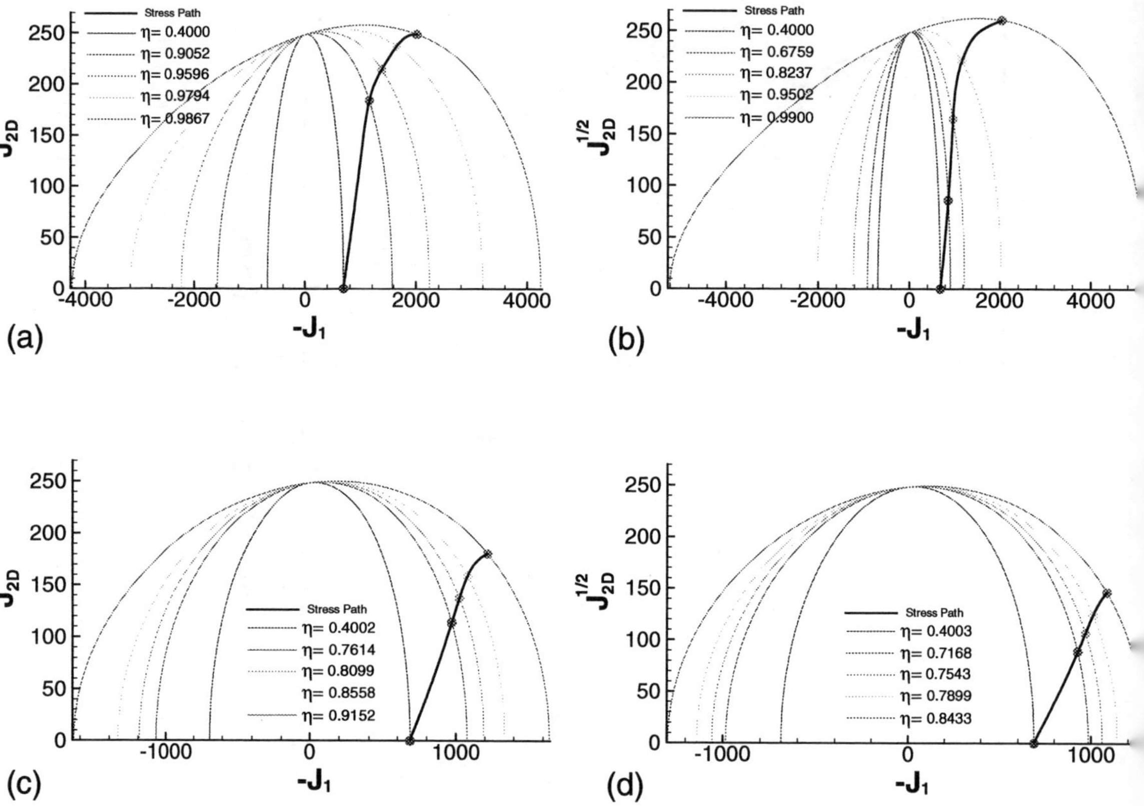

Figure 5.60. A conical shaped-charge liner; 2D stress paths with their associated yield surfaces at different Gauss points; a) A(5.1, 21.5); b) B(21.7, 35.6); c) C(5.6, 11.4); d) D(1.3, 9.1) (see colour plate section at beginning of book)

6 ERROR ESTIMATION AND ADAPTIVITY

For a large class of nonlinear problems the optimal mesh configuration changes continually throughout the deformation process requiring successive mesh adoptions during solution. Considerable benefits may accrue for such problems in terms of robustness and efficiency, realizing that the requirements of computational efficiency are ever increasing. The finite element method has been proven to be a flexible tool, especially when applied to non-linear problems of engineering. Since numerical methods yield approximate solutions, it is essential to introduce some measures of error and to use adaptive mesh refinement to keep this error within prescribed bounds to ensure that the finite element method is effectively used for practical analysis. This is particularly important in powder forming processes in which the mesh is continuously updated and element distortion occurs in the computation.

As metal powder forming analyses involve large deformations where distortion of the mesh is inevitable if an updated Lagrangian coordinate scheme is used, it is important to make use of the error estimate and adaptive remeshing scheme to control the mesh and therefore to improve the solution quality. In this chapter, a simple and reliable error estimator is adopted in the analysis of powder forming processes. The error is then controlled by the adaptive remeshing scheme to reach a given accuracy. This is a posteriori error estimate based on the fact that the first derivatives of the problem variables, after smoothing, (by a simple nodal averaging, least squares method, or SPR technique) are more accurate than those piecewise continuous ones obtained by the finite element method. The difference of these two is measured in terms of the L_2 norm and used as an error estimator. It is proved theoretically that the estimator is asymptotically exact for linear elasticity problems. This naturally gives us a guideline to extend the methodology to elasto-plastic behavior of powder materials.

6.1. Introduction

The finite element method has been widely used in the analysis of powder forming processes (Chenot *et al.* 1990, Brekelmans *et al.* 1991, Haggblad 1994, Oliver *et al.* 1996 and Khoei and Lewis 1998). By using the method, various powder forming conditions of nonlinear behavior, hardening, boundary friction, boundary geometry and loading can be considered and unsteady forming processes can be simulated for very large deformations. The analysis undoubtedly plays an important part in the design of the tool and billet geometry, material properties and the processing conditions, which otherwise rely largely on experience. However, an improperly constructed mesh or a distorted mesh, which often results during the analysis of a large deformation problem, may induce large discretization errors or make further analysis impossible. A great deal of work has been done by employing an error estimation and adaptive remeshing in the analysis of forming processes.

The updated Lagrangian description is generally adopted for non-steady metal forming processes, with dependency on material history. An advantage of the Lagrangian description is that deformation history dependent variables (strain hardening) can easily be taken into account and the changing shape of the formed product can be followed. However, one of the significant limitations of this approach, as described in Chapter 5, is the progressive distortion which can sometimes lead to entanglement of the finite element mesh. This can result in a severe distortion of the finite element mesh where intensive shear takes place. The distortion can cause large approximation errors or make the Jacobian determinant negative and make further analysis impossible. To restart the analysis, and obtain a reasonable solution, mesh regeneration is inevitable. Another reason to change the discretization and regenerate it, is that due to a complicated die geometry, where the initially generated mesh may not have sufficient local degrees of freedom to describe the details of material flow because of the die boundary conditions. This usually happens when a part of the work-piece is in contact with the curved die border and therefore a much finer mesh is necessary.

The process of mesh regeneration often requires human intervention making the analysis difficult in industrial use. In order to solve this problem, we can either set up different remeshing schemes locally as described by Gelten and Konter (1982), Zienkiewicz (1984) and Cheng and Kikuchi (1986) where intensive shear with high values is simulated by dividing the analysis mesh into several parts and updating so as to avoid distortion at the corner, or by use of different coordinates to describe the deformation in the ALE (Arbitrary Lagrangian Eulerian) method as described by Huetink (1982), Ponthot (1989) and Huetink *et al.* (1995). However, the remeshing or rezoning methods, which are most usually adopted in metal forming analyses, can not be regarded as a general panacea for geometrically complex forming problems. The only efficient way is to use error estimates and adaptive remeshing, which not only control the discretization error, but automates the simulation process. This consists in defining a new appropriate finite element mesh within the updated, deformed geometry and interpolating (mapping) the pertinent variables from one mesh to another in order to continue the simulation.

An alternative error estimator was suggested by Cheng in 1986, using the sum of the difference between the effective strains of each element and its adjacent elements, but no application of the method has been observed in the literature. The method was first applied to the simple cylinder upsetting problem by Zienkiewicz *et al.* (1988) and then developed to both compressible and incompressible forming analyses of forging and extrusion problems (Zienkiewicz *et al.* 1989, 1990). Many applications of the Zienkiewicz-Zhu estimator have been seen in the analysis of forming processes. Dyduch *et al.* (1995) applied a local finite difference method to the Zienkiewicz-Zhu error estimator with various smoothed fields and interpolation error indicators for the elasto-viscoplastic analysis of forging problems. They used an error indicator for strain rates and its interpolation for the local refinement during the adaptive remeshing procedure.

An analysis of error estimation was presented in the framework of viscoplasticity problems for incompressible and non-linear materials by Fourment and Chenot (1995). They applied two error estimators, based on the Zienkiewicz-Zhu estimator and a Δ estimator for the velocity error according to the L^2 and L^∞ norms, to an industrial problem of extrusion. It was then developed to 3D forming problems by employing an anisotropic remeshing procedure to optimize the number of elements for highly oriented flows (Fourment *et al.* 1995). An automatic, adaptive remeshing scheme with a new, independent mesh was developed by Dyduch *et al.* (1992) to enable finite element simulations of forming processes with complicated die geometries. Dyduch's mesh generation procedure was based on a variable density mesh generator, where high density points are defined in the centre of grossly distorted elements. An application of Dyduch's method, based on element distortion checks, has recently been applied to forging problems by Dutko *et al.* (1997) where corner angle and element slenderness may also trigger remeshing.

Despite the extensive application of adaptive FEM strategy in the industrial manufacturing and processing operations, powder forming technology has been mostly based on experience and trial and error methodology, rather than methods arising from a rigorous scientific approach. It is only recently that numerical simulation has matured sufficiently to practically influence industrial powder forming operations (see the conference proceedings: NUMIFORM 1995, 1998, 2001 and 2004). However, to date little has been done to extend an adaptive procedure to powder forming processes. An implementation of finite element techniques using adaptive remeshing was reported by Tran *et al.* (1993) for simulating powder compaction processes - however, from their results on a flange component, it can be seen that the refined mesh appeared coarser during the varying stages of compaction which can lead to unrealistic results, especially at the corner region where intensive shear takes place.

In the present study, we intend to employ a-posteriori error estimator developed by Zienkiewicz and Zhu (1987, 1991) using an h-refinement adaptive procedure in the analysis of powder forming processes. This adaptive procedure can be applied as follows; an error and mesh control component, remeshing of the deformed body, mapping of history dependent variables from the old to the new mesh and a restart of the simulation.

6.2. Adaptive FEM strategy

The subject of error estimates for finite element solutions and a consequent adaptive analysis, in which the approximation is successively refined to reach predetermined standards of accuracy, is central to the effective use of finite element codes for practical engineering analysis. Much of the pioneering mathematical work, such as Babuska and Rheinboldt (1978, 1979, 1981), Kelly *et al.* (1983), Gago *et al.* (1983) and Zienkiewicz and Zhu (1987, 1991), has been today translated into engineering usage. However, difficulties still remain in extending such procedures to non-linear analyses. The solution of a non-linear problem is usually carried out by linearization of the governing equation during the iterative procedure. It is interesting to find out if the same procedures of adaptive analysis for linear problems can be extended to non-linear analyses.

For most of the early practical simulations involving history-dependent materials on evolving meshes work has been concentrated on adopting a suitable criterion to control the remeshing procedure. An adaptive mesh technique was developed by Belytschko *et al.* (1985) for nonlinear problems of explicit dynamic analysis of shells. The strategy consisted in splitting a single quadrilateral into four equal-sized smaller elements (fission), and derefining, or combining, four elements into a single element (fusion). An adaptive remeshing scheme was proposed by Ortiz and Quigley (1991), in which the resolution of the mesh is increased in certain regions so that the variation of the solution over each element is within a prescribed tolerance throughout the mesh. A similar procedure was adopted by Jin *et al.* (1989) who use isolines of effective strain as a measure in their adaptive remeshing scheme. In each of the above schemes a suitable transfer operator must be defined to ensure that the state variables are correctly distributed to the new mesh. Lee and Bathe (1994) developed an adaptive strategy for large deformation processes at large inelastic strains, which provides a discussion on important aspects of the adaptive procedure for history-dependent materials on evolving finite element meshes. Their technique analyses various aspects of error indicators, remeshing algorithms and transfer operations. Peric *et al.* (1996) presented a similar procedure based on a general mapping scheme for history-dependent materials and demonstrated aspects of data transfer of the plastic strain tensor and internal variables from one mesh to the other.

The focus of this study is on an error estimation, adaptive mesh refinement and transfer of variables during a mesh adaption procedure in the finite element analysis of history-dependent materials. The history-dependent nature of the process necessitates transfer of all relevant problems variables from the old mesh to the new one, as the successive remeshing is applied during the process simulation. As the mesh is adapted, with respect to an appropriate error estimator, the solution procedure, in general, cannot be re-computed from the initial state, but has to be continued from the previously computed state. In particular, some suitable means for transferring the state variables between meshes, or transfer operators, needs to be defined.

The objective of adaptive techniques is to obtain a mesh which is optimal in the sense that the computational costs involved are minimal under the constraint that the error in the finite element solution is acceptable within a certain limit. Since the computational effort can be linked to the number of unknowns of the finite element mesh the task is to find a mesh with a minimum number of unknowns, or nodes, for a given error tolerance. To obtain an optimal mesh, in the sense of an equal solution quality, it is desirable to design the mesh such that the error contributions of the elements are equally distributed over the mesh. In a practical scheme for an adaptive analysis in plasticity, the mesh must be changed when the error is more than a prescribed value. After changing the mesh, all state variables must be transferred to the integration points of the new mesh. Since, in a conventional displacement formulation of FEM, the state variables are stored at these integration points, then the transfer of these values from the old integration points to the new ones can be carried out by using nodal points as the reference points to store the information. In the case that nodal points are used as reference points, the variables at any arbitrary point can be evaluated by interpolation. In general, the procedure described above can be executed in four parts; an error estimation, an adaptive mesh refinement, an adaptive mesh generator and the mapping of variables.

6.3. Error estimation

The aim of error estimation in any analysis is firstly, to determine the error for the chosen, perhaps distorted, mesh and secondly, to reduce this error to a permissible value by automatic adaptive remeshing. The discretization error represents the difference between the exact and the approximate (e.g. the finite element) solutions for the mathematical model. Since exact solutions for practical, nonlinear problems are usually not available, an estimated error, instead of the real error, is adopted to guide the finite element analysis.

Various methods for the estimation of spatial discretization errors have been proposed and used in practice (Babuska *et al.* 1986). Basically, there are two types of discretization error estimates; *'a-priori'* estimates and *'a-posteriori'* estimates. A-priori estimates are based on a knowledge of the characteristics of the solution and provide qualitative information about the asymptotic rate of convergence as the number of degrees of freedom goes to infinity. A-posteriori estimates employ the solution obtained by the numerical analysis, in addition to some a-priori assumptions about the solution. This method can provide quantitatively accurate measures of the discretization error, while a-priori estimate method cannot. In the present study, a simple yet reliable a-posteriori error estimator developed by Zienkiewicz and Zhu (1987, 1991), which is not only reasonably accurate but whose evaluation is computationally simple and can be readily implemented in existing finite element codes, are applied to the analysis of static and dynamic behavior of metal powder forming processes. For this purpose an error estimation using an L_2 norm of stress by a recovery procedure in which the improved value of finite element solution is the basis of an adaptive procedure, is proposed.

The finite element Galerkin discretization formulation with the independent approximations of $\mathbf{u} \approx \hat{\mathbf{u}} = \mathbf{N}_u^T \overline{\mathbf{u}}$ for the elasto-plastic transient dynamic behavior of metal powder forming problems and its direct time stepping algorithm have been described in Chapter 2, as

$$\int_\Omega \overline{\mathbf{B}}^T \boldsymbol{\sigma}\, d\Omega + \mathbf{M}\ddot{\overline{\mathbf{u}}} - \mathbf{f}_u = 0 \tag{6.1}$$

In the present formulation, C_0 continuity is assumed for the displacement interpolation function $\mathbf{N}_u$. For variables such as the stress $\boldsymbol{\sigma}$, defined as $d\boldsymbol{\sigma} = \mathbf{D}\, d\boldsymbol{\varepsilon}$ with $d\varepsilon = \overline{\mathbf{B}}\, d\overline{\mathbf{u}}$, then the finite element approximation $\hat{\boldsymbol{\sigma}}$ is discontinuous and differs from the exact value $\boldsymbol{\sigma}$. Thus, the error in stress may be defined as

$$\mathbf{e}_\sigma = \boldsymbol{\sigma} - \hat{\boldsymbol{\sigma}} \tag{6.2}$$

Since the exact value of stress $\boldsymbol{\sigma}$ for nonlinear problems is not available, an improved solution is sought instead of the exact solution by creating a better approximation by a suitable projection of the available solution. Thus, if we have value of $\hat{\boldsymbol{\sigma}}$ derived by a finite element solution and $\boldsymbol{\sigma}^*$ represent such an improved solution then the error can be estimated by inserting such values in place of the exact solution. In order to obtain an improved solution, resort is generally made to smoothing or nodal averaging. For instance, a global smoothing for a continuous stress $\boldsymbol{\sigma}^*$ can be formed by using the interpolation function $\mathbf{N}_u$ and nodal parameter $\overline{\boldsymbol{\sigma}}^*$ as

$$\boldsymbol{\sigma}^* = \mathbf{N}_u \overline{\boldsymbol{\sigma}}^* \tag{6.3}$$

Computing parameter $\overline{\boldsymbol{\sigma}}^*$ so that $\boldsymbol{\sigma}^*$ is equal to $\hat{\boldsymbol{\sigma}}$, in the mean sense

$$\int_\Omega \mathbf{N}_u^T (\boldsymbol{\sigma}^* - \hat{\boldsymbol{\sigma}}) d\Omega = 0 \tag{6.4}$$

Substituting $\boldsymbol{\sigma}^*$ by $\mathbf{N}_u \overline{\boldsymbol{\sigma}}^*$ in equation (6.4), we can obtain

$$\overline{\boldsymbol{\sigma}}^* = \mathbf{A}_u^{-1} \int_\Omega \mathbf{N}_u^T \hat{\boldsymbol{\sigma}}\, d\Omega \tag{6.5}$$

where

$$\mathbf{A}_u = \int_\Omega \mathbf{N}_u^T \mathbf{N}_u\, d\Omega \tag{6.6}$$

Alternatively, the nodal parameter $\overline{\boldsymbol{\sigma}}^*$ can be obtained by simply averaging the value over all elements around each node. This procedure is computationally faster and provides equally good results. It is suggested that the smoothed value $\boldsymbol{\sigma}^*$ is an approximation of a higher order than $\hat{\boldsymbol{\sigma}}$ and the differences between these two sets of values can be used to estimate the spatial discretization error, i.e.

$$\mathbf{e}_\sigma \approx \boldsymbol{\sigma}^* - \hat{\boldsymbol{\sigma}} \tag{6.7}$$

This process is easily implemented into any code as these generally already evaluate the appropriate $\boldsymbol{\sigma}^*$ and $\hat{\boldsymbol{\sigma}}$ quantities.

A pointwise definition of errors, as given in equation (6.7) is generally not convenient and occasionally misleading. For instance, under a point load the errors in stresses will be locally infinite but the overall solution may well be acceptable. Similar situations will exist near re-entrant corners where, as is well known, stress singularities exist in elastic analyses and gradient singularities develop in field problems. For this reason various 'norms' representing some integral scalar quantity are often introduced to measure the error, or indeed the function itself. One of the most common of such measures, which can be associated with the errors in any quantity, is the 'L_2 norm' written as

$$\| \mathbf{e} \| = \left(\int_\Omega \mathbf{e}^T \mathbf{e} \, d\Omega \right)^{\frac{1}{2}} \tag{6.8}$$

for a vector $\mathbf{e}$ over the domain Ω. Thus, we can estimate the L_2 norm of the error $\mathbf{e}_\sigma$ for the stress $\boldsymbol{\sigma}$ by

$$\| \mathbf{e}_\sigma \| = \| \boldsymbol{\sigma}^* - \hat{\boldsymbol{\sigma}} \| = \left(\int_\Omega (\boldsymbol{\sigma}^* - \hat{\boldsymbol{\sigma}})^T (\boldsymbol{\sigma}^* - \hat{\boldsymbol{\sigma}}) \, d\Omega \right)^{\frac{1}{2}} \tag{6.9}$$

Although the L_2 norm written above is defined over the whole domain, we note that the square of each can be obtained by summing element contributions. i.e.

$$\| \mathbf{e}_\sigma \|^2 = \sum_{i=1}^{m} \| \mathbf{e}_\sigma \|_i^2 \tag{6.10}$$

where i represents an element contribution and m is the total number of elements. Indeed for an 'optimal' mesh in which the distribution of the estimated error is uniform over the whole domain, we shall generally try to make the contributions to this square of the norm equal for all elements. The uniform distribution of error is usually used as a criterion for calculating the new mesh size, as described in the next section.

6.4. Adaptive mesh refinement

Adaptive remeshing of finite element solutions refers to improving the quality of the solutions by enriching the approximation in some manner so as to achieve the best solution for a given computational effort. An adaptive process is actually a feedback approach, which is optimal with respect to clearly defined objectives. It is clear that all adaptive processes aiming at a specified accuracy solution involve three stages which have to be applied cyclically,

- obtaining a solution with a specified refinement and estimating the error;
- predicting the refinement required for achieving a specified accuracy (error) most economically;
- implementing this refinement and obtaining a new solution and estimating the new error, followed by a return to the second stage if this does not meet the requirements.

The adaptive mesh refinement strategy will of course depend on the nature of the criteria on accuracy which we wish to satisfy. In the early works of adaptive analysis, refinement was generally done in many stages, using 'error indicators' to guide the introduction of additional degrees of freedom in both *h*-refinement and *p*-refinement techniques (Babuska and Rheinboldt 1978 and Kelly *et al.* 1983). Such error indicators were often associated with the desired, local, reduction of error and generally served to introduce refinements which reduced the maximum values of error, thus tending, albeit very slowly, to the optimal solution in which the element errors were equal. This process is in general not practicable, requiring a large number of resolutions of the full problem, and more efficient procedures have been developed by the authors, allowing a direct achievement of the desired (specified) error in one or two resolutions.

For refinement criteria in non-linear adaptive analyses, several methods have been suggested by authors which can be classified into two main categories; refinement criterion based on gradients (or curvature) of displacement which is usually used in localization problems (Ortiz and Quigley 1991 and Zienkiewicz *et al.* 1995) and refinement criterion based on using an error norm on each element, which is usually evaluated through computing an improved stress/strain (or other parameters) field (Samuelsson and Wiberg 1991 and Lee and Bathe 1994). In both cases the optimal mesh is defined to be the one in which the error in the whole domain is uniformly distributed over the elements. In the first category the new mesh size is computed based on the magnitude of the gradient of displacement. This kind of criterion does not indicate the error of the solution. In the second category, usually an error norm is defined for the whole domain. An improved stress/strain field is used to calculate the error norm on each element or on the whole domain.

A very common requirement is to specify the achievement of a certain minimum percentage error in the L_2 norm. Thus, we require that after remeshing, each element will have the same error and the overall percentage error

$$\eta = \frac{\|\mathbf{e}_\sigma\|}{\|\hat{\boldsymbol{\sigma}}\|} \tag{6.11}$$

will be equal to the target percentage error, i.e.

$$\frac{\|\mathbf{e}_\sigma\|_{aim}}{\|\hat{\boldsymbol{\sigma}}\|} = \overline{\eta}_{aim} \tag{6.12}$$

where $\overline{\eta}_{aim}$ is the prescribed target percentage error. From the definition of the L_2 norm, it is noted that the square of the overall error norm over the whole domain is equal to the sum of the squares of the local error norm of each element. If we assume that the error is equally distributed between elements the above requirements can be translated into our placing a limit on the error in each element. Thus, for each element we require that

$$\left(\|\mathbf{e}_\sigma\|_i\right)_{aim} = \frac{1}{\sqrt{m}}\|\hat{\boldsymbol{\sigma}}\|\,\overline{\eta}_{aim} \tag{6.13}$$

For elements which have no singularity present, it is assumed that the rate of convergence of the local error is $O(h^\lambda)$, where h is the size of the element and λ is the polynomial order of the interpolation function. Thus, the new element size can be predicted by the following equation

$$\frac{(h_i)_{new}}{(h_i)_{old}} = \left[\left(\|\mathbf{e}_\sigma\|_i\right)_{aim} \Big/ \left(\|\mathbf{e}_\sigma\|_i\right)_{old}\right]^{\frac{1}{\lambda}} \tag{6.14}$$

In this study, λ is assumed to be 2 when quadratic triangular elements are used and 1 when a linear triangular element is employed. The new element size $(h_i)_{new}$ of the element i can be calculated by equation (6.14). This information is used to compute the appropriate mesh sizes in a particular element for the continuing refinement procedure. To obtain the 'nodal' element size, a simple averaging between elements joining a node is used. In order to avoid too large and too small elements, both the upper and the lower limits for the element size are often specified, i.e.

$$\bar{h}_{\min} \leq (h_i)_{new} \leq \bar{h}_{\max} \tag{6.15}$$

For problems where a singularity exists, then the mesh size predicted by the subdivision requirement of equation (6.14) is invalid. In this case, a finer element subdivision around the singularity is desired by modifying the mesh size at the singular point. The modified mesh size h' can be specified as

$$h' = q\,(h_i)_{new} \tag{6.16}$$

where the grading factor q is less than one and in the present analysis the value of $q = 0.5$ is used for the singular points. As the mesh tends progressively to be optimal with the error uniformly distributed between elements this theoretical rate of convergence appears very effective.

The complete mesh regeneration procedure, according to the rule of equation (6.14), is probably the most efficient method of allowing mesh 'derefinement' where originally too many elements were used. It must be coupled with an efficient mesh generator which allows the new mesh to be constructed according to a predetermined size distribution. Such a generator was developed for triangular elements by Peraire *et al.* (1987) and is described in next section.

6.5. Adaptive mesh generator

Since the early days of finite element analysis much attention has been given to the problem of mesh generation. The initial objective was to avoid the 'manual labour' of creating a mesh which would fit the problem boundaries and roughly provide a 'reasonable' subdivision. In this context many and varied procedures have been developed and numerous papers published. Very broadly these can be grouped into two categories;

- those based on 'mapping' the domain (or various sub-domains into which it is divided) into simple forms in which direct structured mesh forms can be provided; and
- those in which the 'mesh' is constructed to join a set of predetermined nodes or to satisfy prescribed element shape and size requirements.

In the first category such processes as isoparametric, blending or Laplacian function mapping are typical (Zienkiewicz and Phillips 1971, Gordon and Hall 1973 and Thompson *et al.* 1974). These directly provide meshes which are, at least partially, structured. Although such meshes are very convenient for parallel computations and present some economies, the achievement of locally predicted mesh densities is not possible and for this reason we shall not discuss these techniques any further. In the second category many variants are possible - but as the mesh is 'constructed' it can, in general, achieve a prescribed density. Here, techniques such as the Delaunay procedure for triangles or tetrahedral (Cavendish *et al.* 1985), the modified octree scheme (Yerry and Shepard 1984), or the frontal schemes (Lo 1985) are typical. However, a more efficient process directly constructs triangles or tetrahedral of an approximately equilateral shape and of a 'size specified at nodes of a background mesh' which embrace the domain and interpolates these values to any point (Peraire *et al.* 1987). This sophisticated methodology, whose essentials are illustrated diagrammatically in Figure 6.1 for a two dimensional example, has proved to be extremely useful in adaptive mesh generation and is widely used throughout this research.

In adaptive computation, we invariably start from some crudely specified mesh fitting the domain and based on a mesh size h, prescribed as 'a nodal function', with the usual interpolation carried out on such a mesh. Adaptivity requirements predict new mesh sizes in the manner of equation (6.14) for each element. These values, 'interpolated to the nodes of the existing mesh', provide the background mesh necessary for further generation. It should be noted that if a completely new mesh is generated it is possible not only to refine but also to de-refine (or make the mesh locally coarser) if initially it was too fine. This of course ensures economy and also a more efficient progression towards an optimal mesh, where the efficiency of the error measures increases.

In this study, we have employed the mesh generator developed by Peraire *et al.* (1987) which is naturally suited to triangular refinement in two dimensions. This triangular element mesh generator requires a background mesh (Figure 6.1) and mesh size to generate a finer mesh. Usually, the coarse mesh of the last refinement stage can be used as the background mesh. The mesh size h at a node is used to determine the sizes of the regenerated elements around the node. The expected dimensions for elements to be generated within a background element can be obtained simply by linear interpolation. For instance, sizes for generated elements in triangle ΔABC in Figure 6.1 are determined by mesh sizes h_A, h_B and h_C via linear interpolation.

An error estimation and adaptive analysis procedure demonstrated above has been employed with numerical examples in the last section. The computation is first carried out on the initial mesh, then the error is estimated. If the prescribed accuracy is not satisfied, a new mesh is generated, the values of the unknowns are interpolated to the nodes of the new mesh and used as the initial condition for the next computation. Such a refinement process continues until the accuracy requirement is satisfied.

6.6. Mapping of variables

In this section, general aspects of the transfer operation for evolving finite element meshes are provided for the case of a typical elasto-plastic material whose behavior is described by a set of state and internal variables. Once a new mesh is generated, state and internal variables need to be mapped from the old finite element mesh to the new one. The mapping of state and internal variables between two finite element meshes is the essential part of any adaptive strategy that is employed in simulation of history-dependent material processes on evolving general unstructured meshes. The state variables consist of the nodal displacements and the internal variables include the Cauchy stress tensor, the strain tensor, the plastic strain tensor and a vector of internal variables. Several important aspects of the mapping process have to be addressed, i.e.,

i) consistency with the constitutive equations,
ii) requirement of equilibrium,
iii) compatibility of the history-dependent internal variables transfer with the displacement field on the new mesh,
iv) compatibility with evolving boundary conditions,
v) minimization of the numerical diffusion of the transferred state fields.

Let $\boldsymbol{u}_n^{\text{old}}, \boldsymbol{\varepsilon}_n^{\text{old}}, (\boldsymbol{\varepsilon}_n^p)^{\text{old}}, \boldsymbol{\sigma}_n^{\text{old}}, \boldsymbol{q}_n^{\text{old}}$ denote the values of displacement, strain tensor, plastic strain tensor, stress tensor and a vector of internal variables at time t_n for the mesh h. For simplicity of notation, we define a state array $\boldsymbol{\Lambda}_n^{\text{old}} = \left(\boldsymbol{u}_n^{\text{old}}, \boldsymbol{\varepsilon}_n^{\text{old}}, (\boldsymbol{\varepsilon}_n^p)^{\text{old}}, \boldsymbol{\sigma}_n^{\text{old}}, \boldsymbol{q}_n^{\text{old}}\right)$. Furthermore, assume that the estimated error of the solution

$\boldsymbol{\Lambda}_n^{\text{old}}$ respects the prescribed criteria, while these are violated by the solution $\boldsymbol{\Lambda}_{n+1}^{\text{old}}$. In this case, a new mesh $h+1$ is generated and a new solution $\boldsymbol{\Lambda}_{n+1}^{\text{new}}$ is computed by evaluating the plastic strain $(\boldsymbol{\varepsilon}_n^p)^{\text{new}}$ and the internal variables $\boldsymbol{q}_n^{\text{new}}$ for a new mesh $h+1$ at time step t_n. In this way, the state array $\widehat{\boldsymbol{\Lambda}}_n^{\text{new}} = \left((\boldsymbol{\varepsilon}_n^p)^{\text{new}}, \boldsymbol{q}_n^{\text{new}}\right)$ is constructed, where $\widehat{\boldsymbol{\Lambda}}$ is used to denote a reduced state array. It must be noted that the state array $\widehat{\boldsymbol{\Lambda}}$ characterizes the history of the material and provides sufficient information for computation of a new solution $\boldsymbol{\Lambda}_{n+1}^{\text{new}}$. Box 6.1 illustrates a typical transfer operation that includes both the mapping of the internal variables and mapping of the displacement field. In this figure, the mapping of internal variables is denoted by the transfer operator $\mathcal{T}_1$. In addition, a transfer operator is employed that transfers the state variables, i.e. displacement field, from the old to a new mesh. In the context of the backward Euler scheme where solution is sought at time step t_{n+1}, this transfer provides a trial solution. The mapping of state variables is denoted by the transfer operator $\mathcal{T}_2$.

The mapping process of displacement field is stated using the transfer operator $\mathcal{T}_2$ as

$$\boldsymbol{u}^{\text{old}}(\Omega) \rightarrow \boldsymbol{u}^{\text{new}}(\Omega) \tag{6.17}$$

where $\boldsymbol{u}^{\text{old}}$ and $\boldsymbol{u}^{\text{new}}$ denote the displacement fields given by the old and new finite element meshes, respectively, where both finite element models occupy the same domain Ω. The mapping process for the internal variables can be stated in a similar manner using the transfer operator $\mathcal{T}_1$ as

$$\boldsymbol{I}^{\text{old}}(\Omega) \rightarrow \boldsymbol{I}^{\text{new}}(\Omega) \tag{6.18}$$

where $\boldsymbol{I}^{\text{old}}(\Omega)$ and $\boldsymbol{I}^{\text{new}}(\Omega)$ are the list of internal variables of the old and new finite element meshes.

Before the solution variables can be transferred from the integration points of the old finite element mesh to the nodal points of the new finite element mesh, we must first determine which element in the old mesh contains the node n in the new finite element mesh. Since this task has to be performed for every nodal and integration point in the new finite element mesh, it is important that a reliable and computationally efficient procedure be used. In this study, we have employed the 'superconvergent patch recovery' SPR method to project the values from the old Gauss points to the old nodal points (Zienkiewicz and Zhu 1992). In this case, we first obtain the smoothed nodal point values for each of the variables in the list $\boldsymbol{I}^{\text{old}}$, then use the inverse mapping technique to transfer these variables to the nodal points of the new finite element mesh and finally the corresponding values at the new Gauss points can be calculated using interpolation functions in the new finite element mesh.

6.6.1. Mapping of internal variables - Transfer operator $\mathcal{T}_1$

Consider that at time step $t = t_{n+1}$ the error calculated for the mesh h does not satisfy the initially specified error tolerance. In addition, assume that the triangulation has already been performed based on an error estimation resulting in a new mesh denoted by $h+1$. The aim is to transfer the internal variables $\left((\boldsymbol{\varepsilon}_n^p)_{\text{G}}^{\text{old}}, (\boldsymbol{q}_n)_{\text{G}}^{\text{old}}\right)$ stored at the Gauss points of the old mesh h to the Gauss points of new mesh $h+1$. The transfer operator $\mathcal{T}_1$ between meshes h and $h+1$ can be defined as

$$\left((\boldsymbol{\varepsilon}_n^p)_{\text{G}}^{\text{new}}, (\boldsymbol{q}_n)_{\text{G}}^{\text{new}}\right) = \mathcal{T}_1\left[(\boldsymbol{\varepsilon}_n^p)_{\text{G}}^{\text{old}}, (\boldsymbol{q}_n)_{\text{G}}^{\text{old}}\right] \tag{6.19}$$

The variables $\left((\boldsymbol{\varepsilon}_n^p)_{\mathrm{G}}^{\mathrm{old}},(\boldsymbol{q}_n)_{\mathrm{G}}^{\mathrm{old}}\right)$ specified at Gauss points of the mesh h are transferred by the operator $\mathcal{T}_1$ to every point of the domain Ω, in order to specify the variables $\left((\boldsymbol{\varepsilon}_n^p)_{\mathrm{G}}^{\mathrm{new}},(\boldsymbol{q}_n)_{\mathrm{G}}^{\mathrm{new}}\right)$ at the Gauss points of the new mesh $h+1$. The operator $\mathcal{T}_1$ can be constructed in different ways;

i) A simple version of the transfer operator may be constructed by taking constant values over the area associated with every Gauss point.

ii) Another possibility is to construct a solution which is continuous, for instance by a least-squares method, or by a suitable projection of $(\boldsymbol{\varepsilon}_n^p)^{\mathrm{old}}$ which satisfies

$$\int_{\Omega} \boldsymbol{P}\left[\left((\boldsymbol{\varepsilon}_n^{p*})^{\mathrm{old}},(\boldsymbol{q}_n^*)^{\mathrm{old}}\right)-\left((\boldsymbol{\varepsilon}_n^p)^{\mathrm{old}},\boldsymbol{q}_n^{\mathrm{old}}\right)\right]dx=0 \tag{6.20}$$

where $\boldsymbol{P}$ is the so-called projection matrix.

In equation (6.20), the continuous plastic strain tensor $(\boldsymbol{\varepsilon}_n^{p*})^{\mathrm{old}}$ and the internal variable vector $(\boldsymbol{q}_n^*)^{\mathrm{old}}$ are obtained by projection the Gauss point components $(\boldsymbol{\varepsilon}_n^p)_{\mathrm{G}}^{\mathrm{old}}$ and $(\boldsymbol{q}_n)_{\mathrm{G}}^{\mathrm{old}}$ to the nodal points and evaluation the components $(\boldsymbol{\varepsilon}_n^p)_{\mathrm{N}}^{\mathrm{old}}$ and $(\boldsymbol{q}_n)_{\mathrm{N}}^{\mathrm{old}}$. The projection of the Gauss point components to the nodal points can be carried out using the shape function of the element. The nodal point averages of the projected values can be simply performed resulting in $(\boldsymbol{\varepsilon}_n^{p*})_{\mathrm{N}}^{\mathrm{old}}$ and $(\boldsymbol{q}_n^*)_{\mathrm{N}}^{\mathrm{old}}$. The nodal components of the plastic strain tensor $(\boldsymbol{\varepsilon}_n^{p*})_{\mathrm{N}}^{\mathrm{old}}$ and the internal variable vector $(\boldsymbol{q}_n^*)_{\mathrm{N}}^{\mathrm{old}}$ for the mesh h are then transferred to the nodes of the new mesh $h+1$ resulting in components $(\boldsymbol{\varepsilon}_n^{p*})_{\mathrm{N}}^{\mathrm{new}}$ and $(\boldsymbol{q}_n^*)_{\mathrm{N}}^{\mathrm{new}}$. The components of plastic strain tensor and internal variable vector at the Gauss points of the new mesh $h+1$, i.e. $(\boldsymbol{\varepsilon}_n^p)_{\mathrm{G}}^{\mathrm{new}}$ and $(\boldsymbol{q}_n)_{\mathrm{G}}^{\mathrm{new}}$, are finally obtained by using the interpolation of the shape functions of the new finite elements.

In the above procedure, the local coordinates are used to interpolate the variables from the nodes of the mesh h to the nodes of the mesh $h+1$. In the case of the three-noded triangular element, the local coordinates are obtained by resolving a linear system of algebraic equations. In other cases, the problem becomes nonlinear and the Newton-Raphson iterative scheme can be used to obtain the local coordinates, as described by Lee and Bathe (1994). An alternative solution may be to split the higher-order finite elements into a set of simple three noded triangles, and employ a linear interpolation over each triangle. The described transfer operation $\mathcal{T}_1$ is schematically presented in Box 6.2. For meshes composed of constant strain triangles, the general scheme of Box 6.2 can be split into three distinct steps, as demonstrated in Figure 6.2, as follows (Peric *et al.* 1996);

Step 1. The Gauss point components $\left((\boldsymbol{\varepsilon}_n^p)_{\mathrm{G}}^{\mathrm{old}},(\boldsymbol{q}_n)_{\mathrm{G}}^{\mathrm{old}}\right)$ are projected to the nodes $\left((\boldsymbol{\varepsilon}_n^p)_{\mathrm{N}}^{\mathrm{old}},(\boldsymbol{q}_n)_{\mathrm{N}}^{\mathrm{old}}\right)$. For the constant strain triangle we have trivially $\left((\boldsymbol{\varepsilon}_n^p)_{\mathrm{N}}^{\mathrm{old}},(\boldsymbol{q}_n)_{\mathrm{N}}^{\mathrm{old}}\right)=\left((\boldsymbol{\varepsilon}_n^p)_{\mathrm{G}}^{\mathrm{old}},(\boldsymbol{q}_n)_{\mathrm{G}}^{\mathrm{old}}\right)$. In general, the finite element shape functions can be used in projecting the Gauss points values to nodes. The nodal point averages are then performed resulting in $\left((\boldsymbol{\varepsilon}_n^{p*})_{\mathrm{N}}^{\mathrm{old}},(\boldsymbol{q}_n^*)_{\mathrm{N}}^{\mathrm{old}}\right)$.

Step 2. In this step, the nodal components of internal variables $\left((\boldsymbol{\varepsilon}_n^{p*})_{\mathrm{N}}^{\mathrm{old}},(\boldsymbol{q}_n^*)_{\mathrm{N}}^{\mathrm{old}}\right)$ will be transferred from the old mesh h to the new mesh $h+1$ resulting in $\left((\boldsymbol{\varepsilon}_n^{p*})_{\mathrm{N}}^{\mathrm{new}},(\boldsymbol{q}_n^*)_{\mathrm{N}}^{\mathrm{new}}\right)$. This step of the transfer operation is the most complex one and can be carried out into three stages; the construction of the background triangular mesh, the evaluation of the nodal coordinates and finally, the mapping of the nodal values. In order to construct the background triangular mesh for each node A of the new mesh $h+1$ with known coordinates $(\boldsymbol{x}_n)_A^{\text{new}}$, the background element is first determined from the old triangular element $(\Omega^e)^{\text{old}}$, in which $(\boldsymbol{x}_n)_A^{\text{new}} \in (\Omega^e)^{\text{old}}$. The evaluation of the local coordinates of node A within the background element $(\Omega^e)^{\text{old}}$, which corresponds to its position in new mesh $h+1$ can be then computed by

$$(\boldsymbol{x}_n)_A^{\text{new}} = (\mathbf{N}^T)^{\text{old}} (\boldsymbol{x}_n)^{\text{old}} \tag{6.21}$$

where $\mathbf{N}^{\text{old}}$ represents the shape functions of old triangular element $(\Omega^e)^{\text{old}}$. For the background mesh h composed of constant strain triangles, the values of the coordinates for each node A of the new mesh $h+1$ is therefore obtained by solving the algebraic system of (6.21). Finally, the mapping of the nodal values of internal variables $(\widehat{\boldsymbol{\Lambda}}_n)_B^{\text{old}} = \left((\boldsymbol{\varepsilon}_n^{p*})_B^{\text{old}}, (\boldsymbol{q}_n)_B^{\text{old}}\right)$ from the node B of the old mesh h to the nodes A of the new mesh $h+1$ is calculated by using the shape functions of old triangular element $\mathbf{N}^{\text{old}}$. This mapping is expressed as

$$(\widehat{\boldsymbol{\Lambda}}_n)_A^{\text{new}} = (\mathbf{N}^T)^{\text{old}} (\widehat{\boldsymbol{\Lambda}}_n)_B^{\text{old}} \tag{6.22}$$

Step 3. The internal variables at the Gauss points of the new mesh $\left((\boldsymbol{\varepsilon}_n^{p*})_{\text{G}}^{\text{new}}, (\boldsymbol{q}_n^*)_{\text{G}}^{\text{new}}\right)$ can be easily obtained by employing the shape functions of new triangular element $(\Omega^e)^{\text{new}}$, i.e.

$$(\widehat{\boldsymbol{\Lambda}}_n)_{\text{G}}^{\text{new}} = (\mathbf{N}^T)^{\text{new}} (\widehat{\boldsymbol{\Lambda}}_n)_A^{\text{new}} \tag{6.23}$$

In the case of a mesh composed of quadrilateral or higher order elements the background mesh required in the first stage may be constructed by simply subdividing the elements into a local mesh of constant strain triangles. Lee and Bathe (1994) presented an adaptive strategy based on quadrilateral element meshes that can be composed of higher-order elements. The evaluation of local coordinates in element $(\Omega^e)^{\text{old}}$ of the old background mesh h corresponding to each node of the new mesh $h+1$, described in second stage, for quadrilaterals and higher-order elements results in a nonlinear problem. Lee and Bathe (1994) resolved this problem by employing the so-called 'inverse isoparametric mapping technique' which is based on a Newton-Raphson type iterative procedure.

6.6.2. Mapping of displacement field-Transfer operator $\mathcal{T}_2$

In the context of the backward Euler scheme with a strain driven format and Newton-Raphson iterative procedure, the transferred displacement field on a new mesh can be used to provide an initial (trial solution) for the displacements in the first iteration of the Newton-Raphson scheme. In this case, the trial displacement field can be calculated by the transfer of the displacement field $\boldsymbol{u}_{n+1}^{\text{old}}$ obtained from mesh h at time step t_{n+1} as

$$(\boldsymbol{u}_{n+1}^{\text{new}})^{\text{trial}} = \mathcal{T}_2 \left[\boldsymbol{u}_{n+1}^{\text{old}}\right] \tag{6.24}$$

The transfer operator $\mathcal{T}_2$ can be easily constructed by using the nodal values of displacements and finite element shape functions. For convenience, a schematic diagram of the transfer operator $\mathcal{T}_2$ is presented in Box 6.2. As the displacement field over the new mesh $h+1$ is prescribed by the nodal values $\boldsymbol{u}_n^{\text{new}}$ and the shape functions of new element $\mathbf{N}^{\text{new}}$, defined for each element $(\Omega^e)^{\text{new}}$ of the new mesh, the task of transferring displacements (i.e. transfer operator $\mathcal{T}_2$) can be performed by applying the third stage of step

(2), i.e., mapping of the nodal values, in the procedure describing the transfer operator $\mathcal{T}_1$. Applying the shape functions $\mathbf{N}^{\text{old}}$ of the element $(\Omega^e)^{\text{old}}$, the displacements $(\boldsymbol{u}_n)_B^{\text{old}}$ are mapped from the node B of the old mesh h to the node A of new mesh $h+1$. This mapping is expressed as

$$(\boldsymbol{u}_{n+1})_A^{\text{new}} = \mathbf{N}^{\text{old}} (\boldsymbol{u}_n)_B^{\text{old}} \tag{6.25}$$

It must be noted that the same transfer operator $\mathcal{T}_2$ can be used to transfer the displacements and internal variables from the nodal points of the old mesh to the nodal points of a new mesh. However, the numerical diffusion is more pronounced for the described technique than with a possible technique that would define the transfer operator $\mathcal{T}_1$ by a transfer directly from the Gauss points of the mesh h to the Gauss points of the mesh $h+1$.

In the procedure described above, the internal variables $\left((\boldsymbol{\varepsilon}_n^p)^{\text{old}}, (\boldsymbol{q}_n)^{\text{old}}\right)$ are transferred from the old mesh h to the new mesh $h+1$, while stresses and strains $\left((\boldsymbol{\sigma}_n)^{\text{old}}, (\boldsymbol{\varepsilon}_n)^{\text{old}}\right)$ are not transferred. It must be noted that the transfer of the complete state array $\boldsymbol{\Lambda}_n^{\text{old}} = \left(\boldsymbol{u}_n^{\text{old}}, \boldsymbol{\varepsilon}_n^{\text{old}}, (\boldsymbol{\varepsilon}_n^p)^{\text{old}}, \boldsymbol{\sigma}_n^{\text{old}}, \boldsymbol{q}_n^{\text{old}}\right)$ would result in the data that may not be self-consistent. For instance, the yield condition $f\left((\boldsymbol{\sigma}_n)^{\text{new}}, (\boldsymbol{q}_n)^{\text{new}}\right) = 0$ may not be satisfied. In fact, the set of internal variables $\left((\boldsymbol{\varepsilon}_n^p)^{\text{new}}, (\boldsymbol{q}_n)^{\text{new}}\right)$ prescribed at each Gauss point characterizes in full the history of the material on the new mesh $h+1$. These internal variables provide sufficient information for computation of a new solution $\boldsymbol{\Lambda}_n^{\text{new}} = \left(\boldsymbol{u}_n^{\text{new}}, \boldsymbol{\varepsilon}_n^{\text{new}}, (\boldsymbol{\varepsilon}_n^p)^{\text{new}}, \boldsymbol{\sigma}_n^{\text{new}}, \boldsymbol{q}_n^{\text{new}}\right)$.

The transfer operation procedure described above is applicable to problems involving transfer between general unstructured meshes. This is of particular importance since the aim is to apply this procedure to the modeling of complex large deformation processes, where a complete regeneration of the finite element meshes has often proved essential for a successful simulation.

6.6.3. Superconvergent patch recovery technique

The concept of superconvergence is that at some point the approximate solutions are more accurate, or in other words, the rate of convergence at those points is higher than that in other points. The superconvergence was observed by Zienkiewicz and Zhu (1992) for displacement derivatives or stresses evaluated at the Gauss quadrature points of uniform meshes. It was shown that the gradients of the finite element approximation exhibit super convergence at the Gauss integration points of curved isoparametric elements. This concept can be applied for constructing a continuous stress function which plays an important role in data transfer operator, described in preceding sections. In this section, a brief review of superconvergent patch recovery (SPR) technique that uses local discrete and continuous least square smoothing procedures is presented. Attempts are made to recover the nodal values of stresses from the Gauss points and then to evaluate the recovered values of stresses $\boldsymbol{\sigma}^*$ throughout the element by $\boldsymbol{\sigma}^* = \mathbf{N}_u \overline{\boldsymbol{\sigma}}^*$ in the same manner as displacements.

The objective of recovery of finite element solution is to obtain the nodal values of stresses such that the smoothed continues field defined by the shape functions and nodal values is more accurate than that the finite element solution. A procedure for utilizing the Gauss quadrature values is based on the smoothing of such values by a polynomial of order p in which the number of sample points can be taken as greater than the number of parameters in the polynomial. In this case, if we accept the superconvergence of $\hat{\boldsymbol{\sigma}}$ at certain points in each element, the computed values of $\boldsymbol{\sigma}^*$ will have the superconvergent accuracy at all points within the element. Thus, the recovered solution can be obtained for each component of σ_i^* by

$$\sigma_i^* = \mathbf{P}\,\mathbf{a} \tag{6.26}$$

where $\mathbf{P}$ contains the appropriate polynomial terms and $\mathbf{a}$ is a set of unknown parameters. For strain-constant triangular elements, we have

$$\mathbf{P} = \{1,\ x,\ y\}\ , \qquad \mathbf{a} = \{a_1,\ a_2,\ a_3\}^T \tag{6.27}$$

The determination of the unknown parameters of relation (6.26) can be made by performing a least square fit to the values of superconvergent, or sampling points. For this purpose, the following function must be minimized for an element with total sampling points of n

$$\begin{aligned} F(\mathbf{a}) &= \sum_{k=1}^{n}\left[\sigma_i^*(x_k, y_k) - \hat{\sigma}_i(x_k, y_k)\right]^2 \\ &= \sum_{k=1}^{n}\left[\mathbf{P}(x_k, y_k)\,\mathbf{a} - \hat{\sigma}_i(x_k, y_k)\right]^2 \end{aligned} \tag{6.28}$$

where (x_k, y_k) denotes the coordinates of superconvergent points. The minimization of function $F(\mathbf{a})$ implies that

$$\left(\sum_{k=1}^{n}\mathbf{P}^T(x_k, y_k)\,\mathbf{P}(x_k, y_k)\right)\mathbf{a} = \sum_{k=1}^{n}\left[\mathbf{P}^T(x_k, y_k)\,\hat{\sigma}_i(x_k, y_k)\right] \tag{6.29}$$

The parameters $\mathbf{a}$ can be obtained by solving the system equation of (6.29) as

$$\mathbf{a} = \left(\sum_{k=1}^{n}\mathbf{P}_k^T\,\mathbf{P}_k\right)^{-1}\sum_{k=1}^{n}\left[\mathbf{P}_k^T\,\hat{\sigma}_i(x_k, y_k)\right] \tag{6.30}$$

By computing the values of $\boldsymbol{\sigma}^*$, the superconvergent values of $\overline{\boldsymbol{\sigma}}^*$ can be determined at all nodal points. As some nodal points belong to more than one element, the average values can be used as the best estimation. The SPR procedure described above has proved to be a very powerful tool leading to superconvergent results on regular meshes and mush improved results on irregular meshes (Zienkiewicz and Zhu 1992). It has been shown numerically that it produces superconvergent recovery even for triangular elements which do not have superconvergent points within the element.

6.7. Error estimates and adaptive time stepping

In addition to the spatial discretization error described in the preceding sections, there exists in the numerical analysis of dynamic problems another type of error, i.e. the time domain discretization error. This type of error, in a non-linear analysis, is mainly dependent on the magnitude of the increment step. In order to control the time discretization error, we must firstly find some means of estimating the error and then adjust the step size accordingly. The main aim of the adaptive time stepping procedures is to obtain the largest possible step size while maintaining a prescribed accuracy. Such automatic adaptations of the step size during the solution process are of great importance. In this section, a local error estimator and adaptive time stepping strategy based on the Newmark scheme will be introduced for the dynamic analysis of powder forming processes.

An unconditionally stable direct solution procedure, based on an implicit scheme, is described for the solution of the standard dynamic equation in Chapter 2. For these implicit schemes, the step size is governed by accuracy considerations. When single step algorithms are employed, an 'error per step' method is often used for the step size control (Thomas *et al.* 1988) by selecting the step size in such a manner that the local error of each step is roughly equal to a prescribed tolerance. The local error can be estimated either by comparing the results when different step sizes are used (Gear 1971), or by comparing the results when two methods of different orders are used (Thomas *et al.* 1988).

As described in Chapter 2, after the spatial discretization, the standard displacement finite element formulation of dynamic problems has been obtained by the following second order differential equation

$$\mathbf{M}\ddot{\bar{\mathbf{u}}} + \mathbf{K}_{\mathrm{T}}\bar{\mathbf{u}} = \mathbf{f}_u \qquad (6.31)$$

For time domain discretization using the Newmark scheme, it is assumed that the differential equation is to be satisfied at each discrete time station. Typically at time station $t_{n+1} = t_n + \Delta t$ it is required that

$$\mathbf{M}_{n+1}\ddot{\bar{\mathbf{u}}}_{n+1}^{NM} + (\mathbf{K}_{\mathrm{T}})_{n+1}\bar{\mathbf{u}}_{n+1}^{NM} = (\mathbf{f}_u)_{n+1} \qquad (6.32)$$

where $\ddot{\bar{\mathbf{u}}}_{n+1}^{NM}$ represents the Newmark scheme solution. The link between the successive values at t_{n+1} and t_n is established by the following truncated series expansions

$$\ddot{\bar{\mathbf{u}}}_{n+1}^{NM} = \ddot{\bar{\mathbf{u}}}_n + \Delta t\,\Delta\dddot{\bar{\mathbf{u}}}_n \qquad (6.33)$$

$$\dot{\bar{\mathbf{u}}}_{n+1}^{NM} = \dot{\bar{\mathbf{u}}}_n + \Delta t\,\ddot{\bar{\mathbf{u}}}_n + \gamma\Delta t\,\Delta\ddot{\bar{\mathbf{u}}}_n \qquad (6.34)$$

$$\bar{\mathbf{u}}_{n+1}^{NM} = \bar{\mathbf{u}}_n + \Delta t\,\dot{\bar{\mathbf{u}}}_n + \tfrac{1}{2}\Delta t^2\ddot{\bar{\mathbf{u}}}_n + \beta\,\Delta t^2\Delta\ddot{\bar{\mathbf{u}}}_n \qquad (6.35)$$

in which $\Delta\ddot{\bar{\mathbf{u}}}_n$ is as yet an undetermined quantity. The parameters γ and β are related to the accuracy and stability of the scheme and for unconditional stability it is required that $\gamma \geq \frac{1}{2}$ and $\beta \geq \frac{1}{4}$ (Zienkiewicz and Taylor 1989).

6.7.1. Time discretization error

In order to control the time discretization error, it is necessary to find some means of measuring the error. Since the global error is, in general, very difficult to obtain, the local truncation error is usually estimated. To estimate the local error of a time integration scheme in one time step, it is assumed that exact values are used at the beginning of the step t_n. By expanding the exact values of $\bar{\mathbf{u}}_{n+1}^{EX}$ in the Taylor series, we have

$$\bar{\mathbf{u}}_{n+1}^{EX} = \bar{\mathbf{u}}_n + \Delta t\,\dot{\bar{\mathbf{u}}}_n + \tfrac{1}{2}\Delta t^2\ddot{\bar{\mathbf{u}}}_n + \tfrac{1}{6}\Delta t^3\Delta\ddot{\bar{\mathbf{u}}}_n + O(\Delta t^4) \qquad (6.36)$$

On comparing the Newmark scheme solution in equation (6.35) with the exact value, as expressed in (6.36), we will arrive at

$$\overline{\mathbf{u}}_{n+1}^{NM} - \overline{\mathbf{u}}_{n+1}^{EX} = \Delta t^3 (\beta - \frac{1}{6}) \dddot{\mathbf{u}}_n + O(\Delta t^4) \tag{6.37}$$

From the Newmark scheme, the third order derivative $\dddot{\mathbf{u}}_n$ is not available. However, it is possible to find a simple approximation by using a similar procedure demonstrated above for $\overline{\mathbf{u}}_{n+1}^{NM}$ as

$$\ddot{\mathbf{u}}_{n+1}^{NM} = \ddot{\mathbf{u}}_n + \Delta t \, \dddot{\mathbf{u}}_n + O(\Delta t^2) \tag{6.38}$$

Substituting $\dddot{\mathbf{u}}_n$ from equation (6.38) into equation (6.37), we have

$$\overline{\mathbf{u}}_{n+1}^{NM} - \overline{\mathbf{u}}_{n+1}^{EX} = \Delta t^2 (\beta - \frac{1}{6})(\ddot{\mathbf{u}}_{n+1}^{NM} - \ddot{\mathbf{u}}_n) + O(\Delta t^4) \tag{6.39}$$

Therefore, the local errors may be estimated by

$$\mathbf{e} = \Delta t^2 (\beta - \frac{1}{6})(\ddot{\mathbf{u}}_{n+1}^{NM} - \ddot{\mathbf{u}}_n) \tag{6.40}$$

In the above the local error estimator $\mathbf{e}$ is asymptotically correct to the third order of the step size. At the end of each time step, such a posteriori error estimate can be obtained at almost no computational cost. It has to be pointed out that the error estimator (6.40) is invalid when $\beta = \frac{1}{6}$. In this situation, higher order error estimators have to be found. This can be achieved by following the same procedure as given above in which the value of β is now out of the range for unconditional stability.

6.7.2. Adaptive time stepping

A pointwise definition of errors, as given in equation (6.40) is generally difficult to use in the criterion for step size control, and various normalized measures are more conveniently adopted. One of the most common of such measures, which has been also used for spatial discretization in Section 6.3, is the L_2 norm. The norm of the local error approximation of (6.40) is

$$\eta = \|\mathbf{e}\| = \Delta t^2 (\beta - \frac{1}{6}) \left\| \ddot{\mathbf{u}}_{n+1}^{NM} - \ddot{\mathbf{u}}_n \right\| \tag{6.41}$$

In order to control the step size, an 'error per step' method can be employed in which the step size will be chosen in such a way that the local error of each step is to be roughly equal to a prescribed tolerance. In this case, if the tolerance is given as $\overline{\eta}_t$, we can specify an 'upper limit' and a 'lower limit' for the local error as $\gamma_1 \overline{\eta}_t$ and $\gamma_2 \overline{\eta}$, where the parameters γ_1 and γ_2 are chosen in the range of $0.0 \le \gamma_1 \le 1.0$ and $\gamma_2 \ge 1.0$ (Xie *et al.* 1988). Whenever the error norm is between the lower limit and the upper limit, i.e. $\gamma_1 \overline{\eta}_t \le \eta \le \gamma_2 \overline{\eta}$, the solution is accepted and the time integration proceeds to the next step without changing the time step size.

However, when the error norm exceeds the upper limit the step size needs to be reduced. In such circumstances, to insure the accuracy of the numerical solution, the simplest way is to return to the beginning of the current step and do the calculations again with the decreased new step size. Since the rate of convergence of the local error in equation (6.41) is $O(\Delta t^3)$, then the new step size Δt_{new} may be predicted as

$$\Delta t_{new} = \sqrt[3]{\frac{\overline{\eta}_t}{\eta}}\,\Delta t_{old} \tag{6.42}$$

where Δt_{old} is the original step size of the current step. The new step size (6.42) may also be used to increase the step size, when the error norm is smaller than the lower limit. However, it is undesirable to change the step size too frequently, as a step size change usually requires an inversion of a new overall matrix, which is very expensive for implicit schemes.

6.8. Numerical simulation results

A standard feature encountered in the FE simulation of powder forming operations is that the optimal mesh configuration changes continually throughout the forming operation. As described before the application of the finite element method to the metal powder forming process faces difficulties in avoiding severe distorted meshes caused by a conventional updating scheme. The error estimation and adaptive remeshing scheme are therefore provided as a tool to reorganize the mesh and to control the error automatically, which is very important for an industrial code. In that implementation, some additional controls to those previously discussed should be considered,

- The target error $\overline{\eta}_{aim}$ and the acceptable error $\overline{\eta}$ should be prescribed. The target error is for adaptive remeshing and is the error that the new mesh hopes to achieve. The acceptable error is a threshold error, defining a criterion where remeshing is to be concerned. Both errors should be properly chosen and adjusted according to engineering requirements.
- The maximum and minimum sizes, $\overline{h}_{\max}$ and $\overline{h}_{\min}$, of the elements should be controlled during the remeshing process so that a proper mesh can be obtained.
- Whenever a Jacobian determinant fails, which means that some element is severely distorted, the remeshing should be started.

In order to illustrate the applicability of the adaptive FEM strategy described in the preceding sections, a non-linear adaptive analysis has been employed for the non-linear behavior of powder forming problems of a rotational flanged, a shaped tip and a multi-level surface component. The analysis commences with an uniform distribution of elements. The non-linear solution is carried out for various loading steps (or prescribed displacements) until the estimated error exceeds a prescribed value, or the sign of the Jacobian determinant becomes negative. At this time a new mesh is generated using an appropriate refinement criterion. The information at the end of the previous step is transferred to the integration points of the new mesh. Transferring these values from the old integration points to the new ones is carried out by using nodal points as reference points to store the information. For this purpose, the history dependent values in the old mesh are first transferred to the nodes and then, the values at the new nodal points in the new mesh are computed by a simple interpolation of the old nodal values using shape functions. Finally, the non-linear procedure is performed on the new mesh starting from the beginning of the new step.

The iron powder properties chosen in the analysis for all examples are $E = 40.0$ N/mm^2, $\upsilon = 0.35$, $c = 5.0$ N/mm^2 and $\varphi = 30^\circ$. The die wall friction is simulated using a set of interface elements with the Coulomb friction coefficient $\mu = 0.08$. The initial relative density is $\overline{\rho}_0 = 0.4$. The plastic hardening coefficient χ and the initial hydrostatic stress of compaction σ_{c0} are assumed to have values of 0.16 and 0.2 N/mm^2, respectively.

6.8.1. An elasto-plastic L–shape

The first numerical example is of an elasto-plastic solid mechanics problem chosen to illustrate the concepts of adaptive mesh refinement together with the data transfer procedure in plasticity problems. An elasto-plastic L–shape with its geometry and material properties presented in Figure 6.3 is numerically analyzed. Material parameters pertinent to the von-Mises yield criterion. Numerical simulation is performed using three-noded triangular elements in 100 increments. The initial mesh consists of 24 elements, while the number of elements in subsequent meshes varies from 127 to 1711 elements. Adaptive remeshing was performed at three different stages, as shown in Figure 6.4. The adaptive strategy was employed based on the calculation of energy norm for each increment of plastic analysis using SPR technique. A full Newton-Raphson method is employed in computation. The convergence of finite element solution is established with a tolerance of 10^{-3}.

In Figure 6.5, the stress contours in *x*–direction before and after operation of data transfer are presented at load steps of 80 and 90 between the old and new finite element meshes. The similarities between both contours at different steps are remarkable. Finally, the load versus displacement curves are presented in Figure 6.6 for various adapted meshes. It can be observed from this figure that the load-displacement curve obtained by adaptive procedure shows a good performance of data transfer operators and is in good agreement with different adapted meshes.

6.8.2. A rotational flanged component

The next example is of a simple multi-level component which is modeled by an axisymmetric representation as illustrated in Figure 5.25. As described in Chapter 5, a single top punch motion (Figure 5.26), or a single bottom punch motion (Figure 5.30) results in the severe distortion of the finite element mesh which can cause large approximation errors, or make the Jacobian determinant negative, and make further analysis impossible, thus prohibiting further advance of the punch. In this case, a double-ended compaction process (Figure 5.34) was applied to avoid the mesh distortion using a fixed mesh analysis. Since the crucial numerical problem in this more practical geometry is associated with the distorted mesh, an adaptive FEM analysis has been carried out during the complex movements of the top and bottom punches. The whole compaction process is shown in Figure 6.7 with a top punch movement of 3.44 mm to compact the upper part of the section at the first stage, then a bottom punch movement of 7.70 mm to compact the lower part of the section at the second stage, and finally a further top punch movement of 2.62 mm in the third stage of compaction.

The computations begins with an initial uniform mesh as given in Figure 6.8. The first adaptive mesh refinement is carried out on the deformed mesh obtained by the first top punch motion for a specified target error $\overline{\eta}_{aim} = \%\,4.0$ and a new mesh is generated. After changing the mesh, all state variables are transferred from the old mesh to the integration points of the new mesh by using nodal points as the reference points for storing the information. The computation is continued on the new mesh starting from the beginning of the step. A second adaptive analysis is employed after the bottom punch motion takes place as given in Figure 6.8 for a specified target error $\overline{\eta}_{aim} = \%\,2.0$. The analysis is carried out by transferring the information from the old integration points to the new ones in the new mesh. Finally, a further top punch movement is applied to obtain a fairly uniform density field.

The variation of the top and bottom punch forces at different displacements are plotted in Figure 6.9(b). These results can be compared with the predicted compaction forces for the double-ended compaction process as given in Figures 6.9(a). The distribution of relative density on this flanged component is shown in Figure 6.10 for different powder movements as $3D$ plots where the third direction represents the relative density. The results clearly indicate that the algorithm makes it possible to simulate the powder forming problems efficiently and automatically.

6.8.3. An axisymmetric shaped tip component

The second compaction example is that of an engineering tip component with the curved powder-die border as illustrated in Figure 5.43. The initial relative density for this tip component is $\overline{\rho}_0 = 0.2$. The whole compaction process involves a bottom punch movement of 4.0 mm at the first stage, then a top punch movement of 4.0 mm at the second stage, and a further bottom punch movement of 3.0 mm in the third stage of compaction, as shown in Figure 6.11.

The $3D$ distributions of relative densities for this tip component are shown in Figure 6.12 for different powder movements. The computations are started with an initial uniform mesh as given in Figure 6.13. The first adaptive analysis is employed after the first bottom punch motion for a specified target error $\overline{\eta}_{aim} = \%\, 4.0$ and followed by transferring the information from the old integration points to the new ones in the new mesh. The analysis is continued with a second adaptive mesh refinement after the top punch motion takes place as given in Figure 6.13 for a specified target error $\overline{\eta}_{aim} = \%\, 3.0$. After transferring all the state variables from the old mesh to the integration points of the new mesh, a second bottom punch movement is applied to obtain the final stages of compaction.

6.8.4. A modeled multi-level component

The last example chosen demonstrates the performance of the present formulation for the complicated die geometry of a multi-level component as shown in Figure 6.14. The simulation of an adaptive FEM analysis for this complex shape has been done by a top punch movement of 5.0 mm, then a bottom punch movement of 8.0 mm, and a second top punch movement of 3.0 mm as illustrated in Figure 6.15.

As in the last two examples, the computations begin with an initial uniform mesh. The first adaptive remeshing is carried out on the deformed mesh obtained for the first top punch motion for a specified target error of $\overline{\eta}_{aim} = \%\, 4.0$. A second adaptive analysis is employed after the bottom punch motion for a specified target error of $\overline{\eta}_{aim} = \%\, 2.0$. The whole compaction process of this multi-level component is shown in Figure 6.16. In Figures 6.17 and 6.18, the displacement and relative density contours at different stages of compaction are plotted. It can be seen that the proposed adaptive finite element approach is capable of simulating metal powder compaction processes in an efficient and accurate manner.

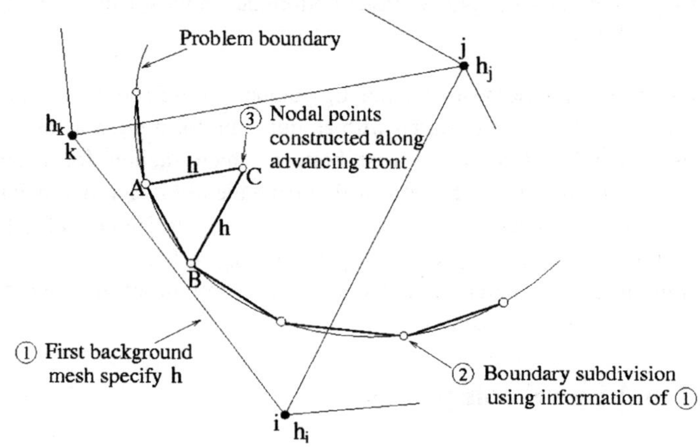

Figure 6.1. Principles of constructive mesh generation of progressive front technique

Step 1

$(\widehat{\boldsymbol{\Lambda}}_n)_\mathrm{G}^\mathrm{old} \rightarrow (\widehat{\boldsymbol{\Lambda}}_n)_\mathrm{N}^\mathrm{old}$

Step 2

$(\widehat{\boldsymbol{\Lambda}}_n)_\mathrm{N}^\mathrm{old} \rightarrow (\widehat{\boldsymbol{\Lambda}}_n)_\mathrm{N}^\mathrm{new}$

Step 3

$(\widehat{\boldsymbol{\Lambda}}_n)_\mathrm{N}^\mathrm{new} \rightarrow (\widehat{\boldsymbol{\Lambda}}_n)_\mathrm{G}^\mathrm{new}$

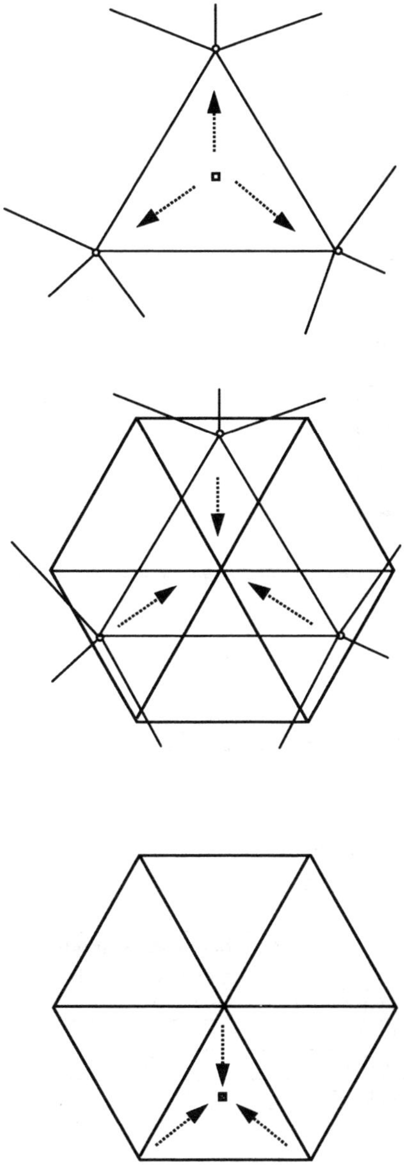

Figure 6.2. A three-steps procedure illustrating the implementation of the transfer operator $\left((\boldsymbol{\varepsilon}_n^p)_\mathrm{G}^\mathrm{new},(\boldsymbol{q}_n)_\mathrm{G}^\mathrm{new}\right)=\mathcal{T}_1\left[(\boldsymbol{\varepsilon}_n^p)_\mathrm{G}^\mathrm{old},(\boldsymbol{q}_n)_\mathrm{G}^\mathrm{old}\right]$ for the finite element meshes composed of constant triangles (Peric *et al.* 1996)

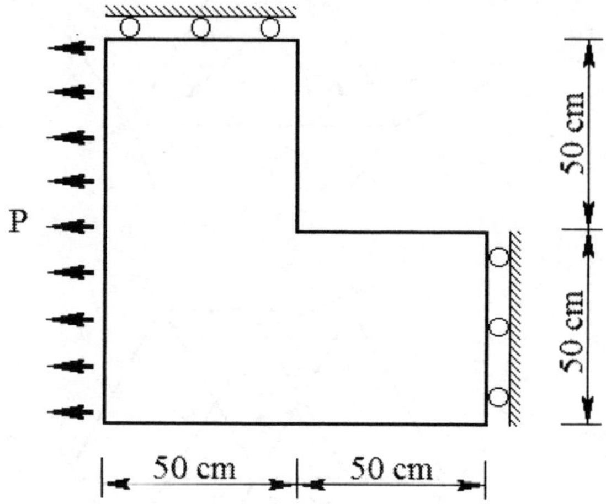

Material Parameters

$E = 1.0 \times 10^4$ Kg/mm^2

$\upsilon = 0.3$

$\sigma_y = 24$ Kg/mm^2

Total number of steps = 100

	Number of steps at which mesh generated	P (Kg)
Mesh (a)	1	
Mesh (b)	70	14
Mesh (c)	80	16
Mesh (d)	90	18

Figure 6.3. Elasto-plastic L–shape; Geometry, material properties and table of loading

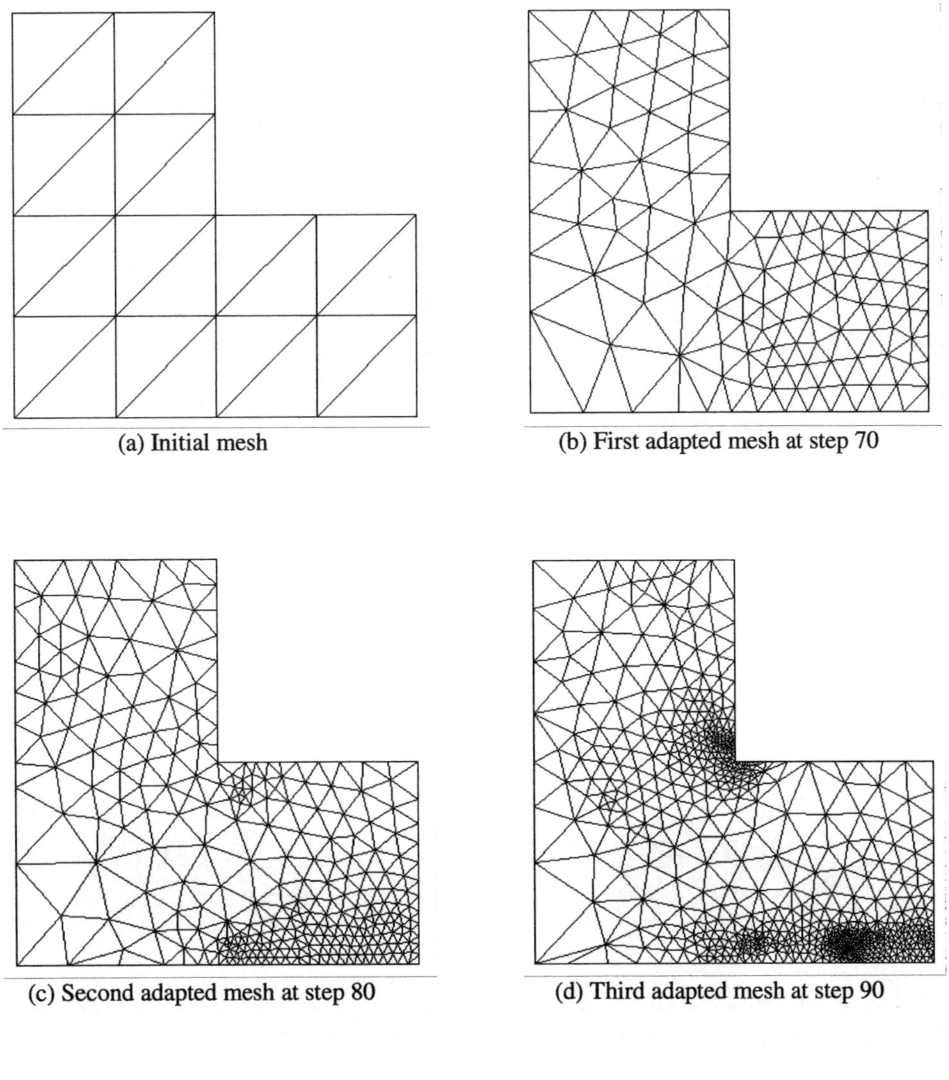

Figure 6.4. Adaptive remeshing for an elasto-plastic L–shape; a) Initial mesh, b) First adapted mesh at step 70, c) Second adapted mesh at step 80, d) Third adapted mesh at step 90

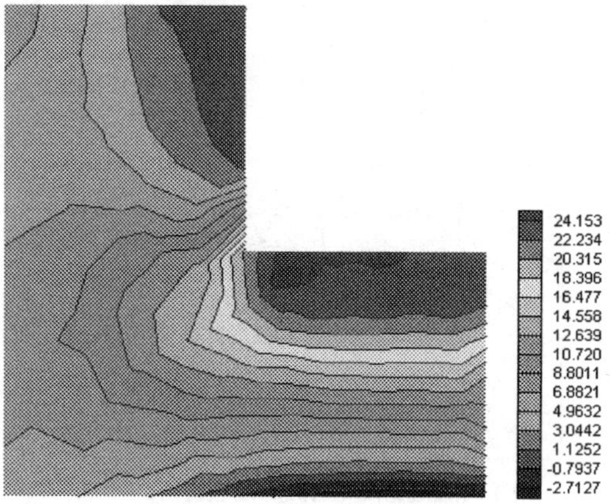

Mesh (b) at step 80 before data transfer

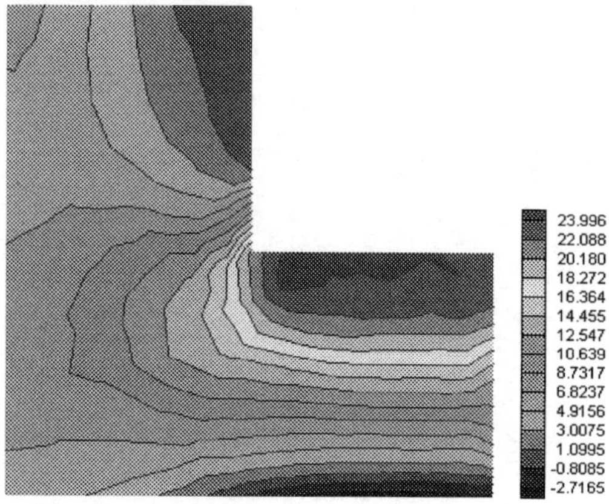

Mesh (c) at step 80 after data transfer

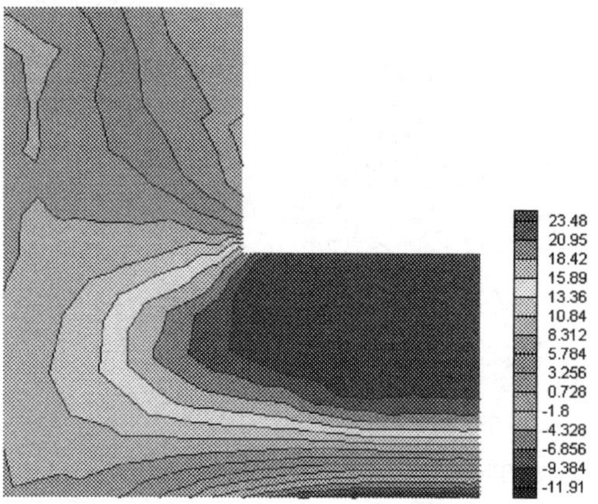

Mesh (c) at step 90 before data transfer

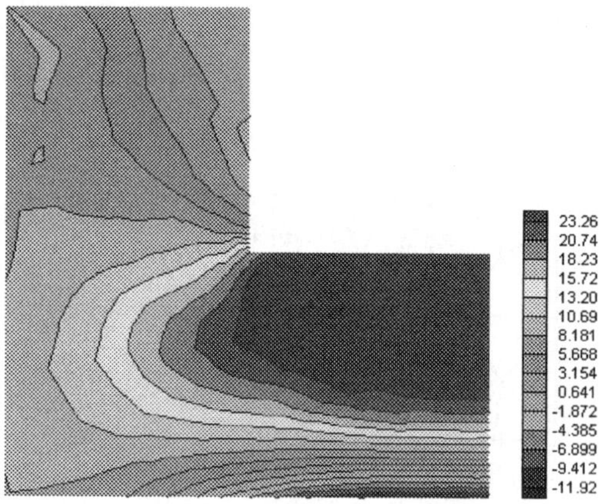

Mesh (d) at step 90 after data transfer

Figure 6.5. Elasto-plastic L–shape; Stress contours in x–direction before and after data transfer operation at load steps of 80 and 90 between the old and new meshes

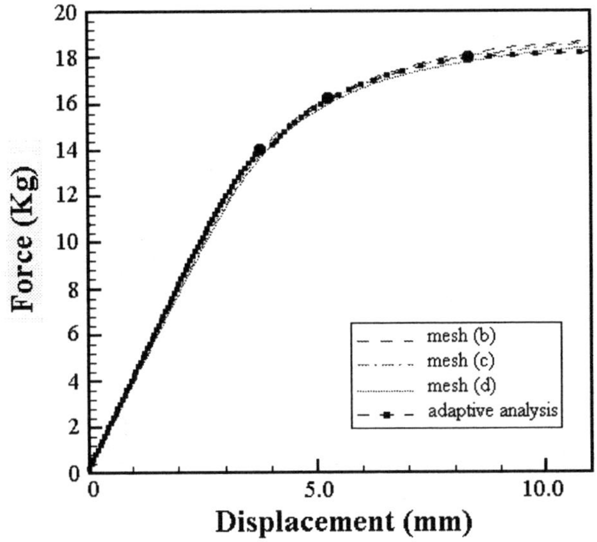

Figure 6.6. Elasto-plastic L–shape; The load versus displacement curves for initial and adapted meshes; ● The places at which the adaptive remeshing and data transfer were performed

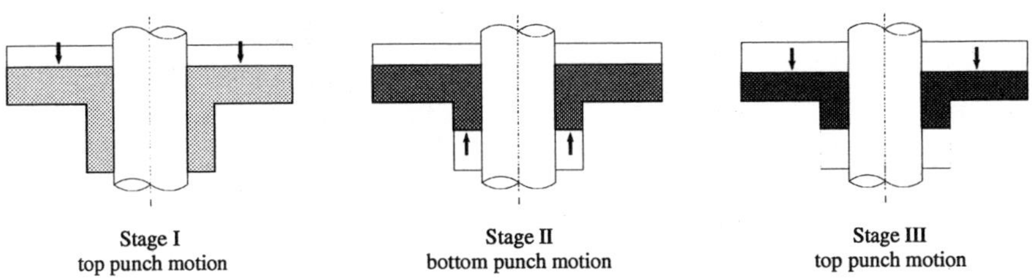

Figure 6.7. Finite element simulation of a flanged component

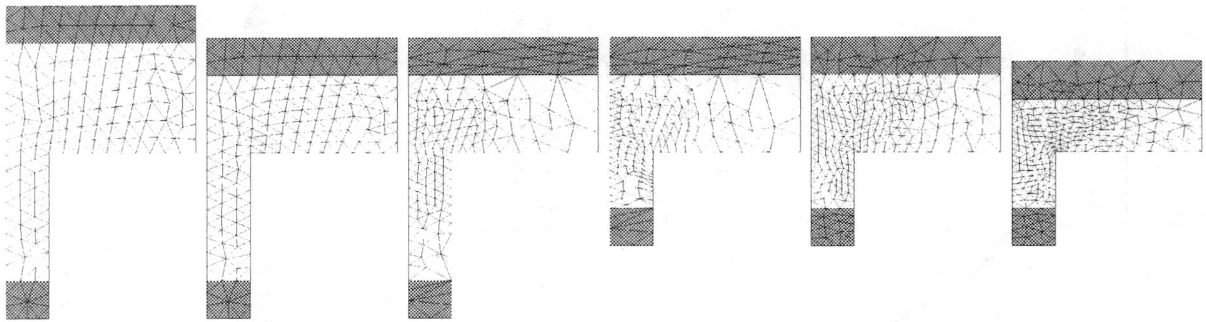

initial mesh	top punch motion	first adaptive mesh	bottom punch motion	second adaptive mesh	top punch motion
267 elements	(d_t = 3.44 mm)	314 elements	(d_b = 7.70 mm)	499 elements	(d_t = 2.62 mm)
		$\bar{\eta}_{aim}$ = % 4.0		$\bar{\eta}_{aim}$ = % 2.0	

Figure 6.8. Adaptive analysis for a rotational flanged component

(a)

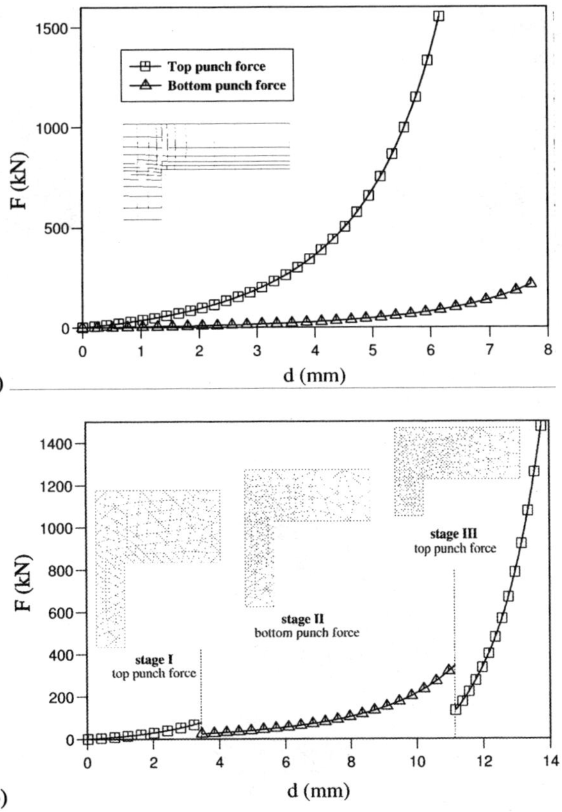

(b)

Figure 6.9. Predicted top and bottom punches compaction forces at different displacements for a flanged component; a) Double punch motion, b) Three stages motion

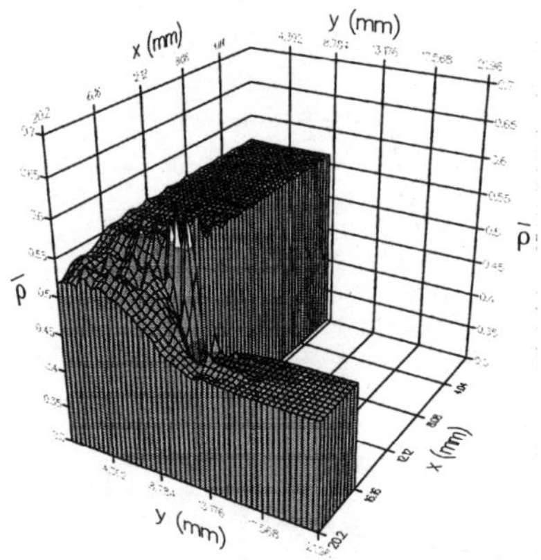

Stage I
Top punch motion
$d_t = 3.44$ mm

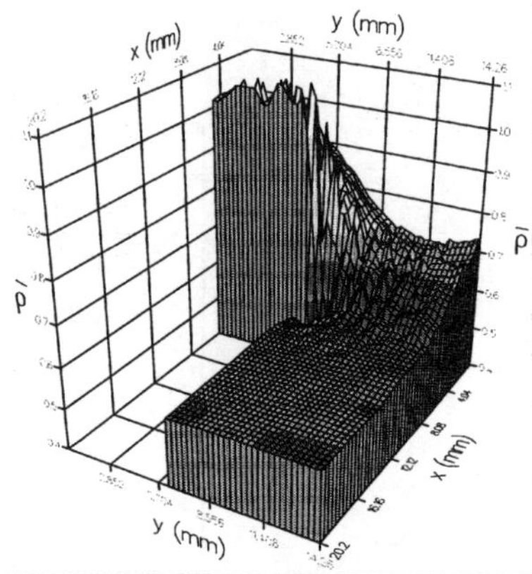

Stage II
bottom punch motion
$d_b = 7.70$ mm

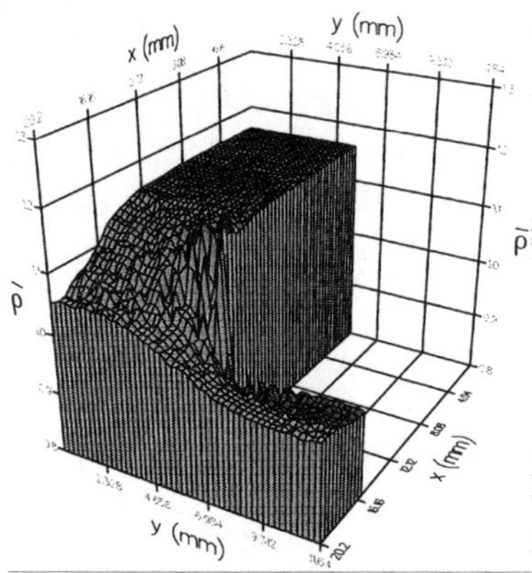

Stage III
top punch motion
$d_t = 2.62$ mm

Figure 6.10. Density distribution at different powder movements for a flanged component

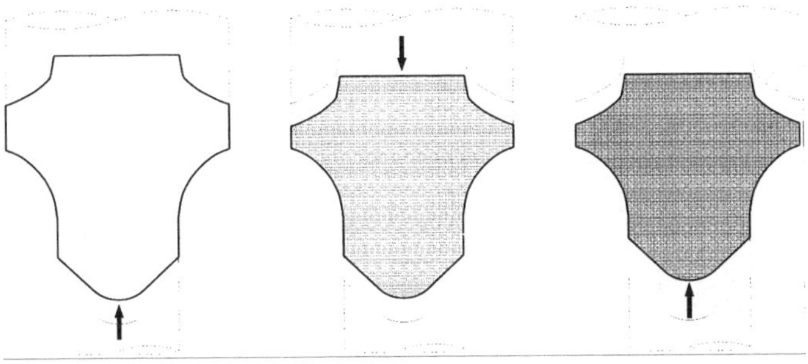

Figure 6.11. Finite element simulation of a tip component

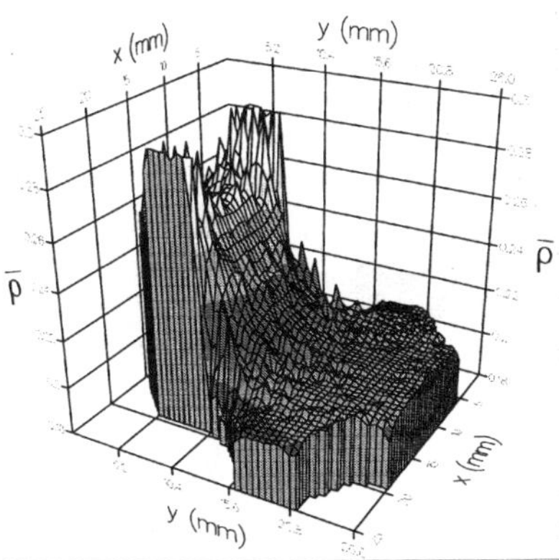

Stage I
bottom punch motion
$d_b = 4.0$ mm

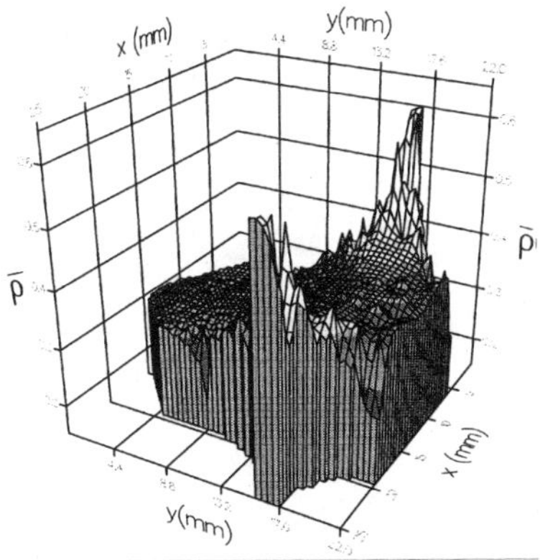

Stage II
top punch motion
$d_t = 4.0$ mm

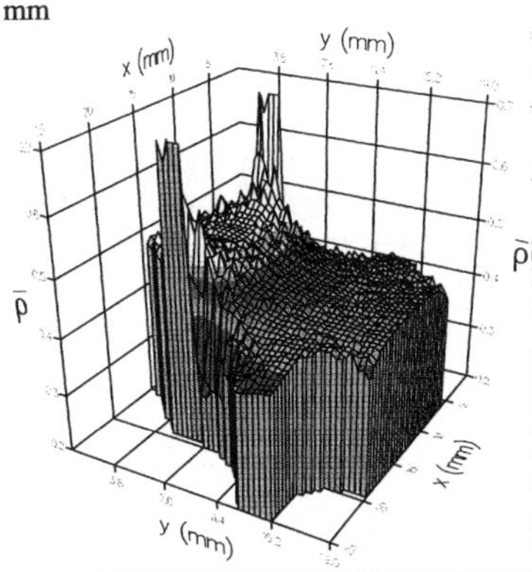

Stage III
top punch motion
$d_b = 3.0$ mm

Figure 6.12. Density distribution at different powder movements for a tip component

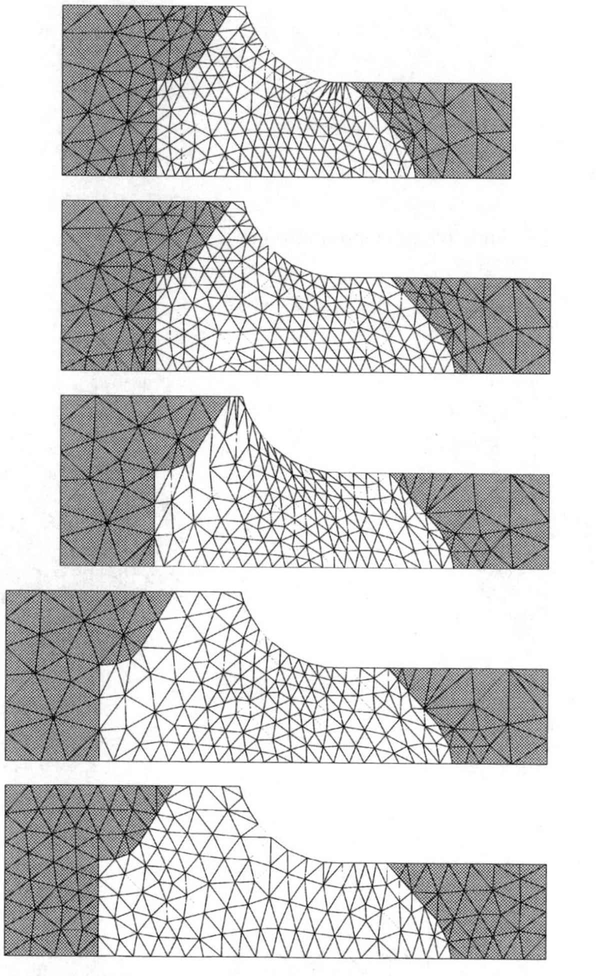

Figure 6.13. Adaptive analysis for a shaped tip component

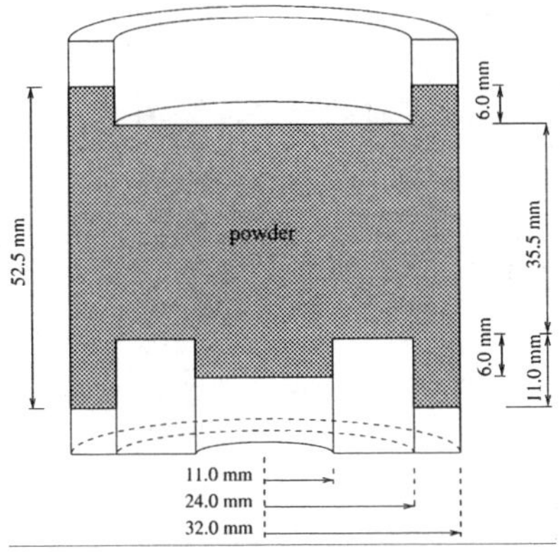

Figure 6.14. Shape, geometry and boundary conditions of the modeled multi-level component

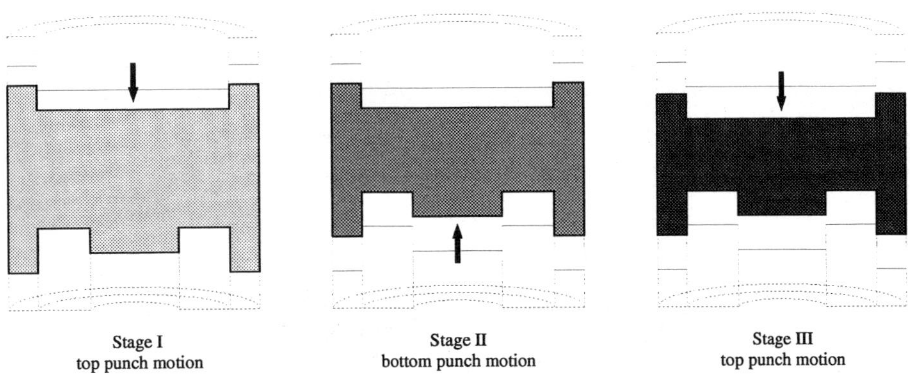

Figure 6.15. Finite element simulation of a multi-level component

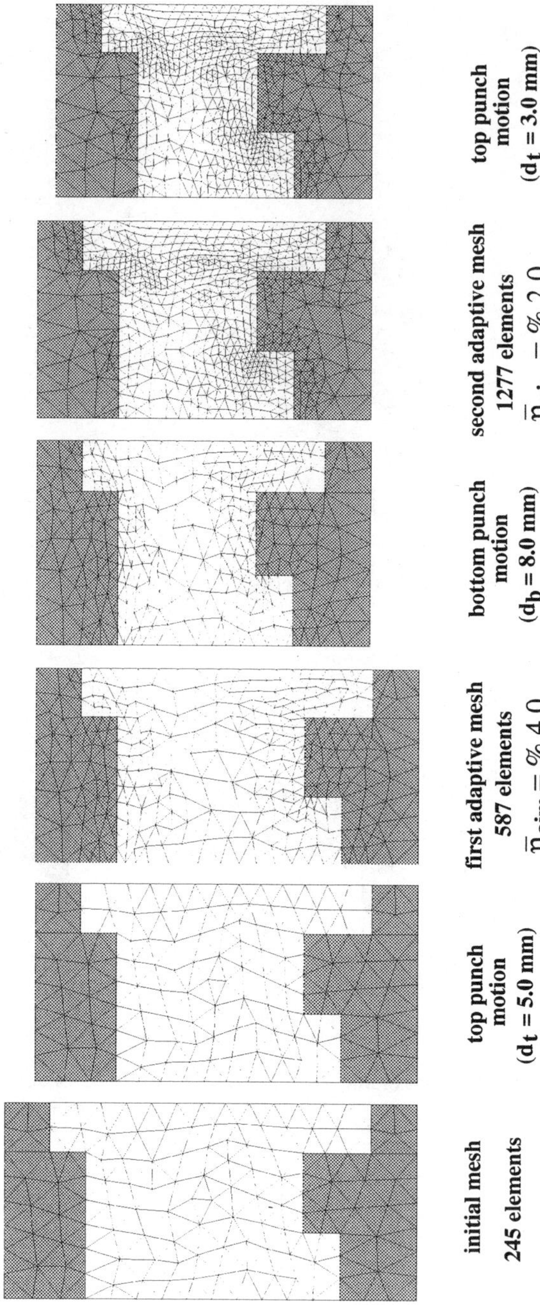

Figure 6.16. Adaptive analysis for a multi-level component

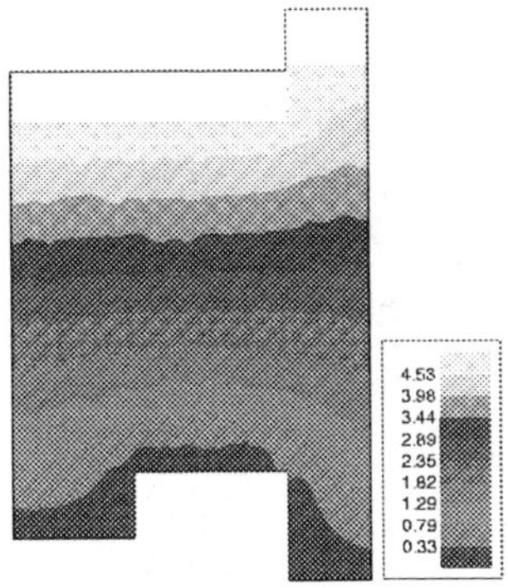

Stage I
Top punch motion

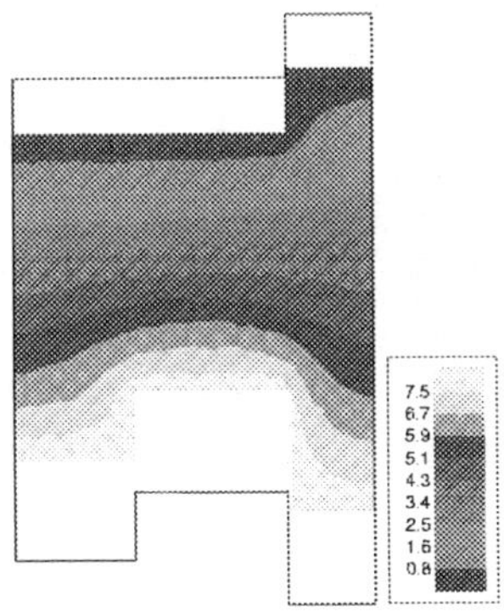

Stage II
Bottom punch motion

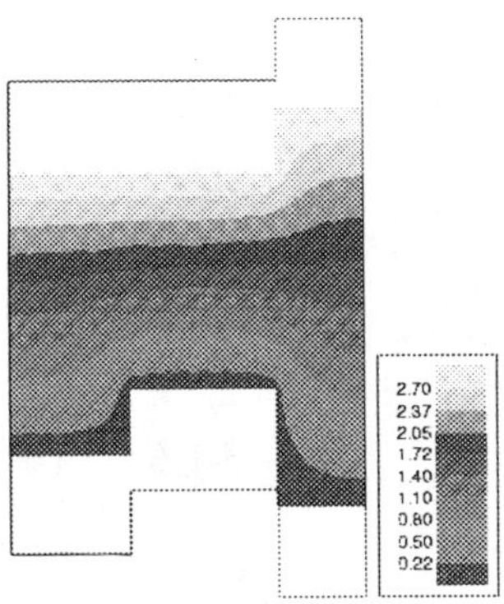

Stage III
Top punch motion

Figure 6.17. Displacement contours at different movements for a multi-level component

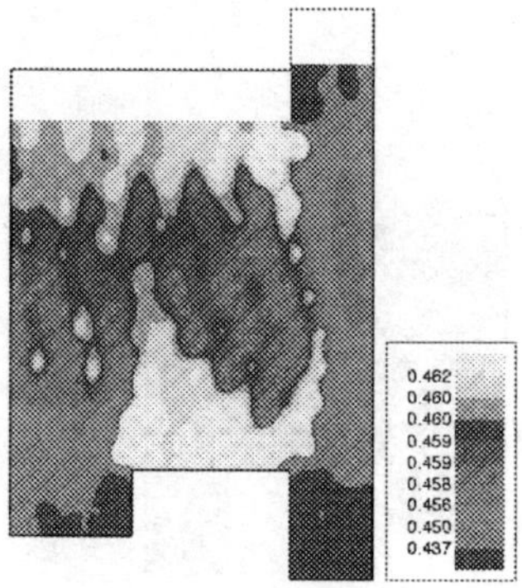

Stage I
Top punch motion

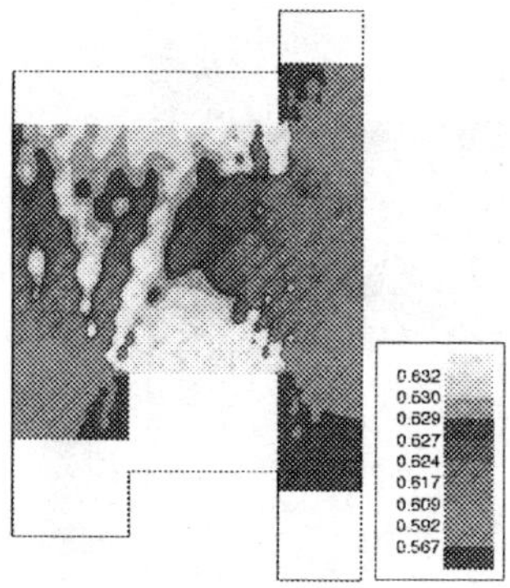

Stage II
Bottom punch motion

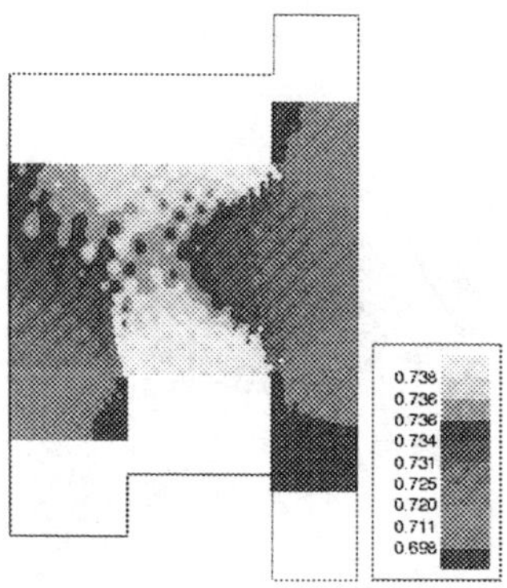

Stage III
Top punch motion

Figure 6.18. Relative density contours at different movements for a multi-level component

Box 6.1. The transfer operators $\mathcal{T}_1$ and $\mathcal{T}_2$

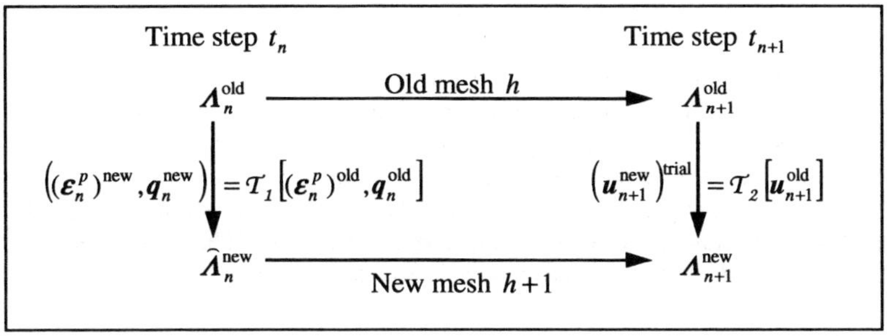

Box 6.2. The schematic diagram of the transfer operator $\mathcal{T}_1$ and $\mathcal{T}_2$

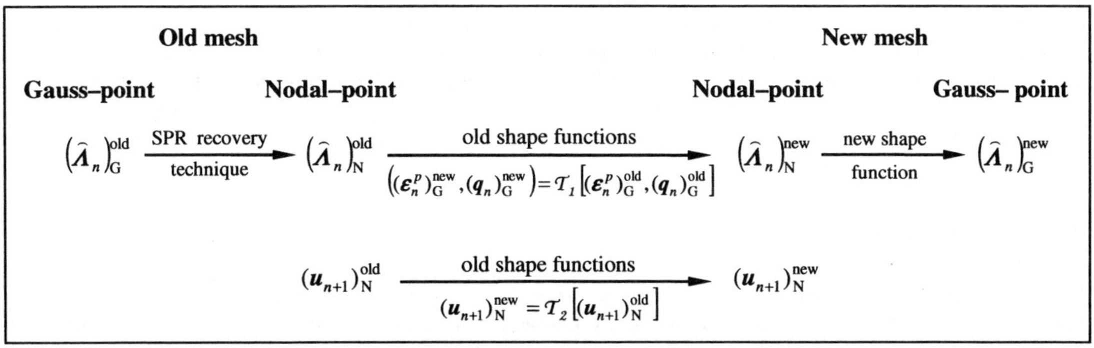

7 ADVANCED PLASTICITY MODELS

In the analysis of powder forming problems, the classical plasticity models suffer from a serious deficiency when the stress-point reaches a yield surface. In the flow theory of plasticity, the transition from an elastic state to an elasto-plastic state appears more or less abruptly. For powder material it is very difficult to define the location of yield surface, because there is no distinct transition from elastic to elastic-plastic behavior. In such materials the domain of the yield surface would collapse to a point, so making the direction of plastic increment indeterminate, because all directions are normal to a point. Thus, the classical plasticity theory cannot deal with such materials and an advanced constitutive theory is necessary.

In the present chapter, a new approach is developed based on an endochronic density-dependent plasticity model for describing the isothermal deformation behavior of metal powder at low homologous temperature. Constitutive equations are established based on coupling between deviatoric and hydrostatic behavior to the endochronic theory. Although the concept of yield surface has not been explicitly assumed in endochronic theory, it has been shown that the cone-cap plasticity and elliptical density-dependent yield surfaces, which are two most commonly used plasticity models for describing the behavior of powder material can be easily derived as special cases of the proposed endochronic theory. In final sections of the chapter, a multi-surface plasticity theory is presented for description of nonlinear behavior of solid mechanics and geotechnical problems. The multi-surface model is formulated using the Mroz translation rule for the direction of movements of the yield surfaces. The fully implicit scheme with radial mapping method is applied and the algorithmic consistent tangent modulus with numerical integration algorithm of constitutive equations is extracted for the mutli-surface model. The model is developed in the class of kinematic hardening models, so the 'Masing' rule is preserved.

7.1. Introduction

Powder metallurgy has been an attractive processing technology for advanced and some conventional materials because it offers net shape capability combined with micro structural control. Cold compaction process plays a fundamental role in powder metallurgy and structural ceramics manufacturing. It consists of the vertical compaction of a powder material through the movement of a set of punches at room temperature. As the magnitude of shrinkage in sintering process is dependent on the relative density of the compacted specimen, a uniform distribution of the relative density is desired in order to avoid micro cracks or fully developed cracks in the sintered specimen. One of the main ingredients of successful quantitative solution possibilities is an appropriate constitutive modeling of powder, which can reproduce powder material behavior under complicated loading conditions and an accurate and stable integration algorithm for the constitutive relation.

The cap plasticity model used in the modeling of geological and frictional materials, is employed to capture the major features of the response of initially loose metal powders to complex deformation histories which are encountered in the manufacture of engineering components by powder metallurgy techniques. The cone-cap model based on a density-dependent Drucker-Prager yield surface and a non-centered ellipse is developed by Aydin *et al.* (1996), Brandt and Nilsson (1999) and Gu *et al.* (2001). A double-surface plasticity model has been recently developed by Lewis and Khoei (2001) for the non-linear behavior of powder materials in the concept of the generalized plasticity formulation for the description of cyclic loading. This model is based on the combination of a convex yield surface consisting of a failure envelope, such as a Mohr-Coulomb yield surface and a hardening elliptical yield cap. The model comprises two surfaces, one to reflect shear failure and the other to capture densification.

In the analysis of metal powder forming problems, the non-linear behavior of powder is adequately described by double-surface plasticity models. However, it suffers from a serious deficiency when the stress-point reaches a yield surface. Results of experimental test on some hard metal powder show that the plastic effects begin immediately upon loading (Haggblad 1991). In this case, the classical plasticity theory cannot describe the behavior of powder materials and advanced constitutive theories are necessary. One way to over come the above difficulties is to use the endochronic plasticity theory which has no yield surface. Here, an endochronic density dependent plasticity model is developed to describe the isothermal deformation behavior of metal powder at low homologous temperature. The fact that a yield surface is not required in the formulation of the endochronic theory is attractive. The theory can be put in a very advanced form which permits the prediction of all main phenomena. Another way to increase the accuracy of the predictions of the plasticity models is to use more than one yield surface in the concept of multi-surface plasticity. In addition to the effects which can be modeled by double-surface plasticity, the multi-surface plasticity approach more easily describes hysteresis effects in cyclic loading. Since additional parameters of the model are obtained by adding extra yield surfaces, close fits to real material behavior can be obtained.

7.2. Endochronic plasticity theory

The endochronic theory deals with the plastic response of materials by means of memory integrals, expressed in terms of memory kernels. Formulation of this theory is based on thermo dynamical concepts and provides a unified point of view to describe the elastic-plastic behavior of material, since it places no requirement for a yield surface and 'loading function' to distinguish between loading and unloading. A key ingredient of the theory is that the deformation history is defined with respect to a deformation memory scale called intrinsic time. In the original version of the endochronic theory, proposed by Valanis (1971), the intrinsic time was defined as the path length in the total strain space. The so-called endochronic theory violates the second law of thermodynamics and leads to constitutive relations, which characterize inherently unstable materials (Rivlin 1981 and Sandler 1978). Aiming at the correction of this deficiency, a new version of the endochronic theory was developed by Valanis (1980) in which the intrinsic time was defined as the path length in the plastic strain space. The new endochronic plasticity was capable of predicting a stress-response to deformation processes, including reversal points in loading in agreement with the experimentally observed mechanical behavior of metals. Also, Valanis demonstrated that an introduction of Dirac delta function into the kernel function leads to a derived result of a yield surface and classical plasticity models of isotropic and kinematic hardening could be derived as a special case of the endochronic theory (Valanis, 1980).

Using original endochronic theory, as a point of departure, a method of extending endochronic theory to concrete, clay and sand was proposed by Bazant and his coworkers (Bazant and Krizek 1976, Bazant and Bhat 1976 and Bazant *et al.* 1979) by introducing sensitivity of intrinsic time to hydrostatic pressure and incorporating inelastic dilatancy due to shear strain. But the proposed approach has shortcomings of the original endochronic theory mentioned above. An extension of the new endochronic theory to concrete and sand was proposed by Valanis *et al.* (Valanis and Read 1986 and Valanis and Peters 1991) and Wu *et al.* (1985). An endochronic rate independent, density-dependent plasticity theory was developed by Khoei *et al.* (2001–2004) to describe the behavior of metal powder in powder compaction processes.

The endochronic theory has also been extended to investigate the rate dependent behavior of materials. Lin and Wu (1976) used original version of the endochronic theory and introduced a rate sensitivity function, which was a function of total strain rate. Wu and Yip (1980) redefined the rate sensitivity function of intrinsic time measure based on the new intrinsic time proposed by Valanis (1980) and discussed the uniaxial stress-strain responses of 1100-O aluminum and mild steel at constant strain rate conditions. Wu and Ho (1995) introduced another functional form for dependency of intrinsic time scale to the equivalent deviatoric plastic strain rate and applied endochronic theory to investigate transient creep of material. A new formulation of the rate sensitivity function was proposed by Pan and Chern (1997) to describe the visco-plastic behavior of material subjected to multi-axial loading. Pan (1997) modified the rate sensitivity function proposed by Pan and Chern (1997) and Pan *et al.* (1996) to incorporate with the finite endochronic constitutive equations. For the case of study, the finite simple torsion of iron and nickle thin-walled tubes were simulated with explicit constitutive equations. Pan *et al.* (1999) proposed a different formulation of the scaling function of the intrinsic time measure, as suggested originally by Valanis (1975), to describe the material behavior under rate dependent elasto-plastic deformation. They used the differential endochronic constitutive equations derived by Valanis (1984) to describe the material responses subjected to rate dependent elasto-plastic deformation.

In the present study, the endochronic plasticity theory developed by Valanis and Peters (1991) is extended for description of metal powder behavior in forming processes. The endochronic plasticity model is developed for rate-independent and rate-dependent materials. Constitutive equations are established based on coupling between deviatoric and hydrostatic behavior to the endochronic theory. The endochronic plasticity describes the behavior of powder material from initial stage of compaction to final stage, in which material behaves as solid metals. A simple procedure for calibrating endochronic material parameters with cone-cap and elliptical yield function is derived. Although the concept of yield surface has not been explicitly assumed in endochronic theory, it has been shown that the cone-cap plasticity and elliptical density-dependent yield surfaces, which are two most commonly used plasticity models for describing the behavior of powder material can be easily derived as special cases of the proposed endochronic theory.

7.3. Endochronic theory of plastic deviatoric deformation

Constitutive equation of the endochronic theory for rate-independent, plastically incompressible, initially isotropic material in the absence of plastic volumetric deformation can be written as

$$\boldsymbol{\sigma}_{dev} = 2\int_0^z \Phi(z - z')\frac{d\boldsymbol{\varepsilon}^p}{dz'}dz' \tag{7.1}$$

where Φ represents the material hereditary function. A time scale 'ς' is introduced which is independent of elapsed time, but intrinsically dependent on the deformation path of material, this memory path is defined in plastic strain space. It is through this parameter that history effects are introduced into constitutive equations of endochronic theory.

$$d\varsigma = (d\boldsymbol{\varepsilon}^p : d\boldsymbol{\varepsilon}^p)^{\frac{1}{2}} \tag{7.2}$$

$$dz = \frac{d\varsigma}{f(\varsigma)} \tag{7.3}$$

where $f(\varsigma)$ denotes a material function called scale function that is dependent to intrinsic time. The role of scale function $f(\varsigma)$ is crucial in behavior of model. By scaling intrinsic time, this function causes hardening or softening plastic behavior as a function of effective plastic strain. In above equations, the total stress tensor is denoted by $\boldsymbol{\sigma}$ and its deviatoric part by $\boldsymbol{\sigma}_{dev}$. The symbol $\boldsymbol{\varepsilon}$ represents the total strain tensor and its deviatoric part is $\boldsymbol{\varepsilon}'$. The superscripts '$e$' and '$p$' indicate the elastic and plastic components, respectively.

As shown by Valanis (1980), the kernel function $\boldsymbol{\Phi}$ can be expressed in term of a Dirichlet series, i.e.,

$$\boldsymbol{\Phi}(z) = \sum_{r=1}^{\infty} C_r e^{-\alpha_r z} \tag{7.4}$$

with the requirements that C_r and α_r are non-negative for all values of 'r', and the conditions

$$\sum_{r=1}^{\infty} \frac{C_r}{\alpha_r} < \infty \tag{7.5}$$

$$\sum_{r=1}^{\infty} C_r = \infty \tag{7.6}$$

The first condition ensures the integrability of $\boldsymbol{\Phi}$ over a finite domain of time scale z and the second condition causes the constitutive equations to reproduce elastic region in onset of loading and elastic unloading. In numerical applications, n-term Dirichlet series can be used as (Hsu *et al.* 1991)

$$\boldsymbol{\Phi}(z) = \sum_{r=1}^{n} C_r e^{-\alpha_r z} \tag{7.7}$$

Substitution of equation (7.7) into equation (7.1) leads to

$$\boldsymbol{\sigma}_{dev} = \sum_{r=1}^{n} (\boldsymbol{\sigma}_{dev})_r \tag{7.8}$$

$$(\boldsymbol{\sigma}_{dev})_r = 2C_r \int_0^z e^{-\alpha_r (z-z')} \frac{d\boldsymbol{\varepsilon}^p}{dz'} dz' \tag{7.9}$$

Taking derivative of above relations results in the following linear first order differential equations

$$d(\boldsymbol{\sigma}_{dev})_r = 2C_r d\boldsymbol{\varepsilon}^p - \alpha_r (\boldsymbol{\sigma}_{dev})_r dz\,, \qquad r = 1, 2, \ldots, n \tag{7.10}$$

and

$$d\boldsymbol{\sigma}_{dev} = \sum_{r=1}^{n} d(\boldsymbol{\sigma}_{dev})_r = 2\boldsymbol{\Phi}(0) d\boldsymbol{\varepsilon}^p - \sum_{r=1}^{n} \alpha_r (\boldsymbol{\sigma}_{dev})_r dz \tag{7.11}$$

In equation (7.11), increment of deviatoric stress is determined based on the unknown value of plastic strain increment (increment of scaled time dz is function of $d\boldsymbol{\varepsilon}^p$) and known value of history variables.

7.4. Endochronic theory of plastic volumetric deformation

Constitutive equations of the endochronic theory for rate-independent, plastically compressible, initially isotropic material due to plastic volumetric deformation can be written as

$$\boldsymbol{\sigma}_{dev} = 2\int_0^{z_d} \boldsymbol{\Phi}_d(z_d - z')\frac{d\boldsymbol{\varepsilon}_{dev}^p}{dz'}dz' \tag{7.12}$$

$$\sigma_h = 2\int_0^{z_h} \Phi_h(z_h - z')\frac{d\varepsilon_{vol}^p}{dz'}dz' \tag{7.13}$$

where $\boldsymbol{\Phi}_d$ and $\boldsymbol{\Phi}_h$ are material functions, called deviatoric and hydrostatic hereditary functions. A time scale 'ς' is introduced which is independent of elapsed time, but intrinsically dependent on the deformation of material. It is through this parameter that history effects are introduced into constitutive equations of endochronic theory.

$$d\varsigma^2 = d\boldsymbol{\varepsilon}_{dev}^p : d\boldsymbol{\varepsilon}_{dev}^p + \kappa^2 d\varepsilon_{vol}^{p\;2} \tag{7.14}$$

$$dz_d = \frac{d\varsigma}{f_d(p)} \tag{7.15}$$

$$dz_h = \frac{d\varsigma}{\kappa\, f_h(\eta, p)} \tag{7.16}$$

where $f_d(p)$ and $f_h(\eta, p)$ denote material functions, called the deviatoric and hydrostatic scale functions. These functions are dependent to hydrostatic pressure p, and relative density η, which is defined as

$$\eta = \frac{\rho_t}{\rho_s} \tag{7.17}$$

where ρ_t is the apparent density in time t and ρ_s denotes the solid density of material. This variable evolve in time by the following relation,

$$\eta = \frac{\eta_0}{J} \tag{7.18}$$

where η_0 is the initial relative density and J is the third invariant of deformation gradient tensor. In equation (7.14), κ is the coupling parameter between the deviatoric and hydrostatic deformations. In above equations, the total stress tensor is denoted by $\boldsymbol{\sigma}$ and its deviatoric and hydrostatic parts by $\boldsymbol{\sigma}_{dev}$ and σ_h, respectively. The symbol $\boldsymbol{\varepsilon}$ represent the total strain tensor by deviatoric and volumetric parts $\boldsymbol{\varepsilon}_{dev}$ and ε_{vol}, respectively. The superscripts 'e' and 'p' indicate the elastic and plastic components, respectively.

As shown by Valanis (1980), the kernel functions $\boldsymbol{\Phi}_d$ and $\boldsymbol{\Phi}_h$ can be expressed in terms of a Dirichlet series, i.e.,

$$\Phi_d = \sum_{r=1}^{\infty} A_r e^{-\alpha_r z_d} \tag{7.19}$$

$$\Phi_h = \sum_{i=1}^{\infty} B_i e^{-\beta_i z_h} \tag{7.20}$$

with the requirements that A_r, B_i, α_r and β_i are non-negative for all values of 'r' and 'i', and the conditions,

$$\sum_{r=1}^{\infty} \frac{A_r}{\alpha_r} < \infty \tag{7.21}$$

$$\sum_{i=1}^{\infty} \frac{B_i}{\beta_i} < \infty \tag{7.22}$$

These conditions ensure the integrability of Φ_d and Φ_h over a finite domain of time scale 'z'. In numerical application, m-term Dirichlet series can be used as

$$\Phi_d = \sum_{r=1}^{m} A_r e^{-\alpha_r z_d} \tag{7.23}$$

$$\Phi_h = \sum_{i=1}^{L} B_i e^{-\beta_i z_h} \tag{7.24}$$

In this section, coupling between deviatoric and hydrostatic behavior is introduced in endochronic theory through intrinsic time 'ς' leading to dilatant deformation, defined by equation (7.14). The role of scale functions $f_d(p)$ and $f_h(\eta, p)$ is crucial in the behavior of model. By scaling intrinsic time, these functions cause hardening or softening plastic behavior as functions of hydrostatic pressure and relative density of material.

The reminder of this section is devoted to describe property of the above endochronic relations and the effects of $f_d(p)$ and $f_h(\eta, p)$ in the behavior of endochronic plasticity. In order to summarize the above discussions, the kernel functions Φ_d and Φ_h are expressed in terms of delta function as,

$$\Phi_d = \phi_d\, \alpha_1\, e^{-\alpha_1 z_d} = \phi_d\, \delta(z_d) \qquad \alpha_1 \to \infty \tag{7.25}$$

$$\Phi_h = \phi_h\, \beta_1\, e^{-\beta_1 z_h} = \phi_h\, \delta(z_h) \qquad \beta_1 \to \infty \tag{7.26}$$

where $\delta(x)$ is Dirac delta function. Substituting the above relations into equations (7.12) and (7.13) and using relations (7.14), (7.15) and (7.16), with some straightforward manipulations leads to the following yield function

$$q^2 + \frac{3}{2}\left(\frac{\phi_d}{\phi_h}\frac{f_d}{f_h}\right)^2 p^2 = \frac{3}{2}\phi_d{}^2 f_d^2 \tag{7.27}$$

where p is the hydrostatic pressure and q is defined as $q=\sqrt{J_2^{'}}$ with $J_2^{'}$ denoting the second invariant of the deviatoric stress tensor. Although yield surface has not been explicitly assumed in endochronic theory, introducing Dirac delta to kernel functions, results to the similar concept in an implicit manner.

7.4.1. Von-Mises yield criterion

Consider the case that the hydrostatic scale function f_h is defined so that $f_h \to \infty$ when $\eta \to 1$, as shown in Figure 7.1. Considering f_h as function of hydrostatic pressure and relative density, it can be happen in final stage of compaction where relative density is high. Introducing this hydrostatic scale function f_h into the yield function (7.27) leads to the von-Mises yield criteria. This property allows endochronic plasticity to describe the behavior of metal powder from initial stage of compaction to final stage, in which material behaves as solid metals.

7.4.2. Elliptical yield criterion

Consider the case where the deviatoric and hydrostatic scale functions are only functions of relative density η. Thus, the yield function (7.27) will be simplified as follows

$$a_1(\eta)\, q^2 + a_2(\eta)\, p^2 = a_3(\eta)\, \sigma_y^{\,2} \tag{7.28}$$

Figure 7.2 shows typically the yield function (7.28) for different values of relative density. This type of yield function has been used efficiently for describing the behavior of powder metals under cold compaction process (Oliver *et al.* 1996 and Perez-Foguet *et al.* 2003).

7.4.3. Cone-cap yield criterion

Consider the case that the deviatoric scale function $f_d(p)$ is a function of hydrostatic pressure, and the hydrostatic scale function $f_h(\eta, p)$ is a function of relative density and hydrostatic pressure. The explicit forms for these functions are considered as follows,

$$f_d = c - \frac{p}{p_0} \tag{7.29}$$

$$f_h = (1-\psi)\, g(\eta) - \psi \frac{p}{\phi_h} \tag{7.30}$$

where c and p_0 are positive parameters, ϕ_d and ϕ_h are defined in equations (7.25) and (7.26), ψ is a positive number in the range of 0–1, and $g(\eta)$ is an increasing function of relative density. Typical evolution of $f_d(p)$ and $f_h(\eta, p)$ are plotted in Figures 7.1 and 7.3.

Introducing relations (7.29) and (7.30) to (7.27) results in a yield function, whose cross sectional shape on meridian plan is plotted in Figure 7.4. This yield surface is very similar to cone-cap yield surface, a combination of Mohr-Coulomb, or Druker-Prager, and elliptical yield surfaces, used extensively to describe the behavior of powder and granular materials (Haggblad and Oldenburg 1994, Khoei and Lewis 1998, 1999, Brandt and Nilsson 1999, Gu *et al.* 2001 and Lewis and Khoei 2001). It is possible to relate the material parameters of this simplified endochronic theory to cone-cap plasticity model. Substituting (7.29) and (7.30) to (7.27) and solving for p at $q=0$ yields to two roots for p, the points of intersection of yield surface with p-axis. The positive root is p_T and the negative root is p_c (Figure 7.4),

$$p_c = -\phi_h \, g(\eta) \tag{7.31}$$

$$p_T = c \, p_0 \tag{7.32}$$

The position of p_T is constant and depends on parameters of deviatoric scale function. Choosing $c > 0$ introduces cohesion to endochronic model. While p_T is constant, p_c increases in absolute value as a function of $g(\eta)$, thus evolution of material relative density, causes hardening behavior of material.

In equation (7.31), a hydrostatic compression test is required to determine $g(\eta)$ in hydrostatic scale function. The stress-strain curve from this test gives the evolution of hydrostatic compression yield stress p_c as a function of material density. Intersection of yield surface with q-axis (q_0 in Figure 7.4) can be obtained from equations (7.27), (7.29) and (7.30) as follows,

$$q_0 = \sqrt{\tfrac{3}{2}} \, c \phi_d \tag{7.33}$$

In above equation, c can be determined by relating q_0 to material cohesion available from standard tests for granular material and considering ϕ_d as a known constant value. It must be noted that ϕ_d and ϕ_h appear as coefficient of f_d and f_h in the yield function (7.27), so they are not independent material parameters and can be given any constant values.

In Figure 7.4, the ratio q_0/p_T is the slope of cone part of the yield surface in meridian plan, which can be determined from the friction angle of material. Thus, p_0 can be determined from equations (7.32), (7.33) and the friction angle of material. In Figure 7.5, the evolution of yield surface with increasing the relative density of material is plotted using deviatoric scale function parameters, $c = 0.093$, $p_0 = 200$, $\phi_d = 15$ and the hydrostatic scale function parameters, $\psi = 0$, $\phi_h = 100$, $g(\eta) = 0.02 \exp(9(1 - \eta_0/\eta))$ and the initial relative density $\eta_0 = 0.4$. It is worth mentioning that because cone-cap yield surface consists of two different yield functions to describe the plastic behavior of material, special treatment should be made to avoid numerical difficulty in intersection of these two surfaces where plastic flow is discontinuous (Gu *et al.* 2001) but the yield surface (7.27) does not have such a drawback.

7.5. Endochronic theory of plastic volumetric and plastic deviatoric deformations

Constitutive equations of the endochronic theory for rate-independent, plastically compressible, initially isotropic material due to plastic volumetric and plastic deviatoric deformation can be expressed as

$$\boldsymbol{\sigma}_{dev} = 2 \int_0^{z_d} \Phi_d (z_d - z') \frac{d\boldsymbol{\varepsilon}^p_{dev}}{dz'} dz' \tag{7.34}$$

$$\sigma_h = 2 \int_0^{z_h} \Phi_h (z_h - z') \frac{d\varepsilon^p_{vol}}{dz'} dz' - 2 \int_0^{z_h} \Psi (z_h - z') \boldsymbol{\sigma}_{dev} : \frac{d\boldsymbol{\varepsilon}^p_{dev}}{dz'} dz' \tag{7.35}$$

where $\Phi_d(z_d)$, $\Phi_h(z_h)$ and $\Psi(z_h)$ are material functions called deviatoric, hydrostatic and coupling hereditary functions. A time scale 'ς' is introduced which is independent of elapsed time, but intrinsically dependent on the deformation of material. It is through this parameter that history effects are introduced into constitutive equations of endochronic theory.

$$d\varsigma^2 = d\boldsymbol{\varepsilon}_{dev}^p : d\boldsymbol{\varepsilon}_{dev}^p + \kappa^2 d\varepsilon_{vol}^{p\ 2} \tag{7.36}$$

$$dz_d = \frac{d\varsigma}{f_d(\eta, p, \theta)} \tag{7.37}$$

$$dz_h = \frac{d\varsigma}{\kappa\, f_h(\eta, p)} \tag{7.38}$$

where $f_d(\eta, p, \theta)$ and $f_h(\eta, p)$ are material functions called deviatoric and hydrostatic scale functions. These functions depend on hydrostatic pressure p, with its positive value in tension, Lode's angle θ and relative density η, defined as

$$\theta = \frac{1}{3}\arccos\left(\frac{3\sqrt{3}\,J_3}{2\,J_2^{3/2}}\right) \tag{7.39}$$

$$\eta = \frac{\rho_t}{\rho_s} \tag{7.40}$$

where J_2 and J_3 are second and third invariants of deviatoric stress tensor, ρ_t is the apparent density in time t and ρ_s denotes the solid density of material. The relative density η evolves in time by the following relation

$$\eta = \frac{\eta_0}{J} \tag{7.41}$$

where η_0 is the initial relative density and J is the third invariant of deformation gradient tensor. In equation (7.36), κ is the coupling parameter between the deviatoric and hydrostatic deformations. In above equations, the total stress tensor is denoted by $\boldsymbol{\sigma}$ and its deviatoric and hydrostatic parts by $\boldsymbol{\sigma}_{dev}$ and σ_h. The symbol $\boldsymbol{\varepsilon}$ represents the total strain tensor by deviatoric and volumetric parts, $\boldsymbol{\varepsilon}_{dev}$ and ε_{vol}, respectively. The superscripts 'e' and 'p' indicate the elastic and plastic components, respectively.

The kernel functions $\Phi_d(z_d)$, $\Phi_h(z_h)$ and $\Psi(z_h)$ can be expressed in terms of a Dirichlet series as (Valanis and Peters 1991)

$$\Phi_d(z_d) = \sum_{i=1}^{\infty} A_i e^{-\alpha_i z_d} \tag{7.42}$$

$$\Phi_h(z_h) = \sum_{i=1}^{\infty} B_i e^{-\beta_i z_h} \tag{7.43}$$

$$\Psi(z_h) = \sum_{i=1}^{\infty} \frac{C_i}{\kappa f_h} e^{-\beta_i z_h} \tag{7.44}$$

with the requirements that A_i, B_i, α_i, β_i and C_i are non-negative for all values of 'i', and the conditions,

$$\sum_{i=1}^{\infty} \frac{A_i}{\alpha_i} < \infty \tag{7.45}$$

$$\sum_{i=1}^{\infty} \frac{B_i}{\beta_i} < \infty \tag{7.46}$$

$$\sum_{i=1}^{\infty} \frac{C_i}{\beta_i} < \infty \tag{7.47}$$

These conditions ensure the integrability of $\Phi_d(z_d)$, $\Phi_h(z_h)$ and $\Psi(z_h)$ over a finite domain of time scale z. In numerical application m-term Dirichlet series can be used as (Hsu *et al.* 1991)

$$\Phi_d(z_d) = \sum_{i=1}^{m} A_r e^{-\alpha_i z_d} \tag{7.48}$$

$$\Phi_h(z_h) = \sum_{i=1}^{L} B_i e^{-\beta_i z_h} \tag{7.49}$$

$$\Psi(z_h) = \frac{1}{\kappa f_h} \sum_{i=1}^{L} C_i e^{-\beta_i z_h} \tag{7.50}$$

In above relations, coupling between deviatoric and hydrostatic behavior is introduced in endochronic theory through intrinsic time ς leading to dilatant deformation, which is defined by equation (7.36) and the rate equation (7.35) through hereditary function $\Psi(z_h)$. The role of scale functions $f_d(\eta, p, \theta)$ and $f_h(\eta, p)$ is crucial in the behavior of model. By scaling intrinsic time, these functions cause hardening or softening plastic behavior as a function of hydrostatic pressure, relative density of material and Lode's angle.

7.5.1. Yield function

This section describes the properties of proposed endochronic formulation and effects of $f_d(\eta, p, \theta)$ and $f_h(\eta, p)$ in behavior of the endochronic plasticity. Consider that the kernel functions $\Phi_d(z_d)$, $\Phi_h(z_h)$ and $\Psi(z_h)$ are expressed by one term of delta function, they can be written as

$$\Phi_d(z_d) = \phi_d\, \alpha_1\, e^{-\alpha_1 z_d} = \phi_d\, \delta(z_d) \qquad \alpha_1 \to \infty \tag{7.51}$$

$$\Phi_h(z_h) = \phi_h\, \beta_1\, e^{-\beta_1 z_h} = \phi_h\, \delta(z_h) \qquad \beta_1 \to \infty \tag{7.52}$$

$$\Psi(z_h) = \frac{1}{\kappa f_h} \psi_h\, \beta_1\, e^{-\beta_1 z_h} = \frac{1}{\kappa f_h} \phi_h\, \delta(z_h) \qquad \beta_1 \to \infty \tag{7.53}$$

where $\delta(x)$ is Dirac delta function and ϕ_d, ϕ_h and ψ_h are positive parameters. Substituting the above relations in equations (7.34) and (7.35) and using relations (7.36), (7.37) and (7.38), with some straightforward manipulations leads to the following relation

$$q^2 + \frac{3}{2}\left(\frac{\phi_d}{\phi_h}\frac{f_d}{f_h}\right)^2 \left(p + \frac{2}{3}\frac{\psi_h}{\phi_d f_d} q^2\right)^2 = \frac{3}{2}\phi_d^{\,2} f_d^2 \tag{7.54}$$

where p is the hydrostatic pressure, with positive value in tension, and q is defined as $q = \sqrt{\frac{3}{2}\boldsymbol{\sigma}_{dev} : \boldsymbol{\sigma}_{dev}}$. Although the yield surface has not been explicitly assumed in endochronic theory, introducing delta function to kernels, results implicitly in this concept. In order to present different aspects of equation (7.54), the following specific forms are proposed for the dependence of deviatoric scale function $f_d(\eta, p, \theta)$ and hydrostatic scale function $f_h(\eta, p)$ on η, p and θ

$$f_d(\eta, p, \theta) = h(\theta)\left(c(\eta) - \frac{p}{p_0}\right) \tag{7.55}$$

$$f_h(\eta, p) = (1-\psi)\, g(\eta) - \psi \frac{p}{\phi_h} \tag{7.56}$$

where p_0 is a positive parameter and ψ is a positive number in the range of 0 – 1. In equation (7.55), $c(\eta)$ is an increasing function of relative density, defined as

$$c(\eta) = A_c + B_c\, \eta^{m_c} \frac{\eta - \eta_m}{1 - \eta_m} \tag{7.57}$$

in which the dependence of $c(\eta)$ on η is a power-law form. In above equation, A_c, B_c, η_m and m_c are material parameters with positive values. In equation (7.55), $h(\theta)$ is defined based on Willam-Warnke function as

$$h(\theta) = \left(\frac{2(e^2-1)\cos\theta + (2e-1)\sqrt{5e^2 - 4e + 4(1-e^2)\cos^2\theta}}{(2e-1)^2 + 4(1-e^2)\cos^2\theta}\right)^{\frac{1}{2}} \tag{7.58}$$

where e is a parameter of the model. In equation (7.56), $g(\eta)$ is an increasing function of relative density with the following specific form

$$g(\eta) = A_\eta\, \eta^{\,m_\eta} \frac{\eta - \eta_m}{1 - \eta_m} \tag{7.59}$$

where A_η and η_m are positive material parameters. Typical evolution of $f_d(\eta, p, \theta)$ and $f_h(\eta, p)$ with η and p are plotted in Figure 7.6. Introducing equations (7.55) and (7.56) into (7.54) results in a yield function, whose cross sectional shape on meridian plan is plotted in Figure 7.4. As demonstrated in previous section, this yield surface is very similar to the cone-cap yield surface, i.e. a combination of Mohr-Coulomb or Druker-Prager and elliptical yield surfaces, which has been extensively used by authors to describe the behavior of powder and granular materials (Haggblad and Oldenburg 1994, Khoei and Lewis 1998, 1999, Brandt and Nilsson 1999, Gu *et al.* 2001 and Lewis and Khoei 2001).

In order to relate the material parameters of the proposed endochronic theory to cone-cap plasticity model, substitute equations (7.55) and (7.56) in (7.54) and solve it for p at $q=0$. It yields to two roots for p, the points of intersection of yield surface with p-axis. The positive root is p_T and negative root is p_c (Figure 7.4)

$$p_c = -\phi_h \, g(\eta) \tag{7.60}$$

$$p_T = p_0 \, c(\eta) \tag{7.61}$$

In equation (7.60), the absolute value of p_c increases as a function of $g(\eta)$. Thus, the evolution of material relative density causes hardening behavior of material. In equation (7.61), the position of p_T is a function of relative density and depends on the parameters of deviatoric scale function. By choosing the non-zero values for parameters of $c(\eta)$, a density dependent cohesion can be introduced to the endochronic model. Typical evolution of yield surface (7.54) is shown in Figure 7.7 for different values of c and p_0 in meridian plan. In equation (7.60), a hydrostatic compression test is required to determine the parameters of function $g(\eta)$. The stress-strain curve obtained from this test gives the evolution of hydrostatic compression yield stress p_c as a function of material density.

As the material parameter ψ_h is generally a small value relative to ϕ_d, the intersection of yield surface with q-axis, defined by q_0 in Figure 7.4, can be obtained with good approximation from equations (7.54), (7.55) and (7.56) as

$$q_0 \cong \sqrt{\tfrac{3}{2}}\, h(\theta)\phi_d \, c(\eta) \tag{7.62}$$

In Figure 7.4, the ratio of q_0/p_T is the slope of cone part of yield surface in meridian plan, which can be related to the friction angle of material. Thus, considering ϕ_d as a constant known value, p_0 can be determined from equations (7.61), (7.62) and friction angle of material, estimated by a standard test, such as hollow cylinder test, simple shear test, or direct shear test. It should be mentioned that ϕ_d and ϕ_h appear as coefficients of f_d and f_h in the yield function, so they are not independent material parameters and can be given any constant values. In Figure 7.7, the effect of variation of p_0 on the shape of yield function is shown in meridian plan. It must be noted that q_0, defined in equation (7.62), can be easily related to the cohesion of material. The parameters of function $c(\eta)$ are determined from the variation of material cohesion with relative density, estimated experimentally by simple compression experiments on samples with different initial relative density. Details of such a procedure for metal powder can be found in Gu *et al.* (2001).

The function $h(\theta)$ controls the shape of yield surface (7.54) in the deviatoric plane. If $e=1$, it results in a circular trace of yield function on deviatoric plan. If $e<1$, it causes triangularity of deviatoric trace along the hydrostatic axis. Figure 7.8 shows the trace of yield function on deviatoric plan for different values of shape parameter e. In Figure 7.9, the shapes of yield surface are depicted in principal stress space.

7.5.2. Dilatancy condition

In this section, the effects of parameters κ and ψ_h, which control coupling between deviatoric and hydrostatic deformations in behavior of the model, are discussed and a procedure for their estimation from experimental data are described. Substituting equations (7.52) and (7.53) in equation (7.35) yields

$$\sigma_h = \phi_h \kappa f_h \frac{d\varepsilon_{vol}^p}{d\varsigma} - \psi_h \boldsymbol{\sigma}_{dev} : \frac{d\boldsymbol{\varepsilon}_{dev}^p}{d\varsigma} \tag{7.63}$$

Eliminating $d\varsigma$ between equations (7.36) and (7.63) and considering $\boldsymbol{\sigma}_{dev} : d\boldsymbol{\varepsilon}_{dev}^p$ as $\|\boldsymbol{\sigma}_{dev}\| \|d\boldsymbol{\varepsilon}_{dev}^p\| \cos\tau$, we obtain the following relation (Valanis and Peters 1991)

$$\kappa \frac{d\varepsilon_{vol}^p}{\|d\boldsymbol{\varepsilon}_{dev}^p\|} = \frac{1 - \left(\psi_h \frac{\|\boldsymbol{\sigma}_{dev}\| \cos\tau}{\sigma_h}\right)^2}{\frac{f_h \psi_h \|\boldsymbol{\sigma}_{dev}\| \cos\tau}{\sigma_h^2} + \sqrt{\left(\frac{f_h}{\sigma_h}\right)^2 + \left(\frac{\psi_h \|\boldsymbol{\sigma}_{dev}\| \cos\tau}{\sigma_h}\right)^2 - 1}} \tag{7.64}$$

In particular case of triaxial compression test, when the loading condition occurs, $\cos\tau = 1$ and the dilatancy condition will be reduced to the following condition

$$\psi_h > -\frac{\sigma_h}{\|\boldsymbol{\sigma}_{dev}\|} \tag{7.65}$$

The two material parameters ψ_h and κ can be determined from the results of triaxial compression test. The constant ψ_h is the ratio of $-\sigma_h / \|\boldsymbol{\sigma}_{dev}\|$ at the initiation of dilatation, where the effect of $d\varepsilon_{vol}^p$ is vanished. After indicating ψ_h, the coupling parameter κ can be computed by substituting the measured dilatancy rate $d\varepsilon_{vol}^p / d\boldsymbol{\varepsilon}_{dev}^p$ into equation (7.64).

7.5.3. Plastic flow rule

In order to investigate the direction of plastic strain increment in stress space, substitute equations (7.37), (7.38), (7.51), (7.52) and (7.53) into equations (7.34) and (7.35). It leads to the following relation for increment of plastic strains

$$d\boldsymbol{\varepsilon}_{dev}^p = \frac{d\varsigma}{f_d \phi_d} \boldsymbol{\sigma}_{dev} \tag{7.66}$$

$$d\varepsilon_{vol}^p = \frac{d\varsigma}{\kappa f_h \phi_h} \left(\sigma_h + \kappa \psi_h \frac{f_h}{\phi_d f_d} \|\boldsymbol{\sigma}_{dev}\|^2 \right) \tag{7.67}$$

The above relations are the flow rule, which is implicitly derived from the proposed endochronic model. The variable $d\varsigma$ can be interpreted as the counterpart of consistency parameter of classical plasticity models. The above relations can be written in (p, q) stress space using the components of flow vector on the meridian plan $\boldsymbol{m} = (m_p, m_q)$ as

$$m_p = \frac{d\varsigma}{\kappa f_h \phi_h}\left(p + \frac{2}{3}\frac{\kappa f_h \psi_h}{\phi_d f_d} q^2\right) \tag{7.68}$$

$$m_q = \frac{2}{3}\frac{d\varsigma}{f_d \phi_d} q \tag{7.69}$$

The direction of the plastic flow vector with components (m_p, m_q) does not, in general, coincide with the gradients of yield function (7.54) and thus, the derived flow rule is non-associative. Applying the higher values for ψ_h, increases the dilatancy in cone region of the yield surface (7.54), as shown in Figure 7.10.

It should be mentioned that the behavior of the proposed endochronic model is investigated with one term kernel function. In fact, the application of more than one term in kernel functions adds kinematic hardening to the endochronic model and improves the behavior of model in cyclic loading condition. Obviously, for calibration of material parameters in this case, additional results of cyclic loading test are necessary. Although one term kernel function is sufficient here for the proposed application, the numerical integration of constitutive relations is implemented for the general n-term kernel functions.

7.6. Endochronic theory of visco-plasticity

Constitutive equations of the endochronic theory for rate-dependent, plastically incompressible, initially isotropic material is as follows,

$$\boldsymbol{\sigma}_{dev} = 2\int_0^z \Phi(z - z')\frac{d\boldsymbol{\varepsilon}^p}{dz'}dz' \tag{7.70}$$

where $\Phi(z)$ is material function, called hereditary function. A time scale ς is introduced which is independent of elapsed time, but intrinsically dependent on the deformation of material. It is through this parameter that history effects are introduced into constitutive equations of endochronic theory.

$$d\varsigma^2 = d\boldsymbol{\varepsilon}^p : d\boldsymbol{\varepsilon}^p \tag{7.71}$$

$$dz = \frac{d\varsigma}{f(\varsigma, \dot{\varsigma})} \tag{7.72}$$

where $f(\varsigma, \dot{\varsigma})$ denotes a material function, called scale function, which is a function of time scale ς, and its rate $\dot{\varsigma}$. In order to describe the visco-plastic material behavior, the rate sensitivity is introduced in scale function through rate of time scale. It should be mentioned that the so-called consistency visco-plastic model can be derived as special case of the proposed model by introducing Dirac delta function to the hereditary function (Wang *et al.* 1997).

In above equations, the total stress tensor is denoted by $\boldsymbol{\sigma}$ and its deviatoric and hydrostatic parts by $\boldsymbol{\sigma}_{dev}$ and σ_h, respectively. The symbol $\boldsymbol{\varepsilon}$ represent the total strain tensor by deviatoric and volumetric parts $\boldsymbol{\varepsilon}_{dev}$ and ε_{vol}, respectively. The superscripts '*e*' and '*p*' indicate the elastic and plastic components, respectively.

As shown by Valanis (1980), the kernel function $\Phi(z)$ can be expressed in terms of a Dirichlet series as

$$\Phi(z) = \sum_{r=1}^{\infty} A_r e^{-\alpha_r z} \tag{7.73}$$

with the requirements that A_r and α_r are non-negative for all values of 'r', and the condition

$$\sum_{r=1}^{\infty} \frac{A_r}{\alpha_r} < \infty \tag{7.74}$$

This condition ensures the integrability of $\Phi(z)$ over a finite domain of time scale z. In numerical application, m-term Dirichlet series can be used as (Hsu *et al.* 1991)

$$\Phi(z) = \sum_{r=1}^{m} A_r e^{-\alpha_r z} \tag{7.75}$$

The role of scale function $f(\varsigma, \dot{\varsigma})$ is crucial in the behavior of model. By scaling intrinsic time, this function causes hardening or softening plastic behavior as functions of time scale and its rate. Although the yield surface has not been explicitly assumed in the endochronic theory, introducing Dirac delta to kernel functions, results implicitly to this concept (Valanis 1980). The above endochronic model contains various isotropic and kinematic hardening rules for special cases (Watanbe and Atluri 1986), depending on the choice of the scale function $f(\varsigma, \dot{\varsigma})$ and the kernel $\Phi(z)$. Prager's linear kinematic hardening rule can be obtained if Dirac delta function and one constant term are used in kernel function. Introducing another exponential term to the kernel function results in the Armstrong and Frederick nonlinear kinematic hardening rule.

7.7. Multi-surface plasticity theory

Complicated behavior of materials has been demanding the implementation of advanced constitutive models in finite element methods. The plastic behavior of most materials shows non-constant hardening function. It means that the uniaxial stress-strain curve has several consecutive line segments with different slopes (hardening parameters). This non-linearity and succession of different material responses motivate the concept of multi-surface plasticity. The yield surfaces may be based on the von-Mises yield surface, or Drucker-Prager criterion. The implementation of the model on the deviatoric plane results in a set of hyper-cylinders in the case of von-Mises yield surfaces. If Drucker-Prager criterion is used (pressure-dependent model) the surfaces have conical shape. The yield surfaces may harden kinematically, or isotropically, or in a combination of both kinds of hardening.

A number of plasticity models have been proposed by researchers for cyclic plasticity. In an alternative approach to the isotropic hardening type models, Iwan (1967) followed the related works of Masing (1926) and developed a one dimensional plasticity model which consists of a collection of perfectly elastic and rigid plastic or slip elements arranged in either a series-parallel or a parallel-series combination. A multi-surface theory was independently proposed by Mroz (1967), in which a collection of nested yield surfaces was used to provide a piecewise linear approximation to the material stress-strain curve. The two-surface theories were developed by Krieg (1975), Dafalias and Popov (1976) and Petersson and Popov (1977). A model based on nonlinear kinematic hardening rules was proposed by Chaboche *et al.* (1979). One difference between these models is the manner in which the generalized plastic modulus is established. In the nonlinear kinematic hardening model, the generalized plastic modulus is derived from the consistency condition, whereas the two- or multi-surface models determine the generalized plastic modulus from a general loading situation, and the consistency condition is then satisfied. Prevost (1978, 1982) extended the Iwan-Mroz model for the undrained behavior of clays under cyclic loading condition, in which the Mroz's translation rule is used to the nested yield surfaces during plastic loading.

Classical isotropic hardening and Prager's kinematic hardening models offer a reasonable explanation of the hardening properties of materials where the load is increasing monotonically and no unloading occurs. Drucker and Palgen (1981) observed that the material behavior under cyclic loading are much more complex than monotonic loading and can not be modeled by isotropic hardening and Prager's kinematic hardening rules. Based on their experimental observations, they suggested several necessary requirements for cyclic plasticity models. These requirements which were further elaborated by Dafalias (1986) are as follows;

i) Unsymmetrical stress cycles will cause cyclic creep or ratcheting effect in the direction of mean stress,

ii) Unsymmetrical strain cycles will cause progressive relaxation of the mean stress to zero,

iii) The model must predict as accurate as possible the variation of the plastic modulus during random cyclic loading which means the constitutive equation should be able to model the smooth elastic-plastic translation, and

iv) Under symmetric stress and strain cycles the material hardens or softens toward a properly defined stabilized state with only kinematic hardening.

One of the most important parts in implementation of multi-surface plasticity models is the yield surface translation rule. The direction of movements of yield surfaces, or kinematic hardening function, has a crucial role in computation of elastoplastic modulus and convergency rate. There are many kinematic hardening rules, which have been employed by researchers, such as: Prager, Ziegler and Saczuk-Baltov hardening rules. Because of overlapping of the surfaces in their translation rule, the typical rule for these kinds of models is considered to be of the Mroz type (McDowell 1989). The Mroz rule allows the multiaxial description of nonlinear hardening functions and preserves the Masing extended behavior, which is important for simulation of cyclic response. Borja *et al.* (1994, 1999) developed a theory of bounding surface model which does not preserve the Masing behavior, but approximate to it. Montans (2000) extended a virtual bounding surface model which consists of a bounding and loading surfaces and preserves the Masing behavior. Recently, a multi-surface plasticity model was developed by Elgamal *et al.* (2003) for capturing the characteristics of cyclic mobility based on the new flow and hardening rules which is originally derived from the traditional Mroz translation rule with some modifications for low confinement levels to gain the desirable numerical convergency. The new flow rule allows for reproducing cyclic shear strain accumulation, and the new hardening rule enhances numerical robustness and efficiency.

In plasticity behavior of material, the Masing rule is an inherent property of kinematic hardening, needs to be preserved. On the other hand, without a modification in hardening rule the model simulates the closed-form stress-strain loops, i.e. zero ratcheting. The cyclic plastic strain accumulation, or ratcheting, is a complicated material behavior, which is important in the design of structural and geotechnical components enduring cyclic loading. The phenomena has been experimentally investigated and numerically modeled by many researchers (Ohno and Wang 1993, Chaboche 1994, Delobelle *et al.* 1995, Khan and Jackson 1999, Abdel-Karim and Ohno 2000, Yoshida 2000, Kobayashi and Ohno 2002 and Feaugas and Gaudin 2004). One of the most well-known models in cyclic plasticity is the non-linear kinematic hardening Armsrtrong-Frederick model. Although the model is highly capable of simulating the stress strain loops in cyclic plastic loading, it suffers from a serious deficiency in simulation of plastic strain accumulation in ratcheting phenomena. In order to solve the problem, the kinematic hardening rule must be modified. Kobayashi and Ohno (2002) presented an implicit kinematic hardening constitutive law to model cyclic plasticity. They derived the consistent tangent modulus used in their model to simulate the multi-linear and nonlinear evolution of back stress. Recently, a constitutive model is proposed by Bari and Hassan (2000, 2002) in simulation of ratcheting behavior. They modified the kinematic hardening rule developed by Delobelle *et al.* (1995) in the framework of the Chaboche model. The new hardening rule introduces only one multiaxial load dependent parameter to the Chaboche model, but performs the best in simulating ratcheting responses. They evaluated the performance of their models by employing the works of Ohno-Wang (1993), Chaboche (1994) and Abdel Karim-Ohno (2000).

In the next sections, an application of the multi-surface plasticity theory is presented for pressure-independent and pressure-dependent materials. The Mroz translation rule is applied to the movements of the yield surfaces, which offers great flexibility in describing the behavior of materials. Fully implicit scheme with radial mapping method is applied in numerical computations. In order to gain the second order convergency, the consistent tangent modulus is extracted. The 'Masing' rule is preserved by employing the kinematic hardening model. In addition, the model is capable to consider the plastic strain accumulation in constant axial stress state, such as ratcheting.

7.8. Multi-surface theory of J_2 plasticity

The multi-surface yield criterion is defined as a function of 'external' and 'internal' variables, in which the stress state is a representative of external variables. The back stress tensor $\boldsymbol{\alpha}$, which is the coordinates of the center of the surface in deviatoric plane, and the radius of yield surface r are the internal variables parameters. The average plastic strain $\bar{\varepsilon}^p$ can be also considered as the internal plastic variable. Thus, the yield surface can be expressed as

$$f^i = \left\| \boldsymbol{\sigma} - \boldsymbol{\alpha}^i \right\| - r^i \tag{7.76}$$

These surfaces have the same shape of yield surface and, initially in the virgin state, they are centered in the origin. However, they can be modeled in an anisotropic hardening rule, as demonstrated in Figure 7.11. In this figure, the configuration of yield surfaces is shown in deviatoric principal stress space, while the hysterisis curve is presented in principal stress-strain space. Note that the change of the slope of hysterisis curve makes active a new yield surface. Three yield surfaces have been considered here and thus, three different slops can be traced out. Figure 7.11(a) presents a virgin state in which the material is inherently isotropic. In this case, no pre-loading/pre-shearing has been experienced by the material in its loading history. Figure 7.11(b) shows the material behavior with no tensile strength, such as granular soils. In this case, the tensile strength and the location of yield surfaces (back stress tensor) can be simply calculated using the typical laboratory tests. Figure 7.11(c) presents the material behavior with different tensile/compression strength, such as concrete. Apart from inherent anisotropy, pre-loading or pre-straining causes anisotropy which can be modeled by the multi-surface plasticity model.

In multi-surface plasticity, there are several yield surfaces. The most inner surface indicates the boundary of elastic region. When the stress state at a Gauss point exceeds the first yield surface, the material behaves plastically and the surface hardens kinematically according to the hardening modulus associated with this yield surface. This process continues until the active yield surface arrives to the next one. At this stage, both yield surfaces, i.e. the 'active' and 'target' surfaces, harden. It means that the stress point will 'tow' both surfaces. In order to formulate different yield surfaces, the super index '*i*' is used for the loading surface, in which the abbreviation '*a*' stands for the 'active surface', and '*a*+1' for the 'target surface'. Thus,

$$f^{a} = \left\| \boldsymbol{\sigma} - \boldsymbol{\alpha}^{a} \right\| - r^{a} = 0 \tag{7.77}$$

$$f^{a+1} = \left\| \boldsymbol{\sigma} - \boldsymbol{\alpha}^{a+1} \right\| - r^{a+1} < 0 \tag{7.78}$$

Note that the abovementioned relations are extracted in accordance with Kuhn-Tucker condition. This condition guarantees that the multiplication of plastic multiplier, which will be introduced later, and the yield surface equation will always be zero. When the stress state is inside the yield surface (elastic state), the plastic multiplier is zero. In the case that the stress state goes beyond the yield surface, the plastic multiplier has a non-negative value, so the yield surface f must be zero. In addition, the active surface can not overlap the target surface which can be guaranteed by the following conditions

$$\begin{aligned} & r^{i} > r^{j} && \forall i > j \\ & \left\| \boldsymbol{\sigma} - \boldsymbol{\alpha}^{i} \right\| > \left\| \boldsymbol{\sigma} - \boldsymbol{\alpha}^{j} \right\| && \forall i > j \end{aligned} \tag{7.79}$$

In order to prevent overlapping of two yield surfaces, the Mroz translation rule is applied. In this method, the translation vector can be obtained by connecting the stress point on active surface to conjugate point, i.e., the point on target surface with the same normal, as shown in Figure 7.12. As the yield surface is circular, the normal to yield surface is similar to radial vector at this point.

$$\boldsymbol{m} = \frac{r^{a+1}}{\left\| \boldsymbol{\sigma} - \boldsymbol{\alpha}^{a} \right\|} (\boldsymbol{\sigma} - \boldsymbol{\alpha}^{a}) - (\boldsymbol{\sigma} - \boldsymbol{\alpha}^{a+1}) \tag{7.80}$$

The Prager's translation rule may be used only for the last surface, i.e. the bounding surface, where no more surfaces are defined. In this rule, the moving direction is normal to yield surface at the stress point, and because all surfaces are circular, the moving direction for bounding surface is in the direction of radial vector at the stress point.

Since an associated flow rule is applied, the plastic strain direction is estimated using the normal vector to the yield surface. The loading condition is defined as

$$\dot{\boldsymbol{\sigma}} : \hat{\boldsymbol{n}} > 0 \quad \text{and} \quad f^{1}(\boldsymbol{\sigma}) = 0 \tag{7.81}$$

where $\hat{\boldsymbol{n}}$ is the unit normal to the yield surface. The total strain is divided into elastic and plastic parts, in which the plastic portion is computed as

$$\dot{\boldsymbol{\varepsilon}}^{p} = \dot{\gamma} \hat{\boldsymbol{n}} \tag{7.82}$$

The stress rate tensor is

$$\dot{\boldsymbol{\sigma}} = \boldsymbol{C} : \dot{\boldsymbol{\varepsilon}}^{e} = \boldsymbol{C} : (\dot{\boldsymbol{\varepsilon}} - \dot{\gamma} \hat{\boldsymbol{n}}) \tag{7.83}$$

where $\boldsymbol{C}$ is the elastic forth order tensor defined as

$$\boldsymbol{C} = K\mathbf{1}\otimes\mathbf{1} + 2\mu\,\boldsymbol{P}_1 \tag{7.84}$$

where μ and K are the shear and bulk moduli. In above equations, the dot product is shown by ':' which denotes the two indices contraction ($\boldsymbol{\alpha}:\boldsymbol{\beta} = a_{ij}b_{ij}$), '$\otimes$' is matrix product and the norm '$\|.\|$' is defined by $\|\boldsymbol{\alpha}\| = \sqrt{\boldsymbol{\alpha}:\boldsymbol{\alpha}}$. '$\boldsymbol{I}$' is the forth order identity tensor and '$\mathbf{1}$' the second order one. The sign '$\hat{\ }$' shows normalized tensors and vectors. All tensors are defined in deviatoric space except those which are shown with superscript '^{f}' which are full tensors ($\boldsymbol{\sigma} = \boldsymbol{\sigma}^d = \boldsymbol{P}_1 : \boldsymbol{\sigma}^f$). The projection to the deviatoric plane is defined by $\boldsymbol{P}_1 = \boldsymbol{I} - (\mathbf{1}\otimes\mathbf{1})/\|\mathbf{1}\|^2$.

Applying the Mroz translation rule, the back stress rate tensor can be obtained in a simple manner as

$$\dot{\boldsymbol{\alpha}} = \dot{\gamma}\,\boldsymbol{H} : \hat{\boldsymbol{m}} \tag{7.85}$$

where $\boldsymbol{H}$ is the hardening tensor, calculated in next sections, and $\hat{\boldsymbol{m}}$ denotes the unitary translation tensor. $\dot{\gamma}$ is called the consistency parameter, or plastic multiplier that can be calculated by enforcing consistency condition ($\dot{f} = 0$), and using equation (7.83) as

$$\dot{\gamma} = \frac{\hat{\boldsymbol{n}} : \boldsymbol{C} : \dot{\boldsymbol{\varepsilon}}}{\hat{\boldsymbol{n}} : \boldsymbol{C} : \hat{\boldsymbol{n}} + \hat{\boldsymbol{n}} : \boldsymbol{H} : \hat{\boldsymbol{m}}} \tag{7.86}$$

Finally, the continuum elasto-plastic matrix can be obtained from relations (7.83) and (7.86) as

$$\boldsymbol{C}^{ep} = \boldsymbol{C} - \frac{\hat{\boldsymbol{n}} : \boldsymbol{C} \otimes \hat{\boldsymbol{n}} : \boldsymbol{C}}{\hat{\boldsymbol{n}} : \boldsymbol{C} : \hat{\boldsymbol{n}} + \hat{\boldsymbol{n}} : \boldsymbol{H} : \hat{\boldsymbol{m}}} \tag{7.87}$$

7.8.1. Computational algorithm

The Mroz translation rule employed in previous section is obviously an explicit rule. In order to implement an implicit manner, the following procedure is applied to modify this translation rule. As the unitary normal tensor needs to be computed iteratively, the translation rule can be defined based on the normal tensor to target surface as

$$\hat{\boldsymbol{t}}_{n+1} = \frac{1}{\left\|\boldsymbol{\sigma}^{tr}_{n+1} - \boldsymbol{\alpha}^{a+1}_{n}\right\|}(\boldsymbol{\sigma}^{tr}_{n+1} - \boldsymbol{\alpha}^{a+1}_{n}) \tag{7.88}$$

Applying the back stress normal tensor, defined in equation (7.88), the final back stress tensor on the active surface will be computed by

$$\overline{\boldsymbol{\alpha}}_{n+1} = \boldsymbol{\alpha}^{a+1}_{n+1} + (r^{a+1} - r^{a})\,\hat{\boldsymbol{t}}_{n+1} \tag{7.89}$$

$$\boldsymbol{m}_{n+1} = \overline{\boldsymbol{\alpha}}^{a}_{n+1} - \overline{\boldsymbol{\alpha}}^{a}_{n} \tag{7.90}$$

In above equations, the sub-index n denotes the last converged step, while n+1 indicates the current state of the analysis. As can be seen from relations (7.88) and (7.89), the normal to active and target yield surfaces are computed based on the current unknown step which demonstrates the fully implicit manner of model. Note that in relation (7.89) $\boldsymbol{\alpha}_{n+1}^{a+1}$ can be replaced by $\boldsymbol{\alpha}_{n}^{a+1}$, as it indicates the back stress of target surface and does not move. In the present algorithm, each surface inside the target surface has its own translation rule such that at the end of step all surfaces contact in the final stress state. The concept of active and target surfaces can be defined as

$$f^{a}(\boldsymbol{\sigma}_{n+1}^{tr},\boldsymbol{\alpha}_{n}^{a})>0 \qquad \text{Active surface(s)} \tag{7.91}$$

$$f^{a+1}(\boldsymbol{\sigma}_{n+1}^{tr},\boldsymbol{\alpha}_{n}^{a})<0 \qquad \text{Target surface} \tag{7.92}$$

The back stress tensor of the active surface can be easily computed by

$$\boldsymbol{\alpha}_{n+1}^{a}=\boldsymbol{\alpha}_{n}^{a}+\Delta\gamma_{n+1}^{a}\overline{H}^{a}\hat{\boldsymbol{m}}_{n+1}^{a} \tag{7.93}$$

where $\overline{H}$ is the modified hardening modulus. The translation vector is computed as

$$\hat{\boldsymbol{m}}_{n+1}^{a}=\frac{\boldsymbol{m}_{n+1}^{a}}{\left\|\boldsymbol{m}_{n+1}^{a}\right\|} \tag{7.94}$$

where

$$\boldsymbol{m}_{n+1}^{a}=(r^{a+1}-r^{a})\hat{\boldsymbol{t}}_{n+1}+\boldsymbol{\alpha}_{n}^{a+1}-\boldsymbol{\alpha}_{n}^{a} \tag{7.95}$$

From the stress state, we will have

$$\begin{aligned}\boldsymbol{\sigma}_{n+1}&=\boldsymbol{\alpha}_{n+1}^{a}+r^{a}\hat{\boldsymbol{n}}_{n+1}\\&=\boldsymbol{\alpha}_{n+1}+r\,\hat{\boldsymbol{n}}_{n+1}\end{aligned} \tag{7.96}$$

Therefore, the back stress tensor of the inner surfaces can be obtained by

$$\boldsymbol{\alpha}_{n+1}^{i}=\boldsymbol{\alpha}_{n}^{i}+\Delta\gamma_{n+1}^{i}\overline{H^{i}}\,\hat{\boldsymbol{m}}_{n+1}^{i} \tag{7.97}$$

where

$$\hat{\boldsymbol{m}}_{n+1}^{i}=(\boldsymbol{\alpha}_{n}-\boldsymbol{\alpha}_{n}^{i})+(r-r^{i})\,\hat{\boldsymbol{n}}_{n+1}+\overline{H}^{a}\Delta\gamma_{n+1}^{a}\hat{\boldsymbol{m}}_{n+1} \tag{7.98}$$

7.8.2. Consistent Tangent Operator

In order to achieve the proper convergency in the implicit scheme, the procedure of iterative solution is performed based on an elastic prediction and plastic correction. The trial stress is therefore calculated using the plastic freezing process as

$$\boldsymbol{\sigma}_{n+1}^{tr}=\boldsymbol{\sigma}_{n}+2\mu\Delta\boldsymbol{\varepsilon} \tag{7.99}$$

where $\Delta\boldsymbol{\varepsilon}$ is the strain increment from iteration n to n+1. Note that the stress state is defined as a function of strain state. In Figure 7.13, the configuration of yield surface is shown for three successive time steps; the 'initial', 'intermediate' and 'final' or 'converged' time steps. The second and third configurations can not be obtained graphically and they are demonstrated for a better description of the scheme. Also plotted in this figure are the normal to target surface, the translation vector of active surfaces, and the trial and converged yield surface configurations.

By differentiating $\Delta\gamma$, defined in the incremental form of equation (7.82) as $\Delta\dot{\boldsymbol{\varepsilon}}^p = \Delta\dot{\gamma}_{n+1}\,\hat{\boldsymbol{n}}_{n+1}$, the residual equation that controls the convergency can be obtained as

$$g(\Delta\gamma_{n+1}^a) = (\boldsymbol{\sigma}_{n+1}^{tr} - \boldsymbol{\alpha}_n^a) : \hat{\boldsymbol{n}}_{n+1} - \overline{H}^a \Delta\gamma_{n+1}^a (\hat{\boldsymbol{m}}_{n+1} : \hat{\boldsymbol{n}}_{n+1}) - 2\mu\Delta\gamma_{n+1} - r \tag{7.100}$$

The non-linear function $\Delta\gamma$ is solved by Newton method in an iterative manner by computing $dg/d(\Delta\gamma_{n+1}^a)$ (Montans 2000). Once the solution is obtained the consistency parameter is extracted and the stress and strain are therefore calculated as

$$\boldsymbol{\alpha}_{n+1}^a = \boldsymbol{\alpha}_n^a + \overline{H}^a \Delta\gamma_{n+1}^a\, \hat{\boldsymbol{m}}_{n+1} \tag{7.101}$$

$$\boldsymbol{\alpha}_{n+1}^i = \boldsymbol{\alpha}_n^a + (r^a - r^i)\,\hat{\boldsymbol{n}}_{n+1} \tag{7.102}$$

$$\boldsymbol{\sigma}_{n+1} = \boldsymbol{\alpha}_{n+1} + r\,\hat{\boldsymbol{n}}_{n+1} \tag{7.103}$$

$$\Delta\boldsymbol{\varepsilon}^p = \Delta\gamma_{n+1}\hat{\boldsymbol{n}}_{n+1} \tag{7.104}$$

$$\boldsymbol{\sigma}_{n+1}^f = \boldsymbol{\sigma}_{n+1} + K(\boldsymbol{\varepsilon}_{n+1}^f : \mathbf{1})\mathbf{1} \tag{7.105}$$

$$\boldsymbol{\varepsilon}_{n+1}^p = \boldsymbol{\varepsilon}_n^p + \sum_{i=1}^{a} \Delta\gamma_{n+1}^i \hat{\boldsymbol{n}}_{n+1} \tag{7.106}$$

Finally, the algorithmic elasto-plastic tangent can be computed as

$$\boldsymbol{C}_{n+1}^{ep} = K\mathbf{1}\otimes\mathbf{1} + \frac{\partial\boldsymbol{\sigma}_{n+1}}{\partial\boldsymbol{\varepsilon}_{n+1}} : \frac{\partial\boldsymbol{\varepsilon}_{n+1}}{\partial\boldsymbol{\varepsilon}_{n+1}^f} \tag{7.107}$$

where $\partial\boldsymbol{\varepsilon}_{n+1}/\partial\boldsymbol{\varepsilon}_{n+1}^f = \boldsymbol{P}_1$.

In order to calculate $\partial\boldsymbol{\sigma}_{n+1}/\partial\boldsymbol{\varepsilon}_{n+1}$ in equation (7.107), the consistency condition $\dot{f} = 0$ must be enforced. By differentiating equation (7.77), which has been rewritten in trial stress state, we obtain

$$\hat{\boldsymbol{n}}_{n+1} : \frac{\partial\boldsymbol{\sigma}_{n+1}^{tr}}{\partial\boldsymbol{\varepsilon}_{n+1}} - \overline{H}^a \Delta\gamma_{n+1}^a \hat{\boldsymbol{n}}_{n+1} : \frac{\partial\hat{\boldsymbol{m}}_{n+1}}{\partial\boldsymbol{\varepsilon}_{n+1}} - \overline{H}^a (\hat{\boldsymbol{m}}_{n+1} : \hat{\boldsymbol{n}}_{n+1}) \frac{\partial\Delta\gamma_{n+1}^a}{\partial\boldsymbol{\varepsilon}_{n+1}} - 2\mu \frac{\partial\gamma_{n+1}}{\partial\boldsymbol{\varepsilon}_{n+1}} = 0 \tag{7.108}$$

The derivative of trial stress, translation tensor, and normal tensor to the active and target surfaces with respect to strain can be calculated as

$$\frac{\partial \boldsymbol{\sigma}_{n+1}^{tr}}{\partial \boldsymbol{\varepsilon}_{n+1}} = 2\mu\, \boldsymbol{I} \tag{7.109}$$

$$\frac{\partial \hat{\boldsymbol{m}}_{n+1}}{\partial \boldsymbol{\varepsilon}_{n+1}} = \frac{(r^{a+1} - r^{a})}{\left\| \boldsymbol{m}_{n+1} \right\|} \boldsymbol{P}_m : \frac{\partial \hat{\boldsymbol{t}}_{n+1}}{\partial \boldsymbol{\varepsilon}_{n+1}} \tag{7.110}$$

$$\frac{\partial \hat{\boldsymbol{t}}_{n+1}}{\partial \boldsymbol{\varepsilon}_{n+1}} = \frac{2\mu}{\left\| \boldsymbol{t}_{n+1} \right\|} \boldsymbol{P}_t \tag{7.111}$$

where $\boldsymbol{P}_m = \boldsymbol{I} - \hat{\boldsymbol{m}}_{n+1} \otimes \hat{\boldsymbol{m}}_{n+1}$ and $\boldsymbol{P}_t = \boldsymbol{I} - \hat{\boldsymbol{t}}_{n+1} \otimes \hat{\boldsymbol{t}}_{n+1}$. Substituting equation (7.111) into (7.110), leads to

$$\frac{\partial \hat{\boldsymbol{m}}_{n+1}}{\partial \boldsymbol{\varepsilon}_{n+1}} = \frac{(r^{a+1} - r^{a})}{\left\| \boldsymbol{m}_{n+1} \right\|} \frac{2\mu}{\left\| \boldsymbol{t}_{n+1} \right\|} \boldsymbol{P}_m : \boldsymbol{P}_t \tag{7.112}$$

The derivative of the increment of consistency parameter can be computed as

$$\frac{\partial \Delta\gamma_{n+1}^{i}}{\partial \boldsymbol{\varepsilon}_{n+1}} = \frac{(r - r^{i})}{\overline{H}^{i}} \hat{\boldsymbol{m}}_{n+1}^{i} : \frac{\partial \hat{\boldsymbol{n}}_{n+1}}{\partial \boldsymbol{\varepsilon}_{n+1}} + \frac{\overline{H}^{a}}{\overline{H}^{i}} (\hat{\boldsymbol{m}}_{n+1} : \hat{\boldsymbol{m}}_{n+1}^{i}) \frac{\partial \Delta\gamma_{n+1}^{a}}{\partial \boldsymbol{\varepsilon}_{n+1}} + \frac{\overline{H}^{a}}{\overline{H}^{i}} \Delta\gamma_{n+1}^{a} \hat{\boldsymbol{m}}_{n+1}^{i} : \frac{\partial \hat{\boldsymbol{m}}_{n+1}}{\partial \boldsymbol{\varepsilon}_{n+1}} \tag{7.113}$$

where

$$\frac{\partial \hat{\boldsymbol{n}}_{n+1}}{\partial \boldsymbol{\varepsilon}_{n+1}} = \frac{1}{\left\| \boldsymbol{n}_{n+1} \right\|} \boldsymbol{P}_n : \left(2\mu\, \boldsymbol{I} - \overline{H}^{a} \Delta\gamma_{n+1}^{a} \frac{\partial \hat{\boldsymbol{m}}_{n+1}}{\partial \boldsymbol{\varepsilon}_{n+1}} - \overline{H}^{a} \hat{\boldsymbol{m}}_{n+1} \otimes \frac{\partial \Delta\gamma_{n+1}^{a}}{\partial \boldsymbol{\varepsilon}_{n+1}} \right) \tag{7.114}$$

where $\boldsymbol{P}_n = \boldsymbol{I} - \hat{\boldsymbol{n}}_{n+1} \otimes \hat{\boldsymbol{n}}_{n+1}$. Thus, the derivative of the consistency parameter can be computed as

$$\frac{\partial \gamma_{n+1}}{\partial \boldsymbol{\varepsilon}_{n+1}} = \frac{\partial \Delta\gamma_{n+1}^{a}}{\partial \boldsymbol{\varepsilon}_{n+1}} + \sum_{i=1}^{a-1} \frac{\partial \Delta\gamma_{n+1}^{i}}{\partial \boldsymbol{\varepsilon}_{n+1}} \tag{7.115}$$

Substituting equations (7.109), (7.112), (7.113) and (7.115) into equation (7.108), leads to

$$\begin{aligned} \frac{\partial \Delta\gamma_{n+1}^{a}}{\partial \boldsymbol{\varepsilon}_{n+1}} = \left(\frac{dg}{d(\Delta\gamma_{n+1}^{a})} \right)^{-1} \Bigg[& 2\mu\, \hat{\boldsymbol{n}}_{n+1} - \overline{H}^{a} \Delta\gamma_{n+1}^{a} \hat{\boldsymbol{n}}_{n+1} : \frac{\partial \hat{\boldsymbol{m}}_{n+1}}{\partial \boldsymbol{\varepsilon}_{n+1}} \\ & - 2\mu \sum_{i=1}^{a-1} \left(\frac{\overline{H}^{a}}{\overline{H}^{i}} \Delta\gamma_{n+1}^{a} \hat{\boldsymbol{m}}_{n+1} : \frac{\partial \hat{\boldsymbol{m}}_{n+1}}{\partial \boldsymbol{\varepsilon}_{n+1}} + \frac{2\mu}{\overline{H}^{i}} \frac{(r - r^{i})}{\left\| \hat{\boldsymbol{n}}_{n+1} \right\|} \hat{\boldsymbol{m}}_{n+1}^{i} : \boldsymbol{P}_n \right) \Bigg] \end{aligned} \tag{7.116}$$

It must be noted that $dg/d(\Delta\gamma_{n+1}^{a})$ has been already computed from the Newton-Raphson algorithm of the previous iteration. Once $\partial \Delta\gamma_{n+1}^{a} / \partial \boldsymbol{\varepsilon}_{n+1}$ is calculated, the value of $\partial \hat{\boldsymbol{n}}_{n+1} / \partial \boldsymbol{\varepsilon}_{n+1}$ can be computed from equation (7.114), and then the strain derivative of the deviatoric stress tensor can be calculated from equations (7.102) and (7.103) as

$$\frac{\partial \boldsymbol{\sigma}_{n+1}}{\partial \boldsymbol{\varepsilon}_{n+1}} = \overline{H}^{a} \Delta\gamma_{n+1}^{a} \frac{\partial \hat{\boldsymbol{m}}_{n+1}}{\partial \boldsymbol{\varepsilon}_{n+1}} + \overline{H}^{a} \hat{\boldsymbol{m}}_{n+1} \otimes \frac{\partial \Delta\gamma_{n+1}^{a}}{\partial \boldsymbol{\varepsilon}_{n+1}} + r \frac{\partial \hat{\boldsymbol{n}}_{n+1}}{\partial \boldsymbol{\varepsilon}_{n+1}} \tag{7.117}$$

Note that the full elasto-plastic tangent must be calculated by adding the non-deviatoric part of equation (7.117), as stated in equation (7.107).

One of the best known schemes for numerical integration of differential equations in return mapping algorithm, is the generalized midpoint rule. Let f: R→R be a smooth function, the initial values problem are defined by $\dot{x}(t) = f(x(t))$ and $x(0) = x_n$ for $t \in [0, T]$, in which $x_{n+1} = x_n + \Delta t\, f(x_{n+\upsilon})$ and $x_{n+\upsilon} = \upsilon\, x_{n+1} + (1-\upsilon)\, x_n$, with x_{n+1} denoting the algorithmic approximation to the exact value of $x(t_{n+1})$ at time $t_{n+1} = t_n + \Delta t$. According to the value of υ, the process is called the forward (explicit) Euler for $\upsilon = 0$, midpoint rule for $\upsilon = 0.5$, and backward (implicit) Euler for $\upsilon = 1$ (Simo and Hughes 1998). This is of particular interest that the second order of convergency can be obtained for $\upsilon = 0.5$, whereas unconditional stability is required for $\upsilon \geq 0.5$ (Hairer *et. al.* 1987). According to the computational algorithm presented here for the normal to active and target surfaces, translation tensor, and backstress tensor with reference to the unknown current step (fully implicit scheme), the proposed model results in second order convergancy which is unconditionally stable (Simo and Hughes 1998).

7.8.3. Parameter Determination

In order to determine the model parameters of proposed multi-surface plasticity, simple shear, uniaxial, or triaxial standard tests may be applied. The parameters of the model are the radius and hardening parameter of surfaces. In the case of anisotropic hardening the location of the yield surfaces (back stress tensor) must be determined. The radius and hardening parameter of the yield surfaces can be evaluated using the shear stress-strain curve ($\tau - \gamma_e$) obtained from a simple shear test. Therefore, the stress tensor is

$$[\sigma] = \begin{vmatrix} 0 & \tau & 0 \\ \tau & 0 & 0 \\ 0 & 0 & 0 \end{vmatrix}, \qquad \|\sigma\| = \sqrt{2}\tau \tag{7.118}$$

or by employing a simple triaxial test and obtaining the deviatoric stress-strain curve ($q - \varepsilon_q$), the stress tensor is

$$[\sigma] = \begin{vmatrix} \sigma^a & 0 & 0 \\ 0 & \sigma^r & 0 \\ 0 & 0 & \sigma^r \end{vmatrix}, \qquad \|\sigma\| = \sqrt{\frac{2}{3}}(\sigma^a - \sigma^r) = \sqrt{\frac{2}{3}} q \tag{7.119}$$

where $q = \sigma^a - \sigma^r$ is the triaxial deviatoric stress, with σ^a and σ^r denoting the axial and radial stresses. If the shear stress-strain and deviatoric stress-strain curves are not available, it can be simulated using a hyperbolic model as

$$\sigma_1 - \sigma_3 = \frac{\varepsilon}{a + b\varepsilon} \tag{7.120}$$

where a and b are the hyperbolic model parameters, with $\frac{1}{a}$ denoting the initial tangent modulus and $\frac{1}{b}$ the maximum deviatoric stress. Applying above procedure, results

$$\dot{\varepsilon} = \dot{\varepsilon}^e + \dot{\varepsilon}^p = \frac{1}{2\mu}\dot{\sigma} + \frac{1}{H}\dot{\sigma} \tag{7.121}$$

Thus, the monotonic tensorial tangent modulus $2\mu^t$ can be calculated as

$$2\mu^t = \frac{2\mu H}{2\mu + H} \tag{7.122}$$

and the hardening parameter can be obtained by

$$H = \frac{2\mu\mu^t}{\mu - \mu^t} \tag{7.123}$$

or

$$H = \frac{2}{3}\frac{E^i E^t}{E^i - E^t} \tag{7.124}$$

where μ is the initial tangent of shear stress-strain curve and μ^t indicates its value at the specified stress point. In the case of pure shear test, equation (7.123) can be employed to evaluate parameters of the model, while equation (7.124) can be applied if the uniaxial tension/compression, or triaxial, test is available. In equation (7.124), E^i denotes the Young modulus and E^t is tangent to the stress-strain curve at the specified stress state. As in computational algorithm of multi-surface plasticity all surfaces inside the active surface move together, the hardening parameter must be modified as follows

$$\frac{1}{\overline{H}^i} = \frac{1}{H^i} - \frac{1}{H^{i-1}} \tag{7.125}$$

where the super index i denotes the number of the surface, and $\frac{1}{H^0} = 0$.

In order to model the plastic strain accumulation and ratcheting, a new parameter is introduced here. In numerical simulation of ratcheting, there are various approaches to enforce the strain accumulation in nearly constant deviatoric stress in successive loops. The main strategy is to define the hardening rule such that the phenomenon can be captured through a modified elastoplastic consistent tangent modulus (Bari and Hassan 2002). Although this model possesses more elaborate formulas and computational algorithms it is able to predict the bi-axial ratcheting. This approach called coupled method is based on the computation of elastoplastic modulus in relation with hardening rule. In this study, the offset parameter is introduced to produce various hardening behavior in compression and tension. This approach which imposes the open stress-strain hysterisis in successive loops is in contrast to Massing rule. The advantage is simplicity the approach in experimental evaluation of the parameter and its implementation in numerical simulation. This parameter can be calculated using the general experimental tests, such as: simple shear test, which is used to estimate other parameters of the model.

Figure 7.14(a) presents the traditional kinematic hardening which preserves the Massing rule. In this figure, the configuration of yield surfaces together with their sizes and hardening parameters do not change throughout the computational simulation, so no extra plastic strain can be produced in successive loops. In Figures 7.14(b) and (c), two different approaches are demonstrated to simulate various hardening behavior in compression and tension. In Figure 7.14(b), the implementation of isotropic and kinematic hardening is presented. In this case, the size of yield surfaces can be modified according to the plastic strain state. It shows an appropriate algorithm for bounding surface model, in which two loading and bounding surfaces are employed. In multi-surface plasticity, the change in size of surfaces makes numerical difficulties. Figure 7.14(c) illustrates an approach proposed here to implement different hardening behavior in tension and compression. In this approach, a multiplier is applied to all surfaces to control the amount of hardening. It must be noted that the size of surfaces do not change during the computation.

7.9. Multi-surface theory of pressure-dependent plasticity

Computational plasticity has been found to be most versatile in modeling pressure-sensitive materials under arbitrary loading conditions. While classical plasticity is based on experimental observations of metal materials, its extension to pressure-dependent materials, such as powders, is drastic advances. The fundamental differences in mechanical behavior between metals and particulate materials are;

1. The bonds between the crystals in metal materials are so strong that metal behavior is not noticeably affected by hydrostatic pressure change in the range of most practical interests. In other words, metals deform independently of confining pressure stress. Conversely, in particulate powders, confining pressure stress is a major or even the only reason that particles hold together.
2. Within a significant deformation range, metal behavior can be treated as purely elastic. However, in particulate systems, such as: powders, elastic deformation is almost always accompanied by plastic deformation, i.e., a purely elastic region is virtually nonexistent.

Consequently, the multi-surface theory of J_2 plasticity, presented in previous section, needs to be extended to pressure-dependent plasticity in order to model general powder behavior.

The multi-surface yield function of pressure-sensitive material can be expressed based on the Drucker-Prager criterion as (Prevost 1985)

$$f = \frac{3}{2}(\boldsymbol{\sigma} - p_a \boldsymbol{\alpha}):(\boldsymbol{\sigma} - p_a \boldsymbol{\alpha}) - M^2 p_a^{\,2} = 0 \tag{7.126}$$

in which the above yield function forms a number of conical yield surfaces in stress space with a common apex and different sizes, as shown in Figure 7.15. In above equation, the orientation of a yield surface is determined by $\boldsymbol{\alpha}$, which is the second-order kinematic deviatoric tensor defining the yield surface coordinates, and M dictates the yield surface size defined by friction angle for the outmost surface. Each surface is associated with a constant plastic modulus. The outmost surface is designated as the failure surface. If $\boldsymbol{\alpha} = 0$, the axis of the cone coincides with the space diagonal and the yield surface presents isotropic behavior at that point. An initially non-zero $\boldsymbol{\alpha}$ indicates material or stress induced anisotropy, which could be due to the material's memory of its past loading history. A deviatoric plane ($p =$ constant) generally intersects an ellipse of a conical yield surface. Only if $\boldsymbol{\alpha} = 0$, it intersects a circle, however – a constant value of M implies that the conical surface deforms in such a way that at any $\boldsymbol{\alpha} \neq 0$, its intersection with a deviatoric plane is always circular of radius Mp_a.

The outer unit normal $\boldsymbol{Q}$ to the yield surface is calculated as

$$\boldsymbol{Q} = \frac{\nabla f}{\|\nabla f\|} \tag{7.127}$$

where $\|\nabla f\|$ is the Euclidean norm of ∇f defined by

$$\nabla f = \frac{\partial f}{\partial \boldsymbol{\sigma}} = 3(\boldsymbol{\sigma} - p_a \boldsymbol{\alpha}) + \left[p_a \left(\boldsymbol{\alpha} : \boldsymbol{\alpha} - \frac{2}{3} M^2 \right) - \boldsymbol{\sigma} : \boldsymbol{\alpha} \right] \boldsymbol{\delta} \tag{7.128}$$

A hardening rule which specifies the modification of yield condition in the course of plastic flow, is adopted for multi-surface plasticity based on the deviatoric kinematic hardening. As mentioned before in the context of multi-surface plasticity, translation of the yield surface is generally governed by the consideration that no overlapping is allowed between yield surfaces (Mroz 1967). Thus, the contact between consecutive similar surfaces f_m and f_{m+1}, shown in Figure 7.15, must occur only at conjugate points with the same direction of outward normal. In this case, the current (deviatoric) stress state $\boldsymbol{\sigma}$ on the active surface f_m can be employed together with its conjugate point on the next outer surface f_{m+1}, to define the translation direction $\boldsymbol{\mu}$ as follows

$$\boldsymbol{\mu} = \frac{\boldsymbol{M}_{m+1}}{\boldsymbol{M}_m} (\boldsymbol{\sigma} - p_a \boldsymbol{\alpha}_m) - (\boldsymbol{\sigma} - p_a \boldsymbol{\alpha}_{m+1}) \tag{7.129}$$

where $p_a \boldsymbol{\alpha}_m$ and $p_a \boldsymbol{\alpha}_{m+1}$ are the centers of f_m and f_{m+1} in deviatoric plane, respectively. With the direction of translation defined above, the amount of translation $d\boldsymbol{\mu}$ can be then obtained by satisfying the consistency condition $\dot{f} = 0$ (Mroz 1967). After updating the active surface f_m, all inner surfaces are translated to be tangential at the updated stress state $(\boldsymbol{\sigma} + d\boldsymbol{\sigma})$ on f_m.

A general scheme for the stress integration algorithm can be followed employing the methodology proposed in Section 7.8, based on the radial return mapping method. In this case, a stress point is only allowed to stay within a half-open space enclosed by the failure surface and $p = 0$ plane. Furthermore, in the case of plastic loading, the stress point must remain on or outside the active yield surface. In a numerical implementation of the radial return mapping procedure, it is possible that after correction, the stress point stays out of the failure surface, or in the $p < 0$ half space, or inside the active surface. However, if the stress point is out of the failure surface, it will be taken back to where its deviatoric component intersects the failure surface. In the case of $p < 0$, the stress point will be taken back to where the $p = 0$ plane intersects the vector connecting the stress point to the trial stress point before correction. Finally, if the stress point is inside the active surface, it will be taken back to the point where the active surface intersects the vector $(\boldsymbol{\sigma} - p_a \boldsymbol{\alpha})$.

It must be noted that the multi-surface yield function of (7.126) is open in the positive direction of hydrostatic axis (Figure 7.15). Thus for a stress path along that direction, the material behavior is nonlinearly elastic. This implies that permanent volumetric deformation may not be simulated accurately to demonstrate the densification behavior of powders. In this case, a cap yield function can be combined together with the multi-surface function to close the open end, as shown in Figure 7.16. However, significant research is necessary for formulation of cap plasticity yield function within the multi-surface context.

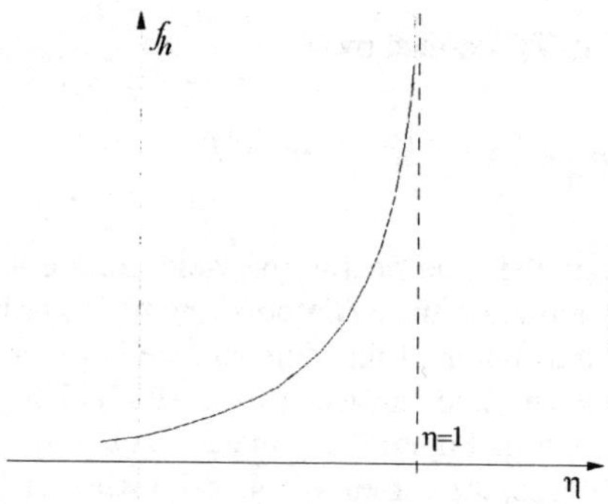

Figure 7.1. Dependence of $f_h(\eta, p)$ on relative density η

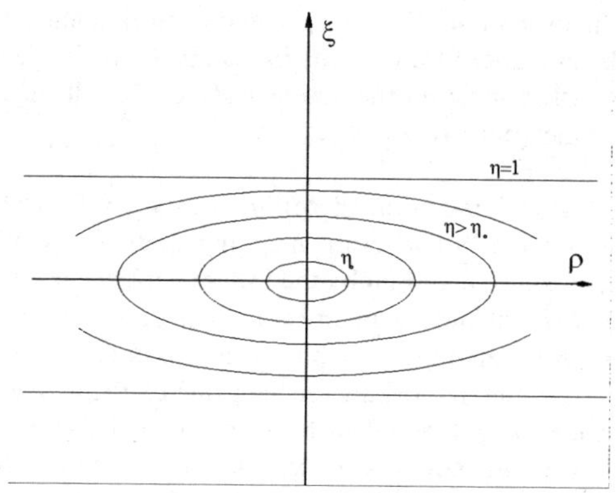

Figure 7.2. Trace of the elliptical yield function on meridian plane

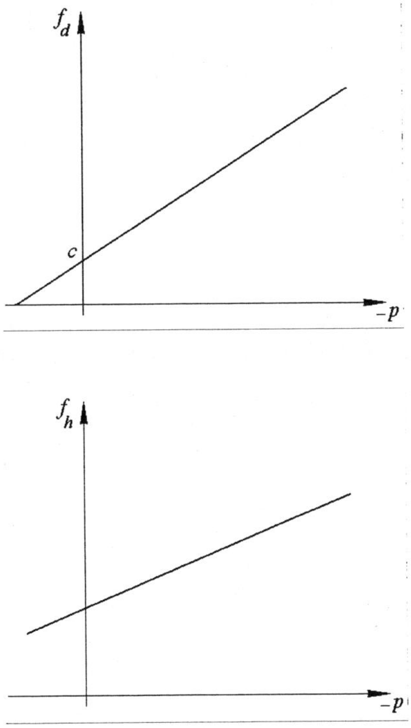

Figure 7.3. Dependence of $f_d(p)$ and $f_h(\eta, p)$ on hydrostatic pressure p

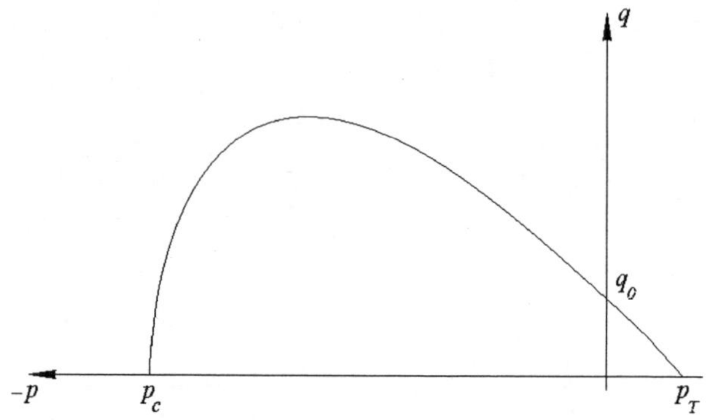

Figure 7.4. Trace of a typical yield function on meridian plane

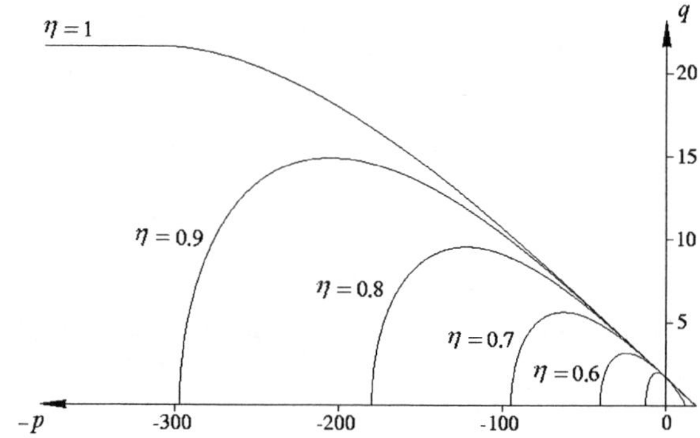

Figure 7.5. Trace of cone-cap yield function on the meridian plane for different relative density η

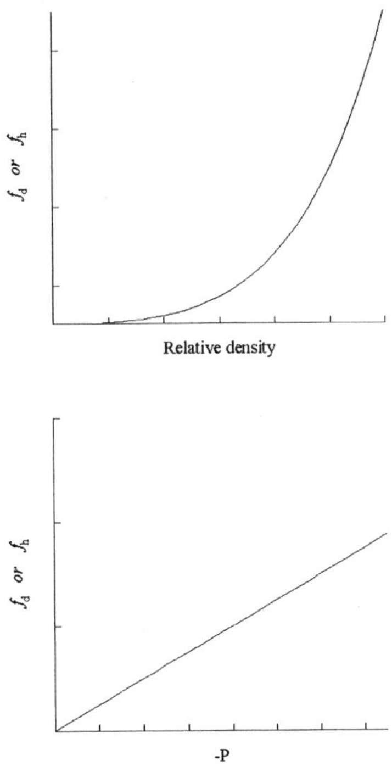

Figure 7.6. Typical evolution of $f_d(\eta, p, \theta)$ and $f_h(\eta, p)$ with η and p

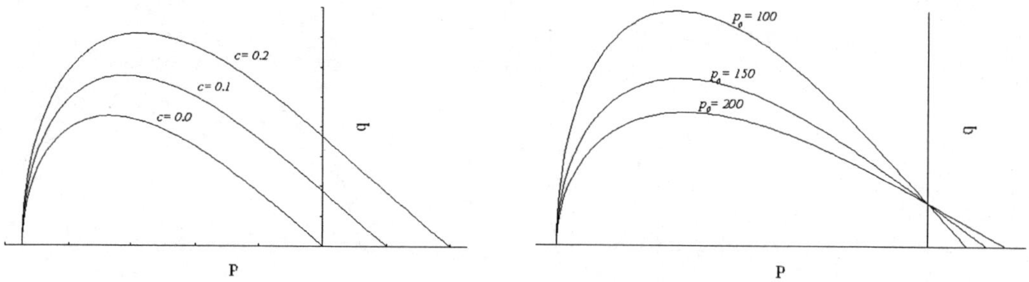

Figure 7.7. Trace of cone-cap yield function on the meridian plane for different values of c and p_0

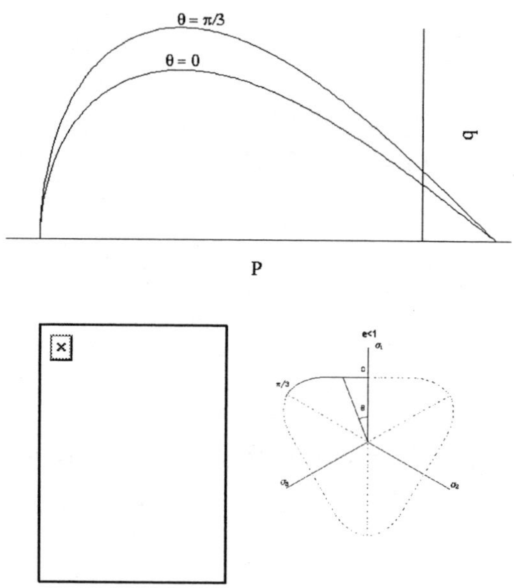

Figure 7.8. The trace of yield function on deviatoric plan for different values of shape parameter e

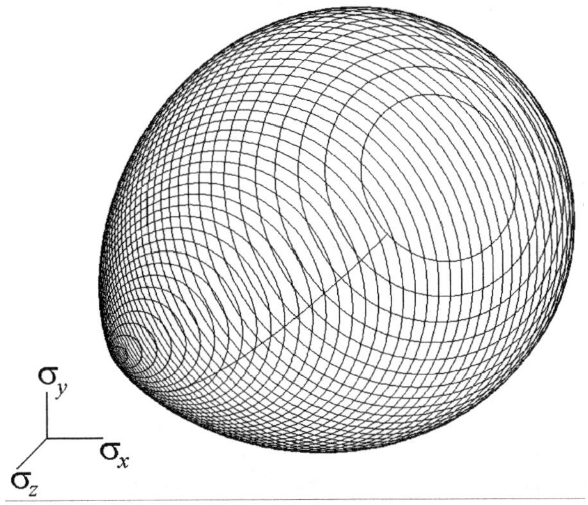

Figure 7.9. The shape of yield function in principal stress space

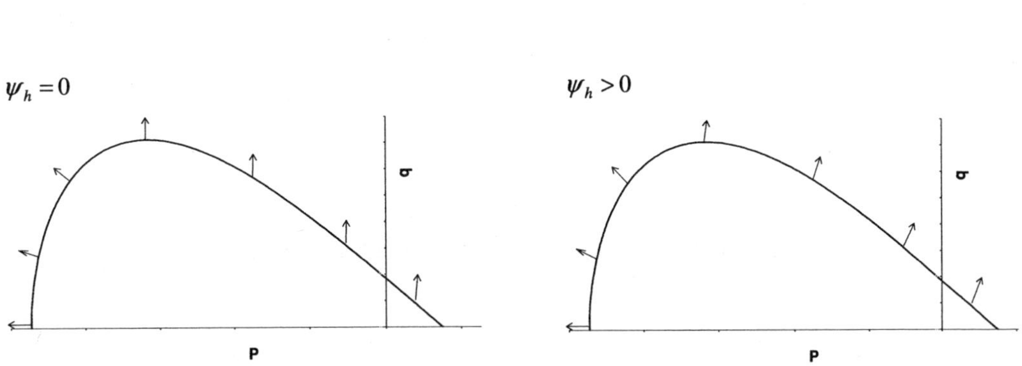

Figure 7.10. The plastic flow rule on the meridian plane for different values of ψ_h

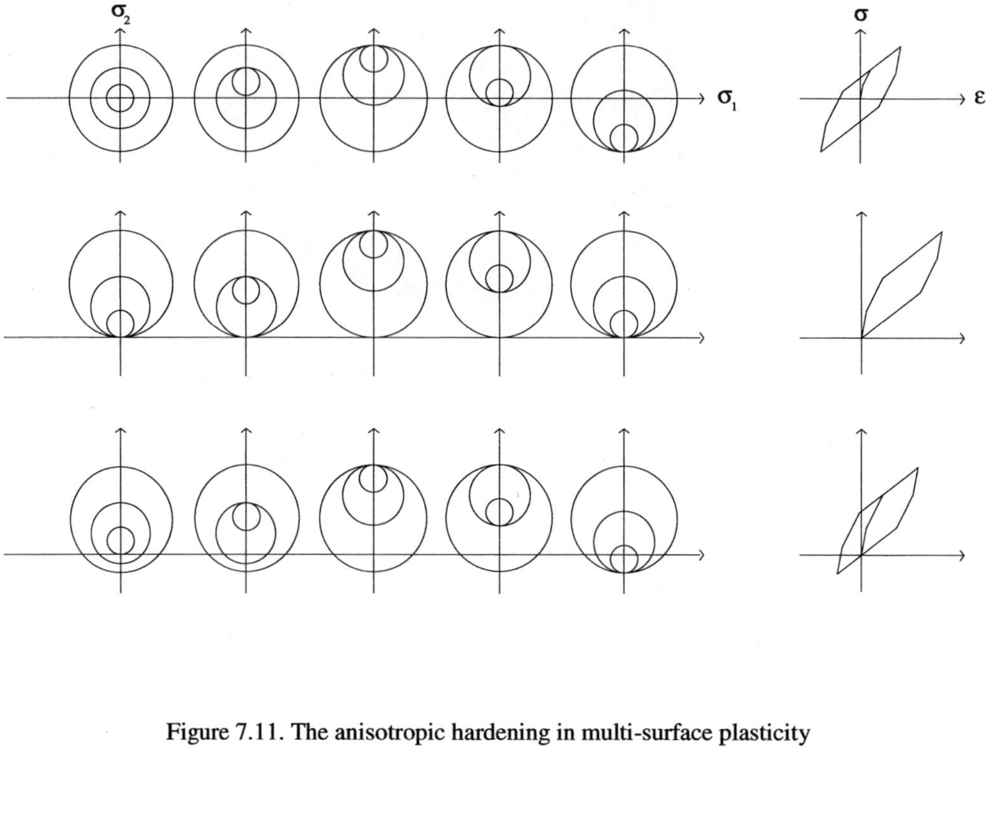

Figure 7.11. The anisotropic hardening in multi-surface plasticity

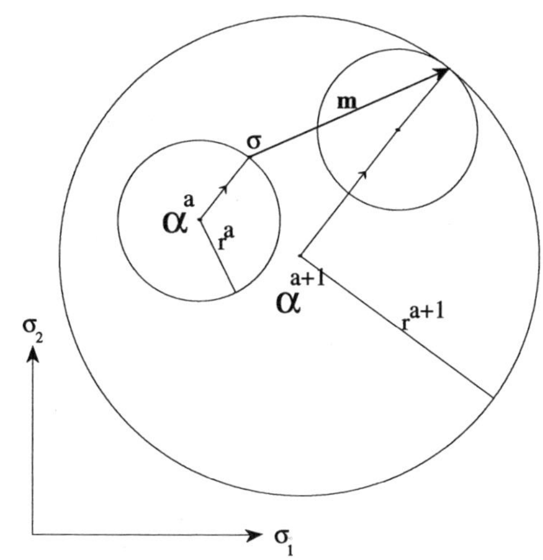

Figure 7.12. The Mroz translation rule

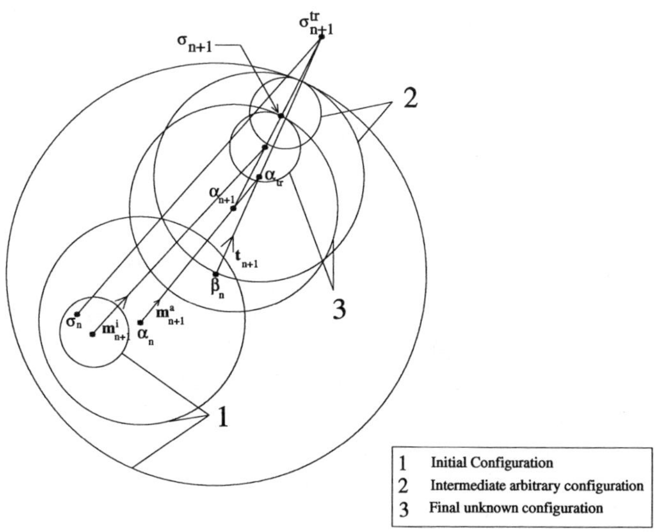

Figure 7.13. Geometrical interpretation of computational algorithm

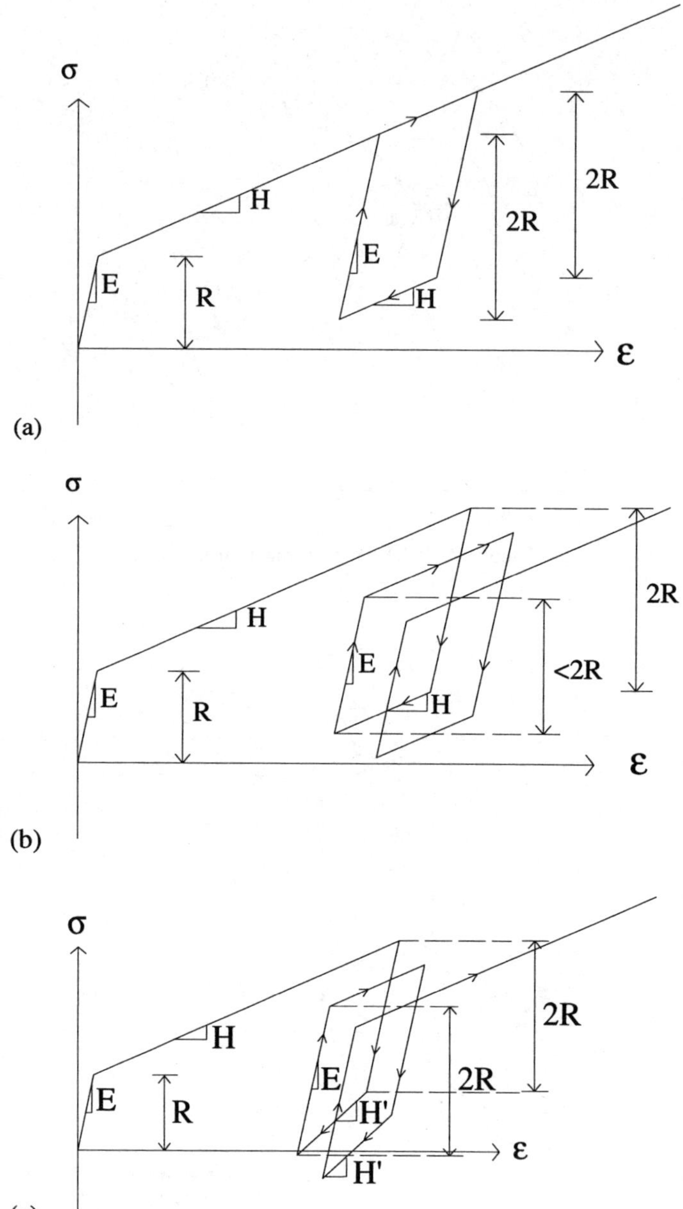

Figure 7.14. The hysterisis stress-strain curve; a) Constant size of surfaces and hardening parameters, b) Constant hardening parameters with different size of surfaces, c) Constant size of surfaces with different hardening parameters

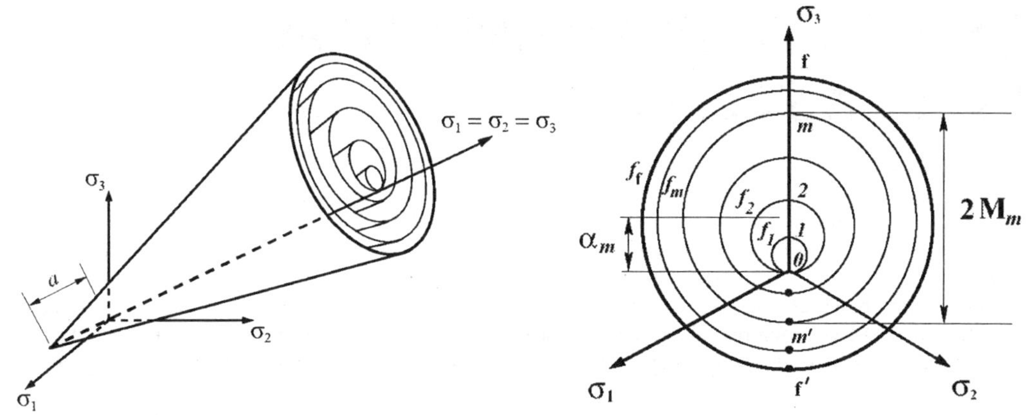

Figure 7.15. Conical multi-surface plasticity in principal stress space and deviatoric plane (Prevost 1985 and Elgamal *et al.* 2003)

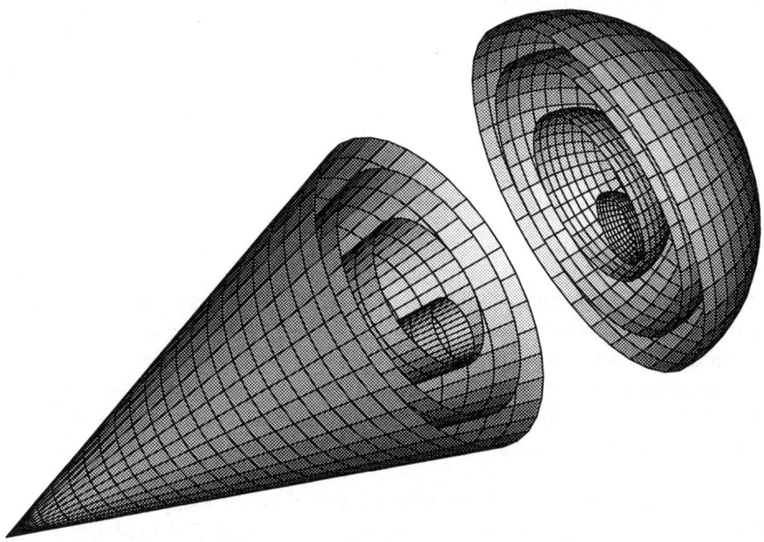

Figure 7.16. A combination of cap model together with the multi-surface plasticity

8 FINITE DEFORMATION PLASTICITY

A common feature of powder compaction simulation with solid mechanics constitutive models is the use of a Lagrangian kinematics formulation. This approach has shown to be adequate for problems that do not exhibit large mass fluxes among different parts of the sample. But in practical problems, as those that appear in realistic design processes, the Lagrangian approach leads to highly distorted and usually useless meshes. In order to solve these problems, the mesh adaptive strategy is employed in Chapter 6, however, it is computationally expensive and information must be interpolated from the old mesh to the new mesh.

In the present chapter, the numerical modeling of powder forming process is simulated by a large strain plasticity. The hypoelasto-plastic formulation is presented in the context of finite deformation problem involving large strain. The constitutive model is stated in the unrotated frame of reference, in which it is simple to achieve incremental objectivity and also in the unrotated reference frame all constitutive models are cast regardless of finite rotations. This greatly simplifies the numerical implementation of endochronic constitutive model. While the explicit treatment of both the integration of constitutive equations and the solution of the equilibrium equation is a common approach in the highly nonlinear behavior of powder compaction simulation, an implicit scheme is employed to present the efficiently and accurately the method for this type of nonlinear problems. Algorithmic modulus consistent with numerical integration algorithm of constitutive equations is extracted. Finally, the implementation of algorithm is validated by means of a set of simple deformation paths.

8.1. Introduction

In finite deformation plasticity, there are generally two approaches. The first class of methods is based on hyperelastic-plastic relations, multiplicative elastic-plastic kinematics and the existence of Helmholtz free energy density governed by either elasto-plastic deformation (Nemat-Nasser 1979 and Simo 1988), or elastic deformation solely (Eve and Reddy 1994). The second class of methods is based on hypoelastic-plastic relations, additive decomposition of rate of deformation and the use of objective stress rates. Even though this formulation is very attractive from the computational point of view (Belytschko 1983 and Ponthot 2002), it is limited to small elastic strain in which the hypothesis of hypoelasticity is valid. This approach is appropriate for most engineering materials, including metals, where elastic strain remains small.

In developing finite strain, rate type elastoplastic models, it is necessary to use objective stress rates in constitutive equations. Until the publication of Nagtegaal and De Jong (1981), which mentioned oscillatory shear response to a monotonically increasing shear strain for a kinematic hardening plasticity model, based on Jaumann rate of back stress, the Jaumann rate (1911) was generally used. That observation triggered a series of investigation to look for objective stress rates appropriate for description of metallic behavior in finite strain range, yielding objective rates such as the Green-Naghdi rate (1965), Sowerby-Chu rate (1984), Cotter-Rivlin rate (1955) and Truesdell rate (1955). In this chapter, the endochronic plasticity model is extended to finite strain scope. Additive decomposition of deformation rate and hypoelastic rate type relations are used in the formulation.

The first implementation of an endochronic theory into a multi-dimensional finite element code was made by Lin *et al.* (1981), who focused on the original endochronic theory with one term exponential for the kernel function. An implicit finite element algorithm for the modern version of endochronic theory without a yield surface was developed by Valanis and Fan (1984), which was incrementally nonlinear. Also, Watanbe and Atluri (1985) presented an implicit finite element algorithm for the modern endochronic theory. They used the endochronic plasticity with yield surface and the resulting constitutive equations were incrementally linear. An unconditional stable integration scheme of endochronic constitutive equations was proposed by Hsu *et al.* (1991, 1992) and its ability is examined in the modeling of random non proportional tests on OFHC copper. Hsu and Griffin (1996) implemented radial return algorithm in integration of endochronic constitutive equations and applied their formulation to finite element micromechanics modeling of a unidirectional metal matrix composite subjected to non-proportional cyclic loading.

The endochronic theory was extended to finite deformation with the concept of the corotational rate and plastic spin by Im and Atluri (1987). They derived the governing equations by using the isoclinic configuration as the intermediate configuration and the corresponding second Piola-Kirchhoff stress tensor. Cases of finite uniaxial compression and torsion were discussed in their work. Wu *et al.* (1995) incorporated the concepts of corotational rate, corotational integral and plastic spin to endochronic theory and applied it to description of rigid-plastic deformation in thin-walled tubes subjected to finite torsion. Pan *et al.* (1996) extended the ordinary differential constitutive equations of endochronic theory to simulate elastoplastic deformation in the range of finite strain using the concept of corotational rate. Different objective rates were incorporated into the theory and cases involving metal tubes under torsion and metal rectangular block under biaxial compression were discussed. An endochronic plasticity theory was developed by Khoei *et al.* (2003) and Bakhshiani *et al.* (2003) to describe the large deformation in finite strain using the concepts of corotational stress rate and the additive decomposition of deformation rate. They derived the constitutive equations for thin-walled tube under torsion to simulate the axial effects for various materials subjected to simple and pure torsional loading. Bakhshiani *et al.* (2002, 2003) and Khoei *et al.* (2001–2004) developed a density-dependent endochronic theory based on coupling between deviatoric and hydrostatic behavior in finite strain plasticity to simulate the compaction process of powder material.

In this chapter, the infinitesimal theory of endochronic plasticity is extended to large strain range on the basis of the additive decomposition of the strain rate tensor and hypoelasticity. This approach is rather different with respect to the multiplicative decomposition of the deformation gradient, which is based on the work of Lee (1969) and used in numerous recent papers (Simo 1988, Eterovic and Bathe 1990, Eve and Reddy 1994, Fish and Shek 2000, Ibrahimbegovic and Chorfi 2000 and Ponthot 2002). The proposed integration algorithm treats the elastoplastic and the elasto-viscoplatic cases in a unified way. Constitutive equations are stated in unrotated frame of reference that greatly simplifies endochronic constitutive relations in finite plasticity and yields the efficiency of the presented algorithm by total uncoupling material and geometrical non-linearities. An implicit scheme is employed, which is an efficient method for the type of nonlinear problems considered here (Ponthot 2002 and Simo 1988). An integration scheme, which is accurate, stable and amenable to consistent linearization, is developed. Although the major challenge in the integration of rate constitutive equations in large deformation analysis is to achieve incremental objectivity, it has been trivially achieved in the proposed algorithm. Algorithmic modulus consistent with numerical integration of constitutive equations for endochronic theory is extracted. The implementation of consistent modulus in global tangent stiffness matrix is essential in preserving the quadratic rate of convergence of

Newton procedure in solving the equilibrium equations (Simo and Taylor 1985). The efficiency of the proposed constitutive model and computational algorithms is demonstrated by several numerical examples.

8.2. Finite deformation of endochronic plasticity

In order to develop the endochronic plasticity model, described in Chapter 7, into the finite strain range, the additive elasto-plastic kinematics is employed using a hypoelasto-plastic formulation. The constitutive model is stated in the unrotated frame of reference (Fish and Shek 2000, Johnson and Bammann 1984 and Ponthot 2002). In unrotated frame, it is simple to obtain the incremental objectivity and all constitutive relations are cast regardless of finite rotations. Thus, it greatly simplifies the numerical implementation of endochronic constitutive model.

A material point in reference configuration Ω_0 with position vector X occupies position x at time t in deformed configuration Ω. So we have $x = \varphi(X,t)$. The motion from the original configuration to the deformed configuration has the deformation gradient $\mathbf{F}$ as

$$\mathbf{F} = \frac{\partial x}{\partial X} \tag{8.1}$$

Applying the polar decomposition theorem to $\mathbf{F}$

$$\mathbf{F} = \mathbf{VR} = \mathbf{RU} \tag{8.2}$$

where $\mathbf{V}$ and $\mathbf{U}$ are the left and right symmetric positive definite stretch tensors, respectively, and $\mathbf{R}$ is a proper orthogonal tensor. The velocity gradient is denoted by $\mathbf{L}$ expressed as

$$\mathbf{L} = \frac{\partial v}{\partial x} = \dot{\mathbf{F}}\,\mathbf{F}^{-1} \tag{8.3}$$

$$v = \frac{d\varphi(X,t)}{dt} \tag{8.4}$$

The velocity gradient can be written in terms of symmetric $\mathbf{D}$ and antisymmetric $\mathbf{W}$ parts, respectively, called rate of deformation and spin tensors, as

$$\mathbf{L} = \mathbf{D} + \mathbf{W} \tag{8.5}$$

The unrotated rate of deformation tensor is defined as

$$\hat{\mathbf{D}} = \mathbf{R}^T \mathbf{D}\,\mathbf{R} \tag{8.6}$$

The finite elasto-plastic kinematics, used in this study is based on the additive decomposition of rate of deformation as (Nemat-Nasser 1982)

$$\mathbf{D} = \mathbf{D}^e + \mathbf{D}^p \tag{8.7}$$

or in unrotated frame as,

$$\hat{\mathbf{D}} = \hat{\mathbf{D}}^e + \hat{\mathbf{D}}^p \tag{8.8}$$

The hypoelastic material law relates the rate of stress to the rate of deformation. A general form of the hypoelastic relation is defined as

$$\boldsymbol{\sigma}^{\nabla} = f(\boldsymbol{\sigma}, \mathbf{D}) \tag{8.9}$$

where $\boldsymbol{\sigma}^{\nabla}$ represents any objective rate of Cauchy stress.

In literature, many objective rates are introduced, such as: Jaumann, Truesdell and Green-Naghdi rates. Diense (1979) has shown that there is spurious oscillation in the stress, which arises directly from the nature of the Jaumann rate. Vanishing of Truesdell rate does not ensure that the invariants of Cauchy stress tensor are constant (Johnson and Bammann 1984), so in this case further plastic flow will exist. In this formulation the constitutive model is posed in terms of Cauchy stress in unrotated configuration as (Fish and Shek 2000, Johnson and Bammann 1984 and Ponthot 2002)

$$\hat{\boldsymbol{\sigma}} = \mathbf{R}^T \boldsymbol{\sigma} \ \mathbf{R} \tag{8.10}$$

The conjugate strain rate to $\hat{\boldsymbol{\sigma}}$ is $\hat{\mathbf{D}}$ defined in equation (8.6). Thus, the hypoelastic part of constitutive equation is

$$\dot{\hat{\boldsymbol{\sigma}}} = \boldsymbol{C}^e : \hat{\mathbf{D}}^e \tag{8.11}$$

$$\hat{\mathbf{D}} = \hat{\mathbf{D}}^e + \hat{\mathbf{D}}^p \tag{8.12}$$

where $\boldsymbol{C}^e$ is the Hook stress-strain tensor given as

$$\boldsymbol{C}^e{}_{ijkl} = K\delta_{ij}\delta_{kl} + 2\mu(\delta_{ik}\delta_{jl} - \tfrac{1}{3}\delta_{ij}\delta_{kl}) \tag{8.13}$$

where K and μ are the bulk and shear modulus of material, respectively.

In order to complete the hypoelasto-plastic constitutive equation in the context of finite deformation plasticity, the endochronic constitutive equations are presented in unrotated frame as

$$\hat{\boldsymbol{\sigma}}_{dev} = 2\int_0^{z_d} \Phi_d(z_d - z')\frac{\hat{\mathbf{D}}^p_{dev}}{\dot{z}'}dz' \tag{8.14}$$

$$\hat{\sigma}_h = 2\int_0^{z_h} \Phi_h(z_h - z')\frac{\hat{D}^p_{vol}}{\dot{z}'}dz' - 2\int_0^{z_h} \Psi(z_h - z')\,\hat{\boldsymbol{\sigma}}_{dev} : \frac{\hat{\mathbf{D}}^p_{dev}}{\dot{z}'}dz' \tag{8.15}$$

$$\dot{z}_d = \frac{\dot{\varsigma}}{f_d(\eta, p, \theta)} \tag{8.16}$$

$$\dot{z}_h = \frac{\dot{\varsigma}}{f_h(\eta, p)} \tag{8.17}$$

$$\dot{\varsigma}^2 = \hat{\mathbf{D}}^p_{dev} : \hat{\mathbf{D}}^p_{dev} + \kappa^2(\hat{\mathrm{D}}^p_{vol})^2 \tag{8.18}$$

where

$$\hat{\boldsymbol{\sigma}} = \hat{\boldsymbol{\sigma}}_{dev} + \hat{\sigma}_h \mathbf{I} \tag{8.19}$$

$$\hat{\boldsymbol{\sigma}}_{dev} = dev[\hat{\boldsymbol{\sigma}}] \tag{8.20}$$
$$\hat{\sigma}_h = \tfrac{1}{3}tr[\hat{\boldsymbol{\sigma}}] \tag{8.21}$$
$$\hat{\mathbf{D}}^p = \hat{\mathbf{D}}^p_{dev} + \hat{\mathrm{D}}^p_{vol}\mathbf{I} \tag{8.22}$$
$$\hat{\mathbf{D}}^p_{dev} = dev[\hat{\mathbf{D}}^p] \tag{8.23}$$
$$\hat{\mathrm{D}}^p_{vol} = tr[\hat{\mathbf{D}}^p] \tag{8.24}$$

and $dev[.]=[.]-\frac{1}{3}tr[.]\mathbf{I}$, with $\mathbf{I}$ denoting the spatial metric tensor.

8.3. Numerical integration of hypoelasto-plastic equations

The major challenge in the integration of the rate constitutive equations in finite strain range is to achieve incremental objectivity. The implementation of objective stress rates in constitutive equations results in objective formulation only in the limit of very small time step (Hughes and Winget 1980). As the standard time discretization procedures cannot be lead to incremental objectivity, one efficient way to overcome this problem is to state constitutive equations in corotational frame. Assuming that the variables of the model at step n and the incremental displacement field $\Delta u = {}^{n+1}x - {}^{n}x$ at load step n+1 are known, the updated value of different variables of the model will be calculated at load step n+1. The left superscript refers to the load step which is omitted for current step.

Applying the polar decomposition $\mathbf{F} = \mathbf{RU}$, relations (8.3) and (8.6) lead to

$$\hat{\mathbf{D}} = \mathbf{R}^T\mathbf{D}\,\mathbf{R} = \frac{1}{2}(\dot{\mathbf{U}}\,\mathbf{U}^{-1} + \mathbf{U}^{-1}\,\dot{\mathbf{U}}) \tag{8.25}$$

The midpoint rule results in

$$\hat{\mathbf{D}}\Delta t = \frac{1}{2}\left(\Delta\mathbf{U}\ {}^{n+\frac{1}{2}}\mathbf{U}^{-1} + {}^{n+\frac{1}{2}}\mathbf{U}^{-1}\ \Delta\mathbf{U}\right) \tag{8.26}$$

where $\Delta\mathbf{U}$ and ${}^{n+\frac{1}{2}}\mathbf{U}$ in the above relation are,

$$\Delta\mathbf{U} = {}^{n+1}\mathbf{U} - {}^{n}\mathbf{U} \tag{8.27}$$
$${}^{n+\frac{1}{2}}\mathbf{U} = \tfrac{1}{2}({}^{n+1}\mathbf{U} + {}^{n}\mathbf{U}) \tag{8.28}$$

The corotational increment of $\mathbf{D}$ is given as

$$\Delta\hat{\mathbf{d}} = \Delta t\hat{\mathbf{D}} = \frac{1}{2}\left(\Delta\mathbf{U}\ {}^{n+\frac{1}{2}}\mathbf{U}^{-1} + {}^{n+\frac{1}{2}}\mathbf{U}^{-1}\ \Delta\mathbf{U}\right) \tag{8.29}$$

Applying the backward Euler scheme to equation (8.11) yields

$$\hat{\boldsymbol{\sigma}} = {}^{n}\hat{\boldsymbol{\sigma}} + \boldsymbol{C}^e : \hat{\mathbf{D}}^e\Delta t \tag{8.30}$$

$$\Delta\hat{\boldsymbol{\sigma}} = \boldsymbol{C}^e : \Delta\hat{\mathbf{d}}^e = \boldsymbol{C}^e : (\Delta\hat{\mathbf{d}} - \Delta\hat{\mathbf{d}}^p) \tag{8.31}$$

One of the most important parts of numerical scheme is numerical integration of endochronic constitutive equations. Substituting equations (7.42), (7.43) and (7.44) in (8.14) and (8.15) give

$$\hat{\boldsymbol{\sigma}}_{dev} = 2\sum_{r=1}^{m} \int_0^{z_d} A_r e^{-\alpha_r (z_d - z')} \frac{\hat{\mathbf{D}}^p_{dev}}{\dot{z}'} dz' \tag{8.32}$$

$$\hat{\sigma}_h = 2\sum_{i=1}^{l} \int_0^{z_h} B_i e^{-\beta_i (z_h - z')} \frac{\hat{\mathbf{D}}^p_{vol}}{\dot{z}'} dz' - 2\sum_{i=1}^{l} \int_0^{z_h} \frac{1}{\kappa f_h} C_i e^{-\beta_i (z_h - z')} \hat{\boldsymbol{\sigma}}_{dev} : \frac{\hat{\mathbf{D}}^p_{dev}}{\dot{z}'} dz' \tag{8.33}$$

In order to integrate equations (8.32) and (8.33) numerically, the loading is divided into n steps, thus

$$\hat{\boldsymbol{\sigma}}^r_{dev}({}^n z_d) = 2\sum_{k=1}^{n} \frac{A_r}{\alpha_r} \frac{{}^k(\Delta \hat{\mathbf{d}}^p_{dev})}{{}^k\Delta z_d} \left[e^{-\alpha_r ({}^n z_d - {}^k z_d)} - e^{-\alpha_r ({}^n z_d - {}^{k-1} z_d)} \right] \tag{8.34}$$

$$\hat{\boldsymbol{\sigma}}_{dev}({}^n z_d) = \sum_{r=1}^{m} \hat{\boldsymbol{\sigma}}^r_{dev}({}^n z_d) \tag{8.35}$$

$$\begin{aligned} \hat{\sigma}^i_h({}^n z_h) = {} & 2\sum_{k=1}^{n} \frac{B_i}{\beta_i} \frac{{}^k(\Delta \hat{\mathbf{d}}^p_{vol})}{{}^k\Delta z_h} \left[e^{-\beta_i ({}^n z_h - {}^k z_h)} - e^{-\beta_i ({}^n z_h - {}^{k-1} z_h)} \right] \\ & - 2\sum_{k=1}^{n} \frac{1}{\kappa f^k_h} \frac{C_i}{\beta_i} \frac{{}^k\hat{\boldsymbol{\sigma}}_{dev} : {}^k(\Delta \hat{\mathbf{d}}^p_{dev})}{{}^k\Delta z_h} \left[e^{-\beta_i ({}^n z_h - {}^k z_h)} - e^{-\beta_i ({}^n z_h - {}^{k-1} z_h)} \right] \end{aligned} \tag{8.36}$$

$$\hat{\sigma}_h({}^n z_h) = \sum_{i=1}^{l} \hat{\sigma}^i_h({}^n z_h) \tag{8.37}$$

It is worth mentioning that the striking feature of the above scheme is the direct integration of hereditary function. Such a procedure, in general leads to more accurate results with much smaller number of increments. Equations (8.34) and (8.35) can be simplified as

$$\hat{\boldsymbol{\sigma}}^r_{dev}({}^n z_d) = \hat{\boldsymbol{\sigma}}^r_{dev}({}^{n-1} z_d) e^{-\alpha_r {}^n\Delta z_d} + 2\frac{A_r}{\alpha_r} \frac{{}^n(\Delta \hat{\mathbf{d}}^p_{dev})}{{}^n\Delta z_d} (1 - e^{-\alpha_r {}^n\Delta z_d}) \tag{8.38}$$

$$\begin{aligned} \hat{\sigma}^i_h({}^n z_h) = {} & \hat{\sigma}^i_h({}^{n-1} z_h) e^{-\beta_i {}^n\Delta z_h} + 2\frac{B_i}{\beta_i} \frac{{}^n(\Delta \hat{\mathbf{d}}^p_{vol})}{{}^n\Delta z_h} (1 - e^{-\beta_i {}^n\Delta z_h}) \\ & - \frac{2}{\kappa\, {}^n f_h\, {}^n\Delta z_h} \frac{C_i}{\beta_i} \left({}^{n-1}\hat{\boldsymbol{\sigma}}_{dev} : {}^n(\Delta \hat{\mathbf{d}}^p_{dev}) \right) (1 - e^{-\beta_i {}^n\Delta z_h}) \end{aligned} \tag{8.39}$$

The incremental form of endochronic constitutive equations can be obtained from equations (8.38) and (8.39) by taking step n+1 as current step. The incremental equations which is necessary in the numerical modeling of initial strain are as follows

$$\Delta \hat{\boldsymbol{\sigma}}_{dev} = \sum_{r=1}^{m} {}^n(\hat{\boldsymbol{\sigma}}^r_{dev})(e^{-\alpha_r \Delta z_d} - 1) + 2\frac{\Delta \hat{\mathbf{d}}^p_{dev}}{\Delta z_d} \sum_{r=1}^{m} \frac{A_r}{\alpha_r} (1 - e^{-\alpha_r \Delta z_d}) \tag{8.40}$$

$$\Delta\hat{\sigma}_h = \sum_{i=1}^{l} {}^n(\hat{\sigma}_h^i)(e^{-\beta_i \Delta z_h} - 1) + 2\frac{\Delta\hat{d}_{vol}^p}{\Delta z_h}\sum_{r=1}^{m}\frac{B_i}{\beta_i}(1 - e^{-\beta_i \Delta z_h})$$
$$- \frac{2({}^n\hat{\boldsymbol{\sigma}}_{dev} : \Delta\hat{\mathbf{d}}_{dev}^p)}{\kappa\ {}^{n+1}f_h\ \Delta z_h}\sum_{r=1}^{m}\frac{C_i}{\beta_i}(1 - e^{-\beta_i \Delta z_h}) \tag{8.41}$$

$$\Delta\hat{\boldsymbol{\sigma}} = \boldsymbol{C}^e : (\Delta\hat{\mathbf{d}} - \Delta\hat{\mathbf{d}}^p) \tag{8.42}$$

$$\Delta\varsigma^2 = \Delta\hat{\mathbf{d}}_{dev}^p : \Delta\hat{\mathbf{d}}_{dev}^p + \kappa^2(\Delta\hat{d}_{vol}^p)^2 \tag{8.43}$$

$$\Delta z_d = \frac{\Delta\varsigma}{f_d({}^{n+1}\eta,\ {}^{n+1}p,\ {}^{n+1}\theta)} \tag{8.44}$$

$$\Delta z_h = \frac{\Delta\varsigma}{f_h({}^{n+1}\eta,\ {}^{n+1}p)} \tag{8.45}$$

A set of seven algebraic equations can be solved for the unknowns values of $\Delta\hat{\mathbf{d}}_{dev}^p$ and $\Delta\hat{d}_h^p$ by applying the Newton-Raphson method. Once $\Delta\hat{\mathbf{d}}^p$ is known, the corotational increment of stress tensor is derived from equation (8.42). The value of stresses will be then calculated from relations $\hat{\boldsymbol{\sigma}} = {}^n\hat{\boldsymbol{\sigma}} + \Delta\hat{\boldsymbol{\sigma}}$ and $\boldsymbol{\sigma} = \mathbf{R}\,\hat{\boldsymbol{\sigma}}\,\mathbf{R}^T$.

It can be seen from equation (8.26) that the proposed scheme is trivially incrementally objective. In the case of rigid body motion, ${}^{n+1}\mathbf{U} = {}^n\mathbf{U}$ and from equation (8.26) $\hat{\mathbf{D}} = 0$, thus the stress tensor will be exactly updated by relation ${}^{n+1}\boldsymbol{\sigma} = \Delta\mathbf{R}\ {}^n\boldsymbol{\sigma}\,\Delta\mathbf{R}^T$. It should be noted that in this scheme the rotation tensor $\mathbf{R}$ is exactly computed from the polar decomposition and not from the numerical integration of rate equation $\dot{\omega} = \Omega\omega$, with Ω denoting a spin tensor and ω indicating an orthogonal rotation tensor.

8.3.1. Consistent tangent operator of hypoelasto-plastic equations

In order to achieve the quadratic convergence rate of Newton method in solving the global system of nonlinear equations, it is essential to use the consistent tangent moduli with integration procedure of constitutive equations in forming tangent stiffness matrix (Simo and Taylor 1985). In this section, the incremental constitutive equations derived in previous section are linearized to yield consistent tangent modulus. Taking material time derivate of equations (8.40) and (8.41) yields

$$\Delta\dot{\hat{\boldsymbol{\sigma}}}_{dev} = \mathbf{H}_1^{dev}\Delta\dot{z}_d + H_2^{dev}\Delta\dot{\hat{\mathbf{d}}}_{dev}^p \tag{8.46}$$

$$\Delta\dot{\hat{\sigma}}_h = H_1^h\Delta\dot{z}_h + H_2^h\Delta\dot{\hat{d}}_{vol}^p + H_3^h\Delta\dot{\varsigma} + H_4^h\boldsymbol{\sigma}_{dev} : \Delta\dot{\hat{\mathbf{d}}}^p \tag{8.47}$$

where

$$\mathbf{H}_1^{dev} = 2\frac{\Delta\hat{\mathbf{d}}_{dev}^p}{\Delta z_d}\sum_{r=1}^{m}A_r e^{-\alpha_r \Delta z_d} - 2\frac{\Delta\hat{\mathbf{d}}_{dev}^p}{\Delta z_d^2}\sum_{r=1}^{m}\frac{A_r}{\alpha_r}(1 - e^{-\alpha_r \Delta z_d}) - \sum_{r=1}^{m}\alpha_r\,{}^n(\hat{\boldsymbol{\sigma}}_{dev}^r)e^{-\alpha_r \Delta z_d} \tag{8.48}$$

$$H_2^{dev} = \frac{2}{\Delta z_d}\sum_{r=1}^{m}\frac{A_r}{\alpha_r}(1 - e^{-\alpha_r \Delta z_d}) \tag{8.49}$$

$$H_1^h = 2\frac{\Delta\hat{d}_{vol}^p}{\Delta z_h}\sum_{i=1}^{l} B_i e^{-\beta_i \Delta z_h} - 2\frac{\Delta\hat{d}_{vol}^p}{\Delta z_h^{\ 2}}\sum_{i=1}^{l}\frac{B_i}{\beta_i}(1-e^{-\beta_i \Delta z_h})$$
$$-\sum_{i=1}^{l}\beta_i{}^n(\hat{\sigma}_h^i)e^{-\beta_i\Delta z_h} + \frac{2\sigma_{dev}:\Delta\hat{\mathbf{d}}^p}{\Delta\varsigma}\sum_{i=1}^{l}C_i e^{-\beta_i\Delta z_h} \tag{8.50}$$

$$H_2^h = \frac{2}{\Delta z_h}\sum_{i=1}^{l}\frac{B_i}{\beta_i}(1-e^{-\beta_i\Delta z_h}) \tag{8.51}$$

$$H_3^h = -\frac{2\boldsymbol{\sigma}_{dev}:\Delta\hat{\mathbf{d}}^p}{\Delta\varsigma^2}\sum_{i=1}^{l}\frac{C_i}{\beta_i}(1-e^{-\beta_i\Delta z_h}) \tag{8.52}$$

$$H_4^h = \frac{2}{\Delta\varsigma}\sum_{i=1}^{l}\frac{C_i}{\beta_i}(1-e^{-\beta_i\Delta z_h}) \tag{8.53}$$

In order to linearize Δz_d and Δz_h, the relative density η, Lode's angle θ and intrinsic time ς, should be determined. For this purpose, taking the time derivative from $\eta = \eta_0/J$ and applying the identity of $\partial J/\partial \mathbf{U} = J\mathbf{U}^{-1}$, we will arrive to

$$\dot{\eta} = -\eta\,\mathbf{U}^{-1}:\dot{\mathbf{U}} \tag{8.54}$$

Using equation (8.29), $\dot{\mathbf{U}}$ can be stated in terms of $\Delta\hat{\mathbf{d}}$ as

$$\Delta t\,\dot{\mathbf{U}} = \boldsymbol{Q}^{-1}:\Delta\hat{\mathbf{d}} \tag{8.55}$$

where the forth order tensor $\boldsymbol{Q}$ is defined as

$$Q_{ijkl} = \tfrac{1}{4}\left(\delta_{ik}{}^{n+\frac{1}{2}}\mathbf{U}_{lj}^{-1} + \delta_{il}{}^{n+\frac{1}{2}}\mathbf{U}_{kj}^{-1} + \delta_{jl}{}^{n+\frac{1}{2}}\mathbf{U}_{ik}^{-1} + \delta_{jk}{}^{n+\frac{1}{2}}\mathbf{U}_{il}^{-1}\right) \tag{8.56}$$

Applying relations (7.39), (8.13) and (8.31) and considering $\dot{J}_2 = -\hat{\boldsymbol{\sigma}}_{dev}:\dot{\hat{\boldsymbol{\sigma}}}_{dev}$ and $\dot{J}_3 = J_3\hat{\boldsymbol{\sigma}}_{dev}^{-1}:\dot{\hat{\boldsymbol{\sigma}}}_{dev}$, it yields

$$\dot{\theta}\,\Delta t = \mathbf{Y}:(\Delta\hat{\mathbf{d}} - \Delta\hat{\mathbf{d}}^p) \tag{8.57}$$

where

$$\mathbf{Y}_{ij} = \frac{-2\sqrt{3}\,\mu\,J_3}{\sqrt{4J_2^3 - 27J_3^2}}\left(\tfrac{3}{2J_2}(\hat{\boldsymbol{\sigma}}_{ij} - \tfrac{1}{3}\hat{\boldsymbol{\sigma}}_{kk}\delta_{ij}) + (\hat{\boldsymbol{\sigma}}_{dev}^{-1})_{ij} - \tfrac{1}{3}(\hat{\boldsymbol{\sigma}}_{dev}^{-1})_{kk}\,\delta_{ij}\right) \tag{8.58}$$

where $\Delta\dot{\varsigma}$ can be obtained by taking time derivative from equation (8.43) as

$$\Delta\dot{\varsigma} = \frac{1}{\Delta\varsigma}\Delta\hat{\mathbf{d}}_{dev}^p:\Delta\dot{\hat{\mathbf{d}}}^p + \frac{\kappa^2}{\Delta\varsigma}\Delta\hat{\mathrm{d}}_{vol}^p\Delta\dot{\hat{\mathrm{d}}}_{vol}^p \tag{8.59}$$

Taking material time derivative of equation (8.44) yields

$$\Delta\dot{z}_d = \frac{\Delta\dot{\varsigma}}{f_d} - \frac{\Delta\varsigma}{f_d^{\ 2}}\left(\frac{\partial f_d}{\partial p}\dot{p} + \frac{\partial f_d}{\partial\eta}\dot{\eta} + \frac{\partial f_d}{\partial\theta}\dot{\theta}\right) \tag{8.60}$$

Applying similar procedure to equation (8.45) yields

$$\Delta\dot{z}_h = \frac{\Delta\dot{\varsigma}}{\kappa f_h} - \frac{\Delta\varsigma}{\kappa f_h^{\,2}}\left(\frac{\partial f_h}{\partial p}\dot{p} + \frac{\partial f_h}{\partial \eta}\dot{\eta}\right) \tag{8.61}$$

Substituting equations (7.47), (8.46), (8.57), (8.59), (8.60) and (8.61) in time derivative equation (8.19), by lengthy but straightforward calculation yields

$$\dot{\hat{\sigma}} = \boldsymbol{N} : \Delta\dot{\hat{\mathbf{d}}}^p + \boldsymbol{P} : \Delta\dot{\hat{\mathbf{d}}} \tag{8.62}$$

where

$$\boldsymbol{N} = \boldsymbol{N}^a + \boldsymbol{N}^b + \boldsymbol{N}^c \tag{8.63}$$

and

$$\boldsymbol{P} = \boldsymbol{P}^a + \boldsymbol{P}^b + \boldsymbol{P}^c \tag{8.64}$$

where tensors $\boldsymbol{N}^a, \boldsymbol{N}^b, \boldsymbol{N}^c, \boldsymbol{P}^a, \boldsymbol{P}^b, \boldsymbol{P}^c$ and $\boldsymbol{I}^{dev}$ are defined as

$$\boldsymbol{N}^a = \left(H_2^h + KH_1^h \frac{\Delta\varsigma}{\kappa f_h^2}\frac{\partial f_h}{\partial p} + \frac{\kappa^2}{\Delta\varsigma}(H_3^h + \frac{H_1^h}{\kappa f_h})\Delta\hat{d}_{vol}^{\,p}\right)\mathbf{I}\otimes\mathbf{I} \tag{8.65}$$

$$\boldsymbol{N}^b = H_2^d\, \boldsymbol{I}^{dev} + H_4^h\,\mathbf{I}\otimes\hat{\boldsymbol{\sigma}}_{dev} + \left(K\frac{\Delta\varsigma}{f_d^2}\frac{\partial f_d}{\partial p} + \frac{\kappa^2}{f_d\Delta\varsigma}\Delta\hat{d}_{vol}^{\,p}\right)\mathbf{H}_1^d\otimes\mathbf{I} \tag{8.66}$$

$$\boldsymbol{N}^c = \frac{\Delta\varsigma}{f_d^2}\frac{\partial f_d}{\partial\theta}\mathbf{H}_1^d\otimes\mathbf{Y} + \frac{1}{\Delta\varsigma f_d}\mathbf{H}_1^d\otimes\Delta\hat{\mathbf{d}}_{dev}^{\,p} + \frac{1}{\Delta\varsigma}(H_3^h + \frac{H_1^h}{\kappa f_h})\mathbf{I}\otimes\Delta\hat{\mathbf{d}}_{dev}^{\,p} \tag{8.67}$$

$$\boldsymbol{P}^a = -\frac{\Delta\varsigma}{f_d^2}\frac{\partial f_d}{\partial\theta}\mathbf{H}_1^d\otimes\mathbf{Y} \tag{8.68}$$

$$\boldsymbol{P}^b = H_1^h\eta\frac{\Delta\varsigma}{\kappa f_h^2}\frac{\partial f_h}{\partial\eta}\mathbf{I}\otimes(\mathbf{U}^{-1}:\boldsymbol{Q}^{-1}) + \eta\frac{\Delta\varsigma}{f_d^2}\frac{\partial f_d}{\partial\eta}\mathbf{H}_1^d\otimes(\mathbf{U}^{-1}:\boldsymbol{Q}^{-1}) \tag{8.69}$$

$$\boldsymbol{P}^c = -KH_1^h\frac{\Delta\varsigma}{\kappa f_h^2}\frac{\partial f_h}{\partial p}\mathbf{I}\otimes\mathbf{I} - K\frac{\Delta\varsigma}{f_d^2}\frac{\partial f_d}{\partial p}\mathbf{H}_1^d\otimes\mathbf{I} \tag{8.70}$$

$$I_{ijkl}^{dev} = \tfrac{1}{2}\left(\delta_{ik}\delta_{jl} + \delta_{il}\delta_{jk}\right) - \tfrac{1}{3}\delta_{ij}\delta_{kl} \tag{8.71}$$

Taking material time derivative from equation (8.42) and substituting the result in equation (8.62), gives

$$\dot{\boldsymbol{\sigma}} = \overline{\boldsymbol{C}}^p : \Delta\dot{\hat{\mathbf{d}}} \tag{8.72}$$

where

$$\overline{\boldsymbol{C}}^p = \boldsymbol{C}^e(\boldsymbol{C}^e + \boldsymbol{N})^{-1}(\boldsymbol{P} + \boldsymbol{N}) \tag{8.73}$$

In order to complete the derivation of consistent modulus, the linearization of ${}^{n+1}\mathbf{R}$ and $\Delta\hat{\mathbf{d}}$ is necessary. For consistent linearization of ${}^{n+1}\mathbf{R}$, we start by relation

$$\mathbf{L} = \dot{\mathbf{F}}\,\mathbf{F}^{-1} = \mathbf{R}\,\mathbf{R}^{T} + \mathbf{R}\,\dot{\mathbf{U}}\,\mathbf{U}^{-1}\,\mathbf{R}^{T} \tag{8.74}$$

Pre-multiplying equation (8.74) by $\mathbf{R}^T$ and post-multiplying by $\mathbf{F}$ yields

$$\mathbf{R}^{T}\mathbf{L}\,\mathbf{F} = \mathbf{R}^{T}\dot{\mathbf{R}}\,\mathbf{U} + \dot{\mathbf{U}} \tag{8.75}$$

Subtracting (8.75) from its transpose, after some manipulation results in

$$\dot{\mathbf{R}} = \boldsymbol{G} : \mathbf{L} \tag{8.76}$$

where

$$G_{ijkl} = (\mathbf{R}_{im}\mathbf{U}_{jn} - \mathbf{R}_{in}\mathbf{U}_{jm})^{-1}(\mathbf{R}_{km}\mathbf{F}_{ln} - \mathbf{R}_{kn}\mathbf{F}_{lm}) \tag{8.77}$$

Substituting equation (8.76) to (8.75) gives $\dot{\mathbf{U}}$ in terms of $\mathbf{L}$ as

$$\dot{\mathbf{U}} = \boldsymbol{H} : \mathbf{L} \tag{8.78}$$

where

$$H_{ijkl} = (\mathbf{R}_{ki}\mathbf{F}_{lj} - \mathbf{R}_{mi}\mathbf{U}_{nj}G_{mnkl}) \tag{8.79}$$

Applying relation (8.29) to evaluate $\Delta\dot{\hat{\mathbf{d}}}$ appeared in equation (8.72), we obtain

$$\dot{\hat{\mathbf{d}}} = \tilde{\boldsymbol{G}} : {}^{n+1}\dot{\mathbf{U}} \tag{8.80}$$

where

$$\tilde{G}_{ijkl} = \tfrac{1}{2}\left({}^{n+\frac{1}{2}}\mathbf{U}^{-1}_{kj}\delta_{il} + {}^{n+\frac{1}{2}}\mathbf{U}^{-1}_{ik}\delta_{jl} - \tfrac{1}{2}\Delta\mathbf{U}_{im}\,{}^{n+\frac{1}{2}}\mathbf{U}^{-1}_{mk}\,{}^{n+\frac{1}{2}}\mathbf{U}^{-1}_{lj} - \tfrac{1}{2}\Delta\mathbf{U}_{mj}\,{}^{n+\frac{1}{2}}\mathbf{U}^{-1}_{ik}\,{}^{n+\frac{1}{2}}\mathbf{U}^{-1}_{lm} \right) \tag{8.81}$$

Substituting equation (8.78) in (8.80) yields,

$$\dot{\hat{\mathbf{d}}} = \tilde{\boldsymbol{H}} : \mathbf{L} \tag{8.82}$$

where $\tilde{\boldsymbol{H}}$ is a forth-order tensor given as

$$\tilde{\boldsymbol{H}} = \tilde{\boldsymbol{G}} : \boldsymbol{H} \tag{8.83}$$

Substituting equation (8.82) into equation (8.72) yields

$$\dot{\hat{\boldsymbol{\sigma}}} = \boldsymbol{C} : \mathbf{L} \tag{8.84}$$

where

$$C_{ijkl} = \overline{C}^{\,p}_{ijmn}\tilde{H}_{mnkl} \tag{8.85}$$

Applying equations (8.76) and (8.84) along with $\dot{\hat{\boldsymbol{\sigma}}} = \mathbf{R}^{T}\dot{\boldsymbol{\sigma}}\,\mathbf{R} + \dot{\mathbf{R}}^{T}\boldsymbol{\sigma}\,\mathbf{R} + \mathbf{R}^{T}\boldsymbol{\sigma}\,\dot{\mathbf{R}}$, yields

$$\dot{\boldsymbol{\sigma}} = \tilde{\boldsymbol{C}} : \mathbf{L} \tag{8.86}$$

where

$$\tilde{C}_{ijkl} = \mathbf{R}_{im} C_{mnkl} \mathbf{R}_{jn} - \mathbf{R}_{im} G_{nmkl} \boldsymbol{\sigma}_{nj} - \boldsymbol{\sigma}_{im} G_{mnkl} \mathbf{R}_{jn} \tag{8.87}$$

8.4. Finite deformation of endochronic visco-plasticity

In order to develop the hypoelasto-viscoplastic constitutive equation, described in section 7.6, in the context of finite deformation plasticity, the endochronic constitutive equations (7.70), (7.71) and (7.72) are presented in unrotated frame as

$$\hat{\boldsymbol{\sigma}}_{dev} = 2\int_0^z \Phi(z - z') \frac{\hat{\mathbf{D}}^p}{\dot{z}'} dz' \tag{8.88}$$

$$\dot{z} = \frac{\dot{\varsigma}}{f(\varsigma, \dot{\varsigma})} \tag{8.89}$$

$$\dot{\varsigma}^2 = \hat{\mathbf{D}}^p : \hat{\mathbf{D}}^p \tag{8.90}$$

Substituting equation (7.75) to (8.88), gives

$$\hat{\boldsymbol{\sigma}}_{dev} = 2\sum_{r=1}^{m} \int_0^z A_r e^{-\alpha_r (z-z')} \frac{\hat{\mathbf{D}}^p}{\dot{z}'} dz' \tag{8.91}$$

In order to integrate equation (8.91) numerically, the loading is divided into n steps, thus

$$\hat{\boldsymbol{\sigma}}^r_{dev}({}^n z) = 2\sum_{k=1}^{n} \frac{A_r}{\alpha_r} \frac{{}^k(\Delta\hat{\mathbf{d}}^p)}{{}^k\Delta z} \left[e^{-\alpha_r({}^n z - {}^k z)} - e^{-\alpha_r({}^n z - {}^{k-1} z)} \right] \tag{8.92}$$

$$\hat{\boldsymbol{\sigma}}_{dev}({}^n z) = \sum_{r=1}^{m} \hat{\boldsymbol{\sigma}}^r_{dev}({}^n z) \tag{8.93}$$

The incremental form of endochronic constitutive equations can be obtained from equation (8.92) by taking step $n+1$ as current step. Thus,

$$\hat{\boldsymbol{\sigma}}^r_{dev}({}^n z) = \hat{\boldsymbol{\sigma}}^r_{dev}({}^{n-1} z) e^{-\alpha_r {}^n\Delta z} + 2\frac{A_r}{\alpha_r} \frac{{}^n(\Delta\hat{\mathbf{d}}^p)}{{}^n\Delta z} (1 - e^{-\alpha_r {}^n\Delta z}) \tag{8.94}$$

The incremental equations which is necessary in numerical modeling of initial strain are

$$\Delta\hat{\boldsymbol{\sigma}}_{dev} = \sum_{r=1}^{m} {}^n(\hat{\boldsymbol{\sigma}}^r_{dev})(e^{-\alpha_r \Delta z} - 1) + 2\frac{\Delta\hat{\mathbf{d}}^p}{\Delta z} \sum_{r=1}^{m} \frac{A_r}{\alpha_r}(1 - e^{-\alpha_r \Delta z}) \tag{8.95}$$

$$\Delta\hat{\boldsymbol{\sigma}}_{dev} = 2\mu(\Delta\hat{\mathbf{d}} - \Delta\hat{\mathbf{d}}^p) \tag{8.96}$$

$$\Delta\varsigma^2 = \Delta\hat{\mathbf{d}}^p : \Delta\hat{\mathbf{d}}^p \tag{8.97}$$

$$\Delta z = \frac{\Delta \varsigma}{f({}^n\varsigma + \beta\,\Delta\varsigma, \Delta\varsigma/\Delta t)} \tag{8.98}$$

where β is a parameter with the condition $\beta \geq \frac{1}{2}$. Although the set of six algebraic equations can be solved for the unknown values of $\Delta\hat{\mathbf{d}}^p$ by applying the Newton-Raphson method, however, more efficient numerical scheme can be devised by making a few algebraic manipulations. Substituting $\Delta\hat{\boldsymbol{\sigma}}_{dev}$ from equation (8.96) into (8.95) yields

$$2\mu\,\Delta\hat{\mathbf{d}} - \sum_{r=1}^{m} {}^n(\hat{\boldsymbol{\sigma}}^r_{dev})(e^{-\alpha_r\Delta z} - 1) = 2\frac{\Delta\hat{\mathbf{d}}^p}{\Delta z}\left(\sum_{r=1}^{m} \frac{A_r}{\alpha_r}(1 - e^{-\alpha_r\Delta z}) + \mu\,\Delta z\right) \tag{8.99}$$

Taking inner product of equation (8.99) with itself, results

$$\mathbf{A} : \mathbf{A} = B^2 \frac{\Delta\varsigma^2}{\Delta z^2} \tag{8.100}$$

where

$$\mathbf{A}_{ij} = \mu\,\Delta\hat{\mathbf{d}}_{ij} - \frac{1}{2}\sum_{r=1}^{m} {}^n(\hat{\boldsymbol{\sigma}}^r_{dev})_{ij}(e^{-\alpha_r\Delta z} - 1) \tag{8.101}$$

and

$$B = \mu\Delta z + \sum_{r=1}^{m} \frac{A_r}{\alpha_r}(1 - e^{-\alpha_r\Delta z}) \tag{8.102}$$

Substituting equation (8.98) into (8.100) gives

$$R(\Delta\varsigma) = \mathbf{A} : \mathbf{A} - B^2 f^2({}^n\varsigma + \beta\,\Delta\varsigma, \frac{\Delta\varsigma}{\Delta t}) = 0 \tag{8.103}$$

The above equation can be solved for $\Delta\varsigma$ by the Newton-Raphson technique using the following expression for derivative of residual with respect to unknown value of $\Delta\varsigma$

$$\frac{\partial R}{\partial \Delta\varsigma} = 2\frac{\partial \mathbf{A}}{\partial \Delta\varsigma} : \mathbf{A} - 2f\,B\left(\frac{\partial B}{\partial \Delta\varsigma} f + B\frac{\partial f}{\partial \Delta\varsigma}\right) \tag{8.104}$$

$$\frac{\partial \mathbf{A}}{\partial \Delta\varsigma} = \frac{1}{2}\frac{\partial \Delta z}{\partial \Delta\varsigma}\sum_{r=1}^{m} \alpha_r\, {}^n(\hat{\boldsymbol{\sigma}}^r_{dev}) e^{-\alpha_r\Delta z} \tag{8.105}$$

$$\frac{\partial B}{\partial \Delta\varsigma} = \frac{\partial \Delta z}{\partial \Delta\varsigma}\left(\mu + \sum_{r=1}^{m} C_r e^{-\alpha_r\Delta z}\right) \tag{8.106}$$

$$\frac{\partial \Delta z}{\partial \Delta\varsigma} = \frac{1}{f} - \frac{\Delta\varsigma}{f^2}\frac{\partial f}{\partial \Delta\varsigma} \tag{8.107}$$

$$\frac{\partial f({}^n\varsigma + \beta\,\Delta\varsigma, \Delta\varsigma/\Delta t)}{\partial \Delta\varsigma} = \beta\frac{\partial f}{\partial \varsigma} + \frac{1}{\Delta t}\frac{\partial f}{\partial \dot{\varsigma}} \tag{8.108}$$

Once equation (8.103) is solved for $\Delta\varsigma$, Δz can be obtained from equation (8.98) and $\Delta\hat{\mathbf{d}}^p$ from (8.99). The corotational increment of stress tensor will be then calculated.

8.4.1. Consistent tangent operator of hypoelasto-viscoplastic equations

In order to achieve the quadratic rate of convergence of Newton method for solution of global nonlinear system of equations, the incremental constitutive equations derived in previous section are linearized to yield the consistent tangent modulus. Taking material time derivative of equation (8.95) yields

$$\Delta\dot{\hat{\boldsymbol{\sigma}}}_{dev} = \mathbf{H}_1 \Delta\dot{z} + H_2 \Delta\dot{\hat{\mathbf{d}}}^p \tag{8.109}$$

where

$$\mathbf{H}_1 = 2\frac{\Delta\hat{\mathbf{d}}^p}{\Delta z}\sum_{r=1}^{m} A_r e^{-\alpha_r \Delta z} - 2\frac{\Delta\hat{\mathbf{d}}^p}{\Delta z^2}\sum_{r=1}^{m}\frac{A_r}{\alpha_r}(1-e^{-\alpha_r\Delta z}) - \sum_{r=1}^{m}\alpha_r{}^n(\hat{\sigma}^r_{dev})e^{-\alpha_r\Delta z} \tag{8.110}$$

$$H_2 = \frac{2}{\Delta z}\sum_{r=1}^{m}\frac{A_r}{\alpha_r}(1-e^{-\alpha_r\Delta z}) \tag{8.111}$$

Taking material time derivative of equation (8.98) yields

$$\Delta\dot{z} = \left(\frac{1}{f} - \frac{\Delta\varsigma}{f^2}\left(\beta\frac{\partial f}{\partial\varsigma} + \frac{1}{\Delta t}\frac{\partial f}{\partial\dot{\varsigma}}\right)\right)\Delta\dot{\varsigma} \tag{8.112}$$

where $\Delta\dot{\varsigma}$ can be obtained by taking time derivative from equation (8.97) as

$$\Delta\dot{\varsigma} = \frac{1}{\Delta\varsigma}\Delta\hat{\mathbf{d}}^p : \Delta\dot{\hat{\mathbf{d}}}^p \tag{8.113}$$

Substituting equations (8.112) and (8.113) into (8.109) yields

$$\Delta\dot{\hat{\boldsymbol{\sigma}}}_{dev} = \boldsymbol{C}^p : \Delta\dot{\hat{\mathbf{d}}}^p \tag{8.114}$$

where

$$\boldsymbol{C}^p = \frac{1}{\Delta\xi}\left(\frac{1}{f} - \frac{\Delta\varsigma}{f^2}\left(\beta\frac{\partial f}{\partial\varsigma} + \frac{1}{\Delta t}\frac{\partial f}{\partial\dot{\varsigma}}\right)\right)\mathbf{H}_1 \otimes \Delta\mathbf{d}^p + H_2\boldsymbol{I} \tag{8.115}$$

Taking time derivative from relation $\hat{\boldsymbol{\sigma}} = \hat{\boldsymbol{\sigma}}_{dev} + \hat{\sigma}_h\mathbf{I}$ and applying the Hook stress-strain tensor along with equations (8.96) and (8.114) by straightforward calculation yields

$$\dot{\hat{\boldsymbol{\sigma}}} = \overline{\boldsymbol{C}}^p : \Delta\dot{\hat{\mathbf{d}}} \tag{8.116}$$

where

$$\overline{\boldsymbol{C}}^p = \boldsymbol{C}^e\left(\boldsymbol{I} - \left(\frac{1}{2\mu}\boldsymbol{C}^p + \boldsymbol{I}\right)^{-1}\boldsymbol{I}^{dev}\right) \tag{8.117}$$

and $\boldsymbol{I}^{dev}_{ijkl} = \frac{1}{2}\left(\delta_{ik}\delta_{jl} + \delta_{il}\delta_{jk}\right) - \frac{1}{3}\delta_{ij}\delta_{kl}$.

8.5. Numerical examples of finite elasto- and visco-plasticity

In this section, several numerical examples are presented to study the performance of the proposed finite elasto- visco-plasticity formulation. The implementation is validated by means of a set of simple deformation paths, including simple shear, extension and rotation, two benchmark tests in nonlinear mechanics, i.e. the necking of a circular bar and expansion of a thick-walled cylinder, a cyclically loaded notched bar test which demonstrates the capabilities of the proposed model in simulation of cyclic loading and ratcheting in finite strain case and finally, the analysis of a tensile test, which presents a shear band with a finite thickness independent of the finite element mesh using endochronic viscoplastic constitutive model.

The finite element mesh employed in all simulations is four-node isoparametric elements with bilinear displacement interpolation and constant pressure to handle the isochoric nature of plastic flow. The convergence criterion used is based on the relative value of norm of residual forces to norm of applied forces. Unless otherwise stated, the convergence tolerance is set to 10^{-4} and β is set to $\frac{1}{2}$.

8.5.1. Simple shear test

The first example is chosen to examine the behavior of model and accuracy of the numerical integration scheme in the case of large deformation. The problem statement for the simple shear test is shown in Figure 8.1. Both elastic and elasto-plastic behaviors are considered. The equations describing the motion of the continuum are

$$x(t) = X + Y\,t\,, \qquad y(t) = Y \tag{8.118}$$

Elastic behavior is represented by Young's modulus of 2000 and Poisson's coefficient of 0.3. Figure 8.2 presents the results for the elastic case. The results of hypoelastic model with Green-Naghdi and Jaumann rates are also depicted for comparison. Both mid-point and fully implicit algorithms yield to sufficiently accurate results for shear strain response in relatively large increment (50 percent shear strain in each increment). Fully implicit algorithm grossly underestimates σ_{xx} component when not enough increments are employed, and demands a smaller strain increment to get accurate solution.

Endochronic plasticity model with kernel function parameters of $C_1 = 2000$, $C_2 = 200$, $r_1 = 100$, $r_2 = 10$ and scale function of $f(\varsigma) = 1 + \varsigma/10$ is employed for simulation of elasto-plastic behavior. The mid-point numerical integration algorithm is used for this case. Figure 8.3(a) shows shear stress response versus shear strain for three different time steps. It can be seen from this figure that the proposed integration algorithm results in acceptable solutions even in the case of large prescribed strain increments. Taking the solution at 5000 time steps, as the exact solution, the error in final shear stress corresponding to shear strain of 5 is computed and plotted versus the number of time increments in Figure 8.3(b). From this figure, the second order accuracy of the integration algorithm is corroborated.

8.5.2. Extension and rotation test

In second example, a combined extension and rotation of an element is considered to illustrate the behavior of the kinematic algorithm and accuracy of the proposed integration scheme in the case of small deformation combined with large rigid body rotation. This example was made by Rashid (1993) to show that the algorithm of Hughes and Winget (1980) is the only weakly objective, which can cause unintended coupling between the rotational part of the motion and the stress update, whenever large rotational increments are present. This characteristic can cause inaccurate results in the case of large rotational increments.

The problem statement for this example is shown in Figure 8.4. A unit square simultaneously undergoes a uniaxial extension of 0.005 and a total superposed rigid rotation of 90°. The equations of motion are

$$\begin{aligned} x(t) &= X(1+0.005t)\cos(\tfrac{\pi}{2}t) - Y\sin(\tfrac{\pi}{2}t) \\ y(t) &= X(1+0.005t)\sin(\tfrac{\pi}{2}t) + Y\cos(\tfrac{\pi}{2}t) \end{aligned} \tag{8.119}$$

and the analytical solution is

$$\begin{aligned} \sigma_{xx} &= \frac{Et}{400(1-\nu^2)}(1+\cos\pi t) \\ \sigma_{yy} &= \frac{Et}{400(1-\nu^2)}(1-\cos\pi t) \\ \sigma_{xy} &= \frac{Et}{400(1-\nu^2)}\sin\pi t \end{aligned} \tag{8.120}$$

The material Young's modulus E is taken as 200000 and the Poisson's ratio is 0.2. The elastic behavior is considered here. The results of this example are shown in Figure 8.5. It is clear that the predictions of the proposed algorithm agree with analytical values for large incremental rotation, even up to 45°.

8.5.3. Expansion of a thick-walled cylinder

The next example is one of the most popular benchmarks for validating finite plasticity formulation (Simo 1988, Fish and Shek 2000 and Ponthot 2002). A thick-walled cylinder with an inner radius of 10 units and an outer radius of 20 units is subjected to internal pressure. This problem is solved using 20 bilinear axisymmetric elements. The convergence tolerance used for this example is 10^{-7}. Figure 8.6 shows the finite element mesh and configuration of the cylinder. The inner radius is driven to a value of 90 in ten load steps. Endochronic plasticity model with kernel function parameters of $C_1 = 180000\,\text{MPa}$, $C_2 = 90000\,\text{MPa}$, $r_1 = 30000$, $r_2 = 10000$ and scale function of $f(\varsigma) = 0.0148$ is employed for simulation of rigid-plastic behavior with the yield stress of 0.5 MPa allowing a comparison with analytical solution. The Young's modulus is 11050 MPa and Poisson's coefficient is 0.454.

Figure 8.7(a) shows the relationship between the inner radius and internal pressure. The radial stress profile versus position through current thickness of the cylinder is shown in Figure 8.7(b). It can be seen that the numerical results are in good agreement with the analytical solution. Newton iterations per load step and values of residual norm during iterations of two typical time steps are given in Tables 8.1 and 8.2, respectively, which demonstrate the quadratic rate of asymptotic convergence.

8.5.4. Cyclically loaded notched bar

This example is chosen to demonstrate the capabilities of endochronic plasticity in simulation of cyclic loading and ratcheting. An axisymmetric notched bar is subjected to cyclic displacement and stress loadings.

The geometry and finite element mesh of the problem are shown in Figure 8.8. This example was considered by Kobayashi and Ohno (2002) to show the behavior of a new form of kinematic hardening plasticity in small displacement context. The Young's modulus and Poisson's coefficient are 165000 MPa and 0.3, respectively. For the proposed example, a constant scale function suppresses is sufficient for the isotropic hardening of model and three terms of kernel function are enough to simulate a nonlinear kinematic hardening behavior. Endochronic material parameters used for simulation are $C_1 = 200000\,\text{MPa}$, $C_2 = 4000\,\text{MPa}$, $C_3 = 800\,\text{MPa}$, $r_1 = 50000$, $r_2 = 1000$, $r_3 = 50$ and scale function of $f(\varsigma) = 35$.

Figure 8.9(a) illustrates the variation of applied stress with the number of cycles. The increment of applied stress is prescribed to allocate ten and fifty steps per half a cycle. Figure 8.9(b) shows the computed axial stress-strain relations at the integration point near the notch root. The capability of the endochronic model to simulate ratcheting is clear from the figure. As can be seen it gives accurate results using only ten stress increments for such a complicated loading history. Figure 8.10(a) shows cyclic loading diagram and the increment of applied displacement at the top of the specimen. Axial stress versus strain relation at the integration point near notch root is shown in Figure 8.10(b). Similar to the previous loading case, ten steps loading per half a cycle is sufficient to obtain an accurate result. It is worth mentioning while the concepts of the yield surface and loading-unloading criteria have not been assumed in the endochronic plasticity, the model can effectively simulate unloading and cyclic loading. Finally, Tables 8.3 and 8.4 show the values of residual norm during iterations of three typical time step for the stress loading and displacement loading cases, respectively. The reported values of residuals are also given corresponding to ten steps loading per half cycle.

8.5.5. Necking of a circular bar

Necking phenomenon in a bar is a well-known test in nonlinear solid mechanics and has been considered by many researchers (Hallquist 1983, Simo 1988, Rodriguez-Ferran *et al.* 1997, Ponthot 2002). A circular bar, with a radius of 6.413 mm and 53.334 mm length, is subjected to uniaxial tension up to total axial elongation of 14 mm, as shown in Figure 8.11(a). For an ideal case of a perfect specimen, necking can start in any section of specimen. In order to replace such a problem with multiple solutions by a problem with unique solution a geometric imperfection (1 percent radius reduction) is introduced to induce necking in the central part of the bar. Figure 8.11(b) shows how the endochronic model can be fitted to experimental data, reported by Simo (1988), which leads to the endochronic material parameters. Endochronic material parameters used for simulation are as follows; $C_1 = 168.6\,\text{GPa}$, $r_1 = 400$ and scale function of $f(\varsigma) = 0.69 - 0.21\exp(-11\varsigma) + 0.16\varsigma$. The Young's modulus and Poisson's coefficient are 206.9 GPa and 0.29, respectively.

Three different meshes, consisting of 50, 200 and 800 quadrilateral bilinear elements, corresponding to one quarter of specimen have been considered, as shown in Figure 8.12, in order to assess the accuracy of discretization. The deformed finite element meshes after total axial elongation of 14 mm corresponding to one quarter of the specimen are shown in this figure. The contours of the Cauchy stress components σ_{rr} and σ_{zz} are shown in Figure 8.13 for 800 elements mesh corresponding to 14 mm axial elongation. They are in good agreement with those reported by Simo (1988) and Ponthot (2002).

A comparison between experimental and some computed results of the ratio of the current to initial radius at the necking section versus axial elongation is shown in Figure 8.14. The result is in good agreement with experimental and previously computed results. The sensitivity of numerical solution to mesh refinement is assessed in Figure 8.15(a), where the necking ratio versus elongation is plotted for three meshes with 50, 200 and 800 elements. This figure corroborates the insensitivity of the numerical results to the mesh refinement. The sensitivity of the solution with respect to time discretization is examined in Figure 8.15(b). It is noted that the results corresponding to 50 and 100 load steps are very similar, which confirms the accuracy of the integration scheme. In Tables 8.5 and 8.6, the computational efforts involved in the calculations are assessed. Table 8.5 summarizes the required number of iteration per load step for 50 element mesh. Also, the values of residual norm during iterations for two typical load steps are shown in Table 8.6.

8.5.6. Strip in tension-shear banding

In the last example, the evolution of the shear band in a plane strain strip with strain softening endochronic plasticity is investigated and the capability of endochronic viscoplastic model in providing the mesh objective results is demonstrated. A strip is constrained at the bottom while a constant velocity of $v = 20\, mm/s$ is imposed at the top, as shown in Figure 8.16. The Endochronic strain softening plasticity is used with material parameters given in Table 8.7. In order to avoid homogeneous solution, the width of specimen is slightly increased toward the top, so that the shear band will be initiated at the bottom right of the specimen. Two different meshes have been used with 200 and 400 bilinear quadrilateral plane strain elements.

It is well known that the strain softening will result in the mathematical ill posedness of field equations (Khoei *et al.* 1997, 2005). Because no length scale is involved in the evolution of shear band, the localization zone is confined to the size of element, which yields to mesh dependent results. Figure 8.17 shows the displacement patterns and load deflection curves corresponding to two different meshes with endochronic softening plasticity. It is evident from the displacement patterns that the width of shear band is determined by the element size and the deformation is localized along a line of integration points. Also, the mesh dependency is obvious from the load deflection curves. It should be noted that the large deformation analysis can not solve the mesh dependency of solution. This lack of objectivity can be overcome by using the rate dependent viscoplastic model introducing a length scale into initial value problem (Needleman, 1988). Endochronic material parameters given in Table 8.7 are used, but additionally a linear viscosity term in the scale function is introduced

$$f(\varsigma,\dot{\varsigma}) = 10(1-10\varsigma+0.0125\dot{\varsigma}) \tag{8.121}$$

Figure 8.18 shows the displacement patterns and load deflection curves of two different meshes using the endochronic viscoplastic model with scale function defined in equation (8.121). The finite width of the shear band and its independency of the finite element size can be observed. Also, similar load-deflection curves can be observed from two different meshes, which corroborates the mesh objectivity of results. Finally, contour plots of the equivalent plastic strain using endochronic plasticity and endochronic viscoplasticity models for two different meshes are shown in Figure 8.19.

8.6. Finite torsion of thin-walled tubes

A general framework of objective corotational rates and spin tensor were developed by Xiao *et al.* (1998). It was proved that the corotational rate of every objective Eulerian symmetric tensor field, defined by spin Ω is objective if and only if the spin tensor is of the form $\Omega = W + \Upsilon(B,D)$, where $\Upsilon(B,D)$ is an anti-symmetric tensor valued isotropic function and B, D, W are left Cauchy-Green, stretching and vorticity tensors, respectively. It was also shown that all commonly used spin tensors are special cases of that general form. Of special interest is the newly discovered spin tensor, defining logarithmic rate (Reinhardt and Dubey 1996 and Xiao *et al.* 1997, 1998), which leads to the self-consistency criterion: the hypoelastic formulation intended for elastic behavior must be exactly integrable to deliver hyperelastic relation.

In this section, the endochronic plasticity model is extended to finite strain scope based on additive decomposition of deformation rate and hypoelastic rate type relations are used in the formulation. The above-mentioned general form of spin tensors is implemented into the endochronic model for problems of thin-walled tube under pure and simple torsion. A new Eulerian material dependent objective rate of Cauchy stress, named here as the combined rate, is developed based on the general form of spin tensors. Logarithmic, Sowerby-Chu, Green-Naghdi, Jaumann and combined rates are incorporated into the theory. Axial effects for various materials subjected to simple torsion and pure torsion are investigated. Experimental data of cast aluminum (Wu *et al.* 1998), Al-1100 (Bailey *et al.* 1972 and Aifantis 1987), 70:30 brass (Swift 1947 and Im and Atluri 1987) and SUS 304 stainless steel (Bammann 1989) are used for comparison with theoretical simulations. It has been shown that the combined rate simulates the axial effect more accurately as compared with the others.

8.6.1. Kinematics for torsion of thin-walled tube

Consider a fixed cylindrical coordinate system (O, e_R, e_Θ, e_Z) with its origin located at the center of one end of tube and the Z-axis coincides with the tube axis. The deformation of tube under torsion can be described by

$$\begin{aligned} r &= \alpha R \\ \theta &= \Theta + \psi Z \\ z &= \lambda Z \end{aligned} \tag{8.122}$$

where (r, θ, z) and (R, Θ, Z) are positions of a point after and before deformation in cylindrical coordinate system, ψ is angle of twist per unite length, α and λ radial and axial extension ratios, respectively. For the above deformation, the deformation gradient F is defined as

$$F = \alpha e_r \otimes e_R + \alpha e_\theta \otimes e_\Theta + \lambda e_z \otimes e_Z + \alpha \psi R \, e_\theta \otimes e_\Theta \tag{8.123}$$

By using equation (8.123) and expressions

$$\begin{aligned} \dot{e}_r &= \dot{\omega} \frac{Z}{R} e_\theta \\ \dot{e}_\theta &= -\dot{\omega} \frac{Z}{R} e_r \\ \dot{e}_z &= 0 \end{aligned} \tag{8.124}$$

where

$$\omega = R \psi \tag{8.125}$$

the velocity gradient, $L = \dot{F} F^{-1}$, can be determined as

$$L = \frac{\dot{\alpha}}{\alpha} e_r \otimes e_r + \frac{\dot{\alpha}}{\alpha} e_\theta \otimes e_\theta + \frac{\dot{\lambda}}{\lambda} e_z \otimes e_z - \dot{\omega} \frac{Z}{R} (e_r \otimes e_\theta - e_\theta \otimes e_r) + \dot{\omega} \frac{\alpha}{\lambda} e_\theta \otimes e_z \tag{8.126}$$

Accordingly, the rate of deformation D and vorticity tensor W are given by

$$D = \frac{1}{2}(L + L^T) = \frac{\dot{\alpha}}{\alpha} e_r \otimes e_r + \frac{\dot{\alpha}}{\alpha} e_\theta \otimes e_\theta + \frac{\dot{\lambda}}{\lambda} e_z \otimes e_z + \dot{\omega} \frac{\alpha}{2\lambda} (e_\theta \otimes e_z + e_z \otimes e_\theta) \tag{8.127}$$

and

$$W = \frac{1}{2}(L - L^T) = -\dot{\omega} \frac{Z}{R} (e_r \otimes e_\theta - e_\theta \otimes e_r) + \dot{\omega} \frac{\alpha}{2\lambda} (e_\theta \otimes e_z - e_z \otimes e_\theta) \tag{8.128}$$

The left Cauchy-Green tensor B is given by

$$B = FF^T = \alpha^2 e_r \otimes e_r + \alpha^2 (1 + \omega^2) e_\theta \otimes e_\theta + \alpha \lambda \omega (e_\theta \otimes e_z + e_z \otimes e_\theta) \tag{8.129}$$

The three eigen-values of B are as follows

$$\begin{aligned} \chi_1 &= \alpha^2 \\ \chi_2 &= \frac{1}{2}(\lambda^2 + \alpha^2 + \alpha^2 \omega^2 + \sqrt{u}) \\ \chi_3 &= \frac{1}{2}(\lambda^2 + \alpha^2 + \alpha^2 \omega^2 - \sqrt{u}) \end{aligned} \tag{8.130}$$

where

$$u = (\lambda^2 + \alpha^2 + \alpha^2 \omega^2)^2 - 4\alpha^2 \omega^2 \tag{8.131}$$

and their corresponding eigen-projections are given by

$$\begin{aligned} B_1 &= e_r \otimes e_r \\ B_2 &= \vartheta_1 e_\theta \otimes e_\theta + \vartheta_2 e_z \otimes e_z + \frac{\alpha \lambda \omega}{\sqrt{u}} (e_\theta \otimes e_z + e_z \otimes e_\theta) \\ B_3 &= \vartheta_2 e_\theta \otimes e_\theta + \vartheta_1 e_z \otimes e_z - \frac{\alpha \lambda \omega}{\sqrt{u}} (e_\theta \otimes e_z + e_z \otimes e_\theta) \end{aligned} \tag{8.132}$$

where

$$\begin{aligned} \vartheta_1 &= \frac{1}{2\sqrt{u}}(-\lambda^2 + \alpha^2 + \alpha^2 \omega^2 + \sqrt{u}) \\ \vartheta_2 &= \frac{1}{2\sqrt{u}}(\lambda^2 - \alpha^2 - \alpha^2 \omega^2 + \sqrt{u}) \end{aligned} \tag{8.133}$$

In order to represent a general formulation of finite strain endochronic hypoelasto-plastic constitutive equations for torsion of thin walled tube, valid for various objective stress rates, the general form of spin tensors, introduced by Xiao *et al.* (1998), is used (See also Reinhardt and Dubey 1996). The general class of spin tensors, defining objective rate of an Eulerian symmetric tensor, is characterized as the following

$$\Omega = W + \sum_{\sigma\neq\tau}^{m} h\left(\frac{\chi_\sigma}{I}, \frac{\chi_\tau}{I}\right) B_\sigma D B_\tau \tag{8.134}$$

$$h: R^+ \times R^+ \to R\,, \qquad h(x,y) = -h(y,x)$$

where I is the first invariant of B, the left Cauchy-Green tensor, χ_i is eigen-value of B, B_i is its corresponding eigen-projection and D is rate of deformation tensor. Now, we derive a general expression for this class of spin tensors specific for torsion problem. First, from equations (8.127) and (8.132), it can be inferred

$$B_1 D B_i = B_i D B_1 = 0\,, \quad i = 2,3 \tag{8.135}$$

Using $h(\chi_2,\chi_3) = -h(\chi_3,\chi_2)$ and equations (8.127), (8.132) and (8.134) results in

$$\begin{aligned}&\sum_{\sigma\neq\tau}^{m} h(\chi_\sigma,\chi_\tau) B_\sigma D B_\tau = \\ &h(\chi_2,\chi_3)\left(\frac{\alpha\lambda\omega}{\sqrt{u}}\left(\frac{\dot\lambda}{\lambda}-\frac{\dot\alpha}{\alpha}\right)+\frac{\alpha}{2\lambda}\dot\omega(\vartheta_1-\vartheta_2)\right)(e_\theta\otimes e_z - e_z\otimes e_\theta)\end{aligned} \tag{8.136}$$

By using equation (8.133), the above expression will be simplified as

$$\begin{aligned}&\sum_{\sigma\neq\tau}^{m} h(\chi_\sigma,\chi_\tau) B_\sigma D B_\tau = \\ &h(\chi_2,\chi_3)\frac{1}{\sqrt{u}}\left(\alpha\lambda\omega\left(\frac{\dot\lambda}{\lambda}-\frac{\dot\alpha}{\alpha}\right)+\frac{\alpha}{2\lambda}\dot\omega\left(-\lambda^2+\alpha^2+\alpha^2\omega^2\right)\right)(e_\theta\otimes e_z - e_z\otimes e_\theta)\end{aligned} \tag{8.137}$$

Hence, for any spin tensor we will have

$$\Omega = W + h(\chi_2,\chi_3)\frac{1}{\sqrt{u}}\left(\alpha\lambda\omega\left(\frac{\dot\lambda}{\lambda}-\frac{\dot\alpha}{\alpha}\right)+\frac{\alpha}{2\lambda}\dot\omega\left(-\lambda^2+\alpha^2+\alpha^2\omega^2\right)\right)(e_\theta\otimes e_z - e_z\otimes e_\theta) \tag{8.138}$$

or

$$\Omega = W + \dot\kappa\, h(\chi_2,\chi_3)(e_\theta\otimes e_z - e_z\otimes e_\theta) \tag{8.139}$$

where

$$\dot\kappa = \frac{1}{\sqrt{u}}\left(\alpha\lambda\omega\left(\frac{\dot\lambda}{\lambda}-\frac{\dot\alpha}{\alpha}\right)+\frac{\alpha}{2\lambda}\dot\omega\left(-\lambda^2+\alpha^2+\alpha^2\omega^2\right)\right) \tag{8.140}$$

Equation (8.139) introduces the general form of spin tensors derived particularly for thin walled tube under torsion. By specifying function $h(x,y)$, corresponding spin tensor will be defined by equations (8.139) and (8.140). In this study, a specific form of function $h(x,y)$ is used for which, equation (8.134) is reduced to

$$\Omega = W + \sum_{\sigma \neq \tau}^{m} h\left(\frac{\chi_\sigma}{\chi_\tau}\right) B_\sigma D B_\tau \tag{8.141}$$

where $h: R^+ \to R$ is a real function of positive real variable with the following property

$$h(z^{-1}) = -h(z), \quad \forall z \in R^+ \tag{8.142}$$

The above subclass is broad enough to include all commonly known spin tensors as particular cases (Xiao *et al.* 1998), including: Zaremba-Jaumann (Jaumann 1911), Logarithmic (Reinhardt and Dubey 1996 and Xiao *et al.* 1998), Green-Naghdi (Green and Naghdi 1965 and Dienes 1979) and Sowerby-Chu (Sowerby and Chu 1984) rates, as shown below.

8.6.1.1. Zaremba-Jaumann rate

Consider $h^J\left(\frac{\chi_2}{\chi_3}\right) = 0$, equation (8.139) is reduced to $\Omega^J = W$, i.e., the well-known Zaremba-Jaumann spin tensor, where W is the vorticity tensor.

8.6.1.2. Logarithmic rate

Consider

$$h^L(z) = \frac{1+z}{1-z} + \frac{2}{\ln(z)} \tag{8.143}$$

From definition of eigen-values of B, equation (8.130), will be derived

$$h^L\left(\frac{\chi_2}{\chi_3}\right) = -\frac{1}{u}\left(\lambda^2 + \alpha^2 + \alpha^2\omega^2\right) + \frac{1}{\sqrt{u}}\left(\ln\frac{\lambda^2 + \alpha^2 + \alpha^2\omega^2 + \sqrt{u}}{2\alpha\lambda}\right)^{-1} \tag{8.144}$$

Then, equation (8.139) yields the newly discovered logarithmic spin tensor (Lehmann *et al.* 1991, Reinhardt and Dubey 1996 and Xiao *et al.* 1998). With the logarithmic spin Ω^{Log}, the logarithmic rate of any Eulerian symmetric tensor, called Ξ, is defined by

$$\overset{\circ}{\Xi}{}^{Log} = \dot{\Xi} - \Omega^{Log}\Xi + \Xi\Omega^{Log} \tag{8.145}$$

In particular, if $\Xi = \ln V$, the Hencky strain measure, the logarithmic rate, defined by equation (8.145) would be rate of deformation D. This result leads to remarkable self-consistency property of hypoelasto-plasticity models based on logarithmic rate.

8.6.1.3. Green-Naghdi rate

Consider

$$h^{GN}(z) = \frac{1-\sqrt{z}}{1+\sqrt{z}} \tag{8.146}$$

Utilizing equation (8.130)

$$h^{GN}\left(\tfrac{\chi_2}{\chi_3}\right) = \frac{-1}{(\alpha+\lambda)^2 + \alpha^2\omega^2} \tag{8.147}$$

Then, equation (8.139) yields the spin tensor defining Green-Naghdi rate (Green and Naghdi 1965 and Dienes 1979)

$$\Omega^{GN} = \dot{R}R^T \tag{8.148}$$

where R is orthogonal tensor in polar decomposition $F = RU$ of F.

8.6.1.4. Sowerby-Chu rate

Consider

$$h^{SC}(z) = \frac{1+z}{1-z} \tag{8.149}$$

Using relations (8.130) and (8.139)

$$h^{SC}\left(\tfrac{\chi_2}{\chi_3}\right) = -\frac{\lambda^2 + \alpha^2 + \alpha^2\omega^2}{u} \tag{8.150}$$

It yields the spin tensor defining Sowerby-Chu rate (Sowerby and Chu 1984 and Reinhardt and Dubey 1996)

$$\Omega^{SC} = \dot{R}_e R_e^T \tag{8.151}$$

That is the twirl tensor of Eulerian triad. In equation (8.151), R_e denotes the diagonal transformation of V, the left stretch tensor ($V = R_e \lambda R_e$), with $\lambda = diag[\lambda_1, \lambda_2, \lambda_3]$ denoting the diagonal tensor containing the eigen-values of V.

8.6.1.5. Combined rate

In this rate, the function $h(z)$ is introduced as a linear combination of $h^{GN}(z)$ and $h^{SC}(z)$, defined by relations (8.146) and (8.149), i.e.

$$h^{Co}(z) = \mu_1 h^{SC}(z) + \mu_2 h^{GN}(z) \tag{8.152}$$

where μ_1 and μ_2 are material constants. In above relation, $h^{Co}(\frac{\chi_2}{\chi_3})$ introduces the combined objective rate, which can be calculated as

$$h^{GN}\left(\frac{\chi_2}{\chi_3}\right)=-\frac{\mu_1}{(\alpha+\lambda)^2+\alpha^2\omega^2}-\frac{\mu_2}{u}\left(\lambda^2+\alpha^2+\alpha^2\omega^2\right) \tag{8.153}$$

This rate can be effectively used to describe the behavior of material in finite strain range, as illustrated in Section 8.6.3.

8.6.2. Finite torsion simulation

In order to extend the endochronic constitutive equations, described in Section 7.3, formulation based upon the additive decomposition of rate of deformation is used, i.e.

$$D=D^e+D^p \tag{8.154}$$

where D^e and D^p are elastic and plastic parts of rate of deformation, respectively. In phenomenological theory of finite plasticity for metals, the following hypoelastic formulation for D^e, the elastic part of rate of deformation and σ an objective rate of Cauchy stress tensor, is often assumed.

$$\overset{\circ}{\sigma}=C^e:D^e \tag{8.155}$$

$$\overset{\circ}{\sigma}=\dot{\sigma}-\Omega\sigma+\sigma\Omega \tag{8.156}$$

$$C^e=2\mu I_4+\frac{2\mu\nu}{1-2\nu}I\otimes I \tag{8.157}$$

where Ω is a spin tensor and C^e is the elastic modulus in which μ is the shear modulus, ν is Poisson's ratio, I and I_4 are the unit second and fourth order tensors. Now, the endochronic constitutive equation of finite deformation can be written as

$$\overset{\circ}{S}=2\Phi(0)D^p-\sum_{r=1}^{n}\alpha_r S_r\dot{z} \tag{8.158}$$

$$\dot{\varsigma}^2=D^p:D^p \tag{8.159}$$

$$f(\varsigma)=\frac{\dot{\varsigma}}{\dot{z}} \tag{8.160}$$

where $\overset{\circ}{S}$ is the deviatoric part of objective stress rate, $\overset{\circ}{\sigma}$. Substituting equation (8.157) into equation (8.155) yields the following expression for S

$$\overset{\circ}{S}=2\mu D'^e=2\mu(D'-D^p) \tag{8.161}$$

where $[]'=[]-\frac{1}{3}tr([])I$ is the deviatoric part of $[]$. Substitution of equation (8.161) into equation (8.158) yields

$$\overset{\circ}{S} = \dot{S} - \Omega\, S + S\, \Omega = \frac{\mu}{\mu + \Phi(0)}\left(2\Phi(0)D' - \sum_{r=1}^{n} \alpha_r S_r \dot{z}\right) \tag{8.162}$$

Considering D^p a traceless tensor, equations (8.156) and (8.157) yield

$$tr(\overset{\circ}{\sigma}) = 3K\, tr(D) \tag{8.163}$$

Using equations (8.128) and (8.139), the general form for spin tensor of different objective rates would be

$$\begin{aligned} \Omega &= W + \dot{\kappa} h(e_\theta \otimes e_z - e_z \otimes e_\theta) \\ &= \left(\dot{\kappa} h + \frac{\alpha}{2\lambda}\dot{\omega}\right)(e_\theta \otimes e_z - e_z \otimes e_\theta) - \frac{Z}{R}\dot{\omega}(e_r \otimes e_\theta - e_\theta \otimes e_r) \end{aligned} \tag{8.164}$$

The latter equation can be written in the form

$$\Omega = \dot{\mathrm{H}}(e_\theta \otimes e_z - e_z \otimes e_\theta) - \tfrac{Z}{R}\dot{\omega}(e_r \otimes e_\theta - e_\theta \otimes e_r) \tag{8.165}$$

where

$$\dot{\mathrm{H}} = H^\alpha \dot{\alpha} + H^\lambda \dot{\lambda} + H^\omega \dot{\omega} \tag{8.166}$$

where H^α, H^λ and H^ω for each objective rate can be derived from equations (8.144), (8.147) and (8.150).

For torsion of a thin walled cylinder, due to loading symmetry, the Cauchy stress σ, is of the form

$$\sigma = \sigma_{\theta\theta} e_\theta \otimes e_\theta + \sigma_{zz} e_z \otimes e_z + \tau e_\theta \otimes e_z + \tau e_z \otimes e_\theta \tag{8.167}$$

Using the identities (8.124) yields the following expression for material time derivative of Cauchy stress tensor

$$\begin{aligned} \dot{\sigma} &= \dot{\sigma}_{\theta\theta} e_\theta \otimes e_\theta + \dot{\sigma}_{zz} e_z \otimes e_z + \dot{\tau} e_\theta \otimes e_z + \dot{\tau} e_z \otimes e_\theta \\ &\quad - \tfrac{Z}{R}\dot{\omega}\left(\sigma_{\theta\theta}(e_r \otimes e_\theta + e_\theta \otimes e_r) + \tau(e_z \otimes e_r + e_r \otimes e_z)\right) \end{aligned} \tag{8.168}$$

Then, from equations (8.156), (8.165) and the above equation, deviatoric part of objective Cauchy stress rate and its trace is deduced

$$\begin{aligned} \overset{\circ}{S} &= -\tfrac{1}{3}(\dot{\sigma}_{\theta\theta} + \dot{\sigma}_{zz}) e_r \otimes e_r + \left(\tfrac{2}{3}\dot{\sigma}_{\theta\theta} - \tfrac{1}{3}\dot{\sigma}_{zz} - 2\tau\dot{\mathrm{H}}\right) e_\theta \otimes e_\theta + \\ &\quad \left(-\tfrac{1}{3}\dot{\sigma}_{\theta\theta} + \tfrac{2}{3}\dot{\sigma}_{zz} + 2\tau\dot{\mathrm{H}}\right) e_z \otimes e_z + \left(\dot{\tau} + (\sigma_{\theta\theta} - \sigma_{zz})\dot{\mathrm{H}}\right)(e_\theta \otimes e_z + e_z \otimes e_\theta) \end{aligned} \tag{8.169}$$

$$tr(\overset{\circ}{\sigma}) = \dot{\sigma}_{\theta\theta} + \dot{\sigma}_{zz} \tag{8.170}$$

Substitution of equations (8.169) and (8.170) into equations (8.162) and (8.163) leads to

$$-\frac{1}{3}\dot{\sigma}_{\theta\theta}-\frac{1}{3}\dot{\sigma}_{zz}=\frac{2}{3}\frac{\mu\Phi(0)}{\mu+\Phi(0)}\left(\frac{\dot{\alpha}}{\alpha}-\frac{\dot{\lambda}}{\lambda}\right)+\frac{1}{3}\frac{1}{\mu+\Phi(0)}\mu\dot{z}\sum_{r=1}^{n}\alpha_r(\sigma_{\theta\theta}^r+\sigma_{zz}^r) \tag{8.171}$$

$$\frac{2}{3}\dot{\sigma}_{\theta\theta}-\frac{1}{3}\dot{\sigma}_{zz}-2\tau(H^{\alpha}\dot{\alpha}+H^{\lambda}\dot{\lambda}+H^{\omega}\dot{\omega})$$
$$=\frac{2}{3}\frac{\mu\Phi(0)}{\mu+\Phi(0)}\left(\frac{\dot{\alpha}}{\alpha}-\frac{\dot{\lambda}}{\lambda}\right)-\frac{1}{3}\frac{1}{\mu+\Phi(0)}\mu\dot{z}\sum_{r=1}^{n}\alpha_r(2\sigma_{\theta\theta}^r-\sigma_{zz}^r) \tag{8.172}$$

$$\dot{\tau}+(\sigma_{\theta\theta}-\sigma_{zz})(H^{\alpha}\dot{\alpha}+H^{\lambda}\dot{\lambda}+H^{\omega}\dot{\omega})=\frac{\mu\Phi(0)}{\mu+\Phi(0)}\frac{\alpha}{\lambda}\dot{\omega}-\frac{1}{\mu+\Phi(0)}\mu\dot{z}\sum_{r=1}^{n}\alpha_r\tau^r \tag{8.173}$$

$$\dot{\sigma}_{\theta\theta}+\dot{\sigma}_{zz}=3K\left(2\frac{\dot{\alpha}}{\alpha}+\frac{\dot{\lambda}}{\lambda}\right) \tag{8.174}$$

Note that the above equations are endochronic constitutive equations derived for finite torsion of thin-walled tube for general objective rate of Cauchy stress tensor.

8.6.2.1. Simple torsion

For the case of simple torsion, axial extension of tube is absent and $\dot{\lambda}=0$. Substituting equation (8.174) into equations (8.171) and (8.172), then eliminating $\dot{\sigma}_{zz}$ from these equations yields

$$\dot{\sigma}_{\theta\theta}=A_{S1}\dot{\omega}+A_{S2}\dot{z} \tag{8.175}$$

where

$$A_{S1}=2\tau H^{\omega} \tag{8.176}$$

and

$$A_{S2}=-\frac{\alpha\tau H^{\alpha}}{3K+3\Phi(0)\frac{K}{\mu}+\Phi(0)}\sum_{r=1}^{n}\alpha_r(\sigma_{\theta\theta}^r+\sigma_{zz}^r)-\frac{\mu}{\mu+\Phi(0)}\sum_{r=1}^{n}\alpha_r\sigma_{\theta\theta}^r \tag{8.177}$$

By similar manipulation $\dot{\sigma}_{zz}$ can be derived from equations (8.171), (8.174) and (8.175) as

$$\dot{\sigma}_{zz}=-A_{S1}\dot{\omega}+A_{S3}\dot{z} \tag{8.178}$$

where

$$A_{S3}=\frac{-3K+\alpha\tau H^{\alpha}}{3K+3\Phi(0)\frac{K}{\mu}+\Phi(0)}\sum_{r=1}^{n}\alpha_r(\sigma_{\theta\theta}^r+\sigma_{zz}^r)+\frac{\mu}{\mu+\Phi(0)}\sum_{r=1}^{n}\alpha_r\sigma_{\theta\theta}^r \tag{8.179}$$

Substituting equations (8.175) and (8.178) into (8.174) yields the following expression for $\dot{\alpha}$,

$$\dot{\alpha}=\frac{\alpha}{6K}(A_{S2}+A_{S3})\dot{z} \tag{8.180}$$

From equations (8.173) and (8.180), $\dot{\tau}$ is derived as

$$\dot{\tau} = A_{S5}\dot{\omega} + A_{S6}\dot{z} \tag{8.181}$$

where

$$A_{S5} = \frac{\mu\Phi(0)}{\mu+\Phi(0)}\alpha + (\sigma_{zz} - \sigma_{\theta\theta})H^{\omega} \tag{8.182}$$

and

$$A_{S6} = -\frac{\mu}{\mu+\Phi(0)}\sum_{r=1}^{n}\alpha_r\tau^r + \frac{\alpha H^{\alpha}}{6K}(\sigma_{zz} - \sigma_{\theta\theta})(A_{S2} + A_{S3}) \tag{8.183}$$

The deviatoric plastic rate of deformation tensor is determined from relation $D^p = D' - \frac{1}{2\mu}\overset{\circ}{S}$, definition of $\overset{\circ}{S}$ and D in equations (8.169) and (8.127) and the derived expression for $\dot{\sigma}_{\theta\theta}$, $\dot{\sigma}_{zz}$, $\dot{\alpha}$ and $\dot{\tau}$ in equations (8.175), (8.178), (8.180) and (8.181), which conclude in the following expression for components of the plastic rate of deformation

$$D^p_{rr} = M_{S2}\dot{z} \tag{8.184}$$

$$D^p_{\theta\theta} = M_{S3}\dot{\omega} + M_{S4}\dot{z} \tag{8.185}$$

$$D^p_{zz} = -M_{S3}\dot{\omega} + M_{S6}\dot{z} \tag{8.186}$$

$$D^p_{z\theta} = M_{S7}\dot{\omega} + M_{S8}\dot{z} \tag{8.187}$$

where

$$M_{S2} = \tfrac{1}{6}\left(\frac{1}{\mu} + \frac{1}{3K}\right)(A_{S2} + A_{S3}) \tag{8.188}$$

$$M_{S3} = \frac{1}{\mu}\left(\tau H^{\omega} - \frac{A_{S1}}{2}\right) \tag{8.189}$$

$$M_{S4} = \tfrac{1}{6}\left(\frac{\alpha}{K}\left(\frac{1}{3\alpha} + \frac{\tau}{\mu}H^{\alpha}\right)(A_{S2} + A_{S3}) - \frac{1}{\mu}(2A_{S2} - A_{S3})\right) \tag{8.190}$$

$$M_{S6} = -\tfrac{1}{6}\left(\frac{1}{K}\left(\frac{2}{3} + \frac{\tau}{\mu}\alpha H^{\alpha}\right)(A_{S2} + A_{S3}) + \frac{1}{\mu}(2A_{S3} - A_{S2})\right) \tag{8.191}$$

$$M_{S7} = \tfrac{1}{2}\left(\alpha - \frac{1}{\mu}A_{S5} - \frac{1}{\mu}H^{\omega}(\sigma_{\theta\theta} - \sigma_{zz})\right) \tag{8.192}$$

$$M_{S8} = -\frac{1}{2\mu}\left(A_{S6} + \frac{1}{6K}\alpha H^{\alpha}(A_{S2} + A_{S3})(\sigma_{\theta\theta} - \sigma_{zz})\right) \tag{8.193}$$

To this point all variables of the model are expressed in term of known values of applied torsion $d\omega$ and the increment of scaled intrinsic time dz, which can be calculate by inserting (8.184) to (8.187) into the following equation

$$(\dot{z}f(\varsigma))^2 = \dot{\varsigma}^2 = D_{rr}^{p\,2} + D_{\theta\theta}^{p\,2} + D_{zz}^{p\,2} + 2D_{z\theta}^{p\,2} \tag{8.194}$$

Substitution of equations (8.184) to (8.187) into equation (8.194) yields the following quadratic equation for dz

$$\upsilon_1 dz^2 + \upsilon_2 dz + \upsilon_3 = 0 \tag{8.195}$$

with coefficients

$$\upsilon_1 = M_{S2}^2 + M_{S4}^2 + M_{S6}^2 + 2M_{S8}^2 - f^2(\varsigma) \tag{8.196}$$

$$\upsilon_2 = 2d\omega\left(M_{S3}M_{S4} - M_{S3}M_{S6} + 2M_{S7}M_{S8}\right) \tag{8.197}$$

$$\upsilon_3 = 2(d\omega)^2\left(M_{S3}^2 + M_{S7}^2\right) \tag{8.198}$$

In equation (8.194), scale function $f(\varsigma)$ is a positive variable and intrinsic time ς is positive or zero. Hence, the quadratic equation (8.195) generates two roots of opposite sign, or zero, the positive root is the desired one. Once the increment of dz is determined, the incremental quantities $d\alpha, d\sigma_{\theta\theta}, d\sigma_{zz}, d\tau$ and $d\varsigma$ are obtained from equations (8.180), (8.175), (8.178) and (8.181), respectively. With updating the value of model variables $\alpha, \omega, \sigma_{\theta\theta}, \sigma_{zz}, \tau$ and $d\varsigma$ in each step, the loading proceeds to next step. It is worth mentioning that the value of history variables $\sigma_{\theta\theta}^r, \sigma_{zz}^r$ and τ^r should be updated in each step with equation

$$\overset{\circ}{S}{}^r = 2C_r D^p - \alpha_r S^r \dot{z} \tag{8.199}$$

8.6.2.2. Pure torsion

For the case of pure torsion, the axial component of stress tensor will be zero, $\dot{\sigma}_{zz} = \sigma_{zz} = 0$. Substituting equation (8.174) into (8.171) yields

$$\dot{\lambda} = A_{P1}\dot{\alpha} + A_{P2}\dot{z} \tag{8.200}$$

where

$$A_{P1} = 2\frac{\lambda}{\alpha}\frac{\Phi(0)(\mu+3K)+3K\mu}{\Phi(0)(2\mu-3K)-3K\mu}\ , \qquad A_{P2} = \frac{\lambda\mu}{\Phi(0)(2\mu-3K)-3K\mu}\sum_{r=1}^{n}\alpha_r\sigma_{\theta\theta}^r \tag{8.201}$$

Using equations (8.171), (8.174) and (8.200) leads to

$$\dot{\alpha} = A_{P3}\dot{\omega} + A_{P4}\dot{z} \tag{8.202}$$

where

$$A_{P3} = \frac{2}{\Lambda}\tau H^{\omega} \tag{8.203}$$

and

$$A_{P4} = \frac{1}{\Lambda}\left(-\frac{2}{\lambda}K A_{P2} + 2\tau A_{P2}H^{\lambda} - \frac{2A_{P2}}{3\lambda}\frac{\mu\Phi(0)}{\mu+\Phi(0)} - \frac{2}{3}\frac{\mu\Phi(0)}{\mu+\Phi(0)}\sum_{r=1}^{n}\alpha_r\sigma_{\theta\theta}^{r}\right) \tag{8.204}$$

where the variable Λ in equations (8.203) and (8.204) is

$$\Lambda = 4\frac{K}{\alpha} + 2A_{P1}\frac{K}{\lambda} - 2\tau\left(H^{\alpha} + A_{P_a}H^{\lambda}\right) - \frac{2}{3}\left(\frac{1}{\alpha} - \frac{A_{P1}}{\lambda}\right)\frac{\mu\Phi(0)}{\mu+\Phi(0)} \tag{8.205}$$

Substituting equation (8.202) into (8.200) yields

$$\dot{\lambda} = A_{P1}A_{P3}\dot{\omega} + (A_{P1}A_{P4} + A_{P2})\dot{z} \tag{8.206}$$

Substitution of $\dot{\alpha}$ and $\dot{\lambda}$ from equations (8.202) and (8.206) into (8.174) leads to

$$\dot{\sigma}_{\theta\theta} = A_{P5}\dot{\omega} + A_{P6}\dot{z} \tag{8.207}$$

where A_{P5} and A_{P6} are defined as

$$\begin{aligned} A_{P5} &= 3K\left(\frac{2}{\alpha}A_{P3} + \frac{1}{\lambda}A_{P1}A_{P3}\right) \\ A_{P6} &= 3K\left(\frac{2}{\alpha}A_{P4} + \frac{1}{\lambda}(A_{P1}A_{P4} + A_{P2})\right) \end{aligned} \tag{8.208}$$

Substituting $\dot{\alpha}$ and $\dot{\lambda}$ from equations (8.202) and (8.206) into equation (8.173) leads to

$$\dot{\tau} = A_{P7}\dot{\omega} + A_{P8}\dot{z} \tag{8.209}$$

where

$$A_{P7} = -\sigma_{\theta\theta}\left(A_{P3}H^{\alpha} + A_{P1}A_{P3}H^{\lambda} + H^{\omega}\right) + \frac{\alpha}{\lambda}\frac{\mu\Phi(0)}{\mu+\Phi(0)} \tag{8.210}$$

$$A_{P8} = -\sigma_{\theta\theta}\left(A_{P4}H^{\alpha} + H^{\lambda}(A_{P1}A_{P4} + A_{P2})\right) - \frac{\mu}{\mu+\Phi(0)}\sum_{r=1}^{n}\alpha_r\tau^{r} \tag{8.211}$$

Like the case of simple shear the amount of components of D^P, plastic rate of deformation tensor, is determined from relation $D^P = D' - \frac{\overset{\circ}{S}}{2\mu}$, definition of $\overset{\circ}{S}$ and D in equations (8.169) and (8.127) and the derived expression for $\dot{\alpha}$, $\dot{\lambda}$, $\dot{\sigma}_{\theta\theta}$ and $\dot{\tau}$ in equations (8.202), (8.206), (8.207) and (8.209), respectively, which yield the following expression for components of the plastic rate of deformation

$$D_{rr}^{p} = M_{P1}\dot{\omega} + M_{P2}\dot{z} \tag{8.212}$$

$$D_{\theta\theta}^{p} = M_{P3}\dot{\omega} + M_{P4}\dot{z} \tag{8.213}$$

$$D_{zz}^{p} = M_{P5}\dot{\omega} + M_{P6}\dot{z} \tag{8.214}$$

$$D_{z\theta}^{p} = M_{P7}\dot{\omega} + M_{P8}\dot{z} \tag{8.215}$$

where variables M_{P1} to M_{P8} in the above equations are

$$M_{P1} = \tfrac{1}{3}\left(\frac{1}{\alpha}A_{P3} - \frac{1}{\lambda}A_{P1}A_{P3} + \frac{1}{2\mu}A_{P5}\right) \tag{8.216}$$

$$M_{P2} = \tfrac{1}{3}\left(\frac{1}{\alpha}A_{P4} - \frac{1}{\lambda}(A_{P1}A_{P4} + A_{P2}) + \frac{1}{2\mu}A_{P6}\right) \tag{8.217}$$

$$M_{P3} = \frac{1}{3\alpha}A_{P3} - \frac{1}{3\lambda}A_{P1}A_{P3} - \frac{1}{3\mu}A_{P5} + \frac{\tau}{\mu}N_1 \tag{8.218}$$

$$M_{P4} = \frac{1}{3\alpha}A_{P4} - \frac{1}{3\lambda}(A_{P1}A_{P4} + A_{P2}) - \frac{1}{3\mu}A_{P6} + \frac{\tau}{\mu}N_2 \tag{8.219}$$

$$M_{P5} = -\frac{2}{3\alpha}A_{P3} + \frac{2}{3\lambda}A_{P1}A_{P3} - \frac{\tau}{\mu}N_1 \tag{8.220}$$

$$M_{P6} = -\frac{2}{3\alpha}A_{P4} + \frac{2}{3\lambda}(A_{P1}A_{P4} + A_{P2}) - \frac{\tau}{\mu}N_2 \tag{8.221}$$

$$M_{P7} = \frac{\alpha}{2\lambda} - \frac{1}{2\mu}A_{P7} - \frac{\sigma_{\theta\theta}}{2\mu}N_1 \tag{8.222}$$

$$M_{P8} = -\frac{1}{2\mu}A_{P8} - \frac{\sigma_{\theta\theta}}{2\mu}N_2 \tag{8.223}$$

In equations (8.218) to (8.223) variables N_1 and N_2 are defined as

$$N_1 = A_{P3}H^{\alpha} + A_{P1}A_{P3}H^{\lambda} + H^{\omega} \tag{8.224}$$

$$N_2 = A_{P4}H^{\alpha} + H^{\lambda}(A_{P1}A_{P4} + A_{P2}) \tag{8.225}$$

Substitution of equations (8.212) to (8.215) into equation (8.194) leads to the following quadratic equation with variable dz

$$\pi_1 dz^2 + \pi_2 dz + \pi_3 = 0 \tag{8.226}$$

The parameters of the above equations are calculated as

$$\pi_1 = M_{P2}^2 + M_{P4}^2 + M_{P6}^2 + 2M_{P8}^2 - f^2(\varsigma) \tag{8.227}$$

$$\pi_2 = 2d\omega\left(M_{P1}M_{P2} + M_{P3}M_{S4} + M_{P5}M_{P6} + 2M_{P7}M_{P8}\right) \tag{8.228}$$

$$\pi_3 = (d\omega)^2\left(M_{P1}^2 + M_{P3}^2 + M_{P5}^2 + 2M_{P7}^2\right) \tag{8.229}$$

Similar to the case of simple torsion, equation (8.226) have two roots of opposite sign or zero and the positive root is the desired one. As discussed in the case of simple torsion, once the increment of dz is determined, the incremental quantities $d\alpha, d\lambda, d\sigma_{\theta\theta}, d\tau$ and $d\varsigma$ are obtained from equations (8.202), (8.206), (8.207) and (8.209), respectively. With updating the variables $\alpha, \omega, \lambda, \sigma_{\theta\theta}, \tau$ and $d\varsigma$ in each step the loading process is completed.

8.6.3. Comparison with experiments

In this section, we compare the simulation results to experimental data found in literature for finite torsion of thin walled tubes. Of greatest interest is comparison between the data obtained from experiments and the results of axial stress and strain, induced by simple and pure finite torsion of tube, derived from simulation. In order to implement the model developed in previous sections, material parameter of endochronic plasticity is required. Material constants include elastic constants, coefficients and exponents of kernel function and scale functions and two material constants μ_1 and μ_2 in the expression of combined rate. In this study, a kernel function is used with three terms and a scale function as

$$f(\varsigma) = 1 - Ce^{-\beta\varsigma} \tag{8.230}$$

Hence, there are two elastic constants and eight endochronic parameters that should be determined. For endochronic model with the combined rate, it is necessary to evaluate two other constants μ_1 and μ_2. These material constants are derived by fitting the theory to the experimental stress-strain curves. Table 8.8 contains constitutive equation parameters for four different materials. Five objective rates, i.e. the logarithmic, Sowerby-Chu, Green-Naghdi, Jaumann and combined rates, are implemented into the model and simulation results are compared with experimental results.

Wu *et al.* (1998) tested the cast aluminum tube subjected to pure and simple torsion in finite strain range. Figure 8.20(a) shows the experimental and simulation result of shear stress versus shear strain curves under pure torsion. It is shown that the shear stress-strain curve predicted by the model with different objective stress rates are similar and are in good agreement with the experiment. The experimental and simulation results of axial stress versus shear strain in simple torsion test are shown in Figure 8.20(b). As can be observed, the behavior of model in predicting induced compressive axial stress depends on objective stress rate. From experimental result, the built up axial compressive stress is about one tenth of shear stress when shear strain is 1. Although the prediction of endochronic model with Green-Naghdi and logarithmic rates agrees fairly well with this experimental observation, the Combined rate with material parameters $\mu_1 = \mu_2 = 0.5$ provides the best prediction for the axial effects.

Figure 8.21(a) shows experimental and simulated shear stress-strain curves for Al-1100 under simple torsion. Experiments were conducted by White and Anand, and the results were reported by Aifantis (1987) and Pan *et al.* (1996). Material parameters of Table 8.8 are used in endochronic model with various stress rates. A comparison between experiment and simulation for induced compressive axial stress is shown in

Figure 8.21(b). The build up axial stress is about one tenth of shear stress in shear strain of 1.5. As for the case of cast aluminum (Figure 8.20-b), the observed experimental axial stress is between prediction of endochronic model with Green-Naghdi and Sowerby-Chu rates, while the best estimation of induced axial stress is provided by the combined rate with $\mu_1 = \mu_2 = 0.5$. The experimental and simulation results of axial strain, developed in pure torsion of Al-1100 tube are shown in Figure 8.21(c). Experiments were conducted by Bailey *et al.* (1972) up to the large shear strain of 10. The induced axial strain is about 0.5 percent of shear strain. Prediction of model with the combined rate agrees very well with the experimental results at large strains. However, endochronic model with Jaumann rate overestimate axial stain up to 1.8 percent of shear strain.

In next step, simulation results are compared with simple and pure torsion experiments conducted by Stout, reported by Swift (1947), Im and Atluri (1987) and Pan *et al.* (1996) on 70:30 brass tubes. Figure 8.22 shows shear stress-shear strain and axial strain-shear strain curves of 70:30 brass tubes under simple and pure torsion, respectively. In Figure 8.22(a), good predictions of shear stress under finite simple torsion are observed for endochronic model with all five rates. However, the combined rate with $\mu_2 = 3\mu_1 = 0.75$ yields the best prediction of axial strain for large pure torsion. Moreover the unacceptable overestimation of endochronic model with Jaumann rate for axial strain can be observed in Figure 8.22(b).

Figure 8.23 shows the behavior of endochronic model and experimental results for SUS 304 stainless steel under simple torsion. The simple torsion experiments are conducted by Ishikawa (1999). Predictions of shear stress in simple torsion by endochronic model with all stress rates are similar and agree well with experimental results. However it can be observed from Figure 8.23(b) that the predictions of axial stresses in simple torsion are different and for the combined rate with $\mu_1 = 0$ and $\mu_2 = 1.2$, the agreement of simulation results with experiment is remarkable.

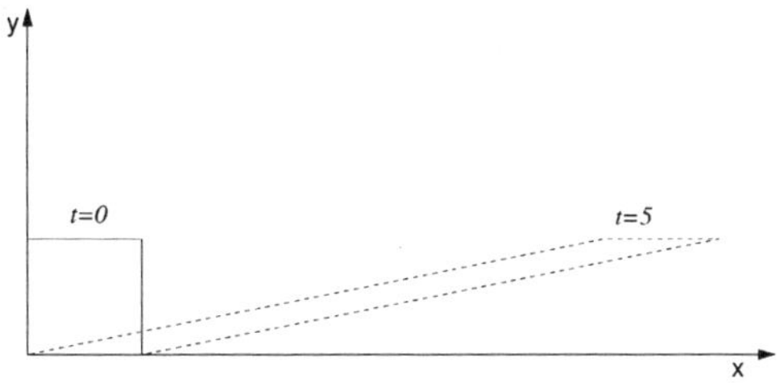

Figure 8.1. Simple shear test; Problem statement

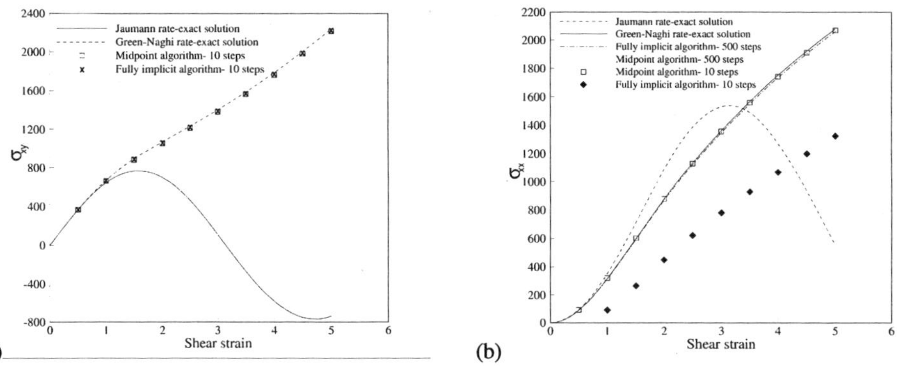

(a) (b)

Figure 8.2. Simple shear test in elastic analysis; a) The stress component σ_{xy} versus shear strain, b) The stress component σ_{xx} versus shear strain

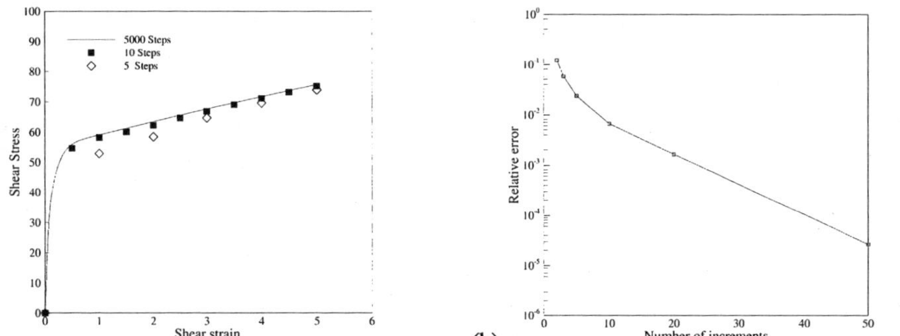

(a) (b)

Figure 8.3. Simple shear test in elasto-plastic analysis; a) The shear stress versus shear strain, b) The error in final value of shear stress versus number of time steps

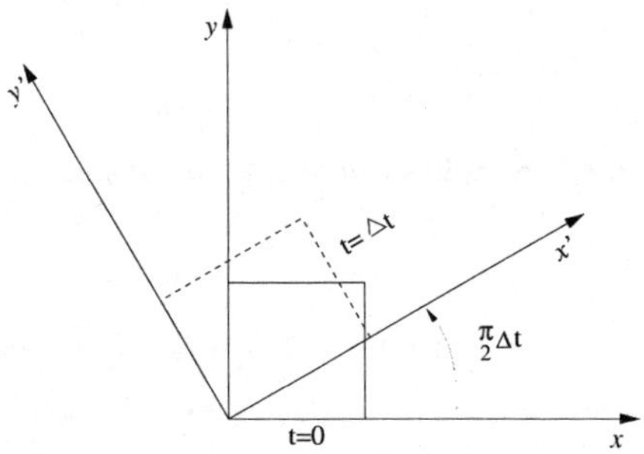

Figure 8.4. Extension and rotation test; Problem statement

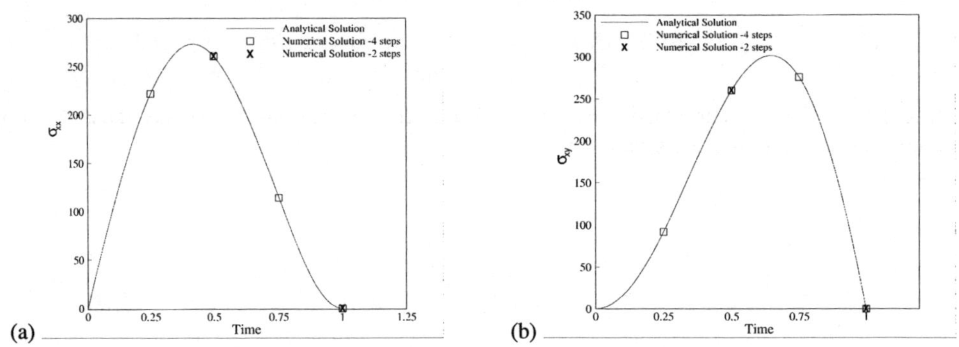

Figure 8.5. Extension and rotation test; a) The stress component σ_{xx} versus time, b) The stress component σ_{xy} versus time

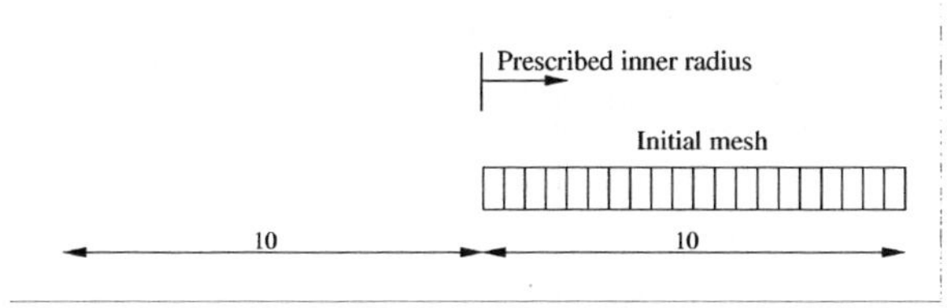

Figure 8.6. Expansion of a thick-walled cylinder; Geometry and initial mesh

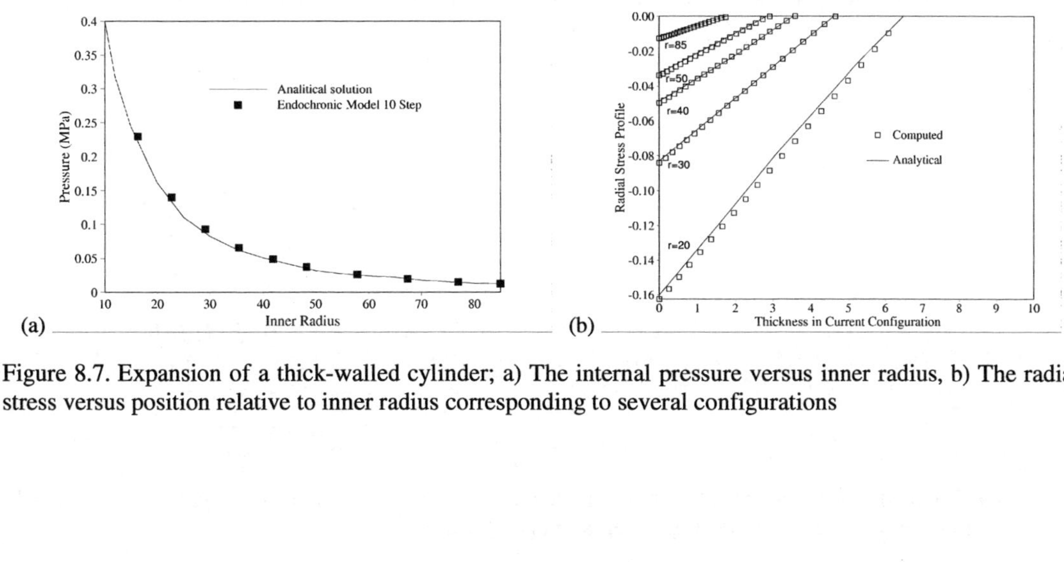

Figure 8.7. Expansion of a thick-walled cylinder; a) The internal pressure versus inner radius, b) The radial stress versus position relative to inner radius corresponding to several configurations

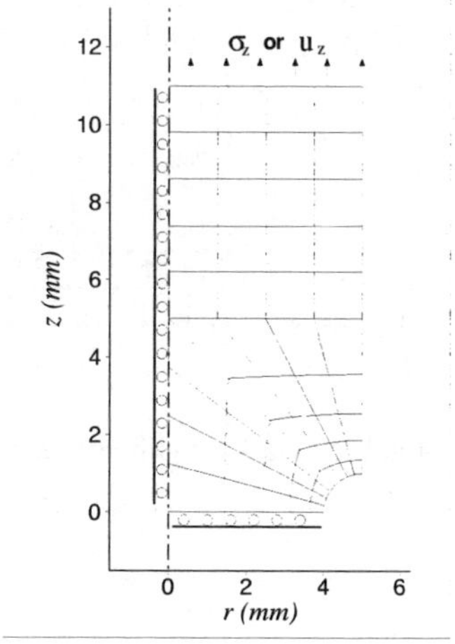

Figure 8.8. Geometry, FE mesh and boundary condition of notched bar subjected to cyclic loading

(a)
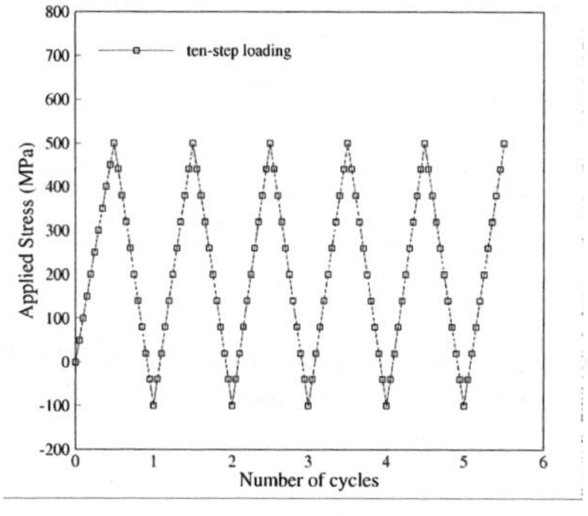

(b)
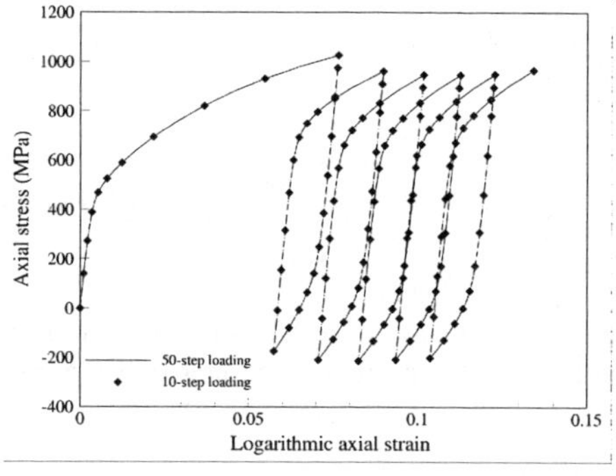

Figure 8.9. Cyclically loaded notched bar; a) Cyclic loading diagram and increment of applied axial stress, b) The axial stress versus strain relation at the integration point near notch root

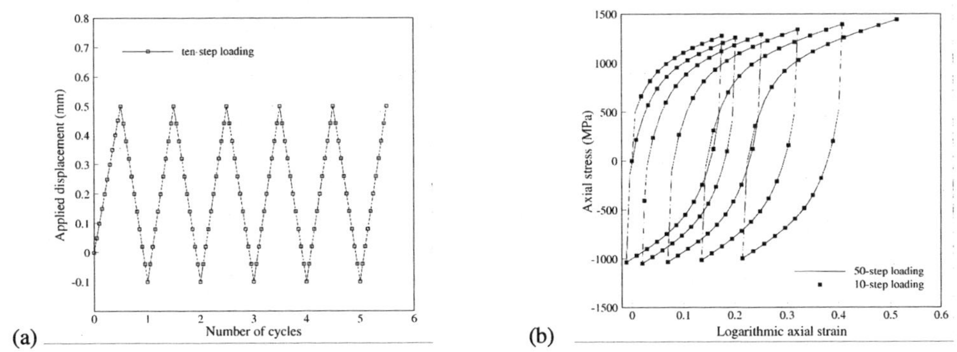

Figure 8.10. Cyclically loaded notched bar; a) Cyclic loading diagram and increment of applied axial displacement, b) The axial stress versus strain relation at the integration point near notch root

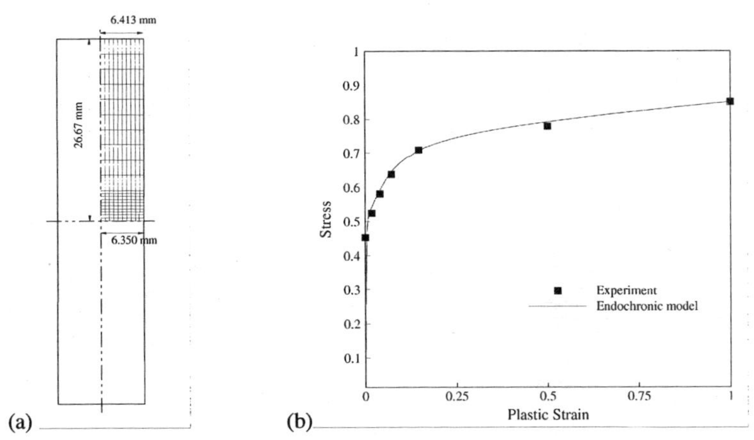

Figure 8.11. Necking of a circular bar; a) Problem description, b) Uniaxial hardening curve

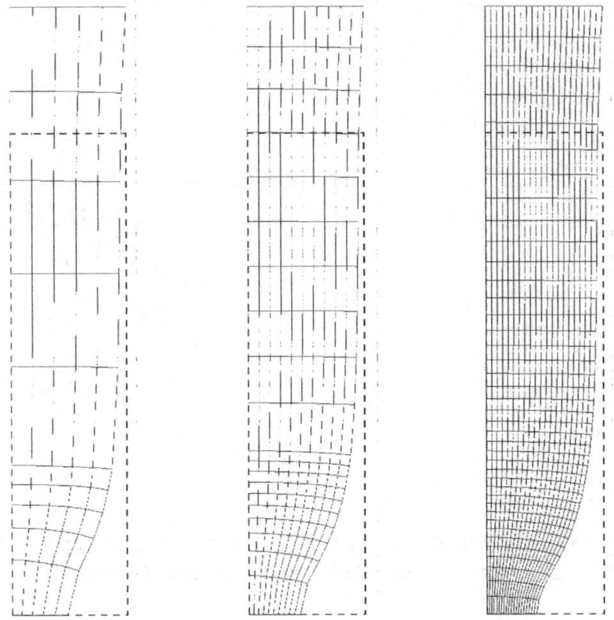

Figure 8.12. Necking of a circular bar; Finite element meshes after the total axial elongation of 14 mm corresponding to one quarter of the specimen, consisting of 50, 200 and 800 elements

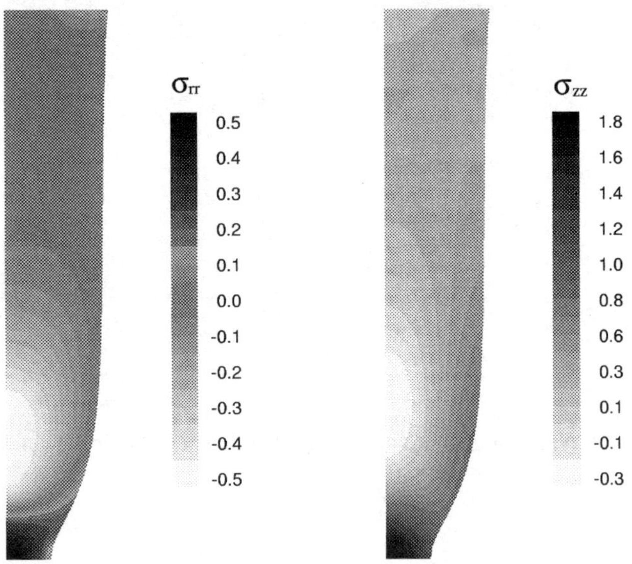

Figure 8.13. Necking of a circular bar; The contours of Cauchy stress components σ_{rr} and σ_{zz} for 800 element mesh

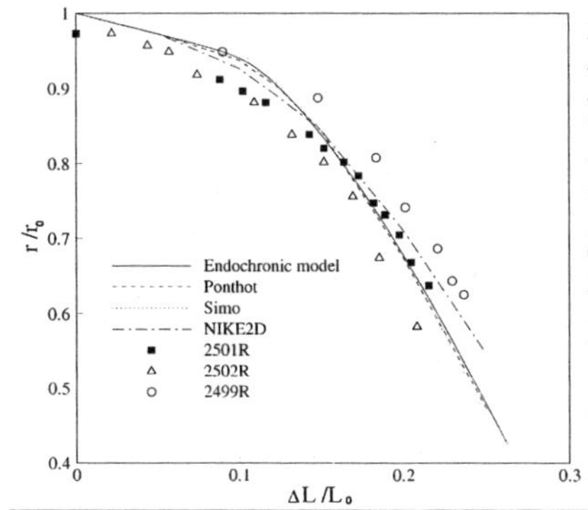

Figure 8.14. Necking of a circular bar; The experimental and computed results of the ratio of current to initial radius at the section undergoing extreme necking versus axial elongation

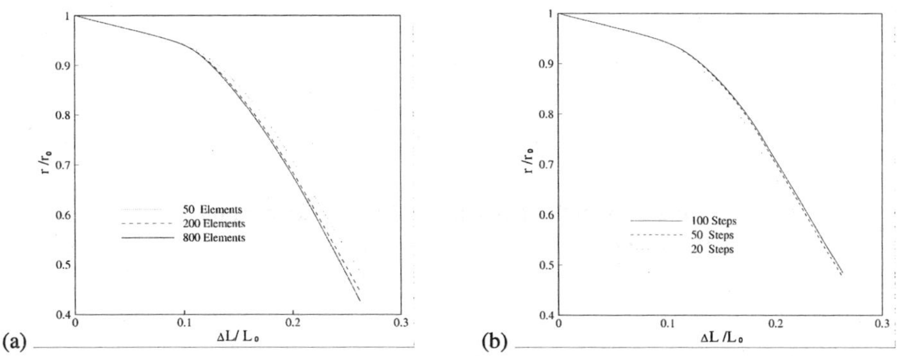

Figure 8.15. Necking ratio versus axial elongation in necking of a circular bar; a) The sensitivity study of numerical calculation with respect to mesh refinement, b) The accuracy of integration algorithm for different load step sizes

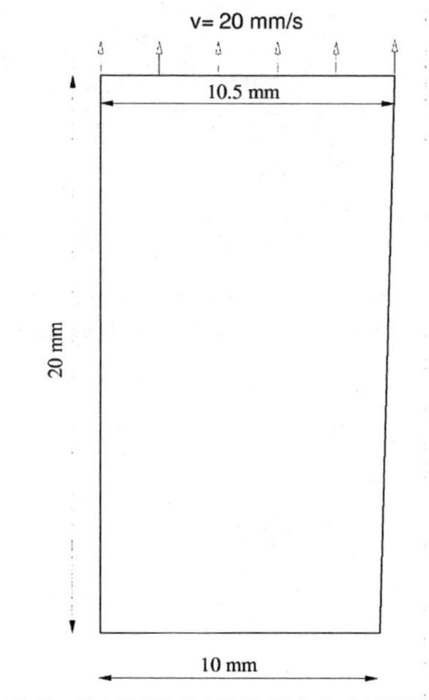

Figure 8.16. Strip in tension; Problem description

(a)

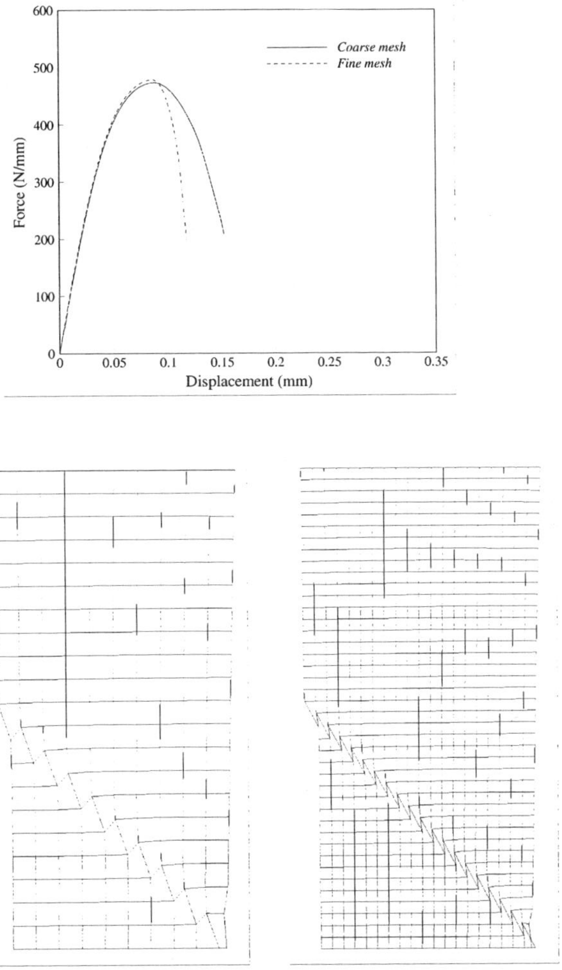

(b)

Figure 8.17. Strip in tension; The mesh-dependent results with endochronic strain-softening plasticity; a) The load-deformation curves, b) The displacement patterns

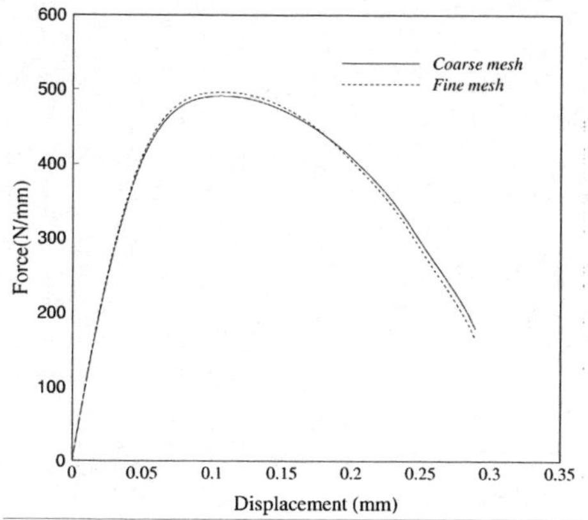

(a)

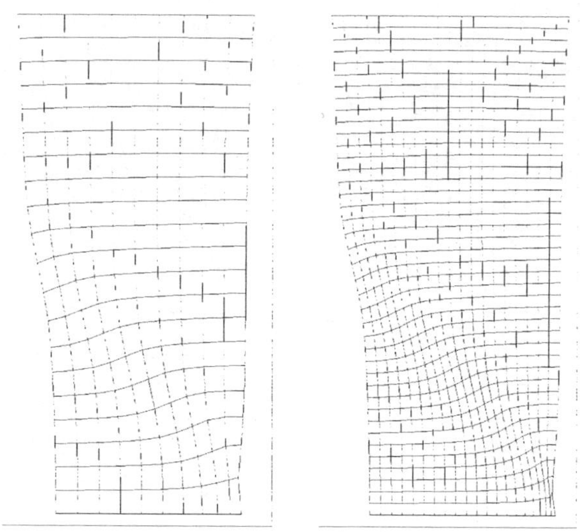

(b)

Figure 8.18. Strip in tension; The mesh objective results with endochronic viscoplastic model; a) The load-deformation curves, b) The displacement patterns

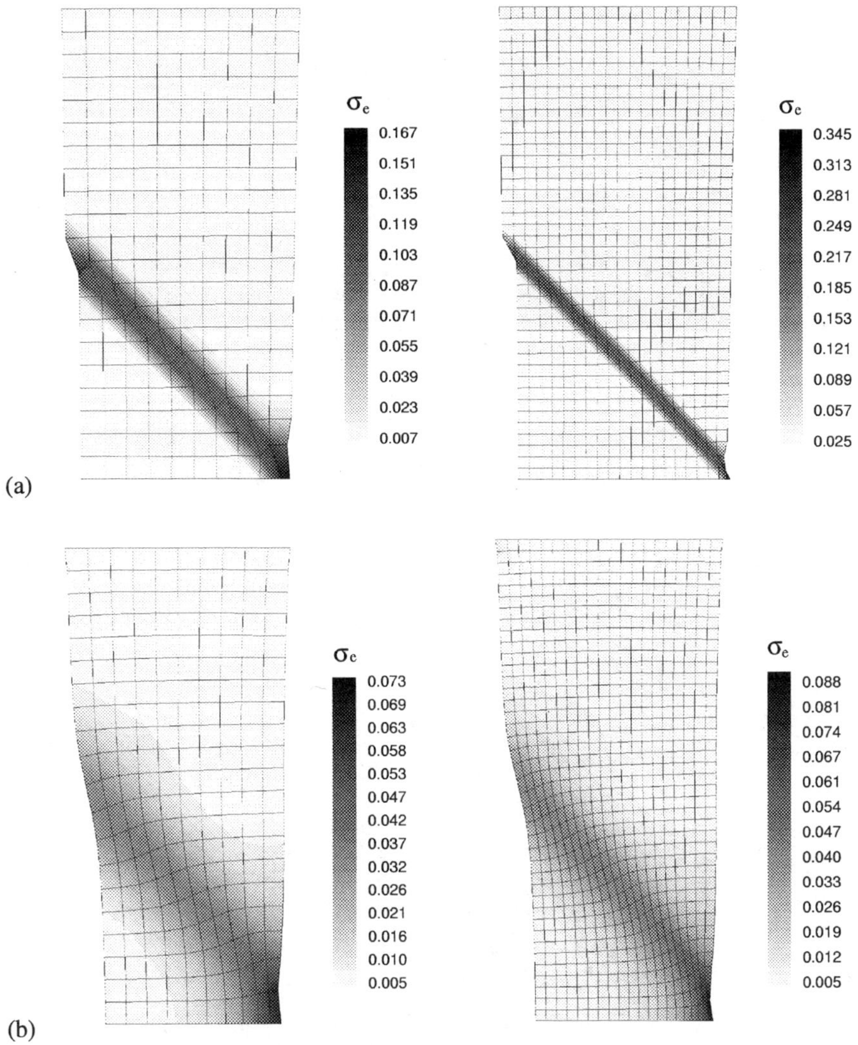

Figure 8.19. Strip in tension; The equivalent plastic strain contours; (a) Endochronic plasticity model for total extension of 20mm, (b) Endochronic viscoplasticity model for total extension of 30mm

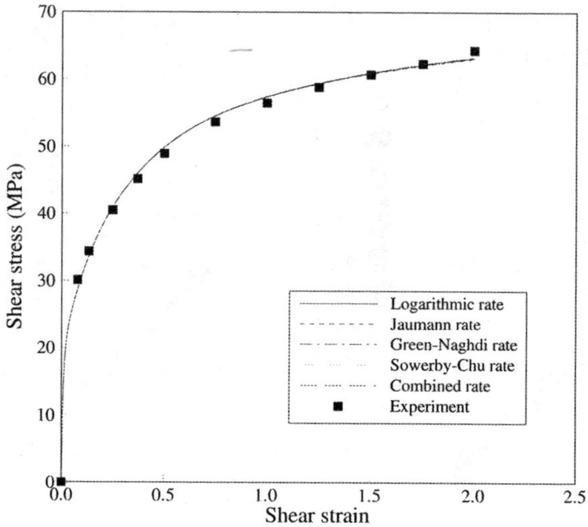

(a)

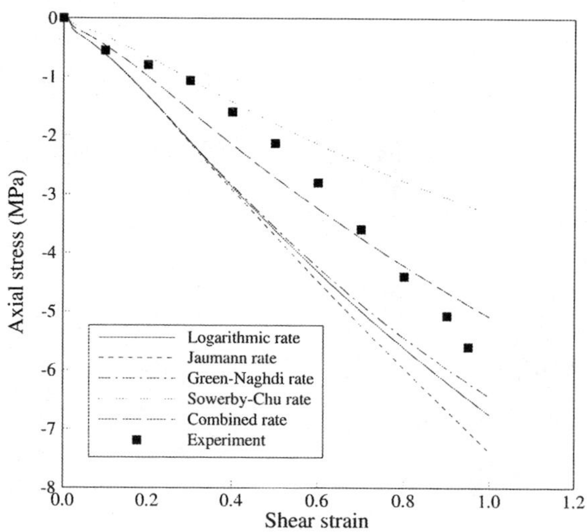

(b)

Figure 8.20. The cast aluminum; a) Experimental and simulation shear stress–strain curves under pure torsion, b) Axial stress versus shear strain curve under simple torsion

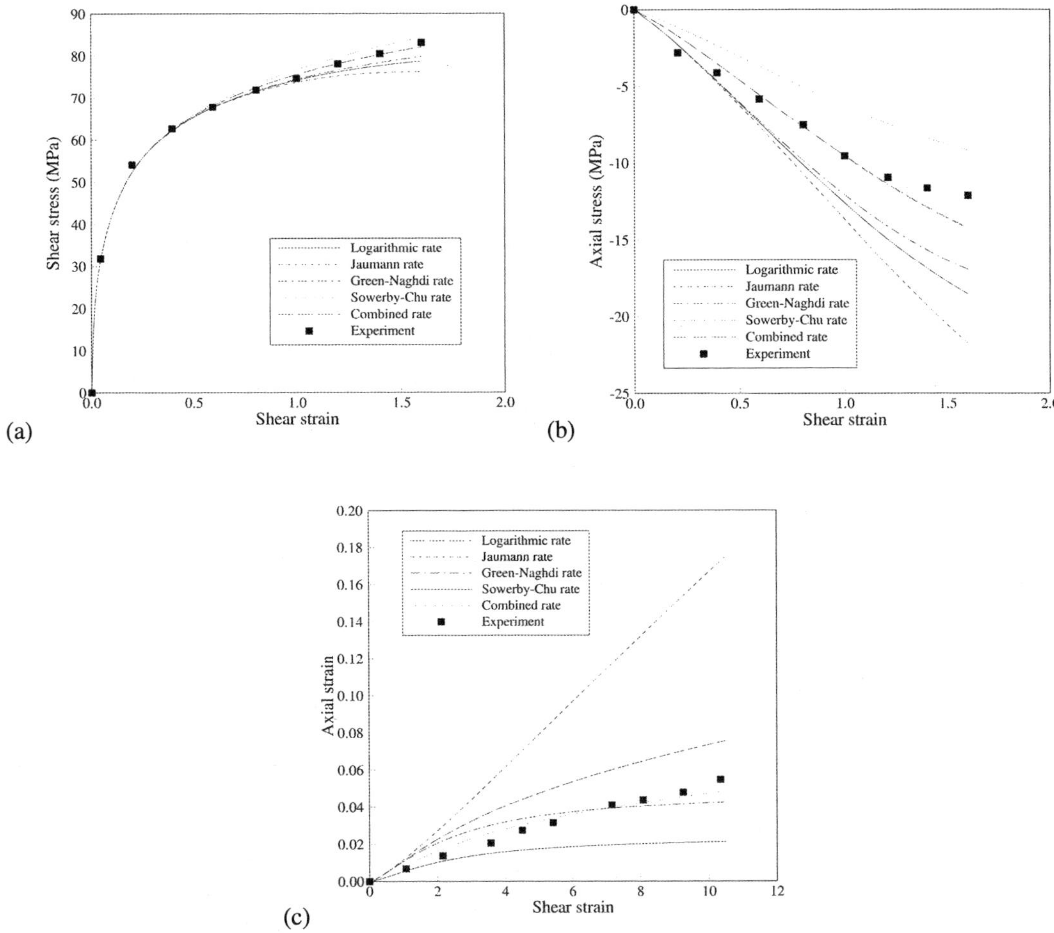

Figure 8.21. Al-1100; a) Shear stress versus shear strain curve under simple torsion, b) Axial stress versus shear strain curve under simple torsion, c) Axial strain versus shear strain curve under pure torsion

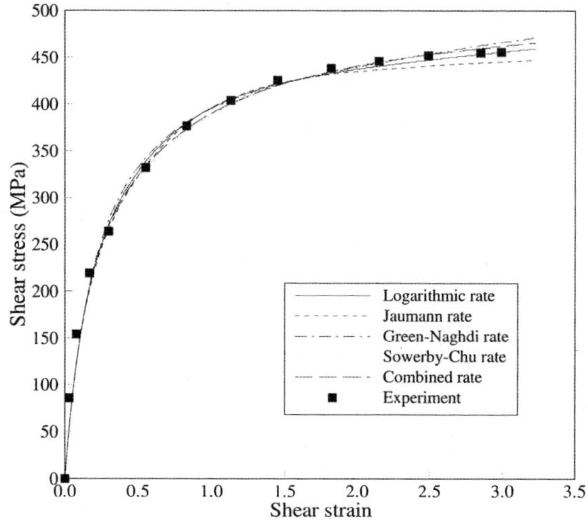

(a)

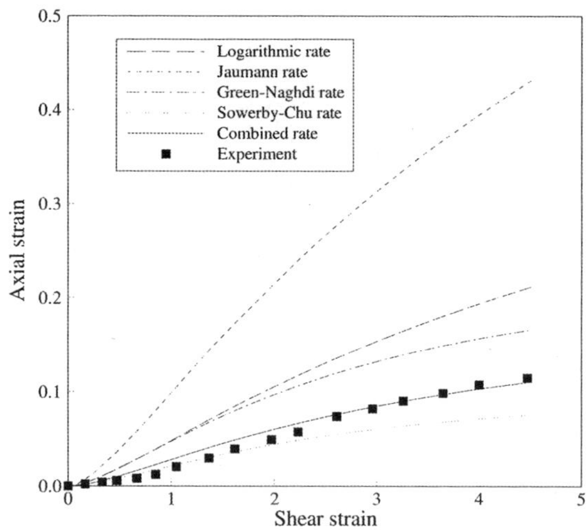

(b)

Figure 8.22. Brass 70:30; a) Shear stress versus shear strain curve under simple torsion, b) Axial strain versus shear strain curve under pure torsion

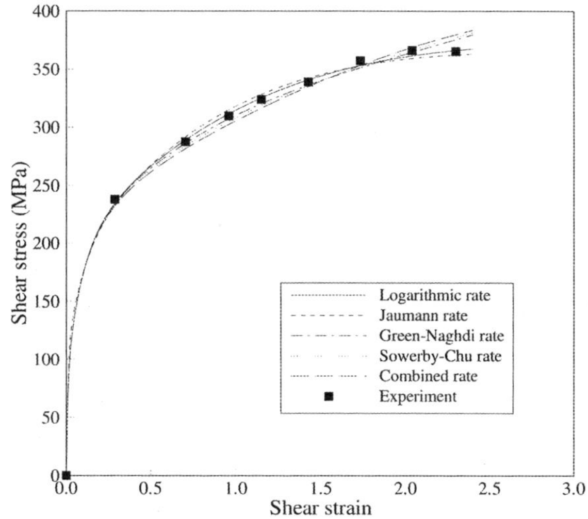

(a)

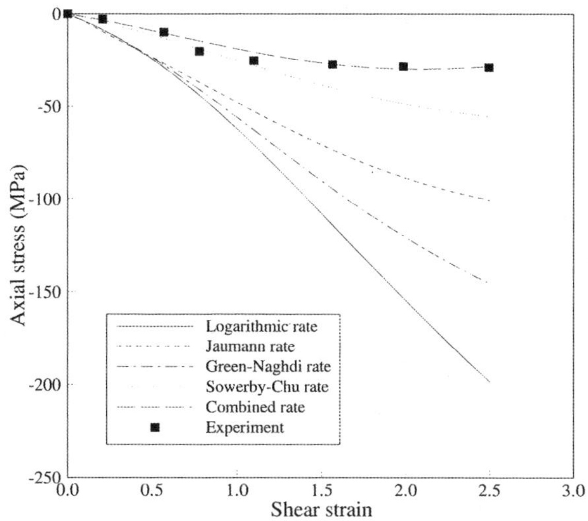

(b)

Figure 8.23. SUS 304 stainless steel; a) Shear stress versus shear strain curve under simple torsion, b) Axial stress versus shear strain curve under simple torsion

Table 8.1. Newton iterations per load step, A thick-walled cylinder

Step	1	2	3	4	5-6	7	8-10
Number of iterations	9	8	7	6	5	6	5

Table 8.2. Residual norm for load steps 4 and 8, A thick-walled cylinder

Iteration	Load step 4	Load step 8
1	1.69E+03	3.20E+03
2	2.15E+01	2.43E+01
3	2.54E-01	1.23E-01
4	2.75E-03	6.23E-04
5	2.94E-05	8.37E-08
6	2.47E-08	

Table 8.3. Residual norm-Stress loading, An axisymmetric notched bar

Iteration	Last increment of 3rd cycle	10th increment of 4th cycle	First increment of 5th cycle
1	1.64E+1	7.29E+1	2.20E+3
2	6.41E-1	3.83E-3	9.34E+2
3	2.08E-2	2.61E-6	2.70E+0
4	1.12E-4		3.83E-3
5			7.40E-7

Table 8.4. Residual norm-Displacement loading, An axisymmetric notched bar

Iteration	Last increment of 3rd cycle	10th increment of 4th cycle	First increment of 5th cycle
1	1.06E+0	2.15E+0	6.84E+4
2	1.06E-2	1.57E-2	5.68E+2
3	9.45E-5	5.60E-5	6.50E+0
4			1.52E-1
5			1.08E-2
6			1.82E-5

Table 8.5. Newton iterations per load step, Necking of a circular bar

Step	1	2	3-4	5-7	8-13	14	15	16	17	18-21	22	23-30
Number of iterations	6	5	4	3	5	4	7	5	6	5	7	5

Table 8.6. Residual norm for load steps 14 and 28, Necking of a circular bar

Iteration	Load step 14	Load step 28
1	2.98E+0	8.41E+0
2	1.71E-2	3.71E-1
3	7.70E-4	8.29E-2
4	2.03E-6	6.42E-3
5		1.41E-5

Table 8.7. Material properties for the strip in tension

Young's modulus	30000	MPa
Poisson ratio	0.3	
C_1	15000	MPa
C_2	1500	MPa
r_1	10000	
r_2	4000	
Scale function	$f(\varsigma)=10(1-10\varsigma)$	

Table 8.8 Endochronic and elastic material constants, Thin-walled tubes

	μ (MPa)	K (MPa)	C_1(MPa)	C_2(MPa)	C_3(MPa)	α_1	α_2	α_3	c	β
Cast aluminum	31000	60500	1830	85	15	80	8	1	0.4	6
Brass 70:30	38000	82300	1600	350	35	11	2.1	0.5	0.1	20
Al-1100	27000	58500	1800	85	35	68	8	1	0.4	20
SUS 304 stainless steel	78000	214000	9000	310	120	80	8	0.6	0.4	20

9 APPLICATION OF ADVANCED PLASTICITY MODELS

An advanced theory of plasticity was presented based on an endochronic plasticity model in Chapter 7 to describe the nonlinear behavior of powder material. In Chapter 8, the endochronic plasticity was developed based on a large strain plasticity using a hypoelastic model in unrotated frame of reference. Constitutive equations for rate-independent, plastically compressible, initially isotropic materials were obtained based on coupling between deviatoric and hydrostatic behavior throughout the endochronic theory. In the present chapter, the numerical modeling of powder cold compaction process is presented using a density-dependent endochronic plasticity model. The material parameters in the constitutive model are calibrated for two samples of metal powder by fitting the model to reproduce data from triaxial compression experiments. A simple procedure for calibrating endochronic material parameters with cone-cap plasticity is derived. The efficiency and accuracy of the material model and numerical schemes are illustrated by the numerical simulation of a set of two-dimensional powder compaction processes.

In Chapter 3, the double-surface density-dependent plasticity model was presented based on a combination of a convex yield surface consisting of a failure envelope and a hardening elliptical cap for nonlinear behavior of powder materials. Also, a three-invariant single-cap plasticity with an isotropic hardening rule was developed, which can be compared with some common double-surface plasticity models proposed for powders in literature. The constitutive elasto-plastic matrix and its components were derived based on the definition of yield surface, hardening parameter and nonlinear elastic behavior for both yield surfaces. In the present chapter, the three-dimensional numerical simulation of powder compaction processes is presented using both double-surface yield function and generalized single-cap plasticity models. It is shown how those plasticity models can describe the behavior of powders from initial stage of compaction to final stage, in which material behaves as solid metals.

9.1. Introduction

In powder metallurgy (P/M) the forming of industrial artifacts requires consolidation of loose powder into dense material leading to near-to-net shape components. The P/M technology being a grain-bulk deformation process has an ability to combine various types of powders with a goal to achieve desired/tailored properties and has much to offer in the campaign of environment-friendly technologies. The process is attractive for the manufacture of parts since it is energy efficient and environmentally friendly. This is attributed to its embodiment of near net shape forming principles and therefore its minimal requirement for finishing operations. In engineering, the process is used to form parts from a range of powder families (e.g. hard metal, ceramic, magnetic, and ferrous) for a wide range of applications, from simple bush families, which are appropriate for bearing applications, through to complex multilevel parts, which are used in automatic transmission systems. Due to near-to-net shaping, the components produced via P/M process result in a quality improvement with less material damage compared to other near-to-net shape processes such as casting; one of the major problem in casting is micro-segregation which leads to dislocation and cracks in the final product. The P/M process also has disadvantages, such as non-uniform densification and presence of voids (pockets of air bubble between a group of particles). However, in automotive industry P/M forgings are being developed for applications requiring good high cycle fatigue properties.

The knowledge of the behavior of powder material undergoing cold compaction is therefore necessary for predicting the final shape and the density distribution within the parts, and for preventing the failures that can occur during the subsequent sintering. Design of a compaction process consists, essentially, in determining the sequence and relative displacements of die and punches in order to achieve this goal. The design process, which has to be done for any new type of piece to be manufactured, could be effectively improved by using a simulation tool, able to predict the mechanical response of the compact along the process. Linking directly with the requirements of simulation, the work on powder forming can be grouped under the headings of input data, simulation, and validation.

Input data: Accurate input data is a key requirement for simulation and in powder forming, yielding of the compressible material bulk and friction between the powder and tool set are dominant parameters. To measure yield at different compaction levels, the powder sample is subjected to loading in up to three independent axes and this induces a complex but controlled and known stress distribution within the sample. This loading may be implemented using a true triaxial facility and the compression of a cubic sample, or more commonly through the compression of a cylindrical sample in which hoop and radial stress components are identical. To establish yield characteristics as a function of density, early types of triaxial tests make use of cylindrical samples precompacted to different density levels and these are then placed in a flexible container and subjected to a combined radial and axial loading path. By monitoring the axial stress–strain history it is then possible to establish the point at which the powder yields. More recent equipment design allows triaxial testing to be carried out starting from a controlled and closely packed fill condition and further allows the simultaneous and direct measurement of stress and strain evolution.

Simulation: The simulation of powder compaction needs to capture the large volumetric changes that are present together with the complex material behavior. Simulation needs to be based on industrial pressing trials where the kinematics are purposely devised to produce both good and defective parts. In spite of its apparent simplicity, powder compaction in a roller press exhibits some behaviors and interactions that are poorly understood from an analytical view. Simulation of cold compaction enables the prediction of the relative density map in the part. This map is very useful for the modeling of sintering, because relative density gradient in the part is the source of heterogeneous deformations. Numerical modeling is also very interesting to predict the presence of cracks in the compacts. The calculated state of stress in the part enables to predict fracture in the green part during ejection. Mathematical models that will allow realistic numerical simulation of powder compaction and appropriate visualization of these results can permit the process engineer to gain a better understanding through the process, leading to its better design and control.

Validation: Relevant validation of simulation is achieved most reliably by comparison with industrial pressing trials. These include conditions under which good compacts are made together with conditions that induce defects, such as large density variation and cracking under extreme circumstances. Generally, validation data is provided in the form of punch forces and density variations. It is evident that a significant wealth of information already exists in connection with the powder forming process. This has been reported mainly in a range of separate studies.

The present chapter includes material characterization, simulation, and validation and it has been focused through the completion of case studies that are relevant to industry. Validation information has been gathered through specific industrial trials. It is focused on two- and three-dimensional complicated die geometries in which wider application requires improvement in compact quality in terms of higher and more closely controlled density. The industry sector is aware of the potential benefits that may be derived through the application of process simulation to assist in achieving this goal, but requires confirmation of the accuracy of such procedures. For wide acceptance, this must also be coupled with low cost techniques for powder characterization in a format that suited to simulation.

9.2. Application of endochronic plasticity

In Chapters 7 and 8, an approach is developed based on an endochronic rate independent, density-dependent plasticity model for describing the isothermal deformation behavior of metal powder at low homologous temperature. As large deformation is observed in powder compaction process, a hypoelastic-plastic formulation is developed in the context of finite deformation plasticity. Constitutive equations are stated in unrotated frame of reference that greatly simplifies endochronic constitutive relation in finite plasticity. While the explicit treatment of both the integration of constitutive equations and the solution of the equilibrium equation is a common approach in the highly nonlinear behavior of powder compaction simulation (Haggblad and Oldenburg 1994, Rednaz 2001, Lewis and Khoei 2001, Gu *et al.* 2001 and Cedergen *et al.* 2002), an implicit scheme is employed to present the efficiently and accurately the method for this type of nonlinear problems (Oliver *et al.* 1996 and Perez-Foguet *et al.* 2001). An integration scheme, which is accurate, stable and amenable to consistent linearization is developed. Although the major challenge in the integration of rate constitutive equations in large deformation analysis is to achieve incremental objectivity, it has been trivially achieved. Algorithmic modulus consistent with numerical integration of constitutive equations for the density-dependent endochronic theory is extracted. The implementation of consistent modulus in global tangent stiffness matrix is essential in preserving the quadratic rate of convergence of Newton procedure in solving the equilibrium equations.

The enodochronic constitutive equations together with the proposed numerical integration and consistent linearization schemes have been implemented in a nonlinear finite element code to evaluate the capability of the model in simulation of powder compaction process. A plain bush, rotational-flanged, shaped tablet, multi-level and shaped tip components are simulated to obtain the density distribution in green compact. The material parameters of endochronic model are calibrated by fitting the model to reproduce data from uni-axial and tri-axial compression experiments. The numerical results are compared with experimental and other simulation results available in literature. It is shown how the endochronic plasticity describes the behavior of powder material from initial stage of compaction to final stage.

9.2.1. A plain bush component

The first example is the uniaxial compaction of a plain bush component, chosen to demonstrate the performance of the present formulation in simulating the metal powder during compaction. Experimental data (Gethin and Lewis 1994) and numerical results (Lewis and Khoei 1998 and Perez-Foguet *et al.* 2001) are available for this example and are used for comparison. The powder material is modeled by endochronic plasticity with the term of deviatoric and hydrostatic hereditary functions and isotropic elastic properties. Simulation is performed with powder material parameters presented in Box 9.1. These material parameters are calibrated by comparing the numerical simulations and experimental results presented by Gethin and Lewis (1994).

The finite element modeling of bush component is performed using an axisymmetric representation, which is illustrated in Figure 9.1. The die wall friction is simulated with Coulomb friction coefficient $\mu = 0.08$ (Lewis and Khoei 1998) and a structured mesh of 160 bilinear elements is used. The problem has been solved with displacement control by increasing top punch displacement up to 12 mm while bottom punch is fixed. The evolution of top punch vertical reaction force with its vertical displacement is depicted in Figure 9.2(a). Also plotted in this figure are the experimental results given by Gethin *et al.* (1994) and numerical results obtained by Lewis and Khoei (1998) and Perez-Foguet *et al.* (2001). It shows a good agreement between the experimental results and present numerical simulation.

Figure 9.2(b) shows the relative density profile at radius of 10.5 mm for top punch displacement of 10 mm. The experimental and numerical results given by Gethin *et al.* (1994) and numerical results obtained by Perez-Foguet *et al.* (2001) are also shown in this figure. The results of all reported simulations for variation of relative density along the profile show the lower value of densities than the experimental results. While the results of present simulation are in good agreement with experimental data up to the deformed height of 7 mm, it deviates from experimental data in upper region. This deviation can be related to treatment of friction between powder and upper punch. The distribution of relative density contours at four different punch displacements are depicted in Figure 9.3. The relative density distributions are in complete agreement with those reported by Perez-Foguet *et al.* (2001), while the range of relative density distribution is a little higher than those presented by Lewis and Khoei (1998). Finally, the normal stress contours σ_y at different stages of compaction are plotted in Figure 9.4. The effect of die wall friction of distribution of stress can be seen in this figure.

9.2.2. A rotational flanged component

The second example refers to the compaction of a flanged component, which is modeled by an axisymmetric representation, as illustrated in Figure 9.5. Also plotted in this figure are the initial geometry and finite element mesh of component. Numerical simulation is performed using 2D structured mesh of 203 bilinear elements. Material parameters are similar to the first example. Both experimental results (Gethin *et al.* 1994) and numerical results (Lewis and Khoei 1998 and Perez-Foguet *et al.* 2001) are available and used for comparative purpose. In order to avoid the mesh distortion, as demonstrated by Lewis and Khoei (1998) and Perez-Foguet *et al.* (2001), simultaneous movements of the top and bottom punches have been applied (the top punch movement of 6.06 mm and the bottom punch movement of 7.7 mm). The friction of powder with punches and die wall are simulated with a Coulomb friction coefficient of $\mu = 0.08$.

The relative density distribution at four different movements of punches is presented in Figure 9.6. The distribution of relative density is almost homogeneous with difference less than 11% in final stage of compaction. The higher value of relative density can be seen in bottom corners of component and the lower value in right bottom corner of flange. Figure 9.7 presents relative density profile along four sections in final stage of compaction. Experimental measured relative density profile (Gethin *et al.* 1994) and results of two other simulations (Lewis and Khoei 1998 and Perez-Foguet *et al.* 2001) are also presented for comparison. All the results are in general agreement.

9.2.3. A shaped tablet component

The next example demonstrates the performance of the present formulation in complicated die geometry and simultaneous high distortional and volumetric deformation of elements. The geometry is used by Lewis and Khoei (2001) to show the applicability of their formulation in tablet pressing process. A shaped tablet is compacted by simultaneous action of top and bottom punches. The finite element mesh, geometry and boundary conditions are shown in Figure 9.8. The component is modeled by an axisymmetric representation, as illustrated in this figure. A 2D mesh of 192 bilinear elements is used. The friction of powder with die wall, bottom and top punch is simulated with a Coulomb friction coefficient of $\mu = 0.08$. The material parameters used in this example are presented in Box 9.1.

The deformed meshes at four different stages of compaction are plotted in Figure 9.9. The relative density contours at different stages of compaction are plotted in Figure 9.10. From the contour of final stage of compaction, it can be seen that the density in the right hand region of tablet is greater than other parts while regions with low density appear between high-density regions. The distribution of density in final stage of compaction is a bit different from results reported by Lewis and Khoei (2001). These discrepancies can be related to different treatments of friction between powder and punches. While in the present simulation contact between powder and punches is modeled with Coulomb friction of $\mu = 0.08$, in this reference the relative movement between powder and punches is restrained.

9.2.4. A multi-level component

The next example is the simulation of a multi-level compaction process chosen to demonstrate the performance of the endochronic formulation for the complicated die geometry of a multi-level component. The scale functions $f_d(\eta)$ and $f_h(\eta)$ are determined from the material properties in terms of relative density. These functions have the highly nonlinear forms to cause the hardening behavior of the deviatoric and volumetric plastic response. The following forms, which are appropriate for metal powder, are proposed for the deviatoric and hydrostatic scale functions

$$f_d(\eta) = \eta + a_1\eta^2 + a_2\eta^3 \tag{9.1}$$

$$f_h(\eta) = \eta^m \frac{\eta - \eta_m}{1 - \eta_m} \tag{9.2}$$

The parameters of hydrostatic scale function (m and η_m) can be calculated by fitting the response of model to the volumetric strain versus hydrostatic pressure curve obtained from the isostatic compression test. Having determined the parameters of $f_h(\eta)$, the response of the model will be fitted to the results of triaxial test, in order to obtain the parameters of $f_d(\eta)$.

9.2.4.1. Determination of Endochronic Parameters

The experimental data are gained from a set of compaction experiments on an iron-based powder (95 % by weight) performed by Doremus *et al.* (1995). These experiments are conducted using the triaxial apparatus schematically plotted in Figure 5.49. Both isostatic compaction and triaxial tests are driven. The compacted specimen has an initial height H_0 of 42 cm and diameter D_0 of 20 cm. The triaxial tests consist of an initial isostatic compaction step up to pressure value of 400 MPa, followed by a subsequent uniaxial compaction step. This step is carried out by keeping pressure constant and increasing the axial stress up to the maximum value of 1250 MPa. The material properties for the simulation of powder are given in Box 9.2. The powder material is modeled by endochronic plasticity with one term for deviatoric and hydrostatic hereditary functions and isotropic elastic properties. It should be noted that the implementation of more than one term in Dirichlet series introduces the kinematic hardening into the endochronic plasticity (Watanabe and Atluri 1986), which is not relevant to the application considered here. A computer program has been developed for the determination of the material parameters.

Figure 9.11 presents the evolution of the density versus the hydrostatic pressure. This evolution is the characteristic of metal powders. The experimental and numerical results are compared for the isostatic compression step. The applicability of the proposed endochronic theory to handle the volumetric terms is evident in this figure. Figure 9.12 corresponds to the complete triaxial compression tests. The density versus axial strain curves are plotted for different values of the hydrostatic pressure attained at the end of isostatic compression step. Remarkable agreements between experimental and numerical results are found.

9.2.4.2. Simulation of a multi-level component

The numerical results are obtained for a multi-level compaction process by Oliver *et al.* (1996). The simulation is performed with the powder material parameters presented in Box 9.2. The initial density at the start of simulation is 2.94 g/cm^3. The compact specimen is a bearing used in the automotive industry whose axisymmetric geometry is shown in Figure 9.13(a). The initial and final shapes are plotted in this figure with dashed and full lines.

The compaction is performed by means of the action of five punches, labeled *a, b, c, d* and *e* in Figure 9.13(a). The component is modeled by 2D axisymmetric structured mesh of 390 bilinear elements. The initial finite element mesh is shown in Figure 9.13(b). Friction effects are considered through a Coulomb's friction model with a friction coefficient of $\mu = 0.1$ (Oliver *et al.* 1996). A numerical model of the frictional contact based on penalty approach is used to simulate sliding resistance at the tool-workpiece interface. A full discussion of the frictional model is given in Chapter 4. The deformed meshes at different stages of compaction are plotted in Figure 9.14. In Figure 9.15, the relative density contours at different stages of compaction process are plotted. The fairly uniform distribution of relative density is also found by Oliver *et al.* (1996). A lower relative density distribution is detected at the top part of the sample. The range of the variation of relative density in the sample is similar to those presented by Oliver *et al.* (1996).

9.2.5. An axisymmetric shaped tip

This example refers to the compaction process of a highway-engineering tip. The endochronic model parameters are determined to predict the experimental results from a set of triaxial loading paths. The experimental data are obtained from experiments on a cylindrical grade tungsten powder specimen (Alm *et al.* 1982, 1983 and Haggblad 1991). A servo-hydraulic testing machine is used to control the axial load. The specimen is placed in a high-pressure loading cell surrounded by fluid. The specimens are cylindrical in shape with a diameter of 30.0 mm and length of 60.0 mm, precompacted to 10 MPa. The loading of the specimens is controlled independently in the radial and axial direction. The loading path corresponds to the conventional triaxial compression test with a pressure range up to 200 MPa. The material properties for simulation of powder are given in Box 9.3.

In Figure 9.16(a), the hydrostatic pressure versus volumetric strain is presented for the isostatic compression step. It shows a good agreement between the endochronic and experimental results. In Figure 9.16(b), the axial stress versus axial and radial strain curves are compared for three different confining pressures corresponding to complete triaxial test. As it can be observed, the endochronic plasticity model captures the behavior of powder in complete triaxial experiment.

In order to investigate the accuracy of the integration scheme, the volume of a unit eight-noded brick element is reduced to 0.6 in different load step sizes. Figure 9.17 shows the hydrostatic pressure versus the volumetric strain curves for different load step sizes. As can be seen from the figure, there is no loss of accuracy even with the extremely unreasonable large load steps. In order to perform a comparison between the endochronic and cap plasticity models, the volume reduction of the unit brick element is simulated using the material parameters described by Khoei and Lewis (1998). The results of simulation for the endochronic and cap plasticity models are given in Table 9.1. The average number of iteration per increment together with the relative CPU time for different load step sizes corresponding to 40 percent volume reduction of the element are given in this table. The tolerance is set to 10^{-8}.

The endochronic material parameters, described in equations (9.1) and (9.2) and presented in Box 9.3, are employed to simulate the compaction process of a highway-engineering tip (Haggblad and Oldenburg 1994, Khoei and Lewis 1998, 2001), whose axisymmetric geometry is shown in Figure 9.18. This example has been simulated by Haggblad and Oldenburg (1994) and Khoei and Lewis (1998, 2001), using a cap plasticity model. The Coulomb friction coefficient is assumed $\mu = 0.2$. This value of friction coefficient is used by Haggblad and Oldenburg (1994), which is based on the sliding experiments, described by Hehenberger *et al.* (1984). The compaction simulation is performed by means of the upper and lower punches. Some problems are caused by discontinuities in geometry, for example the flow around corners and the multi punch set up. The experimental data (Haggblad and Oldenburg 1994) and numerical results (Khoei and Lewis 1998, 2001) are available for this example, both used here for comparative purpose.

In order to illustrate a comparison between the density distribution obtained by this simulation and those given in Chapter 5, the simulation has been performed using the remaining pressing distance from above of 3.0 mm and an under pressing of 1.5 mm. The initial relative density at the start of simulation is 0.25. The finite element modeling of the tip shape component, using the eight-noded quadrilateral elements along with its geometry are illustrated in Figure 9.18. Also plotted in this figure is the deformed mesh at the final stage of compaction. In Figure 9.19, the relative density distributions are shown in 3D plots at the half and final stages of compaction. As can be observed, the results of the present simulation are similar to the measured density distribution given by Haggblad and Oldenburg (1994) and those computed in Chapter 5 using the MCEC plasticity model. These results show how the endochronic theory using finite plasticity model can be effectively used for simulating large deformations in powder forming.

9.2.6. Compaction process of MH-100 iron powder component

The last example of this section describes the applicability of endochronic model in determining the material parameters for dry MH-100 iron powder and predictive capability of the model in simulating the powder compaction process. The material parameters are calibrated by fitting the endochronic model to reproduce the data from the simple compression experiments. Since such experiments on powder materials typically result in non-homogeneous deformations, the fitting procedure requires numerical simulation of the experiments. Figure 9.20 shows a scanning electron micrograph of the powder (Gu *et al.* 2001). The irregular particles tend to deform and interlock during cold compaction. Gu *et al.* (2001) produced the material parameters of the MH-100 iron powder for their cone-cap plasticity using the true triaxial compression experiments, torsion ring-shear experiments, and simple compression experiments.

In order to determine the endochronic material properties in terms of relative density, the following specific forms are proposed for the dependence of deviatoric scale function $f_d(\eta, p)$ and hydrostatic scale function $f_h(\eta, p)$ on η and p

$$f_d(\eta, p) = \left(c(\eta) - \frac{p}{p_0} \right) \tag{9.3}$$

$$f_h(\eta, p) = (1-\psi)\, g(\eta) - \psi \frac{p}{\phi_h} \tag{9.4}$$

where $c(\eta)$ and $g(\eta)$ are the increasing functions of relative density, defined as

$$c(\eta) = A_c\, \eta^{m_c} \frac{\eta - \eta_0}{1 - \eta_0}, \qquad g(\eta) = A_g\, \eta^{m_g} \frac{\eta - \eta_0}{1 - \eta_0} \tag{9.5}$$

where A_c, A_g, m_c and m_g are material parameters with positive values.

9.2.6.1. Material parameters of MH-100 iron powder

In order to determine the material parameters for the consolidation mechanism, the data is obtained from experiments conducted with a high-pressure true-triaxial apparatus developed by Abou-Chedid (1993) and Brown and Abou-Chedid (1994). The design of this apparatus is similar to that developed by Shima and Mimura (1986). Figure 9.21 shows a schematic top view of the apparatus. During an experiment, powder is poured into the cubical cavity created by the blocks. Blocks 1 and 2 are angled sliding-blocks sitting on the corresponding angled guiding-blocks. When the top block (not shown) moves downward in the 3-direction, sliding blocks 1 and 2 move inward in the 1-direction and 2-direction, respectively, and the powder in the initially cubical cavity is compressed in three orthogonal directions. The amount of compression in each of the lateral directions can be controlled by changing the angles of the sliding and guiding angled blocks. The main feature of the apparatus is its ability to load proportionally along different strain paths.

Let $\{\varepsilon_1, \varepsilon_2, \varepsilon_3\}$ denote the absolute values of logarithmic strains in the three directions, $\varepsilon_v \equiv \varepsilon_1 + \varepsilon_2 + \varepsilon_3$ the volumetric strain and $\gamma \equiv \sqrt{\frac{2}{3}}\left((\varepsilon_1 - \varepsilon_2)^2 + (\varepsilon_2 - \varepsilon_3)^2 + (\varepsilon_3 - \varepsilon_1)^2\right)^{1/2}$ the deviatoric strain. A plot of proportional strain paths attainable in the apparatus with the available angles blocks is shown in Figure 9.22. The notation x°/y° denotes the angles of the blocks in the 1- and 2-directions, respectively. For example, a $45^\circ/45^\circ$ arrangement would result in hydrostatic compaction with only ε_v. Gu *et al.* (2001) measured the normal strains $\{\varepsilon_1, \varepsilon_2, \varepsilon_3\}$ and compressive normal stresses $\{\sigma_1, \sigma_2, \sigma_3\}$ in each experiment for the different angled block arrangements, as presented in Figure 9.23. For example, Figure 9.23(a) shows the experimental results with the $45^\circ/45^\circ$ angled block arrangement. The corresponding stress versus strain curves are almost identical in all three directions. Thus, the $45^\circ/45^\circ$ angled block arrangement corresponds to an approximate hydrostatic compression test. Figure 9.23(d) shows the stress versus strain curves for the experiment with the $70^\circ/30^\circ$ angled block arrangement. In this case, the stress-strain curves in three directions are all different from each other, and a true-triaxial stress state is achieved. Figures 9.23(b) and (c) illustrate the stress versus strain curves of the experiments with $30^\circ/30^\circ$ and $70^\circ/70^\circ$ angled block arrangements, respectively. Since ideally the strains in the 1- and 2-directions are the same, this should result in a stress state with two identical principal stresses. However, the $30^\circ/30^\circ$ angled block arrangement shows discrepancy between the stress-strain curves in the 1 and 2 directions.

In order to perform the numerical simulations of the true-triaxial tests on MH-100 iron powder, a single three dimensional element is numerically modeled using endochronic plasticity. The simulations are carried out by subjecting an eight-noded brick element to different principal stretchings D_{11}, D_{22}, D_{33} acting on sides normal to the 1-, 2- and 3-directions. The relative values of the applied stretchings corresponding to the different combinations of angled blocks are given in the Table 9.2. The Young's modulus and Poisson's ratio of MH-100 iron powder compacts have been experimentally determined by Brown and Weber (1988), which are used in the following simulations. The endochronic material parameters are estimated by fitting the response of model to the stress-strain curves obtained from the true-triaxial tests (Figure 9.23). The hydrostatic scale function parameters corresponding to the cap part of cone-cap plasticity are obtained as

$$g(\eta) = \eta^{2.5}\,\frac{\eta - 0.45}{1 - 0.45}\,, \qquad \phi_h = 500\,\text{MPa}\,, \qquad \psi = 0.5 \tag{9.6}$$

Figure 9.24 shows a comparison between the calculated and measured stress-strain curves for the different deformation rate histories. The numerical results for all the different triaxial loading paths are in very good agreement with the experiments.

In order to evaluate the deviatoric scale function parameters corresponding to the cone part of cone-cap plasticity, the data is obtained from experiments conducted with the distortion mechanism of a torsion ring-shear apparatus developed by Kim (1999), as well as by using data obtained from simple compression experiments on unsintered compacts with different initial relative densities. A schematic of the apparatus is shown in Figure 9.25 (Gu *et al.* 2001). In this apparatus a thin hollow cylindrical specimen of the powder of inner and outer diameters of 63.50 and 69.85 mm, respectively, is confined between two floating concentric confining rings, and two annular punches. One of the annular punches is fixed, and the other can be displaced in the axial direction as well as twisted about its axis to subject the powder specimen to various combinations of axial compression and twist.

Gu *et al.* (2001) performed a set of experiments in which ring specimens were first axially compressed under normal pressures of 25, 50, and 75 MPa, to different initial relative densities of 0.51, 0.565 and 0.59, respectively, and the compressed powder rings are then sheared while keeping the corresponding normal pressures constant. Figure 9.26(a) show the nominal friction coefficient versus sliding distance curves obtained from the torsional portion of these experiments. In another set of experiments, the powder ring specimen was first axially compressed under a normal pressure of 50 MPa to a relative density of 0.565. The consolidated MH-100 powder is then sheared under different normal pressures of 12.5, 25, 37.5, and 50 MPa. The nominal friction coefficient versus sliding distance curves obtained from the torsional portion of these experiments are plotted in Figure 9.26(b). In both sets of torsion ring-shear experiments, the nominal friction coefficient starts from zero and go through a peak before decreasing to a substantially lower residual values at large shearing displacements. The residual value for μ_{nom} in both sets of compression-torsion experiments is 0.45. Thus, the values of parameters ϕ_d and p_0 in endochronic model are estimated 90 and 245 MPa, respectively.

In order to evaluate the other deviatoric scale function parameters, i.e. $c(\eta)$, the variation of material cohesion with relative density is obtained using simple compression experiments on samples with different initial relative density given by Gu *et al.* (2001). Cylindrical specimen is formed with a diameter of 25.4 mm and height of 15 mm in a uniaxial strain compaction apparatus. Unconfined simple compression experiments are then performed on these specimens. An estimation of the cohesion with relative density for MH-100 iron powder is presented by Gu *et al.* (2001), as shown in Figure 9.27. The curve plotted in this figure is the fit of our proposed functional form for $c(\eta)$

$$c(\eta) = 1.4\eta^3 \frac{\eta - 0.45}{1 - 0.45} \tag{9.7}$$

Figure 9.28 shows the plots of the absolute values the compressive stress and strain, prior to the catastrophic load drops observed in the experiments. Each stress-strain curve exhibits three typical stages. The non-linear behavior at the beginning of a stress-strain curve may be attributed to initial seating and settling errors during experiments. Strains in the second, linear stage are approximately recoverable, and this response may be modeled as elastic. We shall assume that the inelastic deformation observed in the third non-linear stage of simple compression is due only to the distortion mechanism. It must be noted from Figure 9.28 that the maximum absolute values of the axial stress for the simple compression specimens with different initial relative densities are quite different. The different maximum axial stresses indicate the significantly different values of the cohesion in the specimens. From this figure, the Young's modulus and Poisson's coefficient are obtained 7200 MPa and 0.3, respectively.

In order to illustrate the efficiency and accuracy of the endochronic material model in large strain range, the uniaxial compaction experiment is analysed numerically. The finite element modeling of specimen is performed using 2D axisymmetric structured mesh of 625 bilinear elements, as illustrated in Figure 9.29. As a displacement based formulation is used, the implementation of loading is achieved by the use of prescribed nodal displacements. The direction of this displacement is in a vertical plane which represents the axial punch load. Fixed nodal values are employed to represent the fixed bottom punch. Simulation is performed with powder material parameters presented in this section using different initial relative densities of 0.696, 0.766, 0.813 and 0.850. The evolution of axial stress-strain curves are depicted in Figure 9.30 for different initial relative densities. The numerical results are in very good agreement with the experiments given in Figure 9.28, and the only major difference appears in the non-linear behavior at the beginning of the measured stress-strain curves. This non-linear behavior is attributed to the initial seating and setting errors during the experiments, reported by Gu *et al.* (2001).

9.2.6.2. Simulation of a compaction process

In order to demonstrate the performance of the present model in simulating the metal powder during compaction process, the uniaxial compaction of a cylindrical MH-100 iron powder component is analysed numerically. The experimental data and numerical results are available for this example and are used for comparison. The powder material is modeled by the endochronic plasticity with the material parameters presented in Table 9.3. The finite element modeling of the cylindrical component is performed using an axisymmetric representation and a structured mesh of 144 bilinear elements. The initial geometry, boundary conditions and FE mesh are depicted in Figure 9.31(a). The die wall friction is simulated with Coulomb friction coefficient $\mu = 0.37$ (Gu *et al.* 2001) and the initial relative density of the powder is $\eta_0 = 0.42$. The problem has been solved with displacement control by increasing top punch displacement up to 30 mm while the bottom punch is fixed.

The evolution of top punch vertical reaction force with its vertical displacement is depicted in Figure 9.31(b). Also plotted in this figure, is the experimental result given by Gu *et al.* (2001). The result shows a good agreement between the experimental data and numerical simulation obtained by endochronic model. The relative density distribution over the sample at the end of the simulation is depicted in Figure 9.32. The relative density distribution is in complete agreement with those reported by Gu *et al.* (2001). This example demonstrates that the endochronic model is capable to describe the behavior of MH-100 metal powder in true-triaxial experiments, shear test and axial compression experiments. The results of a powder compaction simulation show how the proposed model can be used to predict the density distribution in the compact and punch forces accurately.

9.3. Application of double-surface plasticity

Up to date, the most computational simulation of powder compaction processes has been presented in two-dimensional cases. A Lagrangian kinematic formulation was developed by Oliver *et al.* (1996) and Lewis and Khoei (1998) to model the 2D multi-level components. The *h*-adaptive FE technique was employed by Khoei and Lewis (1999), in order to avoid the progressive mesh distortion during different stages of compaction. An arbitrary Lagrangian-Eulerian approach was developed by Rodriguez-Ferran *et al.* (2003) and Khoei *et al.* (2003) in 2D powder forming simulation, as the adaptive mesh refinement was computationally expensive and information had to be interpolated from the old mesh to the new mesh. The hypoelasto-plastic endochronic theory was recently proposed by Khoei *et al.* (2003, 2004) in the context of finite deformation plasticity, to evaluate the capability of the endochronic model in simulation of powder compaction process. However, less numerical modeling has been reported for 3D powder compaction simulation. Recently, the three-dimensional compaction simulations of geometries with circular and quadratic cross-sections were performed by Cedergren *et al.* (2002). Also, a three-dimensional FE modeling of multi-level powder components was presented by Khoei and Azizi (2004) using a double-surface plasticity theory.

The double-surface density-dependent plasticity model is developed in Chapter 3 based on a combination of a convex yield surface consisting of a failure envelope and a hardening elliptical cap for nonlinear behavior of powder materials. A general algorithm for the cap plasticity from the viewpoint of efficient numerical modeling is presented. The plasticity model is performed within the framework of large FE deformation, in order to predict the non-uniform relative density distribution during powder die pressing. In this section, the computational simulation of powder forming processes is performed using the three-dimensional double-surface plasticity represented in Figure 9.33. The numerical simulation is illustrated in modeling of 3D rotational-flanged, multi-level and shaped-seal disk components.

The experimental data are gained from a set of compaction experiments on an iron-based powder (95 % by weight) performed by Doremus *et al.* (1995). Both isostatic compaction and triaxial tests are driven. The material model parameters for the double-surface plasticity are given in Table 9.4. The initial relative density is $\bar{\rho}_0 = 0.48$. Figure 9.34(a) presents the evolution of the density versus the hydrostatic pressure. The experimental and numerical results are compared for the isostatic compression step. Figure 9.35(b) corresponds to the complete triaxial compression tests. The density versus axial strain curves are plotted for different values of the hydrostatic pressure attained at the end of isostatic compression step. Remarkable agreements between experimental and numerical results are obtained.

9.3.1. 3D rotational-flanged component

The first simulation example refers to the compaction of a 3D flanged component, which is modeled by an axisymmetric representation, as illustrated in Figure 9.35. Also plotted in this figure are the initial geometry and finite element mesh of component. Numerical simulation is performed with the material parameters given in Table 9.4. On the virtue of symmetry, the flanged component is analyzed for one quarter of component. Both experimental data by Gethin *et al.* (1994) and 2D numerical results by Lewis and Khoei (1998) and Rodriguez-Ferran *et al.* (2002) are available and used for comparative purpose. The loading characteristics are achieved by the use of prescribed nodal displacements for the top and bottom punches. The simulation has been performed using the remaining pressing distance of 6.06 mm from above and an under-pressing of 7.7 mm.

In 3D compaction simulation presented here the friction effects are neglected. Therefore, the results are a qualitative approximation to the real compaction process, where the friction effects have to be taken into account (Lewis and Khoei 1998). Figure 9.36 presents the relative density profile along four sections in final stage of compaction. Experimental measured relative density profile and results of 2D simulations are also presented for comparison. All the results are in general agreement. The relative density distribution at the final movement of punches, i.e. $d_t = 6.06$ and $d_b = 7.7$ mm, is presented in Figure 9.37(a). The distribution of relative density is almost homogeneous with difference less than 10% in final stage of compaction. The higher value of relative density can be seen in bottom corners of component and the lower value in right bottom corner of flange. The result is qualitatively similar to those presented by Lewis and Khoei (1998) and Rodriguez-Ferran *et al.* (2002). The powder displacement contour at final movement of the top and bottom punches of compaction process is shown in Figure 9.37(b). It can be observed that the powder is displaced so that the isolines indicate an almost uniform level.

9.3.2. 3D multi-level component

The second example chosen demonstrates the performance of 3D double-surface plasticity for the complicated die geometry of a multi-level component. The compact specimen is a bearing used in the automotive industry whose axisymmetric geometry is presented in Figure 9.38. The compaction is performed by means of the action of five punches corresponding to one quarter of specimen. Numerical simulation of this complex shape has been employed by a top punch movement of 5.0 mm, then a bottom punch movement of 8.0 mm, and a second top punch movement of 3.0 mm. The 2D numerical modeling of the proposed multi-level compaction process is obtained by Khoei and Lewis (1999) using an adaptive FE analysis. The simulation is performed with the powder material parameters presented in Table 9.4.

In Figure 9.39(a), the relative density contour at final stage of compaction process is plotted. The fairly uniform distribution of relative density is also obtained by Khoei and Lewis (1999) and Perez-Foguet *et al.* (2002). A lower relative density distribution is detected at the top part of the sample. The range of the variation of relative density in the sample is similar to those presented by Khoei and Lewis (1999). In Figure 9.39(b), the displacement contour is depicted at final stage of compaction. It can be seen that the proposed cap plasticity using large deformation formulation is capable of simulating 3D metal powder compaction processes in an efficient and accurate manner.

9.3.3. A shaped seal-disk

The next example demonstrates the applicability of double-surface plasticity model for a complex shaped three-dimensional seal disc, as shown in Figure 9.40. The shape of the compacted powder in its position after compaction is presented in this figure. This component is a challenging example for the proposed approach because it involves sharp geometry boundaries and large mass fluxes between different parts of the sample. The compaction is performed by means of the action of four punches with remaining pressing distance of 1.5 mm from top and bottom punches. The evolution of the relative density fields at final stage of compaction is plotted in Figure 9.40(a). The fairly uniform distribution of relative density can be observed. In Figure 9.40(b), the displacement contour is shown at the final stage of compaction.

9.4. Application of three-invariant single-cap plasticity

A generalized three-invariant single plasticity with an isotropic hardening rule was developed in Chapter 3 for description of powder behavior under cold compaction process. A general form was developed for the cap plasticity which can be compared with some common double-surface plasticity models proposed for powders in literature. The constitutive elasto-plastic matrix and its components were derived based on the definition of yield surface, hardening parameter and nonlinear elastic behavior, as function of relative density of powder. Different aspects of 2D and 3D cap plasticity models are illustrated and the procedure for determination of powder parameters is described. It is shown how the proposed model could generate the elliptical yield surface, double-surface cap plasticity and the irregular hexagonal pyramid of the Mohr-Coulomb and cone-cap yield surface, as special cases demonstrated in Figure 9.41. The procedure for determination of powder parameters was described by fitting the model to reproduce data from true triaxial compression and confining pressure experiments. The single-cap plasticity is performed within the framework of large finite element deformation, in order to predict the non-uniform relative density distribution during powder die pressing. In this section, the numerical schemes are examined for efficiency in modeling a set of three-dimensional compaction processes, including: an automotive component, a conical shaped-charge liner and a connecting-rod.

9.4.1. An automotive component

In the first example, an axisymmetric automotive part which is compacted from iron powder with a mechanical press and a multi-platen die set is simulated numerically using the single cap plasticity. The two-dimensional simulated density distributions were obtained for this component by Khoei and Azami (2004) and compared with those given by Shen *et al.* (2001). The analysis is carried out with the iron powder material parameters presented in Table 3.6. The numerical simulation of automotive part is performed using a 3D finite element analysis, as illustrated by an axisymmetric representation in Figure 9.42. On the virtue of symmetry, the automotive part is analyzed for one quarter of component. The shape of the uncompacted powder in its position before compaction is presented in Figure 9.42(a). In Figures 9.42(b) and (c), the 3D finite element mesh, geometry and boundary conditions of this component are presented. The compaction is employed by means of the action of two bottom punches; a lower inner punch and a lower outer punch labeled 'a' and 'b', respectively, in Figure 9.42(c). The simulation has been performed using the remaining pressing distance of lower inner punch of 9.75 mm and lower outer punch of 31.10 mm.

In the simulation presented here the friction effects are neglected. Therefore, the results are a qualitative approximation to the real compaction process, where the friction effects have to be taken into account. Figure 9.43 presents the final FE mesh and geometry of the component at the final stage of compaction. In Figure 9.44, the predicted relative density and normal stress σ_y distributions are shown at the final stage of compaction. The result of density distribution is similar to those presented by Shen *et al.* (2001) and Khoei and Azami (2004). In Figure 9.45, the 3D representation of the expansion of yield surfaces along with the obtained stress paths are presented at two different Gauss points of an automotive component indicated in Figure 9.42(c). As the densification occurs and stress exceeds, the plastic strain vector, normal to the current yield surface, produces a plastic volumetric decrease which causes the material to harden and the surface expands until the next yield surface is reached. This representation clearly shows how the yield surface grows with densification and expands when the material hardens. The applicability of the single cap plasticity model to handle the compaction simulation of powder is clearly evident in Figure 9.45. The evolution of top punch vertical reaction force with its vertical displacement is depicted in Figure 9.46. Complete agreements can be observed between 2D and 3D results.

9.4.2. A conical shaped-charge liner

The next example is of a shaped-charge liner, which is numerically modeled in two-dimensional analysis in Chapter 5. In the present simulation, the three-dimensional analysis of a conical liner is pressed from the iron powder with the material parameters given in Table 3.6. Due to the symmetry of loading and geometry, only one quarter of axisymmetric component is modeled. The schematic of the process to form a conical shaped charge liner from iron powder along with the geometry and FE mesh before compaction are presented in Figure 9.47. The loading characteristics are achieved by the use of prescribed nodal displacements for the top punch movement. The die and upper punch were modeled as rigid surfaces. The simulation has been performed using the remaining pressing distance of 11.78 mm from above. The two-dimensional simulation of this component was performed by Gu *et al.* (2001) and Azami and Khoei (2004).

Figure 9.48 shows the FE mesh and geometry of the pressed conical liner at the final stage of compaction. The distribution of relative density and normal stress σ_y are presented in Figure 9.49 at final stage of compaction. In the simulation presented here the friction effects are neglected. Therefore, the results are a qualitative approximation to the real compaction process, where the friction effects have to be taken into account. The 3D stress paths with associated yield surfaces are presented in Figure 9.50 at two different Gauss points depicted in Figure 9.47(c). The numerically predicted force on the upper punch versus its vertical displacement is plotted in Figure 9.51. The 2D and 3D force-displacement curves are in complete agreement.

9.4.3. A connecting rod

The last example chosen demonstrates the performance of the proposed plasticity model for the 3D complicated die geometry of a connecting-rod. The compact specimen is a bearing used in the automotive industry whose shape is presented in Figure 9.52. The geometry of component at the initial and final stage of compaction is presented in Figure 9.53. This component is a challenging example for the proposed approach because it involves sharp geometry boundaries and large mass fluxes between different parts of the sample. The compaction is performed by means of the action of five punches from top. The numerical simulation is carried out using eight-noded brick elements. The final shape of component along with the deformed FE mesh are shown in Figure 9.54. The evolution of the relative density fields at final stage of compaction is plotted in Figure 9.55. The fairly uniform distribution of relative density can be observed. In Figure 9.56, the normal stress σ_z contour is shown at the final stage of compaction. The variation of top punch reaction force with its vertical displacement is plotted in Figure 9.57.

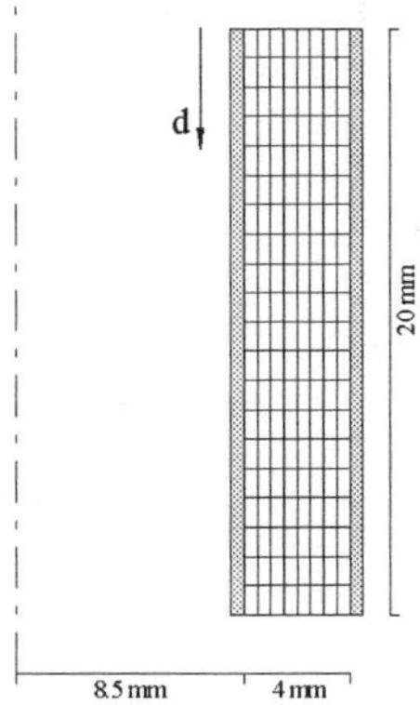

Figure 9.1. Finite element modeling of plain bush component

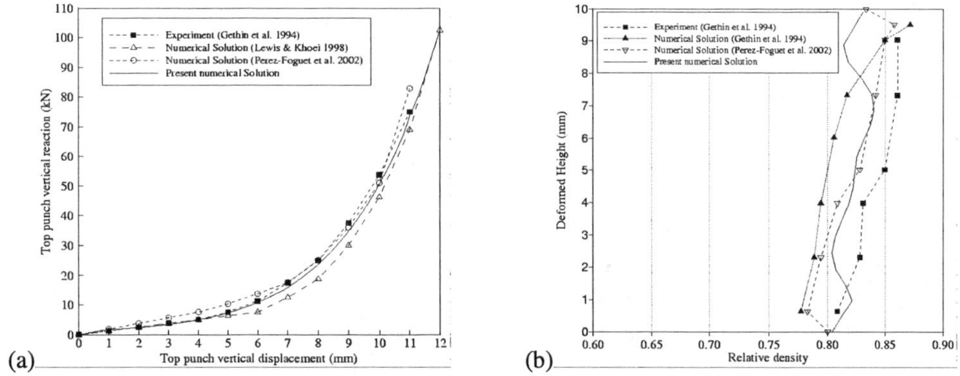

Figure 9.2. A plain bush component; a) The upper punch reaction force at different vertical displacements, b) The relative density profile at radius of 10.5 mm

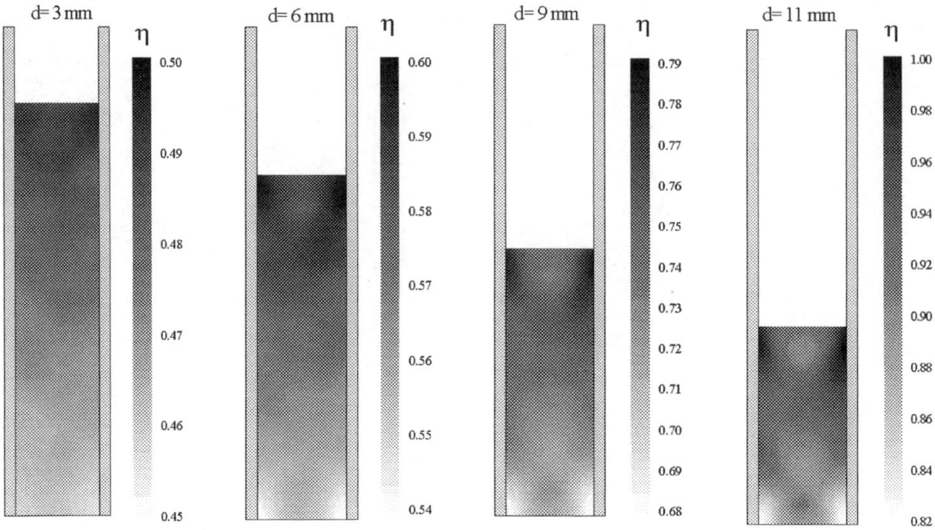

Figure 9.3. The relative density distribution at different top punch movements for plain bush component

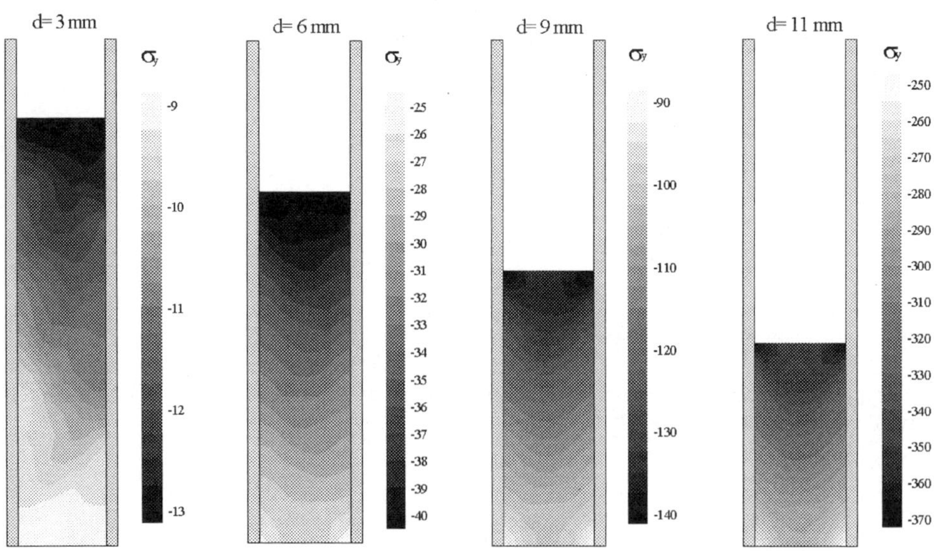

Figure 9.4. The normal stress σ_y contours at different top punch movements for plain bush component

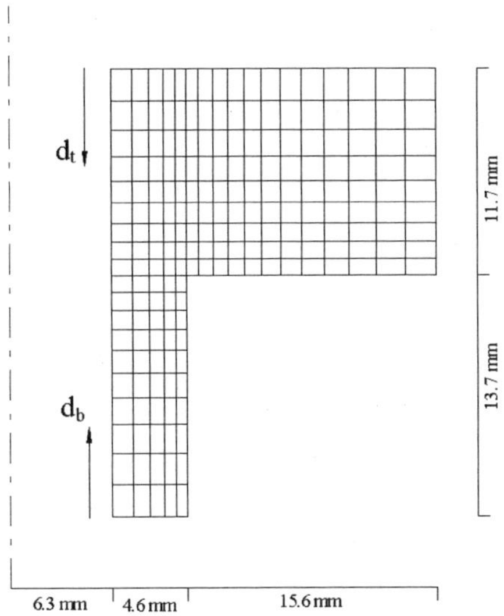

Figure 9.5. Finite element modeling, geometry and boundary condition of a rotational flanged component

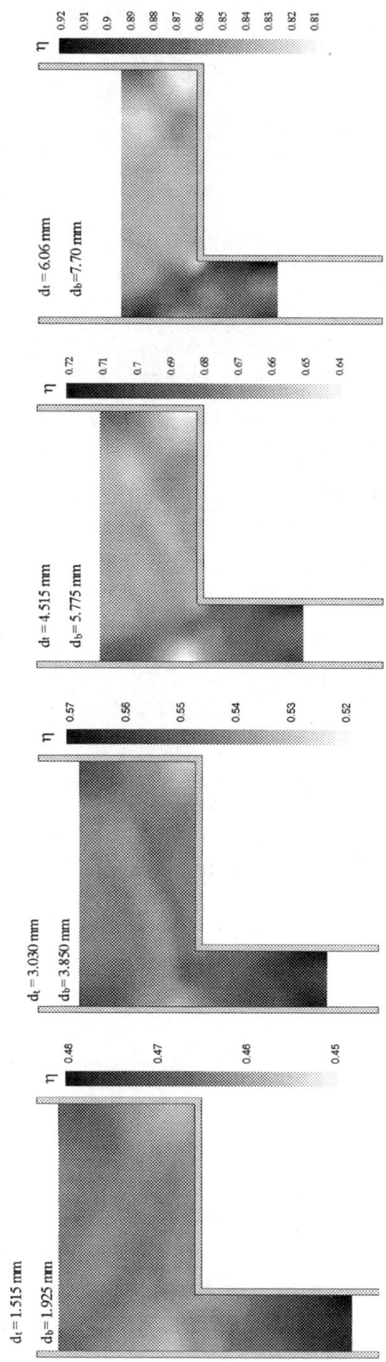

Figure 9.6. Flanged component; the relative density distribution at different compaction stages

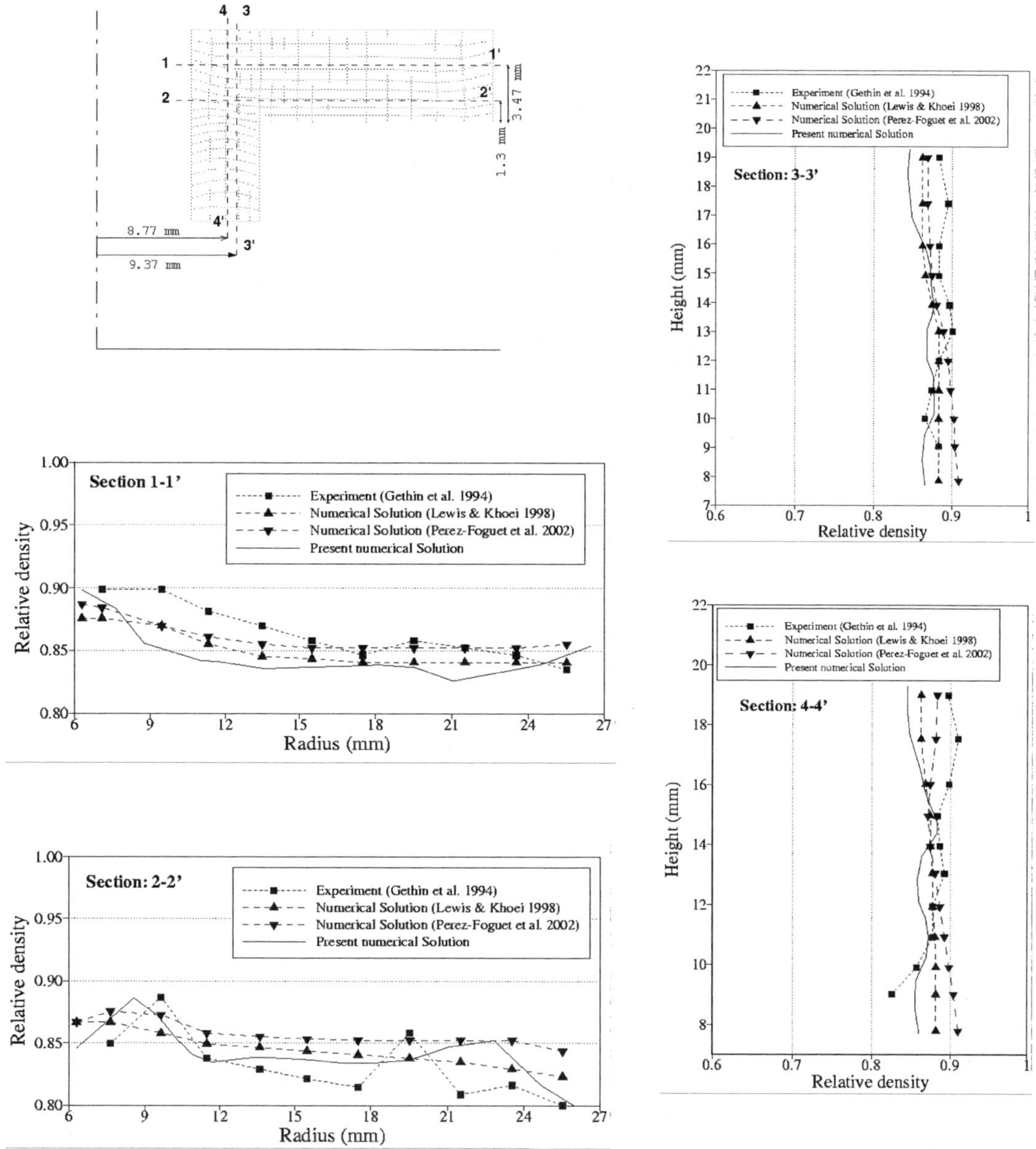

Figure 9.7. Flanged component; the relative density profile at horizontal and vertical sections

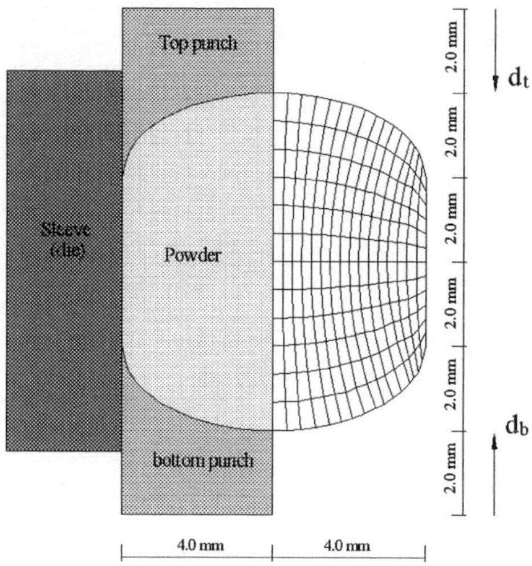

Figure 9.8. Finite element modeling, geometry and boundary condition of an axisymmetric shaped tablet

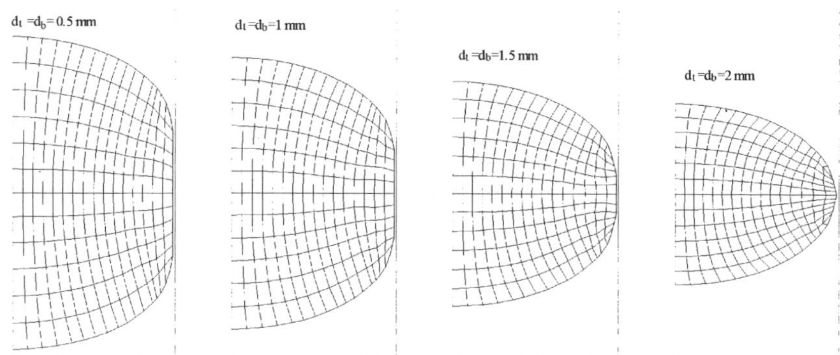

Figure 9.9. Shaped tablet; the deformed mesh at different stages of compaction

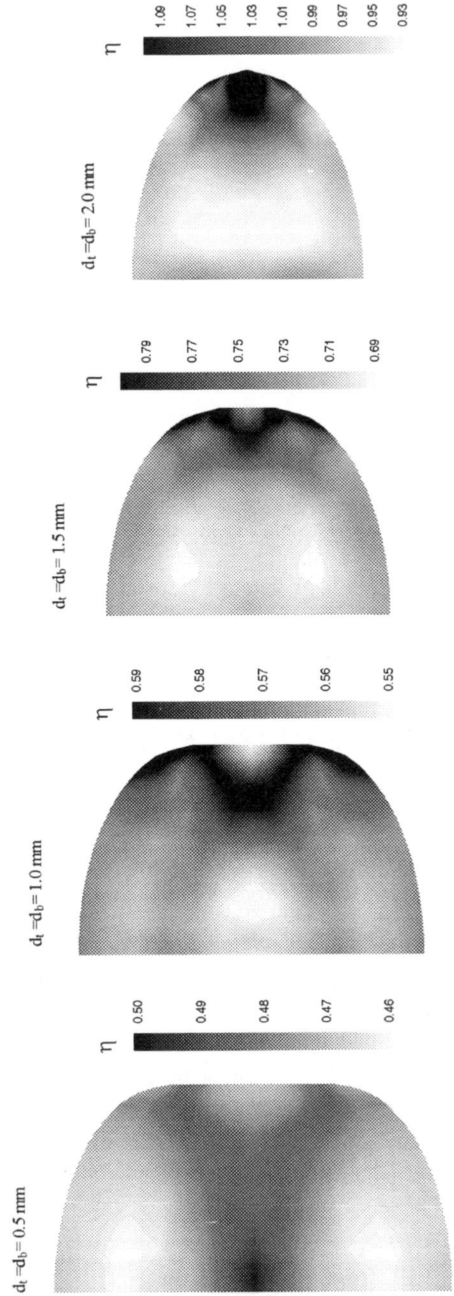

Figure 9.10. Shaped tablet; the relative density distribution at different punch movements

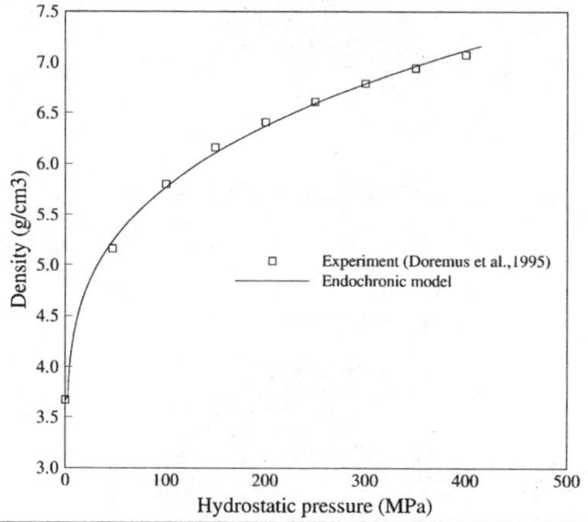

Figure 9.11. The triaxial compression test; Density versus hydrostatic pressure during isostatic compaction

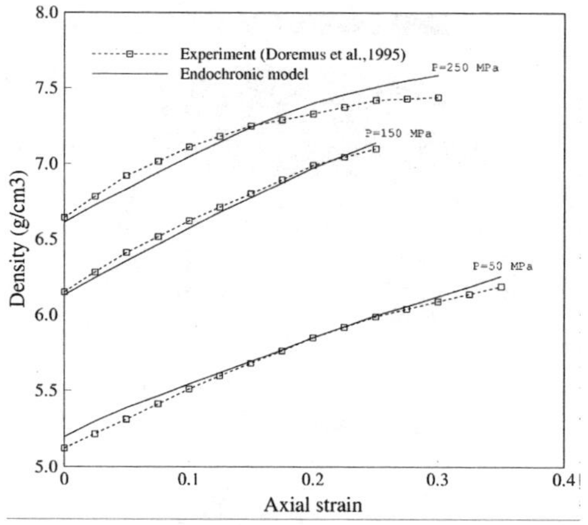

Figure 9.12. The triaxial compression test; Density versus axial strain at given confining pressure

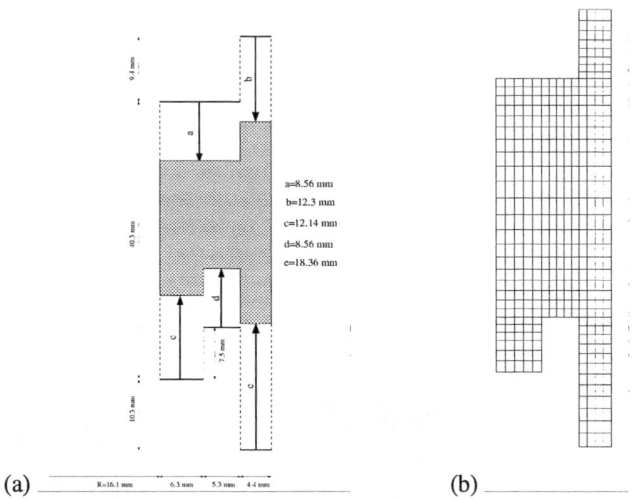

Figure 9.13. A multi-level component; a) Geometry and loading description (after Oliver *et al.* 1996), b) Finite element mesh of the multi-level component

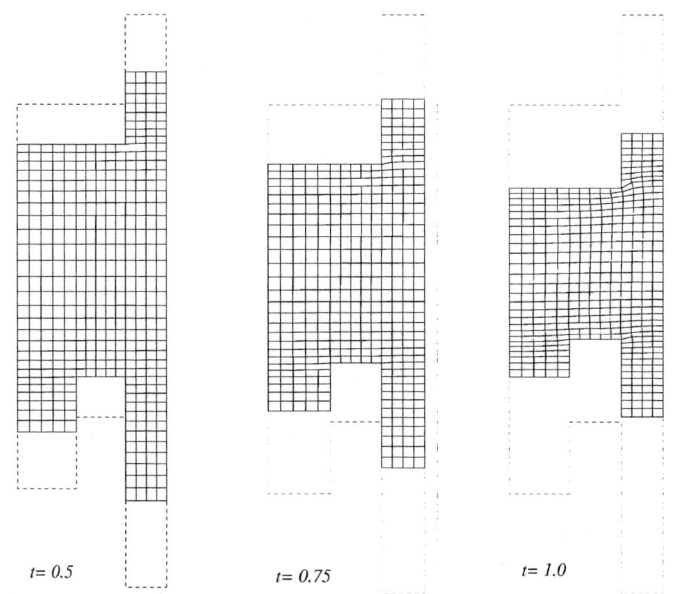

Figure 9.14. The deformed meshes at different stages of compaction process for the multi-level component

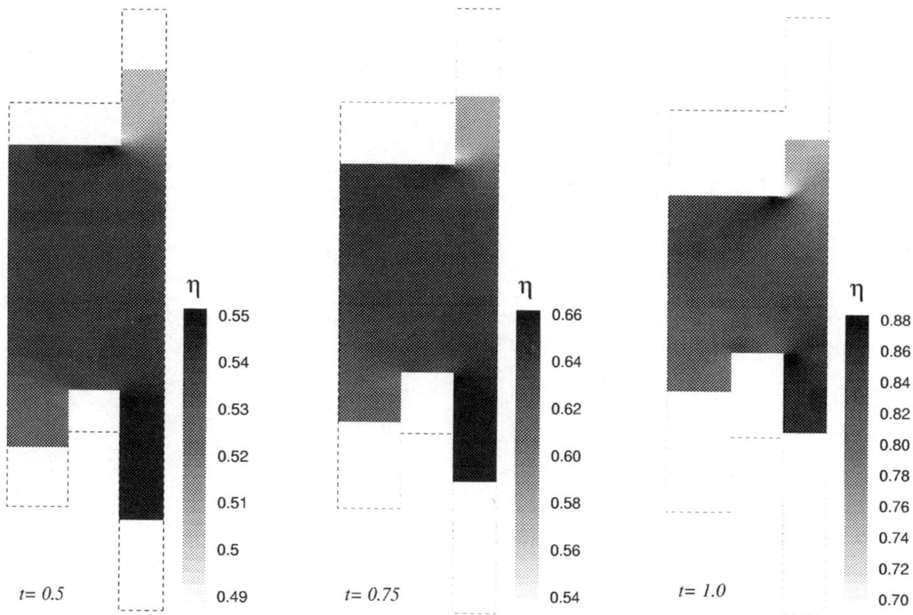

Figure 9.15. The density distribution at different stages of compaction process for the multi-level component

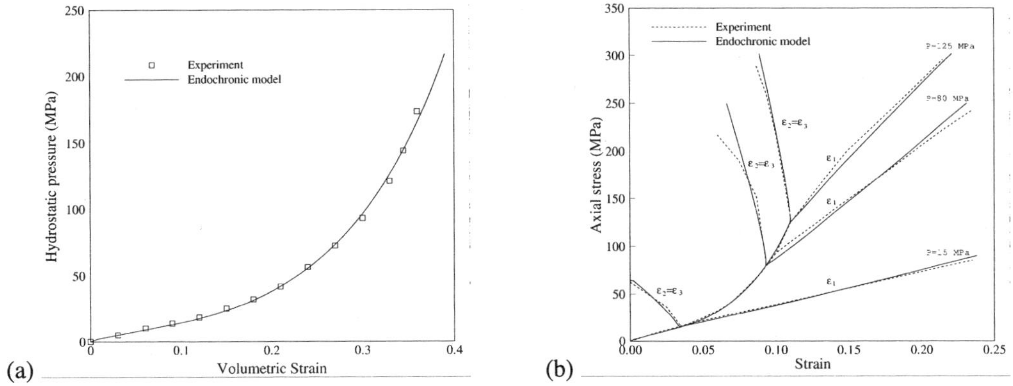

Figure 9.16. The triaxial compression test; a) The hydrostatic pressure versus volumetric strain in isostatic compression test, b) The axial stress versus axial and radial strains at given confining pressure

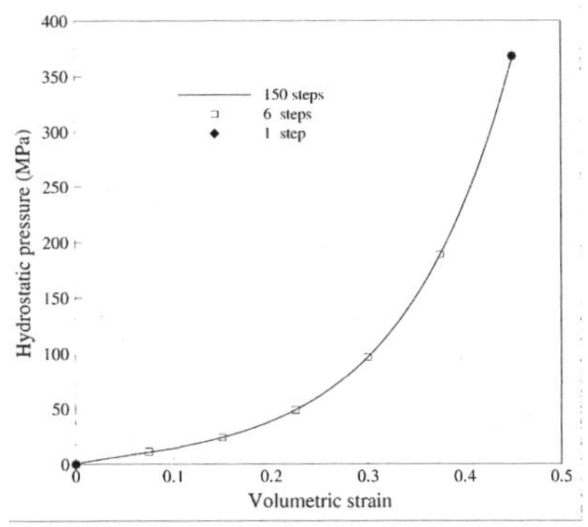

Figure 9.17. The accuracy of the integration algorithm for different load step sizes; The hydrostatic pressure versus volumetric strain

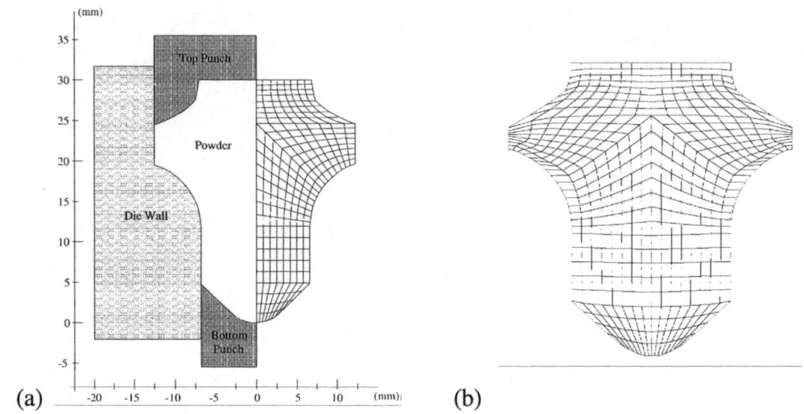

Figure 9.18. An axisymmetric shaped tip component; a) Geometry and finite element mesh, b) The deformed mesh at final stage of compaction

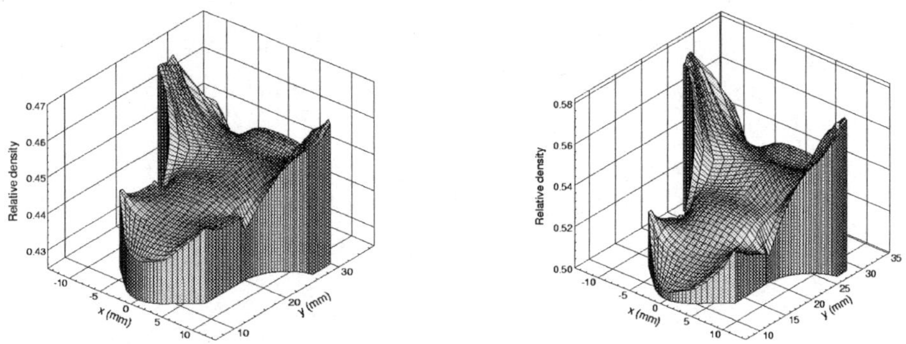

Figure 9.19. A shaped tip component; The density distribution at half and final stages of compaction using endochronic model

Figure 9.20. An electron micrograph of loose MH-100 iron powder scanned by Gu *et al.* (2001)

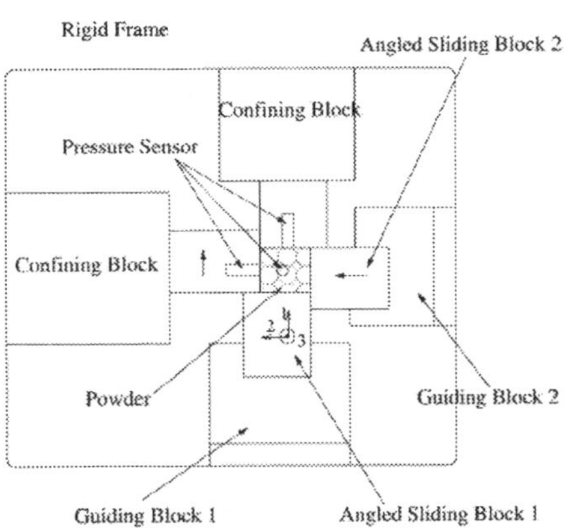

Figure 9.21. A Schematic of the true-triaxial compression apparatus performed by Gu *et al.* (2001)

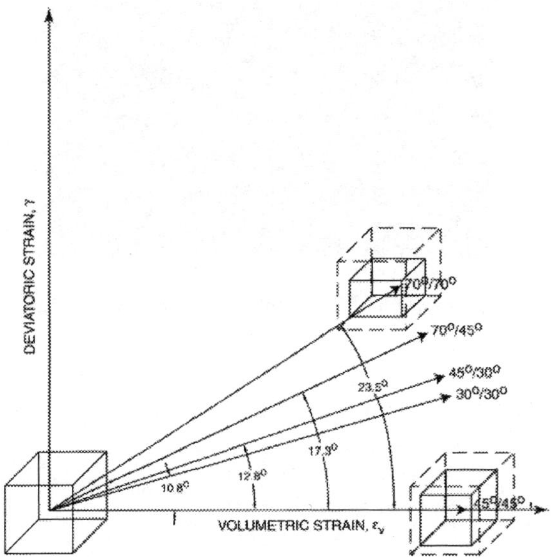

Figure 9.22. Strain paths achievable in the true-triaxial compression experiments

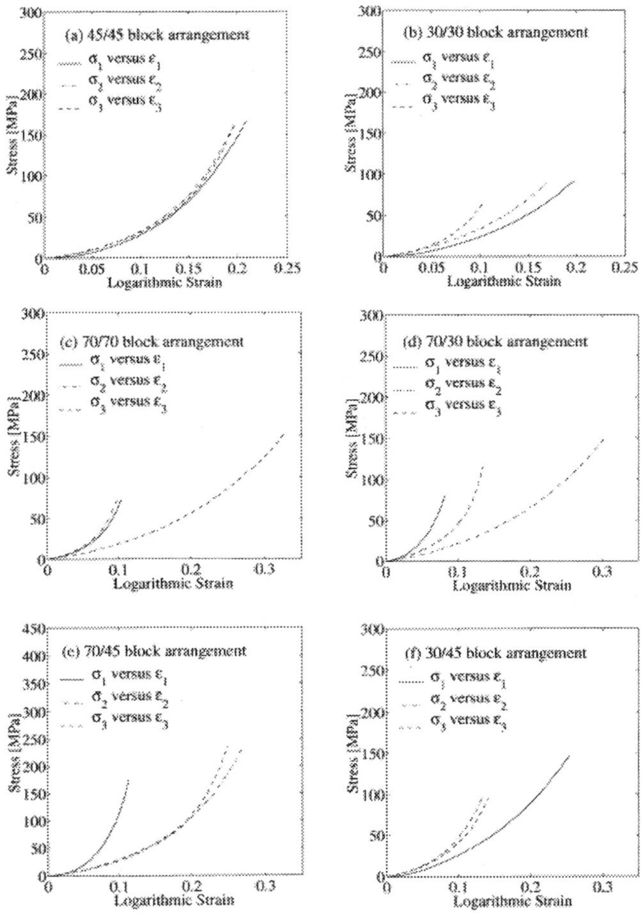

Figure 9.23. The stress versus strain curves from the true-triaxial compression experiment with different angled block arrangements reported by Gu *et al.* (2001)

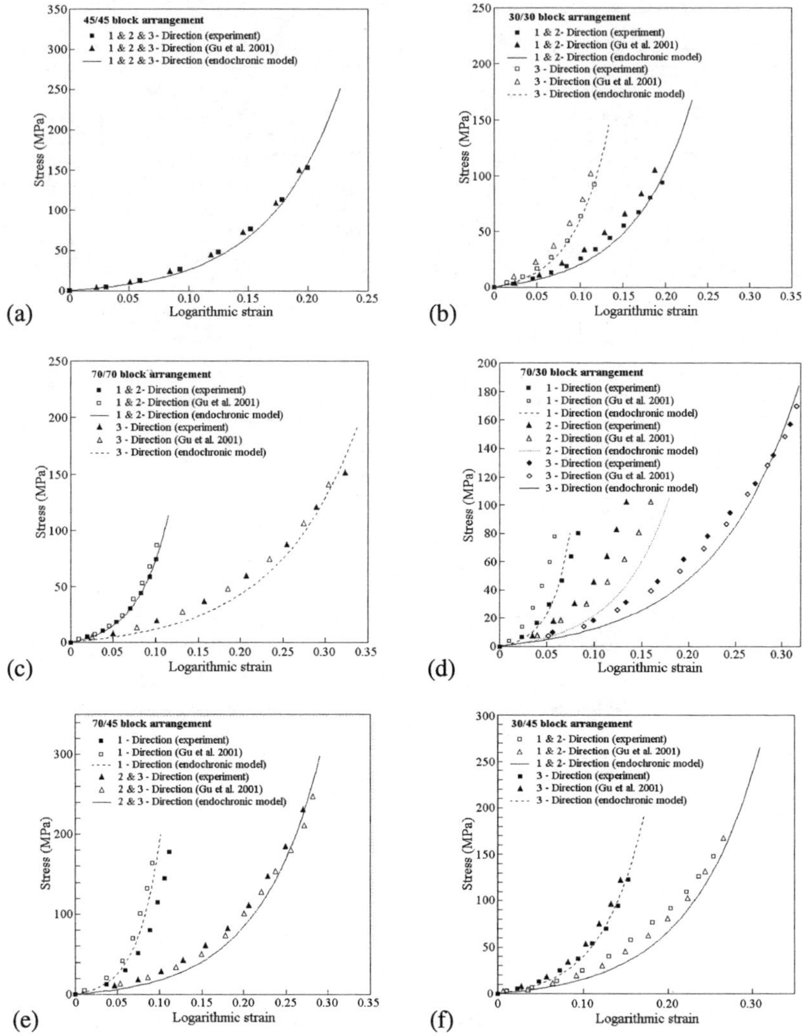

Figure 9.24. Comparison of the calculated stress-strain responses against measured and other simulated results in the true-triaxial compression experiments with different angled block arrangements

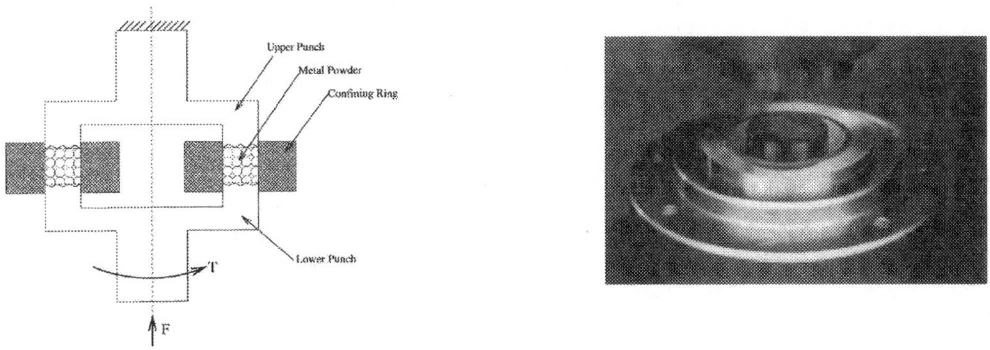

Figure 9.25. Schematic of the torsion ring-shear apparatus performed by Gu *et al.* (2001)

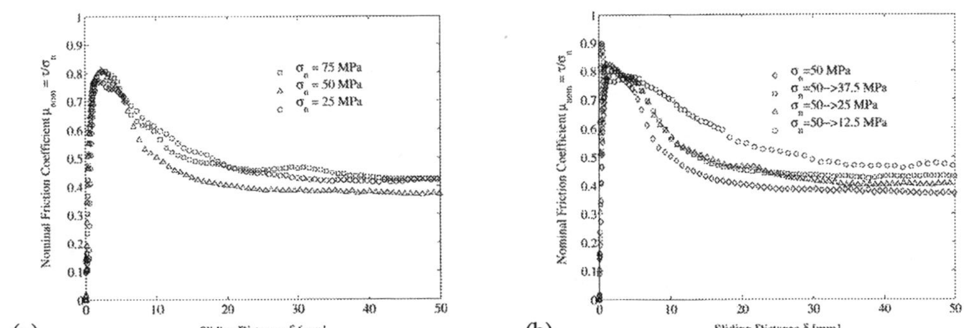

Figure 9.26. Torsion ring-shear experiment on MH-100 iron powder; The nominal friction coefficient versus sliding distance reported by Gu *et al.* (2001)

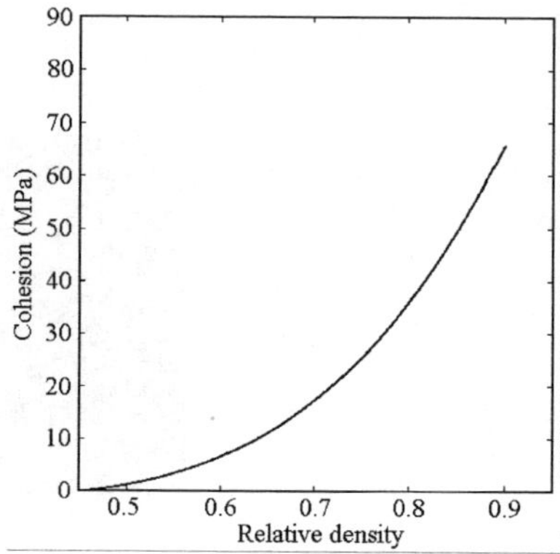

Figure 9.27. Variation of the cohesion with relative density for MH-100 iron powder

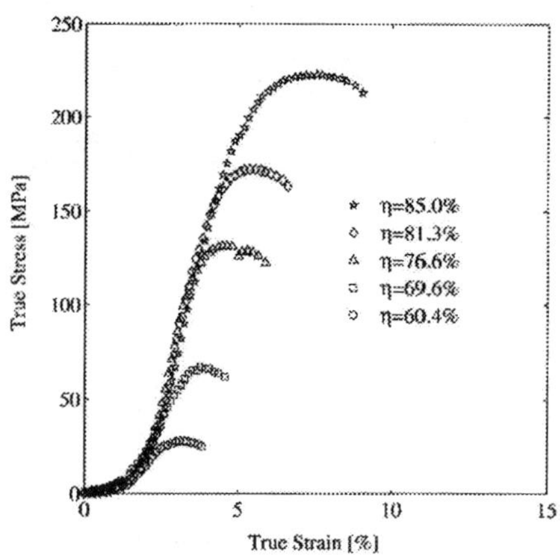

Figure 9.28. Simple compression specimens; The axial stress versus axial logarithmic strain curves obtained by Gu *et al.* (2001)

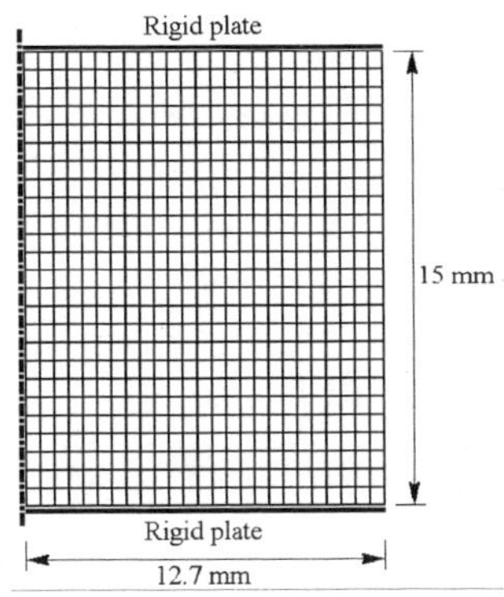

Figure 9.29. Simple compression specimen; An axi-symmetric FE mesh

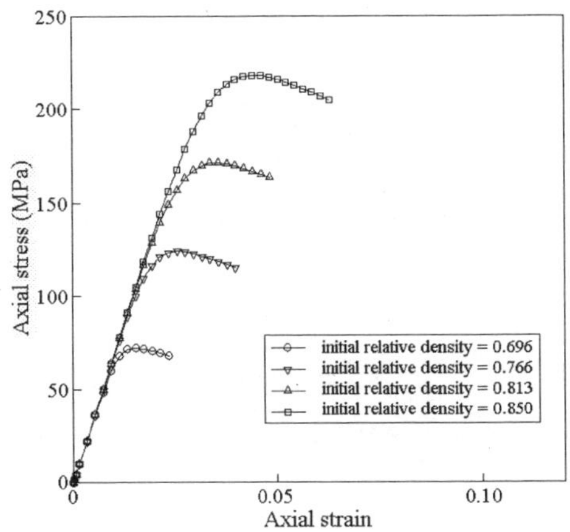

Figure 9.30. Simple compression of compacted specimens; Axial stress versus axial strain curves form numerical simulation of MH-100 powder using endochronic model

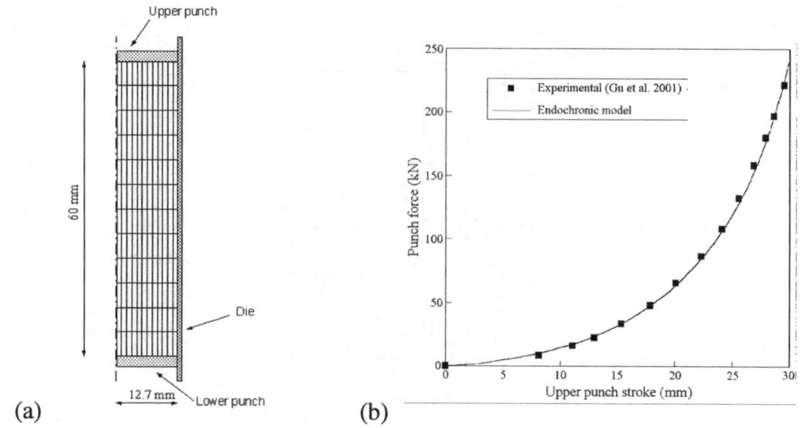

Figure 9.31. The uniaxial compaction of cylindrical MH-100 powder component; a) An axisymetric FE mesh, b) The variation of top punch vertical reaction with displacement

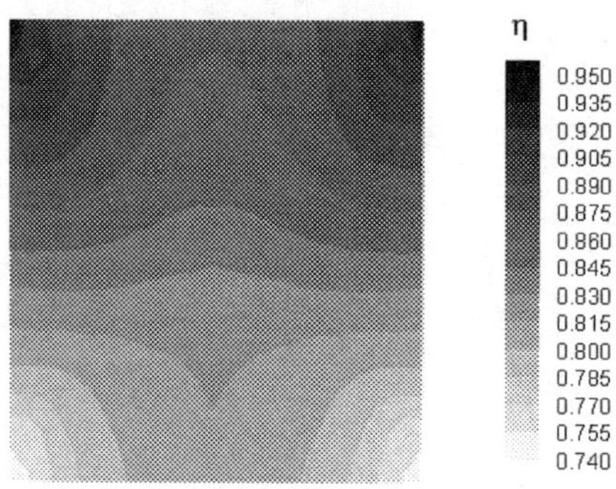

Figure 9.32. Cylindrical MH-100 powder component; The calculated density distribution at final stage of compaction

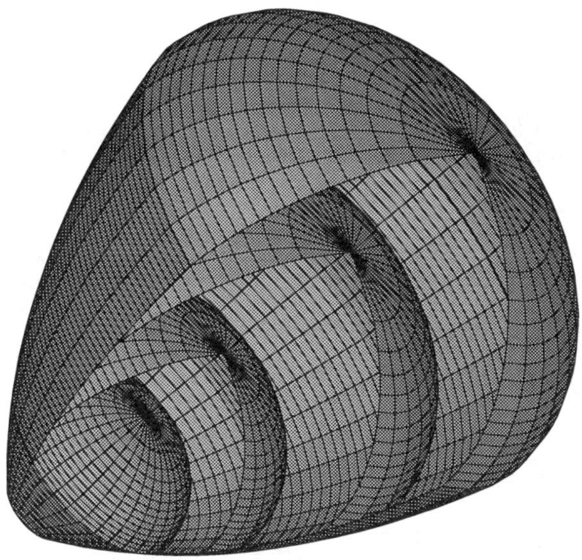

Figure 9.33. 3D representation of the double-surface plasticity; The expansion of moving cap surface with increasing the volumetric plastic strain, or relative density

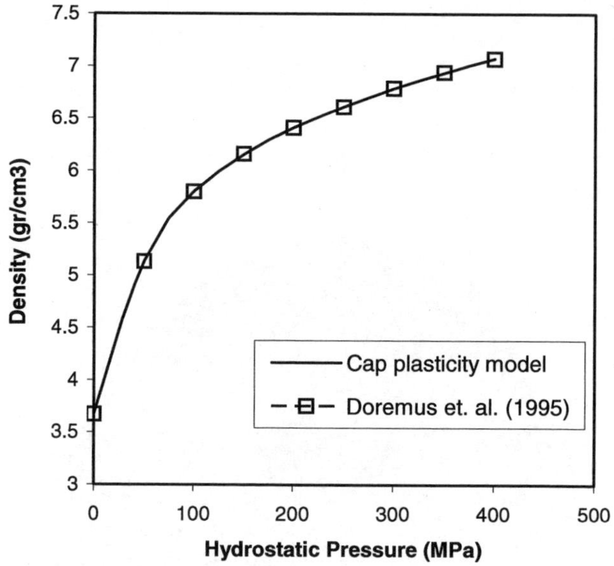

(a)

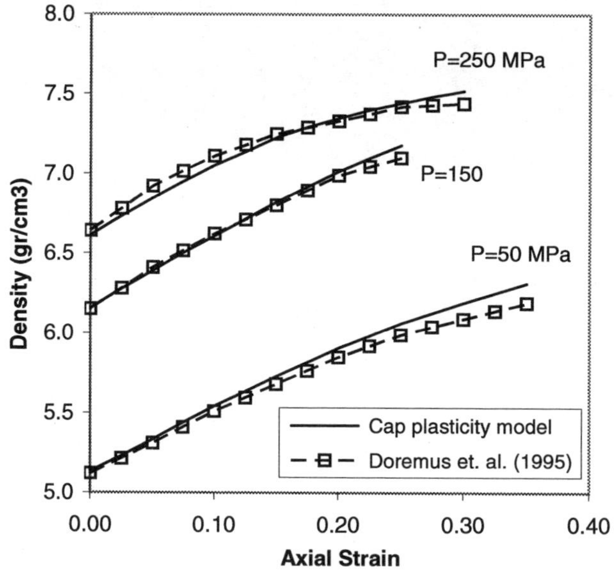

(b)

Figure 9.34. A comparison of numerical and experimental results in confining pressure and triaxial tests; a) The density versus hydrostatic pressure, b) The density versus axial strain

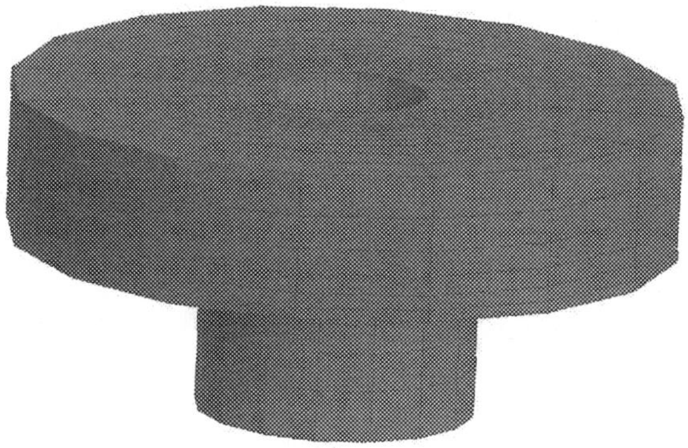

(a)

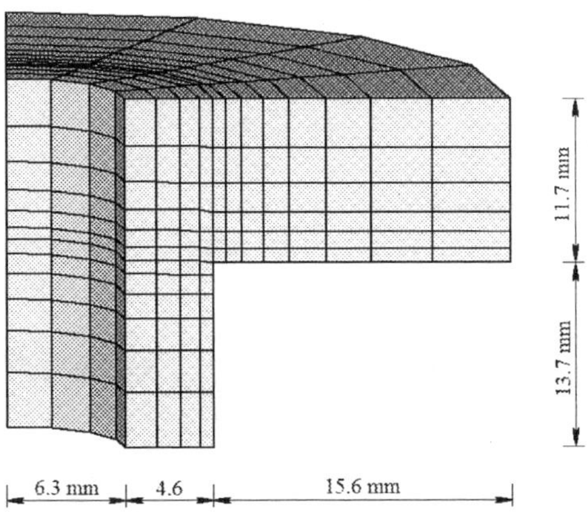

(b)

Figure 9.35. A rotational flanged component; a) 3D finite element mesh, b) Geometry and boundary conditions

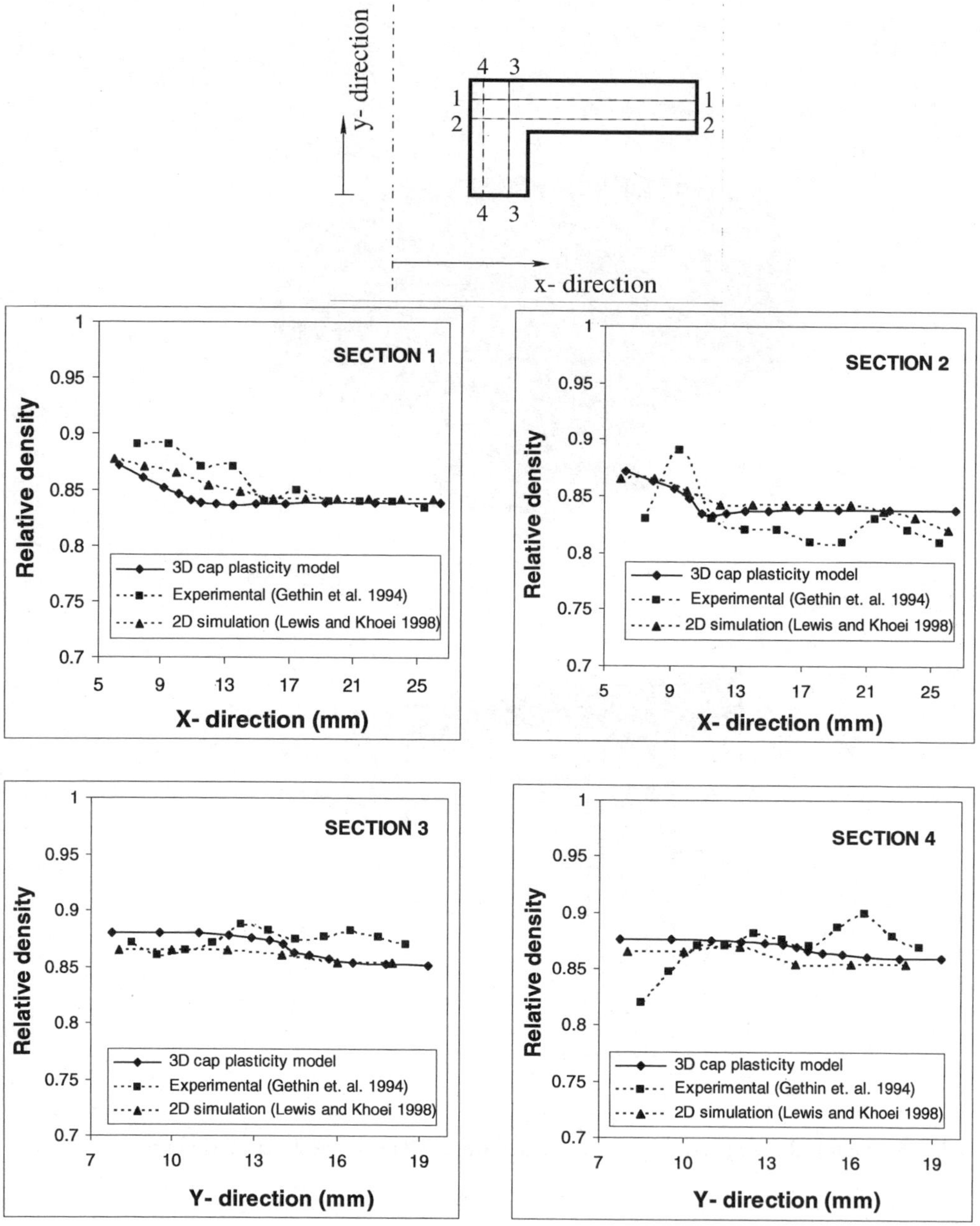

Figure 9.36. A rotational flanged component; Comparison of the relative density profile by the 3D cap plasticity and other numerical and experimental results

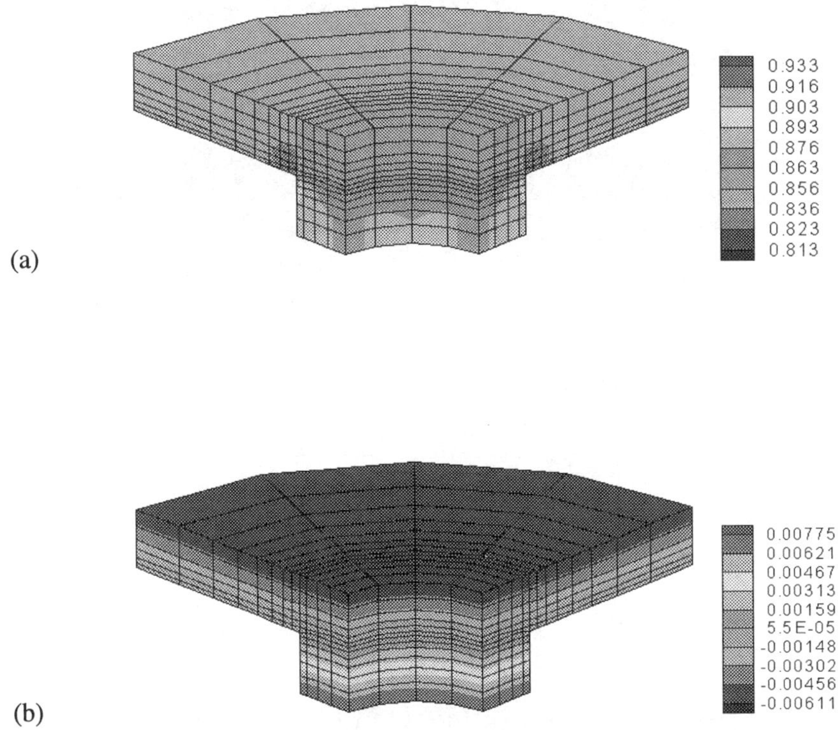

Figure 9.37. A rotational flanged component; a) Relative density contour and, b) Displacement contour at the final stage of powder compaction $d_t = 6.06$, $d_b = 7.7$ mm (see colour plate section at beginning of book)

(a)

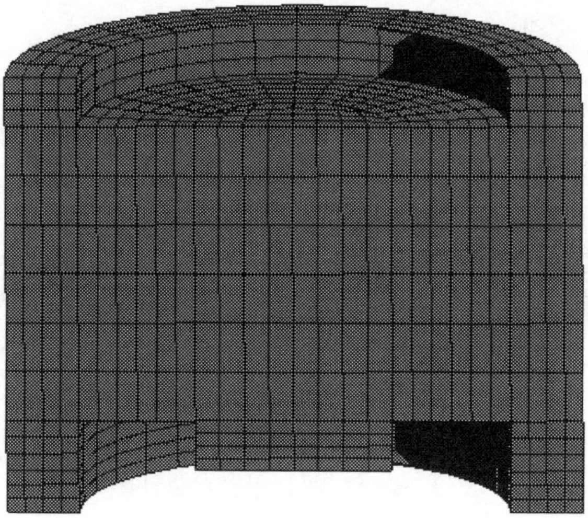

(b)

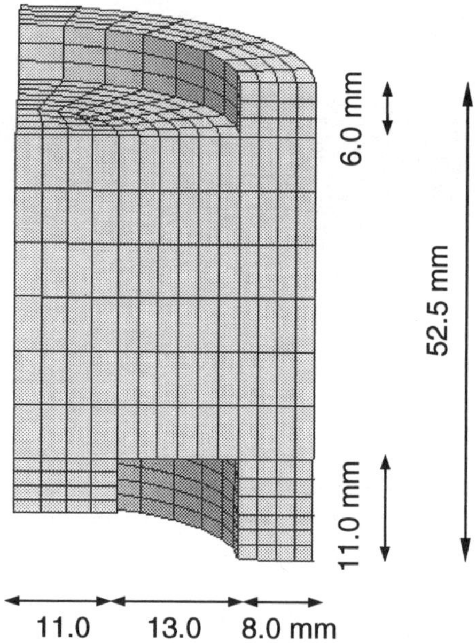

Figure 9.38. A multi-level component; a) 3D finite element mesh, b) Geometry and boundary conditions

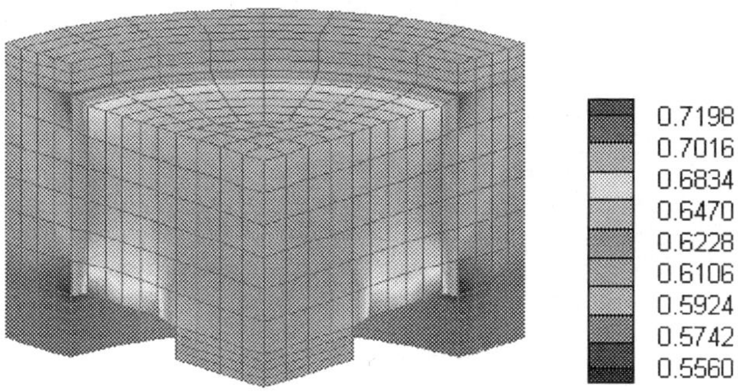

(a)

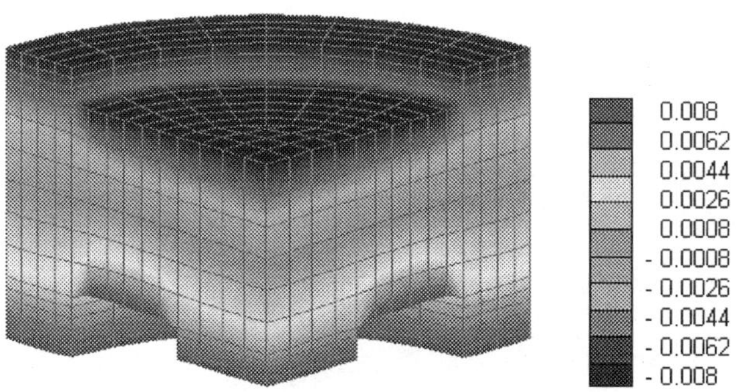

(b)

Figure 9.39. A multi-level component; a) Relative density contour and, b) Displacement contour at the final stage of powder compaction $d_t = 8.0$, $d_b = 8.0$ mm (see colour plate section at beginning of book)

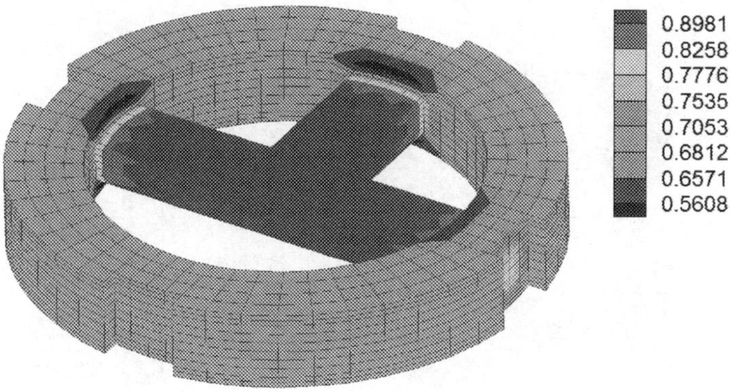

(a)

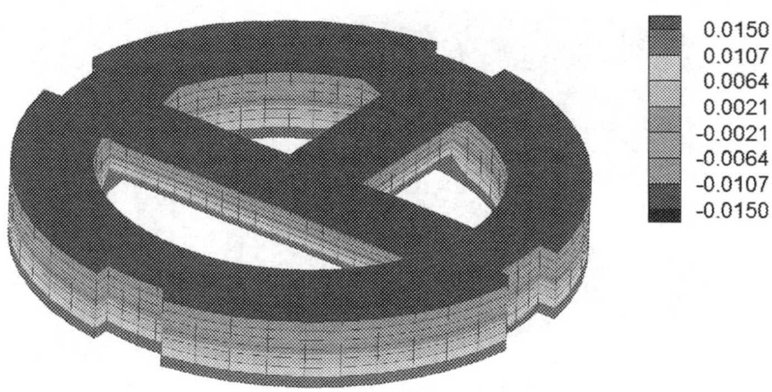

(b)

Figure 9.40. A shaped seal-disk component; a) The relative density contour and, b) The displacement contour at the final stage of compaction (see colour plate section at beginning of book)

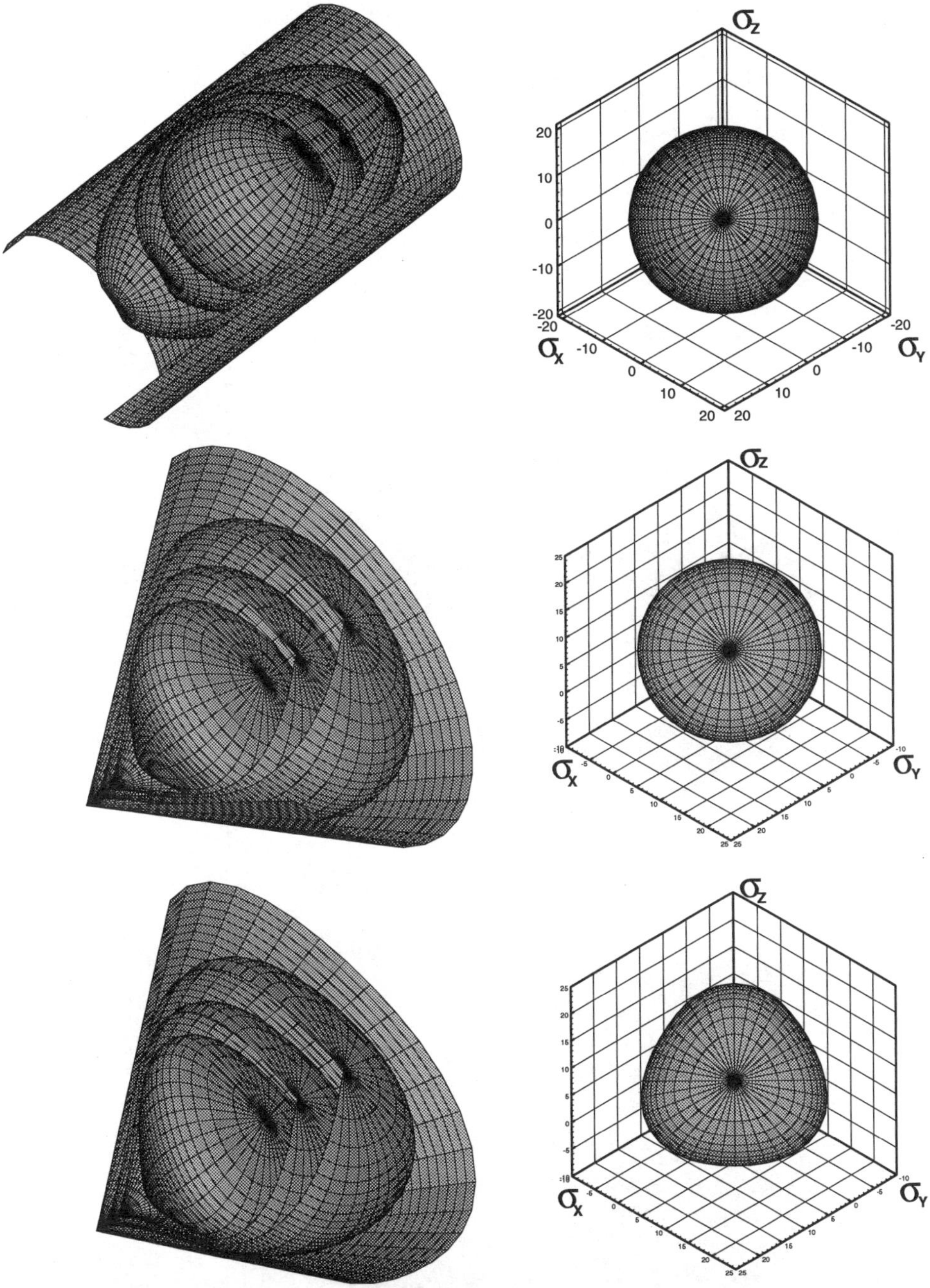

(a)

(b)

(c)

Figure 9.41. Trace of 3D single-cap plasticity in principal stress space for different values of relative density; a) The elliptical yield function, b) The cone-cap yield function, c) The irregular hexagonal pyramid of the Mohr-Coulomb and cone-cap yield function

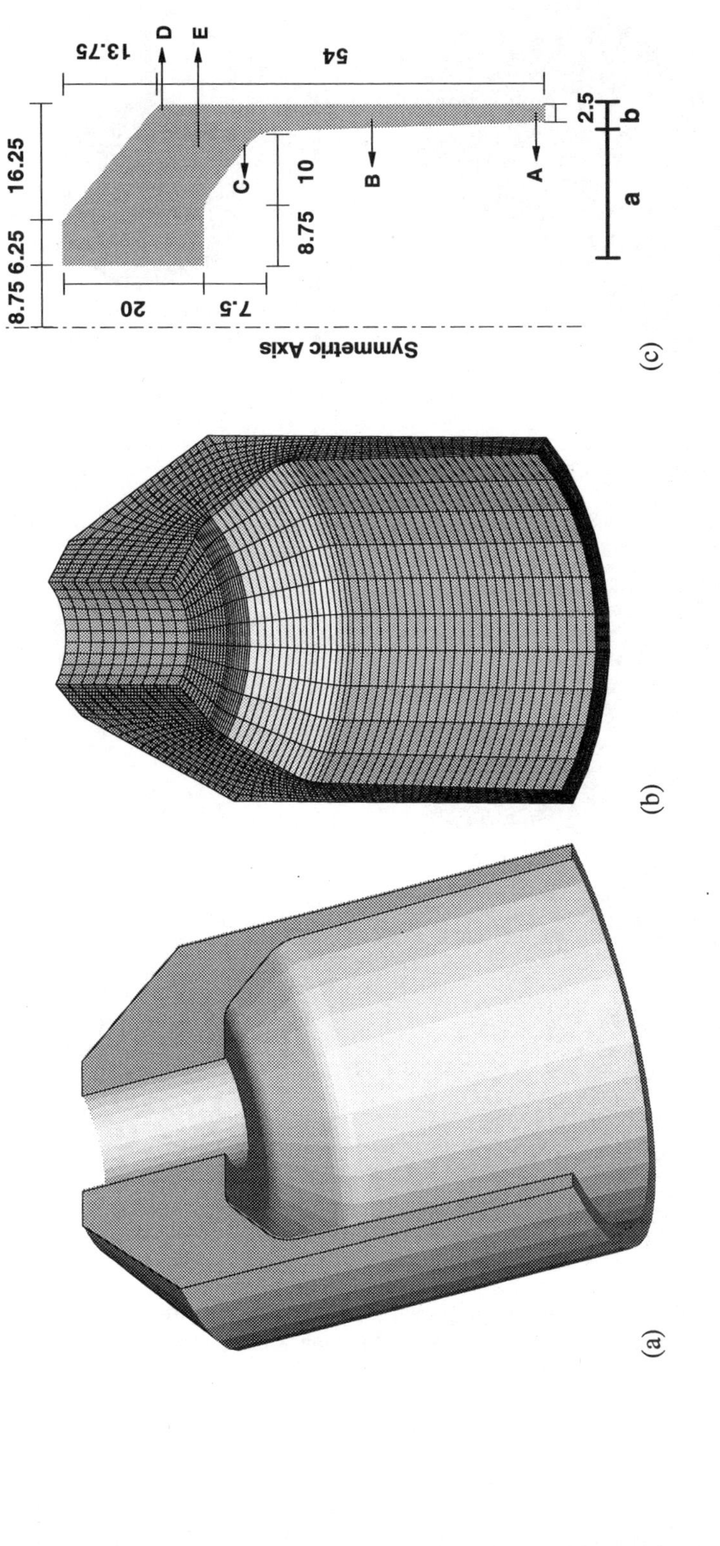

Figure 9.42. An automotive component; a) The uncompacted powder component, b) One quarter of 3D finite element mesh, c) The geometry and boundary conditions

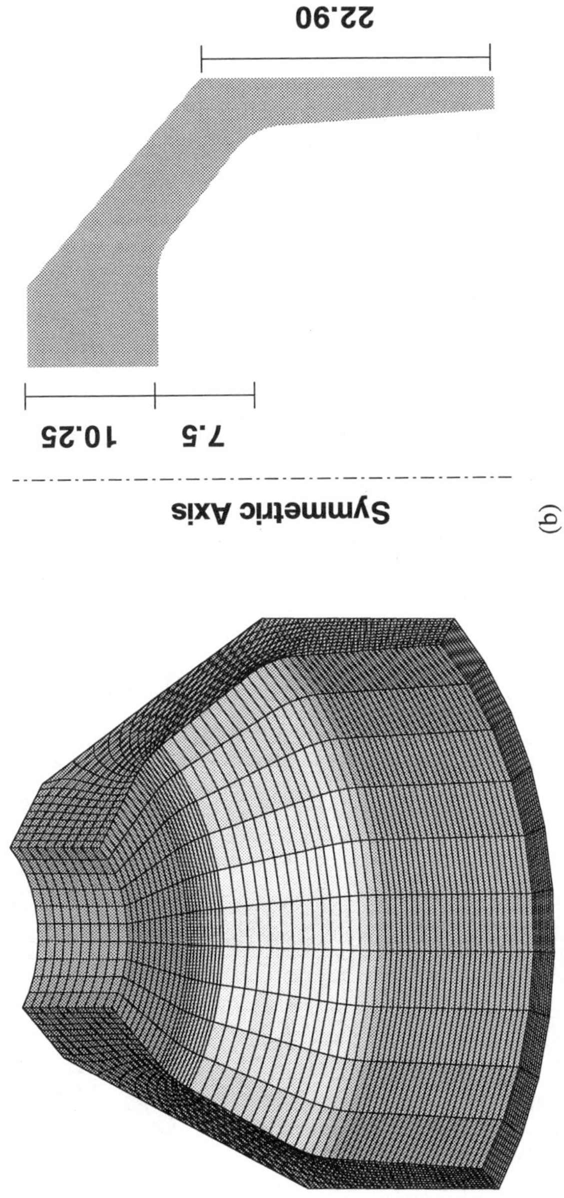

Figure 9.43. An automotive component; a) The final deformed mesh, b) The geometry of compacted specimen

(a)

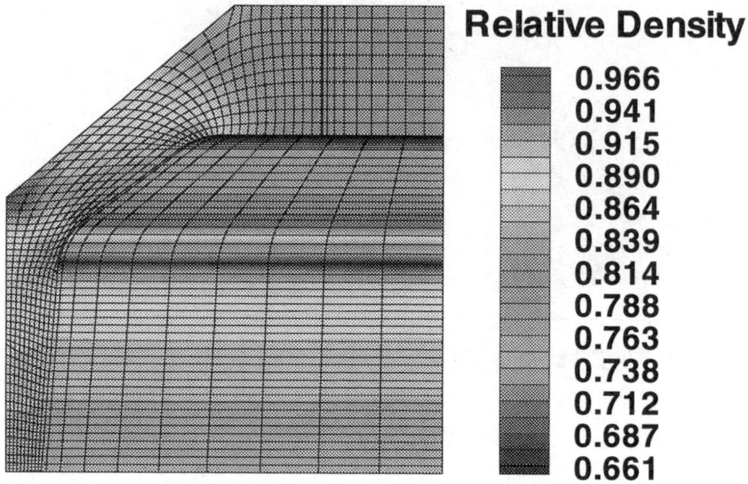

(b)

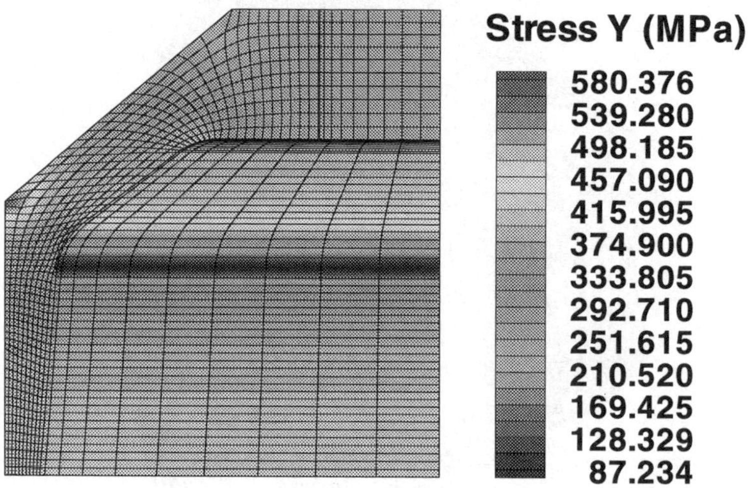

Figure 9.44. An automotive component; a) 3D relative density contour and, b) 3D distribution of normal stress σ_y at the final stage of compaction (see colour plate section at beginning of book)

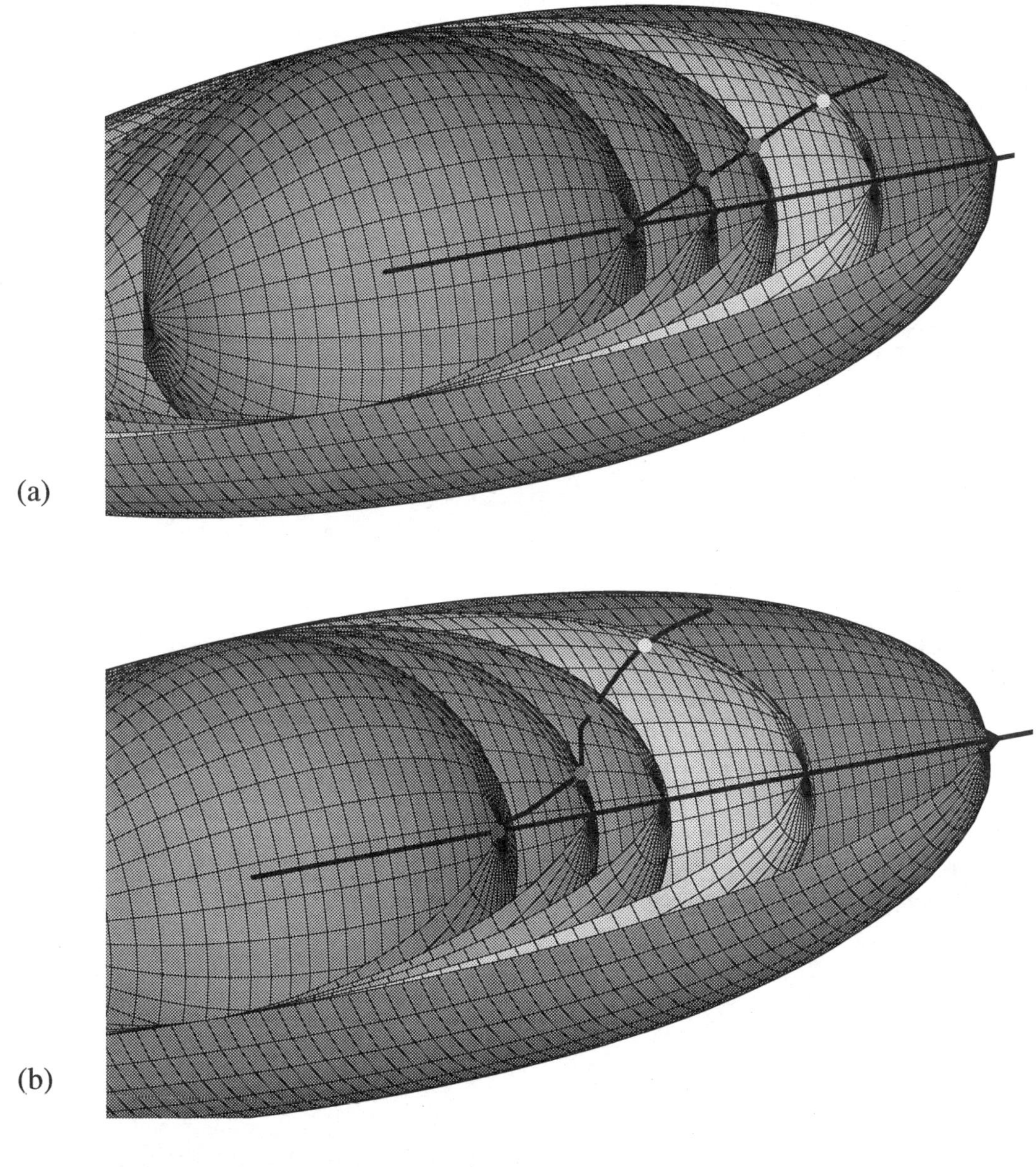

Figure 9.45. An automotive component; 3D representation of stress paths together with the associated yield surfaces at different Gauss points; a) A(30.2, -23.0); b) C(25.6, 17.7) (see colour plate section at beginning of book)

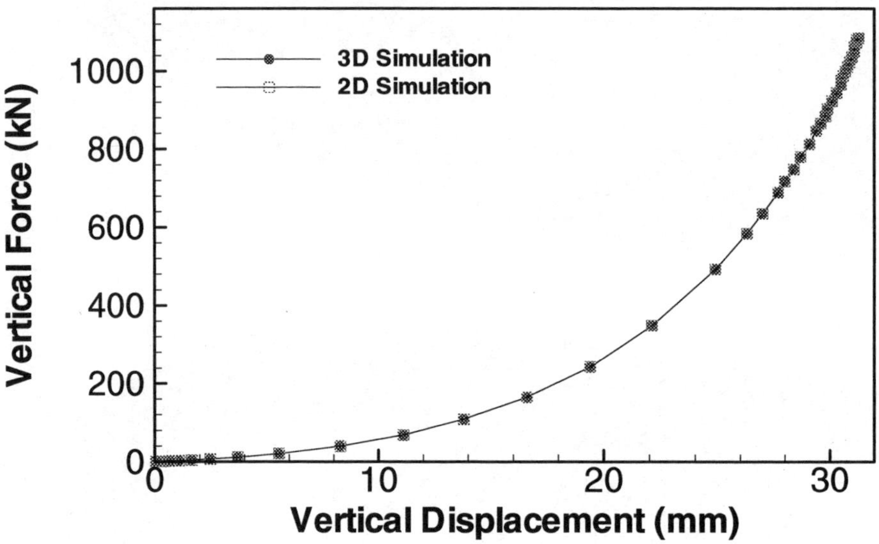

Figure 9.46. The evolution of top punch vertical reaction force with its vertical displacement for an automotive component; A comparison between 2D and 3D results

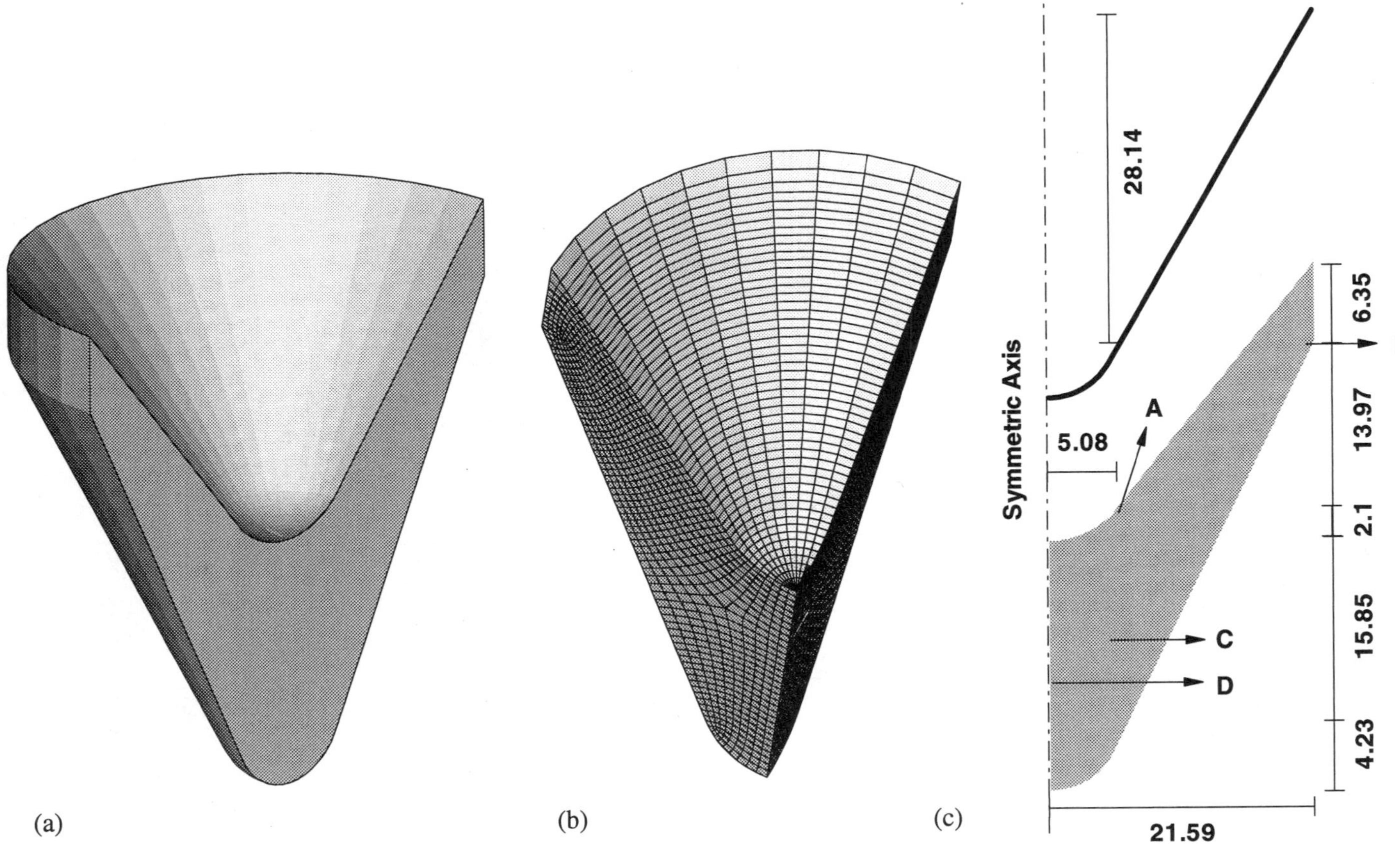

Figure 9.47. A conical shaped-charge liner; a) The uncompacted powder component, b) One quarter of 3D finite element mesh, c) The schematic of the process, geometry and boundary conditions

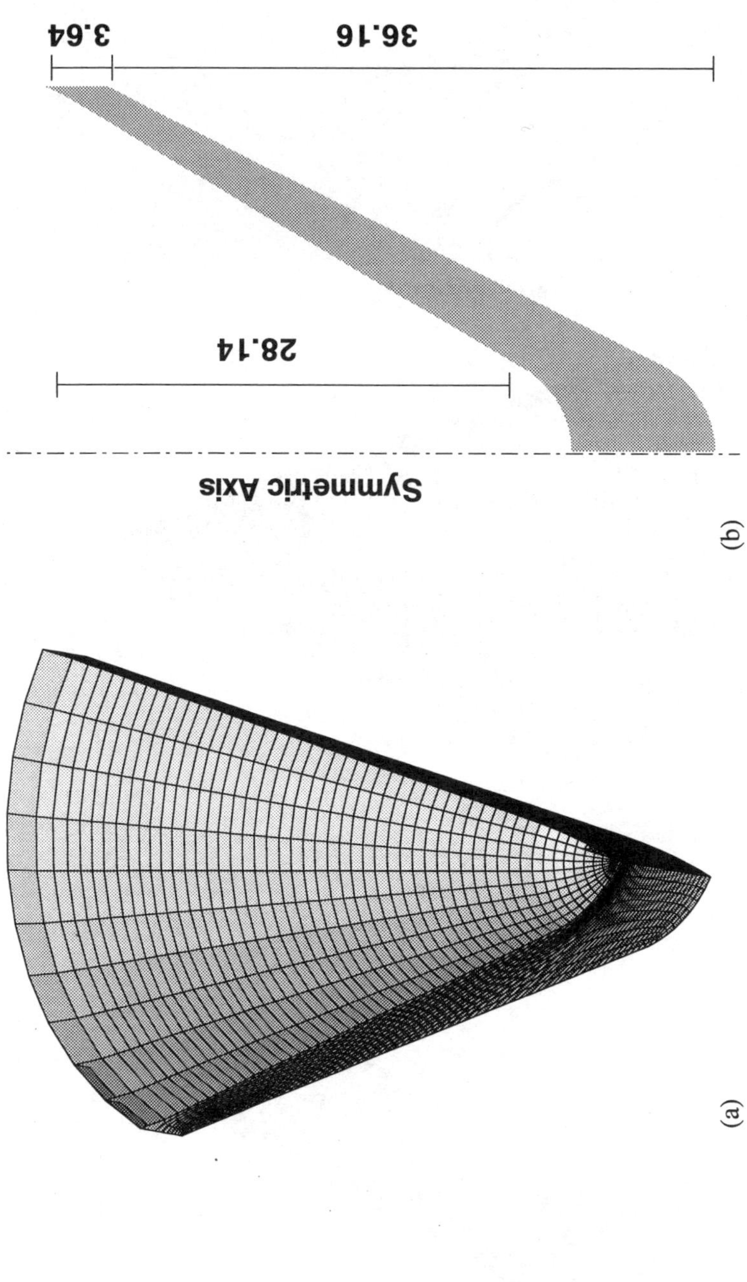

Figure 9.48. A conical shaped-charge liner; a) The final deformed mesh, b) The geometry of compacted specimen

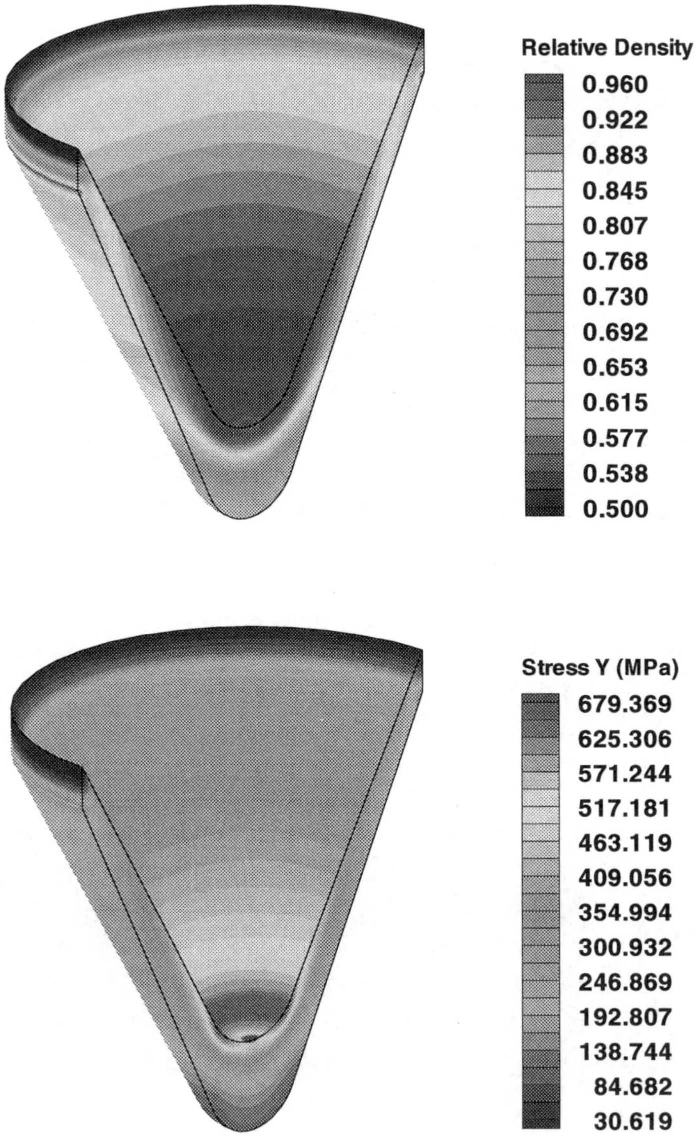

(a)

(b)

Figure 9.49. A conical shaped-charge liner; a) 3D relative density contour and, b) 3D distribution of normal stress σ_y at the final stage of compaction (see colour plate section at beginning of book)

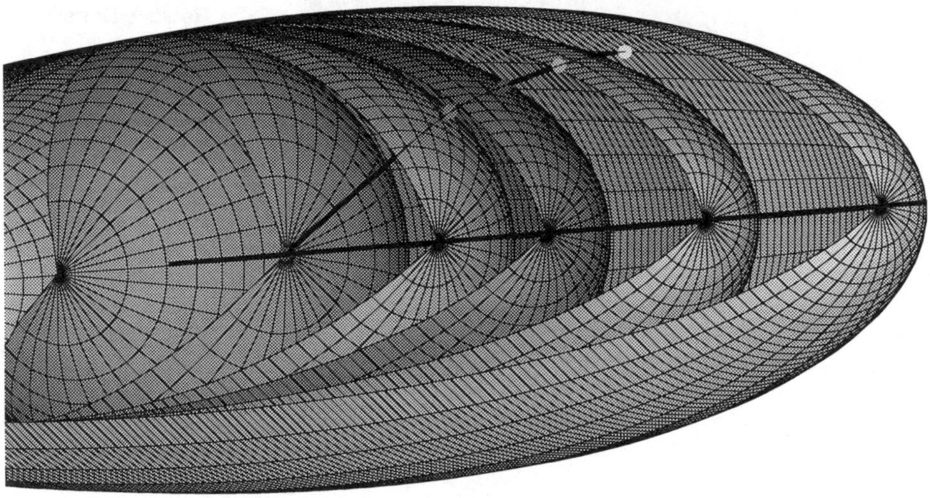

(a)

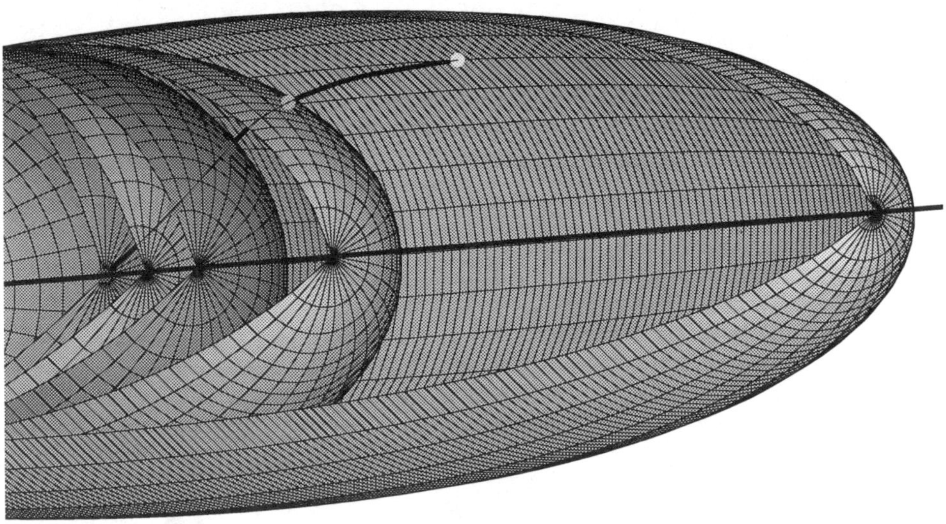

(b)

Figure 9.50. A conical shaped-charge liner; 3D representation of stress paths together with the associated yield surfaces at different Gauss points; a) A(5.1, 21.5), b) B(21.7, 35.6) (see colour plate section at beginning of book)

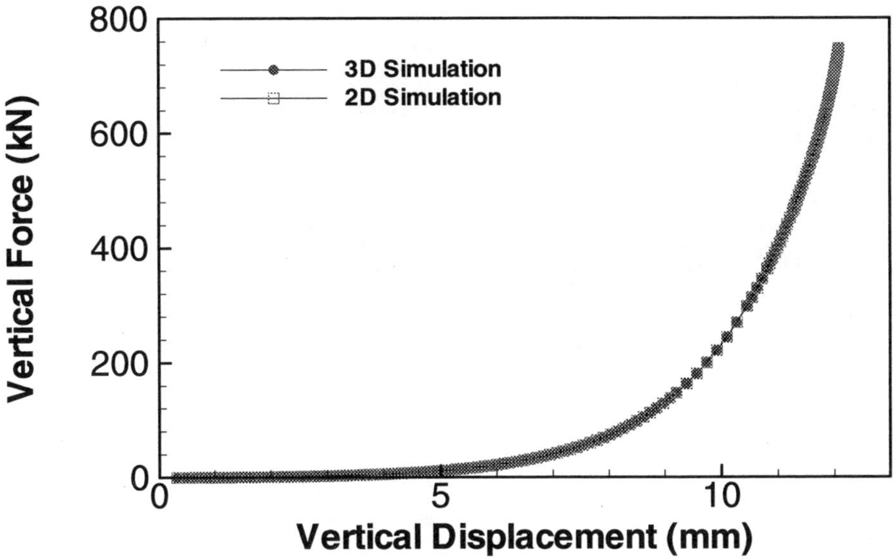

Figure 9.51. The predicted force on the upper punch versus its vertical displacement for a conical shaped-charge liner; A comparison between 2D and 3D results

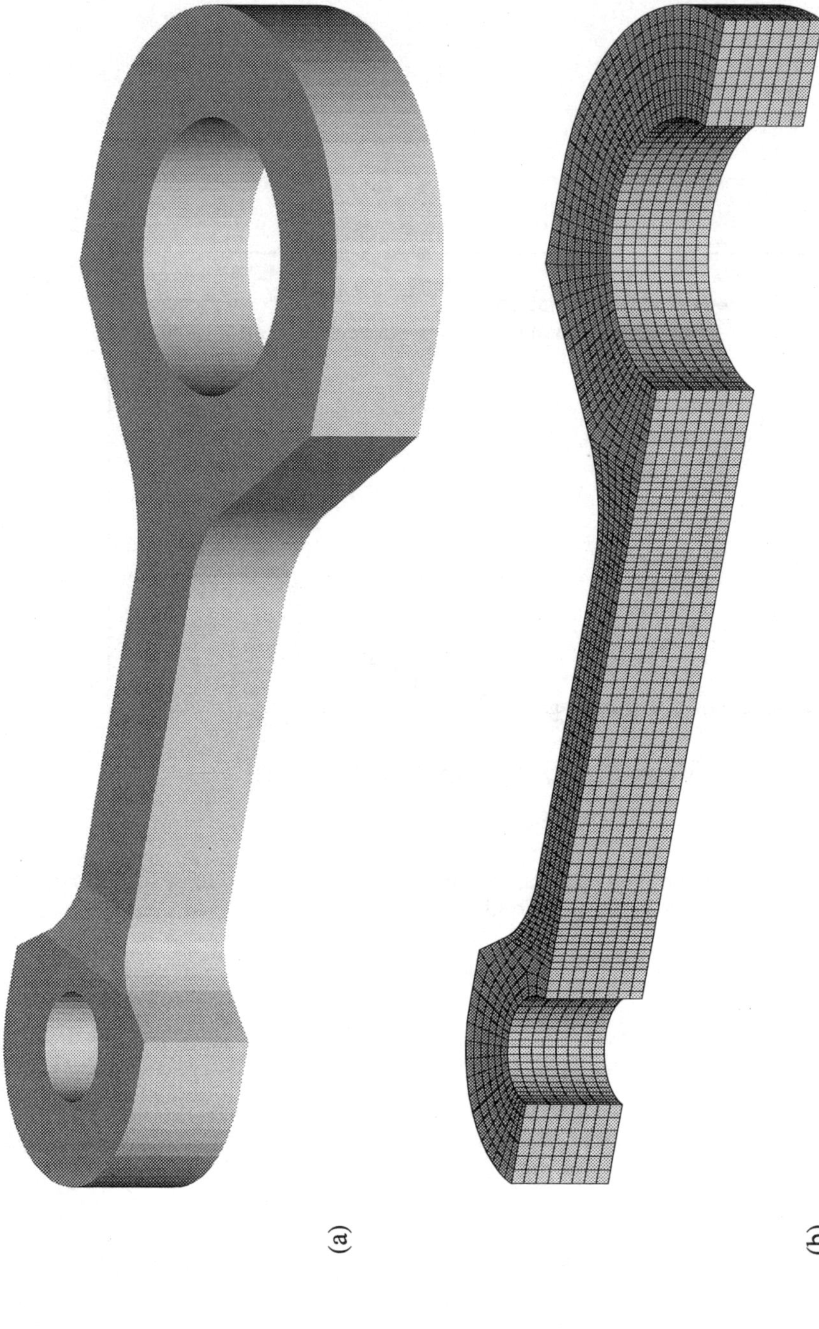

(a)

(b)

Figure 9.52. A connecting-rod; a) The shape of 3D part, b) The initial finite element mesh

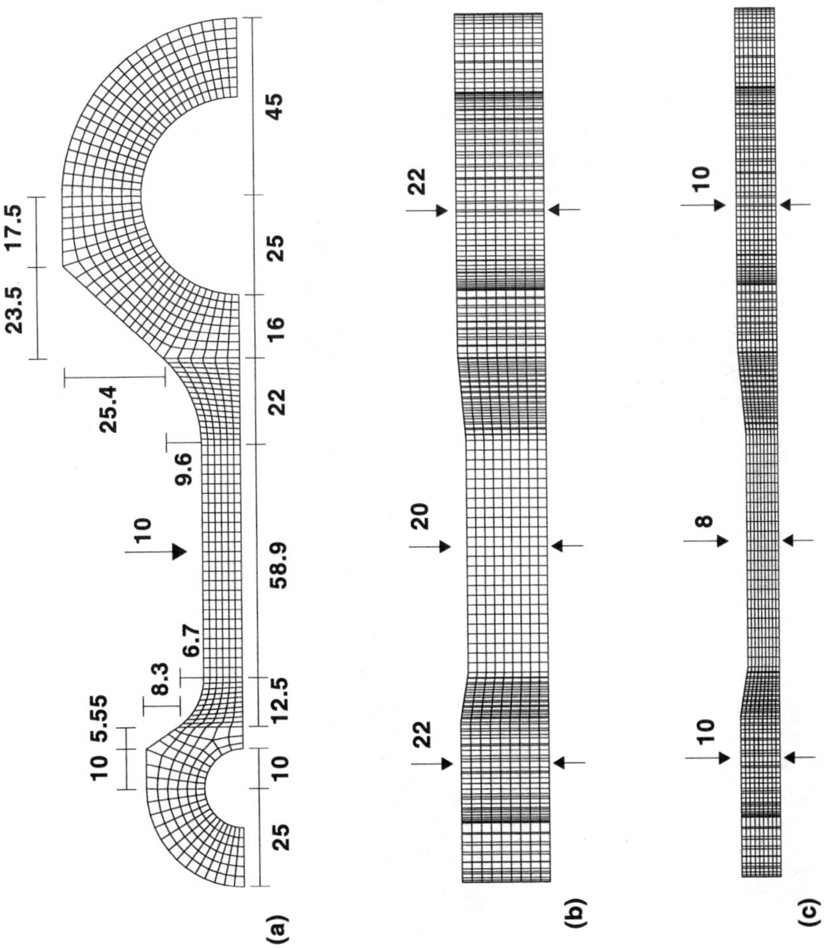

Figure 9.53. A connecting-rod; a) The geometry of component at the initial and final stage of compaction

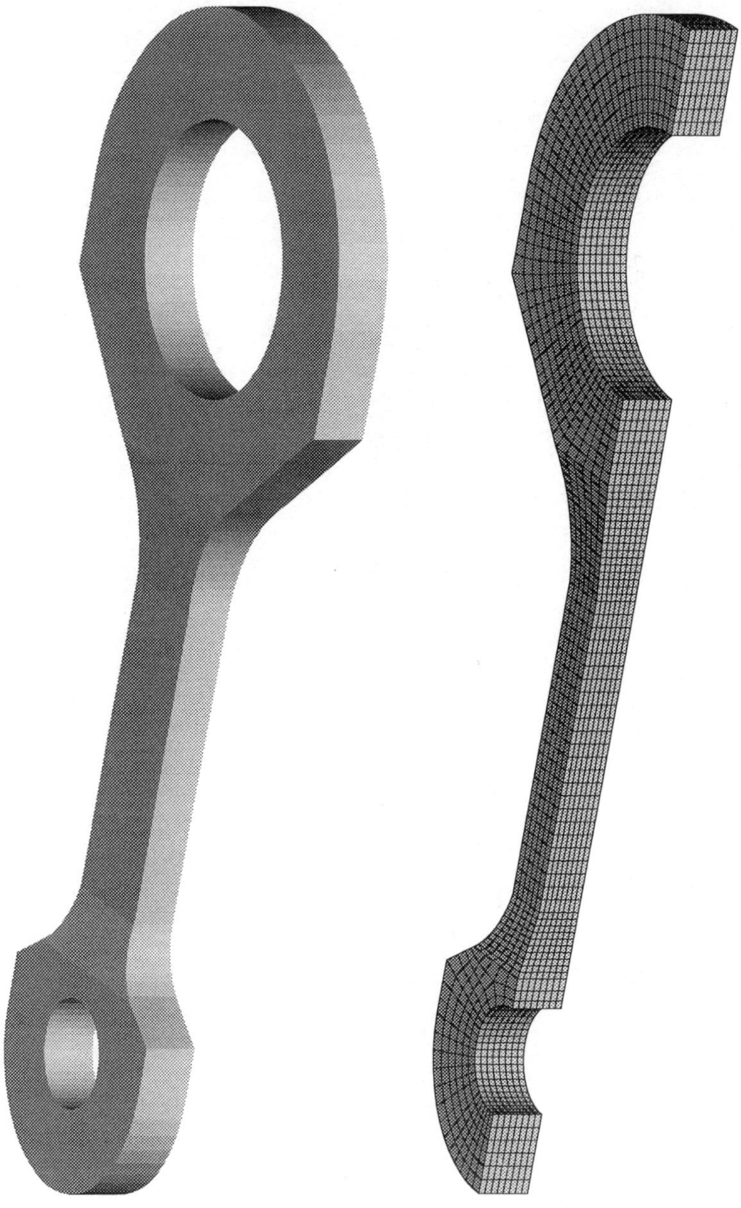

(a) (b)

Figure 9.54. A connecting-rod; a) The shape of compacted specimen, b) The final deformed mesh

Figure 9.55. A connecting-rod; The evolution of the relative density fields at final stage of compaction (see colour plate section at beginning of book)

Figure 9.56. A connecting-rod; The normal stress contour at the final stage of compaction (see colour plate section at beginning of book)

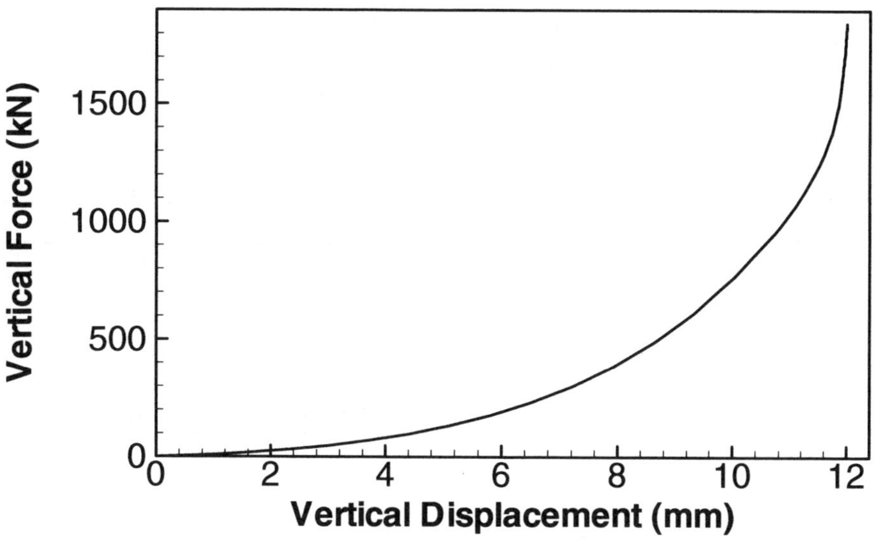

Figure 9.57. The variation of top punch reaction force with its vertical displacement for a connecting-rod

Table 9.1. A unit brick element under 40 percent volume reduction - A comparison of computational costs

	Average number of iteration	Average of CPU time per step (relative value)
Endochronic model- 150 steps	3.0	0.85
Endochronic model- 15 steps	3.5	1.00
Endochronic model- 1 step	4.0	1.16
Cone-cap model- 150 steps	4.8	0.58
Cone-cap model- 15 steps	6.3	0.76

Table 9.2. The values of applied stretchings corresponding to different combinations of angled blocks

	D_{11}	D_{22}	D_{33}
45°/45°	- 1.0	- 1.0	- 1.0
30°/30°	- 1.732	- 1.732	- 1.0
70°/70°	- 0.364	- 0.364	- 1.0
70°/30°	- 0.364	- 1.732	- 1.0
70°/45°	- 0.364	- 1.0	- 1.0
30°/45°	- 1.732	- 1.0	- 1.0

Table 9.3. Endochronic material parameters

Elastic parameters:	$E = 698\exp(5.62\eta)$ MPa , $\nu = 0.3$
Kernel parameters:	$A_1 = 90000$ MPa, $\alpha_1 = 1000$ $B_1 = 90000$ MPa, $\beta_1 = 500$, $\beta_2 = 80$, $\beta_3 = 5$ $C_1 = 0$, $C_2 = 15$, $C_3 = 1.5$ MPa
Initial relative density:	$\eta_0 = 0.45$
Deviatoric scale function parameters:	$c(\eta) = 1.4\eta^3 \dfrac{\eta - \eta_0}{1 - \eta_0}$, $p_0 = 245$ MPa
Hydrostatic scale function parameters:	$g(\eta) = \eta^{2.5} \dfrac{\eta - \eta_0}{1 - \eta_0}$, $\psi = 0.5$, $\phi_h = 500$ MPa
Coupling parameter:	$\kappa = 1.5$

Table 9.4. The material model parameters for the cap plasticity

Fixed yield surface parameters	Cap parameters	Tension cutoff
$\alpha = 225$ MPa	$R = 1.75$	$T = -0.3$ MPa
$\beta = 0.002$ MPa^{-1}	$D = 0.005$ MPa	
$\gamma = 200$ MPa	$W = 0.34$	
$\theta = 0.008$	$X_0 = 1$ MPa	

Box 9.1. Material parameters, A plain bush component

Elastic constants:	Deviatoric scale function parameters:
Young modulus = 2000 MPa	$p_0 = 200$ MPa
Poisson ratio = 0.35	C = 0.093
Kernels function parameters:	Hydrostatic scale function parameters:
A_1 = 15'000 MPa	$\psi = 0$
$\alpha_1 = 1000$	$\phi_h = 100$
B_1 = 100'000 MPa	$g(\eta) = 0.02\, e^{9(1-\frac{\eta_0}{\eta})}$
$\beta_1 = 1000$	
Initial relative density = 0.4	$\kappa = 0.5$

Box 9.2. Material parameters, A multi-level component

Young modulus: $E(\eta) = E_0\left(\dfrac{\eta-\eta_e}{1-\eta_e}\right)^3$, $E_0 = 18000$ MPa, $\nu = 0.35$, $\eta_e = 0.25$

Initial apparent density = 3.67 g/cm^3
Solid density = 7.54 g/cm^3

Hereditary function parameters:

A_1 = 300'000 MPa	B_1 = 116'700 MPa
α_1 = 1000 MPa	$\beta_1 = 300$

Deviatoric scale function parameters: $a_1 = 1.5$, $a_2 = 0$

Hydrostatic scale function parameters: $m = 7.8$, $\eta_m = 0.4$

<u>Box 9.3. Material parameters, A shaped tip component</u>

Young modulus: $E(\eta) = E_0\left(\frac{\eta - \eta_e}{1 - \eta_e}\right)^{2.8}$, $E_0 = 100000$ MPa, $\nu = 0.35$, $\eta_e = 0.2$

Initial relative density = 0.25

Hereditary function parameters:

A_1 = 40'000 MPa $\quad$ B_1 = 100'000 MPa

α_1 = 100 $\quad$ β_1 = 2.9

Deviatoric scale function parameters: $a_1 = -4.0$, $a_2 = 8.5$

Hydrostatic scale function parameters: $m = 5.5$, $\eta_m = 0.21$

10 DISCONTINUOUS DISPLACEMENTS AND LOCALIZATION

An appropriate numerical simulation and computational technique for metal powder compaction processes was described in previous chapters. One of the other major challenges in the computation of powder forming problems is the analysis of failure. Such computations are needed to evaluate the ultimate capacity of new materials, but they are fraught with serious difficulties. One of these difficulties is the process of localization, which is ubiquitous in failure. The standard finite element models, i.e. finite element methods that use standard constitutive models, suffer from excessive mesh dependence when strain-softening models are used in numerical analyses and cannot reproduce the size effect commonly observed in quasi-brittle failure.

In this chapter, a method is presented for dealing with softening plasticity due to material instability. An h-adaptive mesh refinement is developed based on the gradient of deformation with special reference to metal powder forming. As the classical continuum suffer from excessive mesh-dependence in the analysis of strain softening plasticity, a higher order continuum model is used based on Cosserat theory. The governing equations are regularized by adding rotational degrees-of-freedom to the conventional degrees-of-freedom and including the internal length parameter in the model. Adaptive strategy using element elongation is applied to compute the distribution of required element size using the estimated error distribution. Once the new element size and its alignment have been indicated, an automated procedure is used to construct the mesh according to a predetermined size and elongation distribution. The efficiency of the proposed model and computational algorithms is demonstrated by several numerical examples, particularly a compacted multi-level component. Clearly, a finite shear bandwidth is achieved and the load-displacement curves are uniformly converged upon mesh refinement.

10.1. Introduction

Localization of deformation refers to the emergence of narrow regions in a structure where all further deformation tends to localize, in spite of the fact that the external actions continue to follow a monotonic loading program. The remaining parts of the structure usually unload and behave in an almost rigid manner. Indeed such localization is almost certain to occur if strain softening or non-associated behavior exists, though it can be triggered even when ideal plasticity is assumed. The phenomenon has a detrimental effect on the integrity of the structure and often acts as a direct precursor to structural failure. It is observed for a wide range of materials, including rocks, concrete, soils, metals, alloys and polymers, although the scale of localization phenomena in the various materials may differ by some orders of magnitude: the band width is typically less than a millimeter in metals and several meters for crystal faults in rocks.

Finite element techniques together with a suitable constitutive model provide a valuable tool to simulate how the powder deforms during the compaction process (Khoei and Lewis 1998, 1999). However, the standard approach presents some important limitations when failure surfaces are to be obtained. These limitations are caused by the dependence of the solution on the mesh (Zienkiewicz *et al.* 1995 and Khoei *et al.* 1997).

- The first problem is similar to the numerical diffusion observed in fluid dynamics problems (Peraire *et al.* 1987). Here, the plastic strain is spread over several element widths, and therefore, the localized zone does not correspond to the physical phenomena. Of course, use of a much finer mesh would improve the solution but the cost of solving the problem would also increase.
- A second part of the mesh objectivity problem consists on the fact that meshes having a grid size much finer than the physical shear bandwidth produce much narrower shear bands. In the limit, the width tends to zero with the grid size leading to unrealistic results (Pietruszczak and Stolle 1985, 1987 and Needleman 1988).
- Finally, the orientation of the mesh plays also an important role, and the solutions obtained in meshes having different orientations may differ, affecting the predicted mechanism of failure (Zienkiewicz *et al.* 1995 and Khoei *et al.* 1997). In fact, to accurately capture a shear band, finite elements should be aligned along the localization band.

Therefore, the use of conventional elements for problems with localization result in both a size and directional dependence on mesh topology. Furthermore, the numerical solutions may be physically inappropriate because energy dissipation in regions of localization decreases with mesh refinement.

The capturing of such local phenomena in a finite-element analysis is difficult as discontinuities of displacement occur and these are generally precluded in a standard displacement analysis. Although it is possible to include discontinuities in the analysis (by use of special finite elements) this is complex as the position of such discontinuities has to be assumed a priori. Early efforts were devoted to obtaining failure surfaces and their associated safety factors. Later, finite element techniques allowed a more precise analysis of the stress and strain fields, using more sophisticated material models. However, the problem is no longer mathematically well posed after the onset of localization in strain-softening materials, because local continuum allows for an infinitely small band width in shear or in front of a crack tip (Benallal and Tvergaard 1995). At the numerical level, these difficulties translate in mesh dependence of solutions.

Various techniques have been implemented to provide a physically acceptable solution. Some impose restrictions on the constitutive moduli in the post-localization regime, while others artificially restrict the size of finite element by comparison to the localization zone. The former is based on enriching the continuum with non-conventional constitutive relations in such way that an internal or characteristic length scale is introduced. Bifurcation analysis techniques based on the early work of Thomas (1961) and Rice (1977) were adopted by many researchers to determine the shear band localization (de Borst 1988, Runneson *et al.* 1991, Pijaudier-Cabot and Benallal 1993 and Simo *et al.* 1993). Non-local theories are the Cosserat continuum (Fleck and Hutchinson 1993 and de Borst *et al.* 1993), the higher gradient theories (Triantafyllidis and Ainfantis 1986), and the integral theory or the gradient theory (Muhlhaus and Ainfantis 1991 and de Borst and Mulhlaus 1992). The later is based on a suitable mesh refinement using normal, continuous, approximations to all the variables (Pastor *et al.* 1991, Belytschko and Tabbara 1993, Zienkiewicz *et al.* 1995, Lewis and Khoei 2001 and Khoei and Lewis 2002). The main function of this approach is a progressive mesh refinement, which can well model the discontinuity in the limit by using standard continuous functions as shown in Figure 10.1 for a one-dimensional case. In this figure, it has also been illustrated how the discontinuity could be exactly captured by a discontinuous shape function approximation providing the position of the discontinuity was known.

The present chapter is concerned mainly with a combination of both methods based on the Cosserat continuum using adaptive finite element analysis to better simulate strain localization in elasto-plastic solids. The Cosserat continuum theory, first proposed by Cosserat brothers in 1909, has attracted many attentions in the last decade. The first implementation of the theory into a finite element code was made by de Borst (1991). It was then extended to associated and non-associated materials by Peric *et al.* (1994), Iordache and Willam (1998) and Cramer *et al.* (1999). This theory has two main characteristics. Firstly, the rotational degree of freedom is taken into account in addition to translation degrees of freedom. In fact, the introduction of rotational degree of freedom lead to the existence of moment stresses (moment per area) in addition to the stresses of classic continuum. Secondly, an internal length scale is introduced in the field of constitutive equation. This parameter, which plays the most important role in controlling shear bandwidth, relates couple stresses to micro-curvature.

Adaptive remeshing is one of the various computational procedures suggested by authors for solving problems involving strain localization. Peraire *et al.* (1987) developed an adaptive remeshing algorithm for compressible flow computations in the capturing of discontinuous shocks. Ortiz and Quigley (1991) developed an *r*-adaptive method with an error criterion based on interpolation for strain localization problems. Pastor *et al.* (1991, 1992) applied an adaptive remeshing algorithm to solve the inception and development of shear bands on both homogeneous and non-homogeneous stress fields. An *h*-adaptive finite element method used by Belytschko and Tabbara (1993) for dynamic problems, to capture the details of the shear band by refining the mesh at appropriate locations. An automatic adaptive remeshing using element elongation was proposed by Zienkiewicz *et al.* (1995) in the capturing of discontinuous solutions. Deb *et al.* (1996) developed a mesh enrichment procedure which can be adaptively refined in the shear band of localized zone for elasto-visco-plastic solids. Recently, Diez and Huerta (1999) proposed a unified approach to remeshing strategies for finite element *h*-adaptivity, based on a priori error estimates and the standard principles of finite precision computation. More recently, Khoei and Lewis (2002) presented an adaptive remeshing method for capturing the localization phenomenon in powder forming processes.

10.2. Causes of Localization in Solid Mechanics

In many physical phenomena, whose behavior is described by differential equations in a well posed manner, abrupt discontinuities in the main solution variables may develop. Such phenomena are frequently noted in solid mechanics where plastic failure or fracture can localize in surfaces on which discontinuities of stresses and displacements occur. This phenomenon was observed as early as 1972 in an early work describing a general approach to finite element solution of plasticity problems by Nayak and Zienkiewicz (1972). They experienced that even with a very coarse finite element discretization, quite narrow plastic zones occur under a punch when plastic softening is assumed. On the contrary, when the material is hardening the plasticity zone extends practically throughout the whole domain. Though not noted at the time, it became clear some years later, that material softening of the general form assumed by Nayak and Zienkiewicz, will always lead to localization with an abrupt discontinuity.

From a mechanical point of view the driving forces behind localization phenomena are material instabilities, that is, the constitutive relationship violates the stability criterion that the inner product of the stress rate $\dot{\boldsymbol{\sigma}}$ and the strain rate $\dot{\boldsymbol{\varepsilon}}$ is positive (i.e. $\dot{\boldsymbol{\varepsilon}}\,\dot{\boldsymbol{\sigma}} > 0$ Obviously, this inner product becomes negative when, in a uniaxial tension or compression test, the slope of the homogenized axial stress - axial strain curve is negative. We call this phenomenon 'strain softening'. By using the terminology 'homogenized' we refer to the fact that initial flaws and boundary conditions necessarily induce a non-homogeneous stress state in a specimen during testing. In particular during progressive failure of the specimen these flaws and local stress concentrations will cause strongly inhomogeneous deformations of the specimen. The procedure that is normally utilized to derive stress-strain relations, namely dividing the force by the virgin load-carrying area and dividing the displacement of the end of the specimen by the original length so as to obtain stress and strain respectively then no longer reflects what happens at a micro-level and loses physical significance.

In order to illustrate the phenomena for a finite element solution, a simple tension bar is shown divided into several elements and a material having plastic softening properties, as given in Figure 10.2, is assumed. It is statistically highly improbable that the maximum yield stress is uniform throughout and if this is just slightly lower in a single element, yielding will be confined to that location and all other elements will simply unload elastically. As displacements are increased, the 'average strain' (u/L) for the whole bar is shown in Figure 10.2(c) indicates an ultimate discontinuity, or brittle behavior, for small but finite values of h. Obviously the unloading characteristics shown in Figure 10.2 will not occur along the elastic path and it is readily seen that the model results in brittle behavior, i.e. an immediate drop of strain from its maximum value to zero whenever strain softening and localization occur. The capture of this discontinuity as it actually develops can be achieved approximately with finite elements using a continuous interpolation. However, by sufficient refinement, the exact solution can be approached using a continuous interpolation as closely as possible. How far the refinement should proceed is a question which is difficult to answer precisely and here we will need to revert to the notion of error tolerance.

The purpose of this chapter is to develop a numerical solution in indicating clearly the localization and failure surfaces in compacted powder components. The algorithm was firstly applied by Peraire *et al.* (1987) to compressible fluid dynamics problems to capture discontinuities such as shocks, and then developed by Pastor *et al.* (1991, 1992) and Zienkiewicz *et al.* (1995) to incompressible materials of soils to solve inception and development of shear bands. This study is concerned mainly with the manner in which the numerical discretization process has to be devised so as to capture the localization phenomenon. For this purpose, an application of the mixed u-π formulation is firstly presented for incompressible materials. A numerical solution based on the gradient of deformation is then developed in the modeling of shear band localization. An h-adaptive finite element strategy, using element elongation, is applied to indicate the failure surfaces and evaluate the ultimate capacity of material. It is investigated under which conditions the continuum theory permits localization of deformation without losing ellipticity for static problems and hyperbolicity for dynamic problems.

10.3. Governing equations of incompressible plasticity

It is well known that the standard displacement formulation fails when the Poisson's ratio ν approaches 0.5 or when the material becomes incompressible. In particular, if plasticity of an isochoric (volume-preserving) type is used, typified by classical Tresca or von-Mises models, difficulties will arise with a simple, linear triangle (3C) finite element. The main problem in the application of standard displacement formulation to incompressible or nearly incompressible problems lies in the determination of the mean stress or pressure which is related to the volumetric part of the strain. These difficulties can be avoided to some extent by using higher-order elements with reduced integration or simply by use of very high-order elements (Sloan and Randolph 1982). The most transparent is the equivalence of the displacement form with a mixed formulation involving both the displacement u and pressure π as independent variables (Zienkiewicz and Taylor 1991). It is simply found here that failure or success depends on the satisfaction of the so-called 'Babuska-Brezzi' condition by the approximation when full incompressibility occurs (Zienkiewicz *et al.* 1986).

Problems of incompressible plasticity, such as: Tresca or von-Mises material, can be presented in two-field u–π mixed form as

$$\mathbf{S}^T(\boldsymbol{\sigma}_d + \mathbf{m}\pi) + \rho\,\mathbf{b} - \rho\,\ddot{\mathbf{u}} = \mathbf{0} \tag{10.1}$$

$$\mathbf{m}^T\boldsymbol{\varepsilon} - \frac{\pi}{K} = \mathbf{0} \tag{10.2}$$

where $\mathbf{u}$ is the displacement vector, $\mathbf{b}$ is the body force acceleration, ρ is the density, $\boldsymbol{\sigma}$ is the stress and $\mathbf{S}$ is the strain operator relating displacements and strain $(\boldsymbol{\varepsilon} = \mathbf{S}\,\mathbf{u})$. $\boldsymbol{\sigma}_d$ is the deviatoric stress vector defined as $\boldsymbol{\sigma}_d = \boldsymbol{\sigma} - \mathbf{m}\pi$, the mean stress π is $\pi = \frac{1}{3}\mathbf{m}^T\boldsymbol{\sigma}$ with $\mathbf{m}$ denoting a vector which has a form of $\mathbf{m} = [1,\ 1,\ 1,\ 0,\ 0,\ 0]$ for the general three-dimensional stresses, and K is the bulk modulus of the material defined as $K = E/[3(1-2\nu)]$, where E is the elastic Young's modulus and ν is the Poisson's ratio.

In order to develop a finite element formulation for localization phenomena, we need to solve equations (10.1) and (10.2), numerically. Following the standard procedure of the finite element method, we can write the above mixed u–π formulation in the weak form as

$$\int_\Omega \delta\boldsymbol{\varepsilon}^T(\boldsymbol{\sigma}_d + \mathbf{m}\pi)d\Omega - \int_\Omega \delta\mathbf{u}^T\rho\,\mathbf{b}\,d\Omega + \int_\Omega \delta\mathbf{u}^T\rho\,\ddot{\mathbf{u}}\,d\Omega - \int_{\Gamma_t} \delta\,\mathbf{u}^T\bar{\mathbf{t}}\,d\Gamma = 0 \tag{10.3}$$

$$\int_\Omega \delta\pi^T\left(\mathbf{m}^T\boldsymbol{\varepsilon} - \frac{\pi}{K}\right)d\Omega = 0 \tag{10.4}$$

Applying the standard finite element Galerkin discretization process to equations (10.3) and (10.4), with the independent approximations of $\mathbf{u}$ and π defined as, $\mathbf{u} = \mathbf{N}_u\bar{\mathbf{u}}$ and $\pi = \mathbf{N}_\pi\bar{\boldsymbol{\pi}}$, we obtain the following algebraic equations

$$\int_\Omega \mathbf{B}^T\boldsymbol{\sigma}_d\,d\Omega + \mathbf{Q}\bar{\boldsymbol{\pi}} + \mathbf{M}\ddot{\bar{\mathbf{u}}} = \mathbf{f}_u \tag{10.5}$$

$$\mathbf{Q}^T\bar{\mathbf{u}} - \mathbf{C}\bar{\boldsymbol{\pi}} = \mathbf{f}_\pi \tag{10.6}$$

where $\mathbf{B}$ is the strain matrix relating the increments of strain and displacement (i.e. $d\boldsymbol{\varepsilon} = \mathbf{B}d\bar{\mathbf{u}}$). $\mathbf{M}$, $\mathbf{Q}$ and $\mathbf{C}$ are the mass, coupling and compressibility matrices, defined as

$$\begin{aligned} \mathbf{M} &= \int_\Omega \mathbf{N}_u^T\,\rho\,\mathbf{N}_u\,d\Omega \\ \mathbf{C} &= \int_\Omega \mathbf{N}_\pi^T\,\frac{1}{K}\,\mathbf{N}_\pi\,d\Omega \\ \mathbf{Q} &= \int_\Omega \mathbf{B}^T\,\mathbf{m}\,\mathbf{N}_\pi\,d\Omega \end{aligned} \tag{10.7}$$

and $\mathbf{f}_u$ and $\mathbf{f}_\pi$ vectors are defined as

$$\mathbf{f}_u = \int_\Omega \mathbf{N}_u^T \rho\, \mathbf{b}\, d\Omega + \int_{\Gamma_t} \mathbf{N}_u^T\, \bar{\mathbf{t}}\, d\Gamma \qquad \mathbf{f}_\pi = 0 \tag{10.8}$$

The definition of the spatial problem is complete when the constitutive law for use in the first term of equation (10.5) is defined. This term represents the internal force, and for non-linear problems, can be written as

$$\begin{aligned}\int_\Omega \mathbf{B}^T \boldsymbol{\sigma}_d\, d\Omega &= \int_\Omega \mathbf{B}^T (\boldsymbol{\sigma} - \mathbf{m}\,\pi)\, d\Omega \\ &= \int_\Omega \mathbf{B}^T (\mathbf{D_T} - \mathbf{m}\, K\, \mathbf{m}^T)\, \mathbf{B}\, \bar{\mathbf{u}}\, d\Omega = \mathbf{K_T}\bar{\mathbf{u}}\end{aligned} \tag{10.9}$$

where $\mathbf{D_T}$ is a function of the displacement. The consistent tangential stiffness matrix can be obtained by performing a full differentiation on the internal force term. Assuming rate-independent material behavior, the incremental stress-strain relations take the form $\dot{\boldsymbol{\sigma}} = \mathbf{D_T}\dot{\boldsymbol{\varepsilon}}$, with the deviatoric stress rate being defined as $\dot{\boldsymbol{\sigma}}_d = \dot{\boldsymbol{\sigma}} - \mathbf{m}\,\dot{\pi}$. In this study, we refer frequently to the 'effective' plastic strain $\bar{\varepsilon}^p$ which in a rate form may be defined as $\dot{\bar{\varepsilon}}^p = \sqrt{\frac{2}{3}\dot{e}_{ij}^p \dot{e}_{ij}^p}$, where e_{ij}^p are the deviatoric strains, e.g. $e_{ij} = \varepsilon_{ij} - \frac{1}{3}\delta_{ij}\varepsilon_{ij}$.

After the spatial discretization, the mixed formulation is now represented by two sets of ordinary differential equations (10.5) and (10.6). In order to complete the numerical solution, the differential equations (10.5) and (10.6) are integrated in time by a generalized Newmark scheme (*GNpj* method), an unconditionally stable direct solution procedure. The *GNpj* method stands for generalized Newmark p^{th} order interpolation for j^{th} order differential equation. Applying the *GN22* scheme for displacement and then the Newton-Raphson procedure for non-linear iteration, the matrix of the solution can be given by the following expression

$$\begin{bmatrix} \mathbf{M} + \mathbf{K_T}\beta\,\Delta t^2 & \mathbf{Q} \\ \mathbf{Q}^T \beta\,\Delta t & -\mathbf{C} \end{bmatrix}_{n+1}^{i} \begin{Bmatrix} \delta(\ddot{\bar{\mathbf{u}}}_{n+1}) \\ \delta(\bar{\boldsymbol{\pi}}_{n+1}) \end{Bmatrix}^{i} = -\begin{Bmatrix} \mathbf{G}_u \\ \mathbf{G}_\pi \end{Bmatrix}_{n+1}^{i} \tag{10.10}$$

where the tangential stiffness matrix $\mathbf{K_T}$ is defined in equation (10.9), and $\mathbf{G}_u$ and $\mathbf{G}_\pi$ stand for the vectors that can be evaluated from the known initial values. The parameters β is chosen in the range of 0–1, but for unconditional stability of the recurrence scheme, it is required that $\beta \geq \frac{1}{4}$ (Zienkiewicz and Taylor 1991).

With incompressible behavior it is necessary to ensure that the equations do not become singular. The permissible interpolation can be verified by the Babuska-Brezzi conditions or the patch test. The following triangular and quadrilateral elements, presented in Figure 10.3, have been used with good effect and satisfy the stability requirements (Zienkiewicz *et al.* 1989);

i) *T3C1B/3C* triangular linear continuous displacement element with a higher order bubble function and continuous linear pressure;
ii) *T6C/1D* triangular quadratic continuous displacement element with discontinuous constant pressure;
iii) *T6C/3C* triangular quadratic continuous displacement element with continuous linear pressure;
iv) *T6C1B/3D* triangular quadratic continuous displacement element with a higher order bubble function and discontinuous linear pressure;

v) *T6C1B/3C* triangular quadratic continuous displacement element with a higher order bubble function and continuous linear pressure;

vi) *Q4C/1D* four-noded quadrilateral continuous displacement element with discontinuous constant pressure;

vii) *Q9C/4D* nine-noded quadrilateral continuous displacement element with four-noded discontinuous pressure;

viii) *Q9C/4C* nine-noded quadrilateral continuous displacement element with four-noded continuous pressure.

Although the elements with continuous pressure perform best these are more expensive to use than the discontinuous pressure elements. If we use pressure discretizations that are discontinuous between elements, it is efficient to statically condense out the pressure degrees of freedom at the element level providing $C \neq 0$. In other words, we can write equation (10.10) for each element as

$$\begin{bmatrix} \mathbf{M}_e + (\mathbf{K}_\mathrm{T})_e \beta \Delta t^2 & \mathbf{Q}_e \\ \mathbf{Q}_e^T \beta \Delta t & -\mathbf{C}_e \end{bmatrix}_{n+1}^i \begin{Bmatrix} (\delta \ddot{\mathbf{u}})_e \\ (\delta \overline{\boldsymbol{\pi}})_e \end{Bmatrix}_{n+1}^i = -\begin{Bmatrix} (\mathbf{G}_u)_e \\ (\mathbf{G}_\pi)_e \end{Bmatrix}_{n+1}^i \tag{10.11}$$

and on eliminating the pressure variables, we will have

$$\left(\mathbf{M}_e + (\mathbf{K}_\mathrm{T})_e \beta \Delta t^2 + \mathbf{Q}_e \mathbf{C}_e^{-1} \mathbf{Q}_e^T \beta \Delta t\right)_{n+1}^i \left(\delta \ddot{\mathbf{u}}_e\right)_{n+1}^i = -\left((\mathbf{G}_u)_e + \mathbf{Q}_e \mathbf{C}_e^{-1} (\mathbf{G}_\pi)_e\right)_{n+1}^i \tag{10.12}$$

The equivalence of mixed methods with discontinuous pressure approximations and displacement methods employing selective reduced integration techniques is now well understood and either can be used with good results.

10.4. Theory of Cosserat continuum

Conventional continuum mechanics approaches cannot incorporate any intrinsic material length scale. However real materials often have a number of important length scales, which must be included in any realistic model. So-called nonlocal theories can be used to account for size effects in the mechanical behavior of materials. The departure from local theories begins with the Cosserat continuum. A Cosserat (or micropolar) medium is a continuous collection of particles that behave like rigid bodies. Accordingly, each material point is endowed with translational and rotational degrees of freedom, that describe its displacement and the rotation of an underlying microstructure. The material may then oppose couple stresses to the development of curvature i.e., gradient of microrotation. In what follows, the essential concepts of Cosserat continuum theory are briefly reviewed for the two-dimensional elasto-plastic solids.

10.4.1. Cosserat elasticity

During deformation, the displacements of each element is displaced by (u_x, u_y) while it is rotated by the angle ω_z around the z axis. The strain components in Cosserat continuum are defined as

$$\begin{aligned} \varepsilon_{xx} &= \partial u_x / \partial x & \varepsilon_{yy} &= \partial u_x / \partial y \\ \varepsilon_{xy} &= \partial u_y / \partial x - \omega_z & \varepsilon_{yx} &= \partial u_x / \partial y + \omega_x \end{aligned} \tag{10.13}$$

In addition, the micro-curvatures are introduced in Cosserat continuum, which do not appear in classical continuum as

$$\kappa_{xz} = \partial\omega_z / \partial x \qquad \kappa_{yz} = \partial\omega_z / \partial y \tag{10.14}$$

where micro-curvatures (κ_{xz}, κ_{yz}) are conjugate of couple stresses (m_{xz}, m_{yz}) components of stress vector. For convenience, the above equations can be rewritten as

$$\varepsilon_{ij} = u_{j,i} - e_{ijk}\omega_k \tag{10.15}$$

$$\kappa_{ij} = \omega_{j,i} \tag{10.16}$$

where e_{ijk} is the permutation symbol. The displacement vector u is defined as u = $[u_x, u_y, \omega_z]^T$ and the extended strain vector can be represented as

$$\boldsymbol{\varepsilon} = [\varepsilon_{xx}, \varepsilon_{yy}, \varepsilon_{zz}, \varepsilon_{xy}, \varepsilon_{yx}, \kappa_{xz}\ell, \kappa_{yz}\ell]^T \tag{10.17}$$

where ℓ is the internal length parameter, taken into the strain tensor to make all components dimensionless. The existence of rotational degree of freedom and couple stresses lead to unequal shear stresses and therefore, the stress tensor will not be symmetric. The stress vector can be defined as

$$\boldsymbol{\sigma} = [\sigma_{xx}, \sigma_{yy}, \sigma_{zz}, \sigma_{xy}, \sigma_{yx}, m_{xz}\ell, m_{yz}\ell]^T \tag{10.18}$$

where m denotes the couple stress, which is the conjugate of micro-curvature in the strain tensor. The stress and strain vectors can be decomposed into the symmetric and skew symmetric parts, in which the symmetric tensors $[\sigma]_{sym}$ and $[\varepsilon]_{sym}$ introduce the stress and strain tensors in classical continuum and the skew symmetric tensors $[\sigma]_{skw}$ and $[\varepsilon]_{skw}$ indicate the effects of couple stress and rotation in Cosserat continuum.

The constitutive relation for an elastic Cosserat continuum can be defined as $\boldsymbol{\sigma} = \mathbf{D}^e\boldsymbol{\varepsilon}$, where $\mathbf{D}^e$ is the linear elastic operator defined as

$$\mathbf{D}^e = \begin{bmatrix} \lambda+2\mu & \lambda & \lambda & 0 & 0 & 0 & 0 \\ \lambda & \lambda+2\mu & \lambda & 0 & 0 & 0 & 0 \\ \lambda & \lambda & \lambda+2\mu & 0 & 0 & 0 & 0 \\ 0 & 0 & 0 & \mu+\mu_c & \mu-\mu_c & 0 & 0 \\ 0 & 0 & 0 & \mu-\mu_c & \mu+\mu_c & 0 & 0 \\ 0 & 0 & 0 & 0 & 0 & 2\mu & 0 \\ 0 & 0 & 0 & 0 & 0 & 0 & 2\mu \end{bmatrix} \tag{10.19}$$

where λ and μ are the Lame constants, defined as $\lambda = E\nu/[(1+\nu)(1-2\nu)]$ and $\mu = E/[2(1+\nu)]$ and μ_c is an additional shear modulus. If μ_c takes to zero, the classic continuum will be recovered as a particular case of Cosserat continuum. The above elastic operator can be used for both two dimensional problems, i.e. plane strain and plane stress. However, for plane stress problems, zero out-of-plane stress iteration is performed outside the constitutive driver. In this way, both two dimensional problems (i.e. plane stress and plane strain) are solved using the same constitutive laws.

10.4.2. Cosserat elasto-plasticity

In order to introduce the internal length scale into the set of constitutive equations, the classical model of J_2-elastoplasticity is generalized by introducing additional degrees of freedom within the Cosserat continuum. The yield criteria can be presented in a similar manner to classical continuum as

$$F = (3J_2)^{0.5} - \overline{\sigma}(\overline{\varepsilon}^p) = 0 \tag{10.20}$$

where $\overline{\sigma}$ denotes the yield stress, which is a function of the effective plastic strain $\overline{\varepsilon}^p$. In equation (10.20), J_2 is the second invariant of deviatoric stresses defined as

$$J_2 = \frac{1}{4} S_{ij} S_{ij} + \frac{1}{4} S_{ij} S_{ji} + \frac{1}{2} m_{ij} m_{ij} / \ell^2 \tag{10.21}$$

where S_{ij} is the deviatoric stress tensor defined as $S_{ij} = [\sigma_{ij}]_{sym} - \frac{1}{3} I_1 \delta_{ij}$, with δ_{ij} denoting the Kronecker delta and $I_1 = [\sigma_{ii}]_{sym}$. Equation (10.21) can be rewritten in a matrix format as

$$J_2 = \frac{1}{2} \boldsymbol{\sigma}^T \mathbf{P} \boldsymbol{\sigma} \tag{10.22}$$

where **P** is defined as

$$\mathbf{P} = \begin{bmatrix} 2/3 & -1/3 & -1/3 & 0 & 0 & 0 & 0 \\ -1/3 & 2/3 & -1/3 & 0 & 0 & 0 & 0 \\ -1/3 & -1/3 & 2/3 & 0 & 0 & 0 & 0 \\ 0 & 0 & 0 & 1/2 & 1/2 & 0 & 0 \\ 0 & 0 & 0 & 1/2 & 1/2 & 0 & 0 \\ 0 & 0 & 0 & 0 & 0 & 1 & 0 \\ 0 & 0 & 0 & 0 & 0 & 0 & 1 \end{bmatrix} \tag{10.23}$$

As in classical plasticity, the plastic strain is related to yield criteria as follow

$$\varepsilon_{ij}^p = \lambda \frac{\partial F}{\partial \sigma_{ij}} \tag{10.24}$$

where λ is the plastic multiplier. Substituting equation (10.20) into (10.24), the plastic strain will be obtained as

$$\boldsymbol{\varepsilon}^p = \frac{3\lambda}{2\overline{\sigma}(\overline{\varepsilon}^p)} \mathbf{P} \boldsymbol{\sigma} \tag{10.25}$$

The effective plastic strain, defined in equation (10.20), can be evaluated by

$$\overline{\varepsilon}^p = \left[\frac{1}{3} e_{ij}^p e_{ij}^p + \frac{1}{3} e_{ij}^p e_{ji}^p + \frac{2}{3} \kappa_{ij}^p \kappa_{ij}^p \ell^2 \right]^{1/2} \tag{10.26}$$

where e_{ij}^p is the plastic deviatoric strain tensor. Using matrix **P**, equation (10.26) can be rewritten in the following form

$$\bar{\varepsilon}^p = \left[\frac{2}{3}(\mathbf{e}^p)^T \mathbf{P}\, \mathbf{e}^p\right]^{1/2} \tag{10.27}$$

10.4.3. Computational algorithm

The algorithm presented here is completely similar to classical continuum algorithm when von-Mises yield criteria is adopted. The trial stress at the beginning of the new step can be calculated as

$$\boldsymbol{\sigma}_t = \boldsymbol{\sigma}_0 + \mathbf{D}^e \Delta\boldsymbol{\varepsilon} \tag{10.28}$$

In above equation, it is assumed that the stress increment is related to strain increment through the elastic material property matrix (i.e. $\Delta\boldsymbol{\sigma} = \mathbf{D}^e \Delta\boldsymbol{\varepsilon}^e$). The new stress state at the end of the increment is the sum of the stress at the beginning of the step and the stress increment, i.e.

$$\boldsymbol{\sigma}_p = \boldsymbol{\sigma}_0 + \Delta\boldsymbol{\sigma} \tag{10.29}$$

Substituting equations (10.25) and (10.28) into (10.29), we obtain the following explicit expression

$$\boldsymbol{\sigma}_p = \boldsymbol{\sigma}_t - \frac{3\Delta\lambda}{2\bar{\sigma}(\bar{\varepsilon}^p)} \mathbf{D}^e \mathbf{P} \boldsymbol{\sigma}_p \tag{10.30}$$

or in the simpler form as

$$\boldsymbol{\sigma}_p = \left(\mathbf{I} + \frac{3\mu\Delta\lambda}{\bar{\sigma}(\bar{\varepsilon}^p)} \mathbf{P}\right)^{-1} \boldsymbol{\sigma}_t \tag{10.31}$$

in which the only unknown parameter in equation (10.31) is $\Delta\lambda$.

If plasticity occurs (i.e. $F(\boldsymbol{\sigma}_t, \bar{\varepsilon}_0^p) > 0$), a correction for plastic flow must be applied. To this end the yield condition $F(\boldsymbol{\sigma}_n, \bar{\varepsilon}_n) = 0$, where the subscript n denotes the value of a quantity after correction for plastic flow, is developed in a truncated Taylor series around $(\boldsymbol{\sigma}_t, \bar{\varepsilon}_0^p)$ as (de Borst and Muhlhaus 1992)

$$F(\boldsymbol{\sigma}_t, \bar{\varepsilon}_0^p) = \Delta\lambda\left(H + \frac{\partial F^T}{\partial \boldsymbol{\sigma}} \mathbf{D}^e \frac{\partial F}{\partial \boldsymbol{\sigma}}\right) \tag{10.32}$$

where H is the hardening parameter and can be defined similar to that of the classic continuum as

$$H = \frac{\partial \bar{\sigma}}{\partial \bar{\varepsilon}_0^p} \tag{10.33}$$

Substituting $\partial F / \partial \boldsymbol{\sigma}$ from equations (10.24) and (10.25) into (10.32), we obtain an explicit expression for the plastic multiplier $\Delta\lambda$ as

$$\Delta\lambda = \frac{F(\boldsymbol{\sigma}_t, \bar{\varepsilon}_0^p)}{H + 3\mu} \tag{10.34}$$

With expression (10.34) for the magnitude of the plastic strain increment, the computation for elasto-plasticity in Cosserat continuum is completed in analogy to a conventional plasticity model. The plastic strain increment in each step can be determined from equations (10.25) and (10.34) and then, the corrected value of stress can be obtained from equations (10.28) and (10.31).

In order to linearize the set of tangential moduli, a consistent tangent operator is introduced for Cosserat continuum model. By differentiating equation (10.30) and using relation (10.33), we will obtain

$$\dot{\boldsymbol{\varepsilon}} = \widehat{\mathbf{D}}_T^{-1} \dot{\boldsymbol{\sigma}} \tag{10.35}$$

where $\widehat{\mathbf{D}}_T$ is the consistent tangent operator for a Cosserat continuum defined as

$$\widehat{\mathbf{D}}_T = \mathbf{K} - \frac{\left(\frac{3\bar{\sigma}}{2}\right)^2 \mathbf{K}(\mathbf{P}\boldsymbol{\sigma})(\mathbf{P}\boldsymbol{\sigma})^T \mathbf{K}}{\left(\frac{H\bar{\sigma}}{\bar{\sigma} - \Delta\lambda}\right) + \left(\frac{3\bar{\sigma}}{2}\right)^2 (\mathbf{P}\boldsymbol{\sigma})^T \mathbf{K}(\mathbf{P}\boldsymbol{\sigma})} \tag{10.36}$$

where $\mathbf{K}$ is given by

$$\mathbf{K} = (\mathbf{D}^e)^{-1} + \frac{3\Delta\lambda}{2\bar{\sigma}} \mathbf{P} \tag{10.37}$$

10.5. Adaptive strategy for discontinuous displacements

Adaptive mesh refinement in strain softening problems has received important attention in last two decades. The presence of two well-differentiated length scales in such problems seems to indicate that adaptive remeshing strategies, in a general sense, are the natural approach. Recall that the spatial interpolation of the primitive variables must describe both the macroscopic scale associated to the solid geometry and the micro-scale related to the shear band.

Adaptivity in finite element computations requires three main ingredients. The first one is an algorithm for increasing/decreasing the richness of the interpolation in a particular area of the computational domain. Second, an error estimator or error indicator must be employed to locate where there is a need for refinement/de-refinement. And third, a remeshing criterion must be used to translate the output of the error analysis into the input of the mesh generator, for instance, the distribution of desired mesh sizes. These three steps are fundamental in adaptivity. In strain localization problems, the difficulties that have deterred an extensive application of adaptive methods are the path dependent constitutive equations, and the error estimation relying, in statics, on the ellipticity of the equations, which is lost at the inception of localization. Here, these two difficulties are overcome by employing;

a) an error estimator mathematically sound for path dependent constitutive equations, and
b) a well known regularized model which precludes the loss of ellipticity.

The main goal is to show that adaptive mesh refinement based on estimating the actual error is now possible for regularized problems. Any model maintaining ellipticity after the inception of the localization could be used. Moreover, the numerical examples show that adaptivity based on the error estimation is essential for accurate computations and, in certain problems, for capturing a realistic physical behavior.

The use of h-adaptivity generates a sequence of meshes that are expected to converge to a proper mesh. Strategies based on h-adaptivity consist on building a new mesh, using the same type of elements, and 'adapting' the element size to the requirements of the solution. That is, reducing their size where the interpolation must be enriched (i.e. more accuracy is needed) and enlarging the elements where it is already accurate enough. The number of degrees of freedom is increased (or decreased) in every zone of the computational domain in order to obtain some prescribed accuracy. An error indicator is needed to design a new mesh verifying the accuracy prescriptions. In fact, adaptive computations can be viewed as an iterative procedure that generates a convergent sequence of approximations to the exact solution. It is important to notice that convergence is a key issue in adaptive computations that must be ensured in every step of the scheme.

Here, the main interest is the design of finite element grids which are able to resolve sharp shear bands. It has to be remarked that location of the band is not known a priori, and therefore it is necessary to begin with an initial mesh which provides a first solution which is used to obtain an improved mesh. After few iterations the position of the shear band is accurately determined. For this purpose, the mesh corresponding to iteration j is firstly used to obtain a solution of the problem. This mesh is characterized by a set of mesh parameters $\mathbf{P}^{(j)}$, describing element size, aspect ratio, *etc.* Once the new solution is obtained, it is possible to define a new $\mathbf{P}^{(j+1)}$ characterizing an improved mesh $(j+1)$. Of course, the determination of $\mathbf{P}^{(j+1)}$ depends on the objective searched by the analyst.

10.5.1. Error indicator

The error assessment can be performed using error estimators or error indicators. While error estimators provide objective error measures in a suitable error norm, error indicators use heuristic criteria, mainly based on interpolation error estimates, to obtain qualitative information about the error. Error indicators are usually explicitly computed, they may be used to identify the zones where the mesh must be finer and they are cheaper than error estimators, which typically require to solve a set of local problems. However, the information given by error indicators cannot be used as an objective error assessment tool because the selection of the error indicator is very sensitive to the analyzed problem. In discontinuity problems, the available energy norm based error estimators strongly rely on the ellipticity of the problem thus preventing their extension to strain localization, which is accompanied by the loss of ellipticity. In fact, for the analysis of localized failure, there does not appear to exist a global error norm similar to the energy norm in self-adjoint elliptic problems due to the loss of ellipticity. Therefore, attention must be focused on the error occurring in the values of the variables of interest in each individual element.

Failure of elasto-plastic solids is very frequently accompanied by development of surfaces or bands at which high gradients of displacement are present. Thus, we refine the element size h in a manner ensuring that

$$h\left|\frac{\partial \varphi}{\partial x}\right| = C \tag{10.38}$$

where C is a predetermined constant and φ is a scalar variable of the absolute displacement scalar, defined as $\varphi=\left(\mathbf{u}^T\mathbf{u}\right)^{1/2}$, where $\mathbf{u}$ is the displacement vector.

In a two-dimensional context the refinement criterion of equation (10.38) provides a useful indicator, though not an error measure, and leads to a successful capture of discontinuities. Of course now, the maximum gradient of the scalar function φ is used and h refers to the minimum element size. In general, in the vicinity of the discontinuity the behavior is nearly one-dimensional and elements can therefore be elongated, as shown in Figure 10.4. The indicator defined by (10.38) can be interpreted as the maximum value of the first term of the Taylor expansion defining the local error of scalar quantity φ. By ensuring that the mesh is generated so that the quantity C is constant throughout all elements, we achieve a solution which captures well all local discontinuities and which is efficient in achieving the progression which gives overall accuracy. The use of above indicator allows element elongation to be included in the refinement, as on many occasions the feature occurring at high gradients is almost one-dimensional.

In order to measure and indicate the error occurring in the values of displacements in each individual element, we shall estimate its magnitude based on the maximum gradient of absolute displacements, defined in equation (10.38), as

$$e=\varphi-\varphi^h \le \hat{e}=ch\left|\frac{\partial \varphi^h}{\partial \hat{x}}\right|_{\max} \tag{10.39}$$

where φ and φ^h are the absolute values of displacements for the exact and finite element solutions, $\hat{x}$ is any direction chosen, h is the element size, c is a positive constant and the displacement gradient is computed from the numerical solution.

In order to indicate the amount of elongation, the curvatures will be considered by two orthogonal directions, maximum and minimum values, in which the element will be elongated in the direction of the minimum curvature, as shown in Figure 10.5. Thus, if in any element the principal second derivatives and their directions can be found, then the element could be stretched with sizes

$$s=\frac{h_{\max}}{h_{\min}}=\left(\left|\frac{\partial^2 \varphi^h}{\partial \hat{x}^2}\right|_{\max} \Big/ \left|\frac{\partial^2 \varphi^h}{\partial \hat{x}^2}\right|_{\min}\right)^{1/2} \tag{10.40}$$

10.5.2. Adaptive mesh refinement

In practical applications we shall try to devise a procedure in which the error indicator in each element is reduced to, or below, a specified value. For this purpose, a solution including displacements and its first and second derivatives will be obtained from an initial mesh. In order to project the values of derivatives from the Gauss points to the nodal points, the method of 'superconvergent patch recovery' (SPR) described in Chapter 6 is employed here. In this case, the smoothed nodal point values will be obtained for the first and second derivatives.

In order to compute the maximum gradients of displacement at each nodal point, the relative values of displacement are calculated using the sampling nodes surrounding the particular nodal point, as shown in Figure 10.6. The absolute values of displacements are obtained by $\varphi=\left(\overline{\mathbf{u}}^T\overline{\mathbf{u}}\right)^{1/2}$, in which $\overline{\mathbf{u}}=\overline{\mathbf{u}}_i-\overline{\mathbf{u}}_o$. The directions of maximum and minimum curvatures can be computed using the gradients of the first derivatives, as shown in Figure 10.7, by

$$\left(\frac{\partial^2 \varphi}{\partial \zeta^2}\right)_{ii'} = \frac{\theta_i - \theta_{i'}}{(\ell_i - \ell_{i'})/2} \tag{10.41}$$

where

$$\begin{aligned} \theta_i &= \left(\frac{\partial \varphi}{\partial \zeta}\right)_i = \frac{\varphi_i}{\ell_i} \\ \theta_{i'} &= \left(\frac{\partial \varphi}{\partial \zeta}\right)_{i'} = \frac{\varphi_{i'}}{\ell_{i'}} \end{aligned} \tag{10.42}$$

As the constant c in equation (10.39) is not known a priori and the specification of maximum levels for the error indicators $\hat{e}$ is not easy, the maximum value of r will be computed in the whole domain by

$$r = h_{old} \left|\frac{\partial \varphi^h}{\partial \hat{x}}\right|_{\max} \tag{10.43}$$

The new mesh size will be obtained by choosing an appropriate value for r in the new mesh. The new value of r can be considered to be proportional to the absolute maximum value of it in the old mesh as

$$r = \beta\, r_{\max} \tag{10.44}$$

and thus

$$h_{new} = (\beta\, r_{\max}) \Big/ \left(\left|\frac{\partial \varphi^h}{\partial \hat{x}}\right|_{\max}\right) \tag{10.45}$$

where β is a prescribed value suggested to be between 0.1–0.4 (Zienkiewicz *et al.* 1995). After indicating the minimum size of elements from equation (10.45) and determining the elongation s and its direction by expression (10.40), a mesh satisfying the requirements will be finally generated by an efficient mesh generator which allows the new mesh to be constructed according to a predetermined size and elongation distribution. For convenience, the algorithm of adaptive mesh refinement is summarized in Table 10.1.

10.6. Numerical Simulation Results

In preceding sections, a method has been presented for dealing with softening plasticity to study the prediction of localization phenomenon in plasticity problems. The finite element method was applied to the mixed u–π formulation for the spatial discretization, a generalized Newmark scheme was used for the time domain discretization and finally, a Newton Raphson procedure was employed to solve the nonlinear equations. As the classical continuum suffer from excessive mesh-dependence in the analysis of strain softening plasticity, a higher order continuum model based on Cosserat theory was developed and the internal length parameter was included in the model. Adaptive strategy using element elongation was applied to compute the distribution of required element size using the estimated error distribution. Once the new element size and its alignment were indicated, an automated procedure was used to construct the mesh according to a predetermined size and elongation distribution.

In order to demonstrate a part of the wide range of problems that can be solved by the proposed approaches, we have illustrated, the effective performance of the elongated elements using the mixed u–π formulation and Cosserat continuum model on the plastic flow in localization analysis. A serious problem with adaptive analysis of non-linear problems of plasticity in which the results are path dependent is that of data transfer between the various stages of analysis. In principle, the control of the error should be achieved at each load increment separately and this, of course, necessitates the transfer of history dependent data such as stresses, strains, *etc.* from the mesh of the previous step to that used in the next increment. To avoid difficulties we have re-analysed the problem from the start of loading for the adapted mesh.

10.6.1. Vertical cut

The first example is chosen to demonstrate that if a correct approximation is used then both the uniform and non-uniform mesh refinements will converge to the correct answer and clearly indicate the localization phenomenon. It will be shown that the robust elements give good answers even with 'bad' meshes in which the original mesh was not suitably aligned. A vertical cut with a footing resting on its crest, as shown in Figure 10.8, is analyzed for two meshes; 'bad mesh' and a 'lucky mesh'. Material parameters pertinent to the von-Mises yield criterion, along with the strain softening case, are used. The material properties chosen for the analyses are; $E = 2.1 \times 10^5$ N/m^2, $\nu = 0.49$, $\sigma_y = 86.6$ N/m^2. The plastic hardening/softening modulus, H, is -5000 N/m^2 for softening plasticity. The rigid and rough footings were assumed to be elastic, with an elastic modulus one hundred times higher than that of the soil. The problems have been solved with displacement control by increasing the footing settlement δ in an incremental manner.

The effective plastic strain contours and the deformed mesh using the standard displacement formulation (equation 10.1), and applying the triangular quadratic continuous displacement elements *T6C* for the 'lucky' and 'bad' meshes, are shown in Figure 10.9 for softening plasticity $H=-5000$. Obviously, *T6C* displacement-based elements fail to predict the localization behavior of incompressible plastic materials, when the 'bad mesh' model is used. In Figure 10.10, the results of the effective plastic strain contours and final deformed mesh using the mixed u–π formulation (equations 10.1 and 10.2) and applying the triangular quadratic continuous displacement elements with the triangular continuous linear pressure elements *T6C/3C* are presented for both the 'lucky mesh' and the 'bad mesh'. It can be observed that both the uniform and non-uniform mesh refinements converge to the correct answer. In these results, we experienced occasional failures if the original mesh was not suitably aligned. However, it must be emphasized that the 'lucky' and 'bad' meshes with different mesh alignment result in a different width of shear band, because of the fact that alignment is not the only consideration.

In order to attain a definite width of shear band in numerical analysis of strain localization, the vertical cut is analyzed by introducing the internal length parameter within Cosserat continuum. The internal length parameter chosen for the analyses is $\ell = 3$ mm. The strategy of adaptive remeshing, discussed in Section 10.5, was performed in this example starting from an initial uniform mesh, as shown in Figure 10.11. Also plotted in this figure are the adapted mesh refinements at two different stages. The same value of r was used in each remeshing, with this value being determined at the first adaptive step. In this analysis the mesh was adaptively refined with elements not only being reduced in size near the displacement discontinuity, but also stretched in the direction of this discontinuity, which is indicated by the material deformation pattern. It can be observed that the adaptive strategy gives a clear picture of the failure even starting with a coarse mesh.

In Figure 10.12, the effective plastic strain contours are presented for the initial and two adapted meshes. In order to compare the performance of various meshes, the reaction of foundation is obtained for different mesh refinements. The foundation reaction versus displacement plots for softening plasticity corresponding to initial and adapted meshes are shown in Figure 10.13. It can be observed that the adapted meshes improve significantly the results and the reactions become close between different meshes. This result demonstrates the capability of the method for predicting localized failure due to a non-robust plasticity analysis in an automatic manner.

10.6.2. Strip in Tension

The second example is of a plane strain strip with strain softening Cosserat plasticity, as shown in Figure 10.14. The strip is restrained at the bottom edge in the vertical direction and a uniform vertical displacement is imposed on the upper nodes while the other two degrees of freedom are set free. In order to avoid the homogeneous solution, the shaded areas of the bar is taken as the weak inclusion, as shown in Figure 10.14. If the localization condition reaches, this part will be the first part that triggers localization. The Cosserat strain softening plasticity is used with material parameters given in this figure. The internal length parameter is taken equal to 3 mm. The adaptive mesh strategy is performed in the plane strain strip starting from an initial uniform mesh and the adapted meshes are obtained at two different stages, as shown in Figure 10.15. It can be observed that the adaptive strategy gives a clear picture of the failure. In Figure 10.16, the effective plastic strain contours are presented for initial and two adapted meshes. Clearly, a definite width of shear band can be observed for the adaptive mesh refinements.

10.6.3. Plate with a Hole

In the next example, a plate is considered with a hole in its center subjecting to uniaxial tension. On the virtue of symmetry, the plate is analysed for one quarter of specimen, as shown in Figure 10.17. The Cosserat material parameters chosen for the analyses are as follows; $E = 2.1 \times 10^8$ N/mm^2, $\nu = 0.3$, $\sigma_y = 86.6$ N/m^2 and $\ell = 3$ mm. The plastic hardening/softening modulus, H, is zero for ideal plasticity. The initial mesh is refined twice and the adapted meshes are shown in Figure 10.18. It can be seen from the figure that the elements in the vicinity of shear band are elongated in the direction of displacement discontinuity. The effective plastic strain contours for three different meshes are shown in Figure 10.19. Obviously, the adaptive mesh strategy using the Cosserat continuum captures the shear band with a finite thickness independent of finite element mesh.

10.6.4. Slope and foundation

In next two examples, the evolution of the shear band in a slope and foundation with Cosserat plasticity is investigated and the effective performance of the h-adaptive strategy for predicting the localized failure is demonstrated. The Cosserat material parameters for both examples are as follows; $E = 2.1 \times 10^8$ N/mm^2, $\nu = 0.49$, $\sigma_y = 86.6$ N/m^2 and $\ell = 3$ mm. The plastic hardening/softening modulus, H, is zero for ideal plasticity.

The geometry and boundary conditions of the slope are shown in Figure 10.20. The adaptive remeshing has been performed twice for this example. The initial and adapted meshes are shown in Figure 10.21. Obviously, a clear picture of localization can be observed by successive performing of the mesh refinement process in an automatic manner. The results display that the adapted meshes improve significantly the shear band zone.

In Figure 10.22, a foundation soil under central loading, exerted by a rigid strip footing is analyzed numerically. The geometry and boundary conditions of the foundation are shown in this figure. The adaptive mesh strategy is performed three times for this example. The initial and adapted meshes are shown in Figure 10.23. This example shows that the *h*-adaptive mesh refinement using Cosserat continuum captures effectively the shear band and failure surface in an automatic manner.

10.6.5. A compacted multi-level component

In last example, we have illustrated the effective performance of the elongated elements in localization analysis of a compacted powder component. In Chapters 5 and 6, the compaction process of a rotational-flanged component was modeled numerically by an axisymmetric representation, as shown in Figure 5.25. The whole compaction process was performed by a top punch movement of 3.44 mm at the first stage, then a bottom punch movement of 7.70 mm at the second stage, and finally a further top punch movement of 2.62 mm in the third stage of compaction. In this study, we intend to evaluate the ultimate capacity of the new compacted materials by an analysis of the failure. The proposed flanged component, after final stage of compaction where the top and bottom punch movements are d_t = 6.06 and d_b = 7.70 mm, is numerically analyzed, as shown in Figure 10.24. At this stage, the new compacted powder behaves as an incompressible material. Material parameters pertinent to the von-Mises yield criterion, along with the strain softening case, are used. The material properties chosen for the analyses are as follows; $E = 2.1 \times 10^5$ N/m^2, $\nu = 0.49$, σ_y = 86.6 N/m^2 and H = −5000 N/m^2. The rigid and rough punches were assumed to be elastic, with an elastic modulus one hundred times higher than that of the compacted powder.

In Figure 10.25, the effective plastic strain contours at different displacement are presented for softening plasticity ($H = -5000$). As expected the prediction of localization behavior of incompressible plastic material for a uniform mesh using T6C/3C is almost identical. Adaptive remeshing was performed in this example starting from an initial uniform mesh at different displacement as shown in Figure 10.26. The same value of r was used in each remeshing - with this value being determined at the first adaptive step. In Figure 10.27, the final displacement contours together with their vectors are plotted for different adapted meshes. In order to compare the performance of various meshes in predicting limit loads, the bottom punch reaction is obtained for the initial and adapted meshes. In Figures 10.28 and 10.29, the variation with displacement of the reaction of the bottom punch for two values of the plastic hardening/softening modulus, i.e. (H = 0) for ideal plasticity and (H = −5000) for softening plasticity, are plotted. Obviously, all uniform and adapted meshes gave similar accuracy, as illustrated in Figure 10.28, when ideal plasticity ($H = 0$) is used.

The bottom punch reaction versus displacement plots for softening plasticity (H = −5000) corresponding to the uniform mesh, as well as adapted meshes are shown in Figure 10.29. It can be observed that although the uniform mesh using T6C/3C elements improves the predictions of the localization zone and limiting loads as already discussed, the softening response of this mesh and adapted meshes shows large differences. In this figure, we have found that the adapted meshes can improve significantly the results and the reactions become close between different meshes. The strategy of adaptive remeshing was performed in this example using $\bar{r} = r/2$ for the adapted mesh 7, as shown in Figure 10.30. This result demonstrates the capability of the method for predicting localized failure due to a non-robust plasticity analysis in an automatic manner.

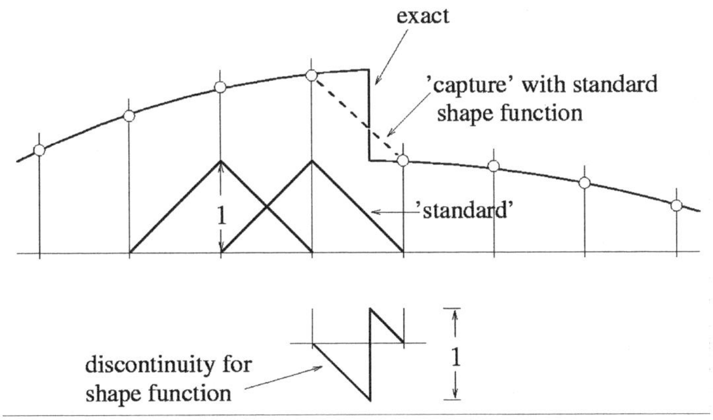

Figure 10.1. Discontinuous one-dimensional phenomenon 'captured' with standard shape function and fitted by additional discontinuous functions

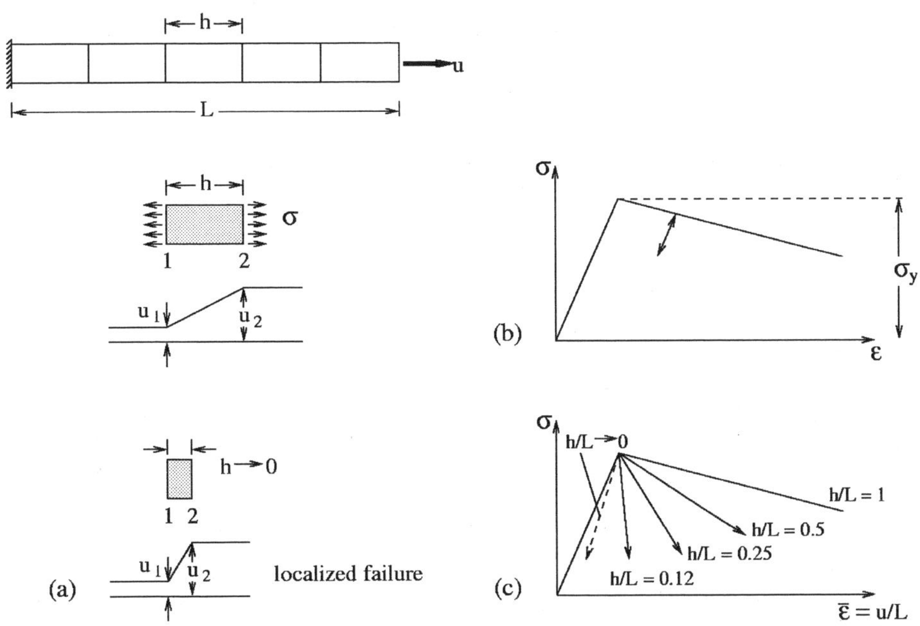

Figure 10.2. A simple tension bar with a strain softening material; a) Discrete failure as $h \to 0$, b) Stress σ versus strain ε for material, c) Stress σ versus average strain $\bar{\varepsilon} = u/L$ assuming yielding in a single element of length h

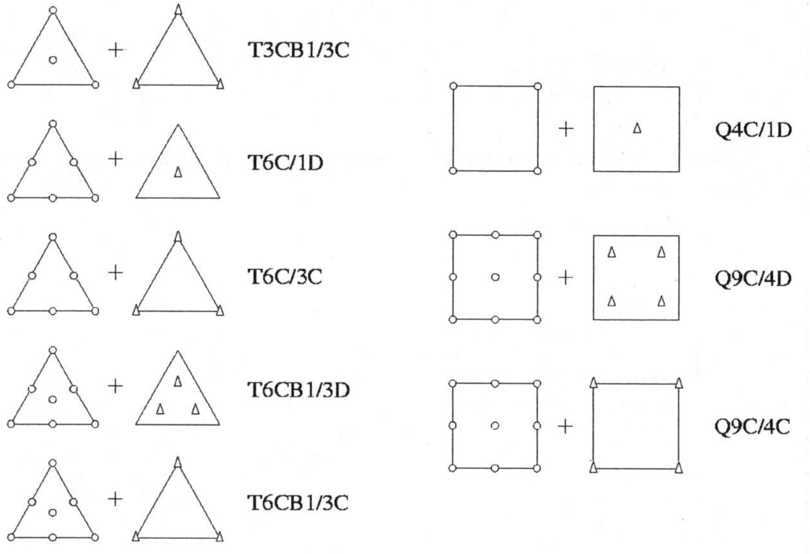

Figure 10.3. Mixed elements for incompressibility; (Δ) pressure variables, (O) displacement variables, (T) triangle, (Q) quadrilateral, (D) discontinuous, (C) continuous, (B) bubble function

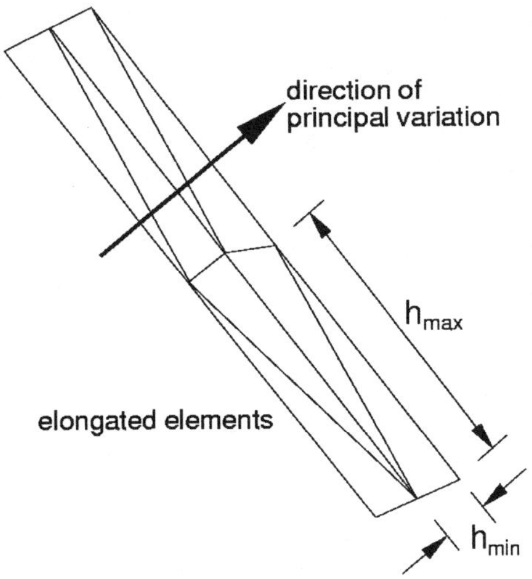

Figure 10.4. Elongation of element to model nearly 1D behavior

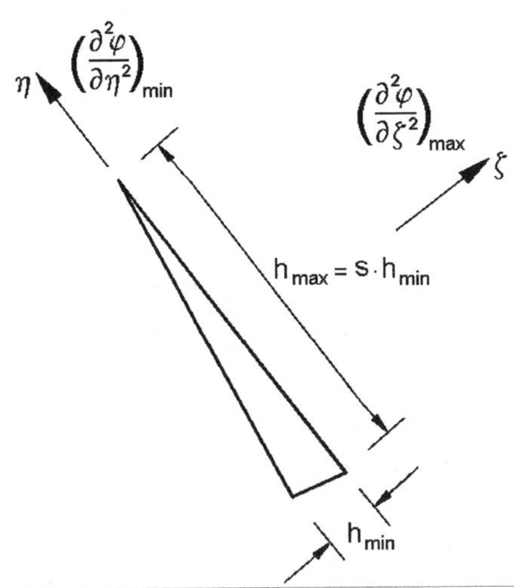

Figure 10.5. The directions of maximum and minimum curvatures for an elongated element

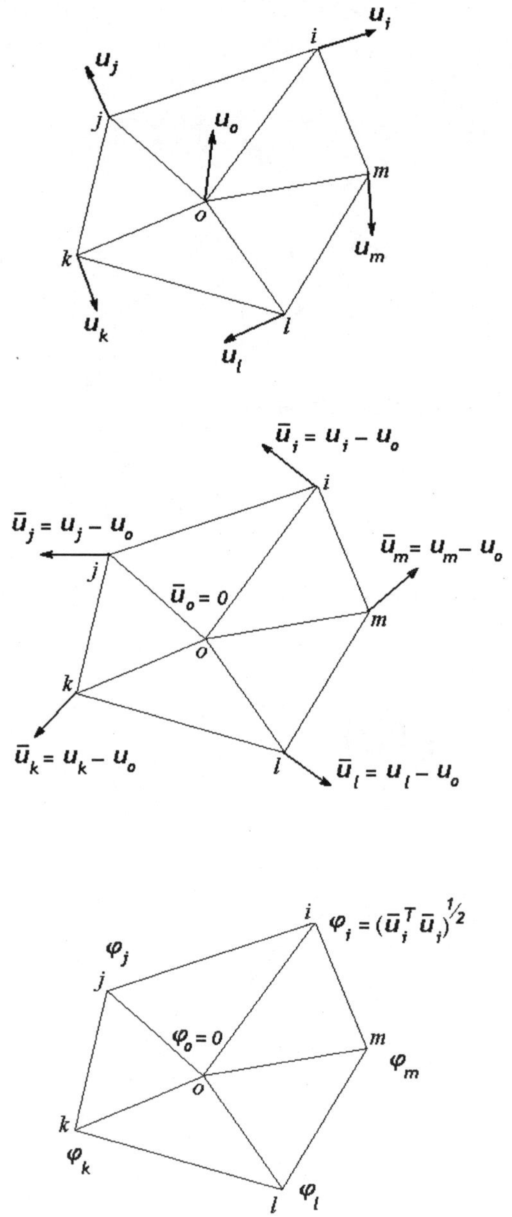

Figure 10.6. Computation of the maximum gradient of displacement at each nodal point

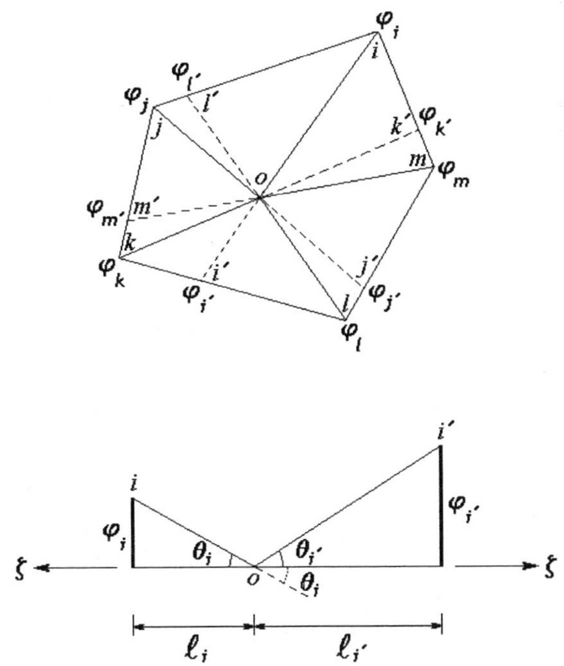

Figure 10.7. Computation of the second derivatives of displacement at each nodal point

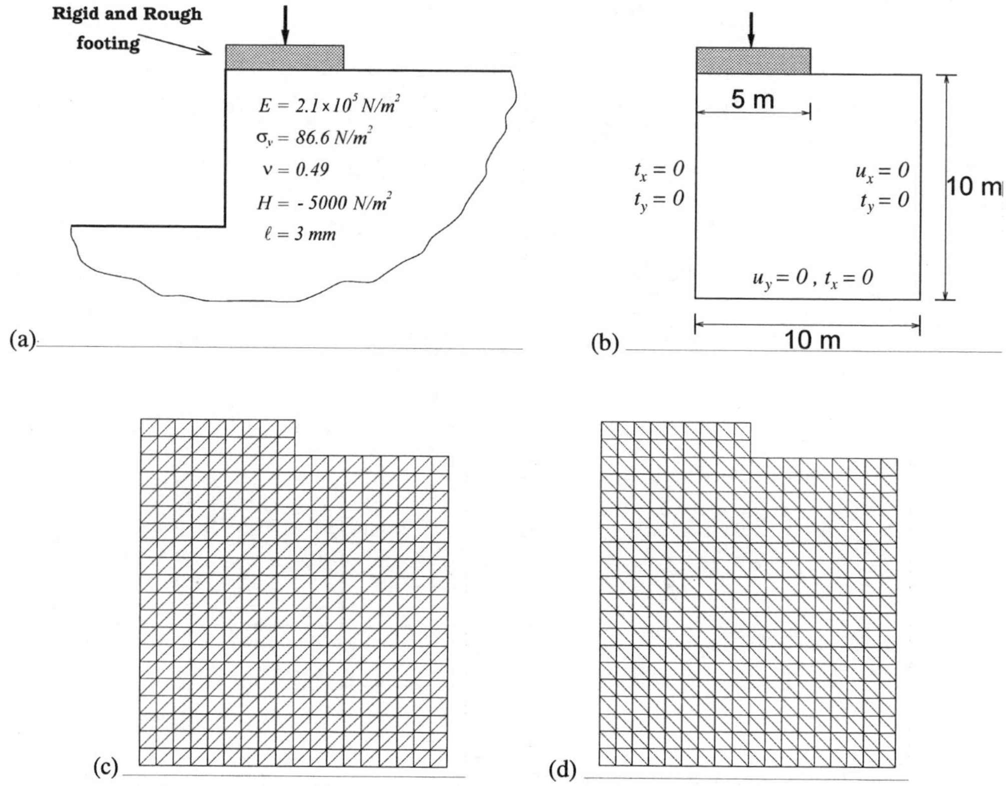

Figure 10.8. A vertical-cut; a) Problem description and elasto-plastic softening material, b) Geometry and boundary conditions, c) 'lucky' mesh, d) 'bad' mesh

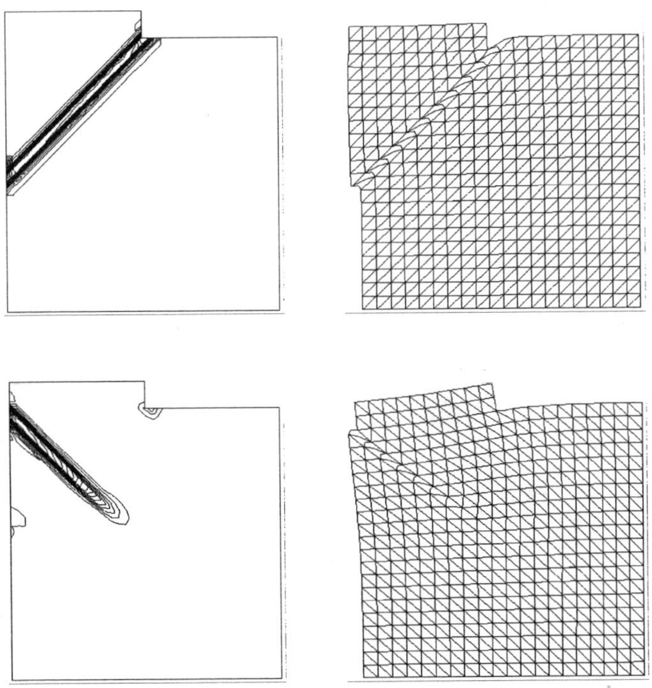

(a)

(b)

Figure 10.9. The effective plastic strain contours and deformed meshes for a vertical-cut using *T6C* elements ($H = -5000$); a) 'lucky' mesh, b) 'bad' mesh

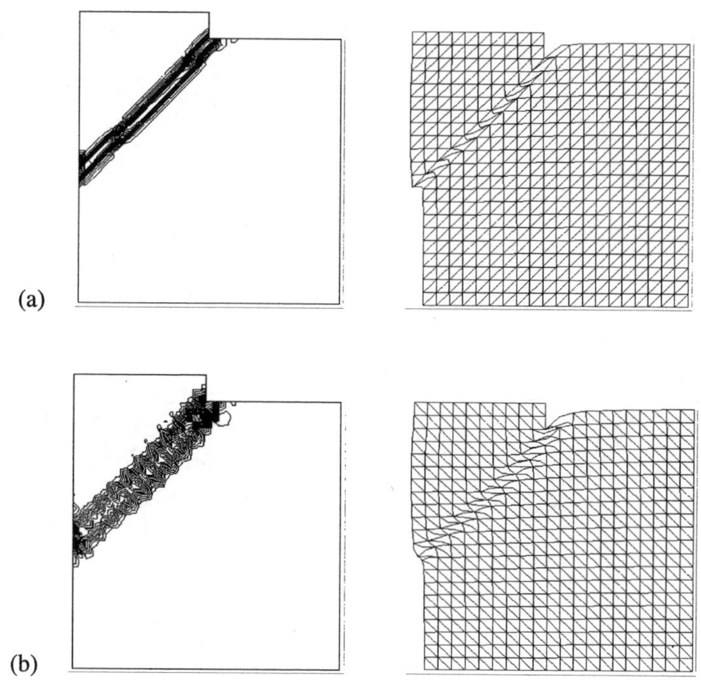

Figure 10.10. The effective plastic strain contours and deformed meshes for a vertical-cut using *T6C/3C* elements ($H = -5000$); a) 'lucky' mesh, b) 'bad' mesh

(a)

(b)

(c)

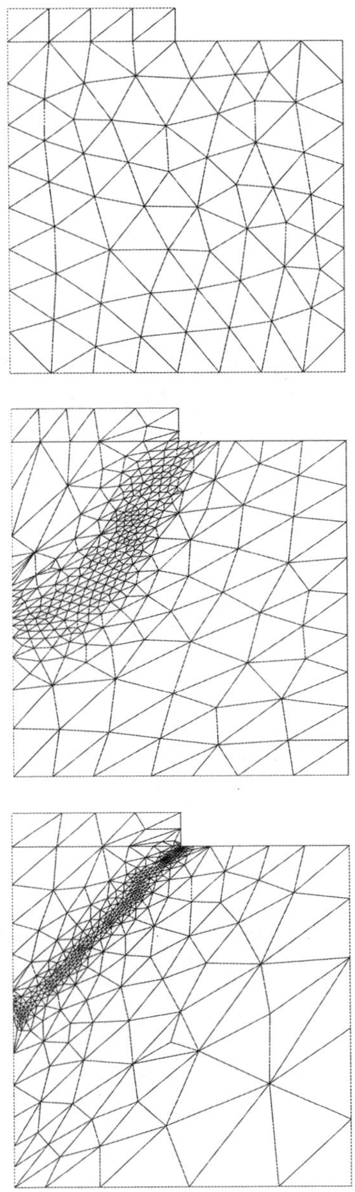

Figure 10.11. Adaptive remeshing for a vertical-cut from initial mesh at $\delta = 12$ mm; a) Initial mesh (mesh 1), b) First adaptive remeshing (mesh 2), c) Second adaptive remeshing (mesh 3)

Figure 10.12. A vertical-cut; The effective plastic strain contours for initial and adapted meshes at $\delta = 25$ mm (see colour plate section at beginning of book)

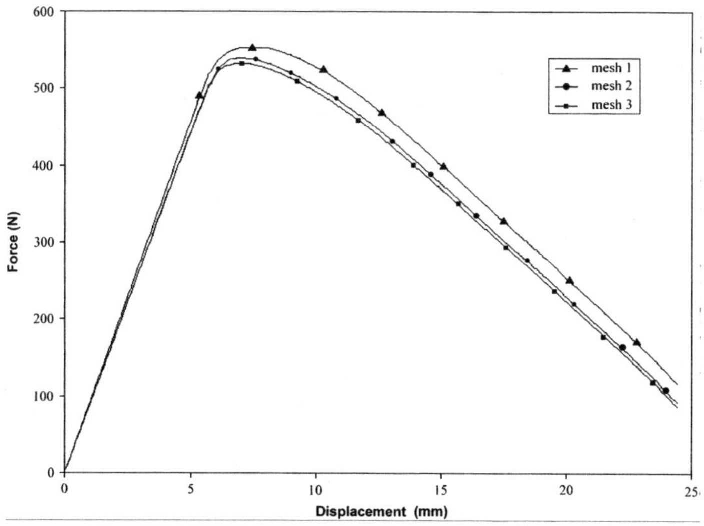

Figure 10.13. A vertical-cut; Comparison of foundation reaction versus prescribed displacement curves using various meshes

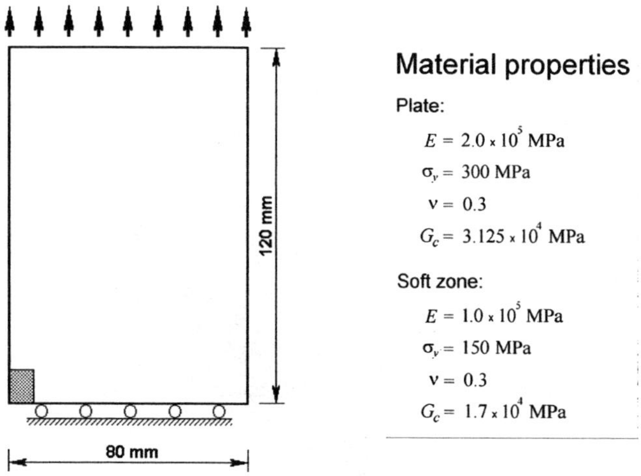

Figure 10.14. A plane strain strip in tension; Geometry, boundary conditions and Cosserat material properties

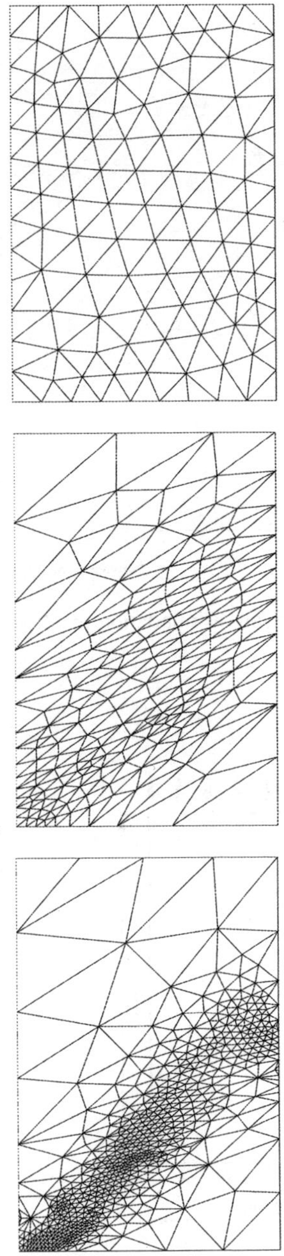

(a)

(b)

(c)

Figure 10.15. Adaptive remeshing for a strip in tension from initial mesh at δ = 30 mm; a) Initial mesh (mesh 1), b) First adaptive remeshing (mesh 2), c) Second adaptive remeshing (mesh 3)

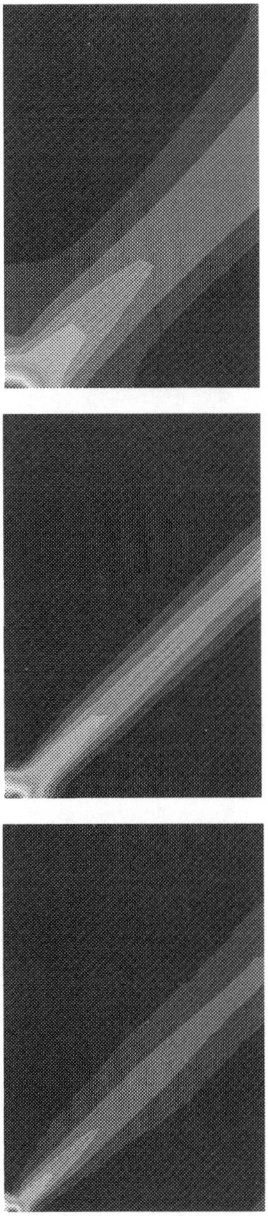

Figure 10.16. A strip in tension; The effective plastic strain contours for initial and adapted meshes at $\delta = 30$ mm. (see colour plate section at beginning of book)

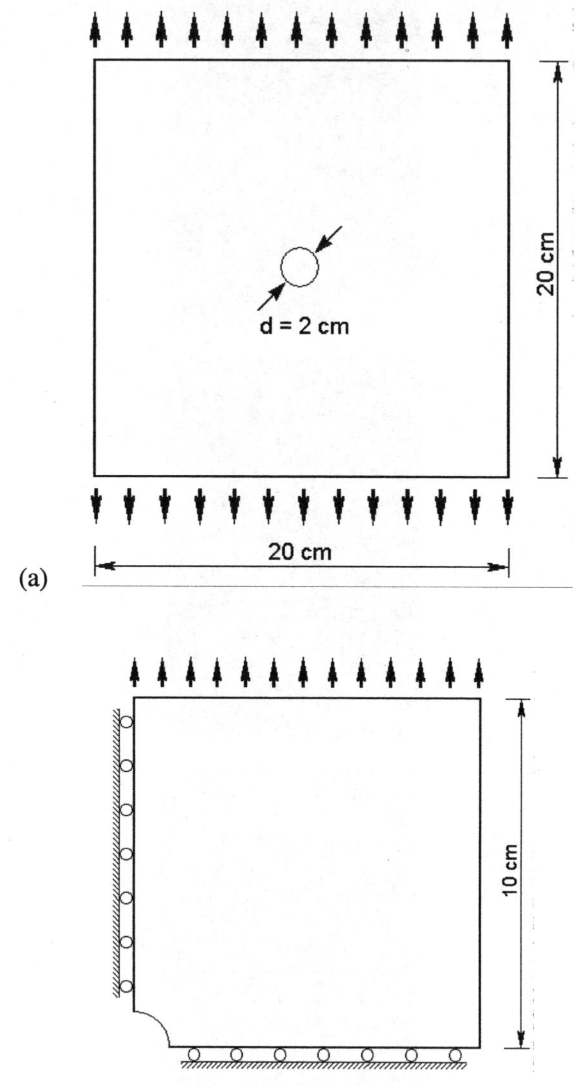

Figure 10.17. A plate with a hole subjecting to uniaxial tension; a) Geometry and boundary conditions, b) One quarter of speciemen

(a)

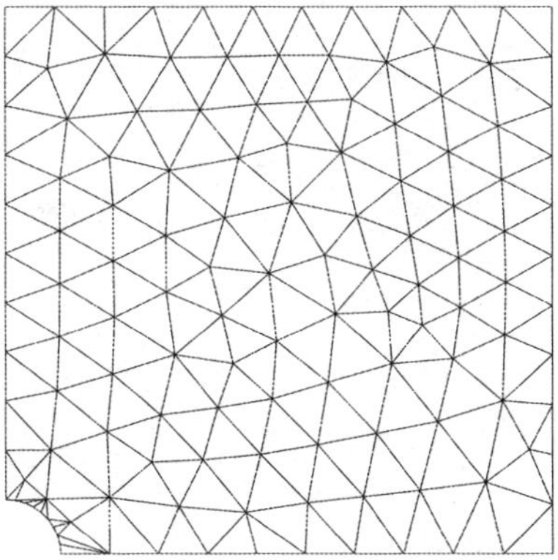

(b)

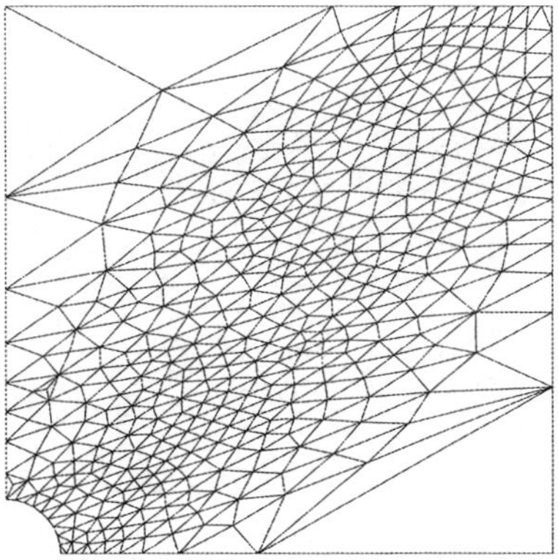

(c)

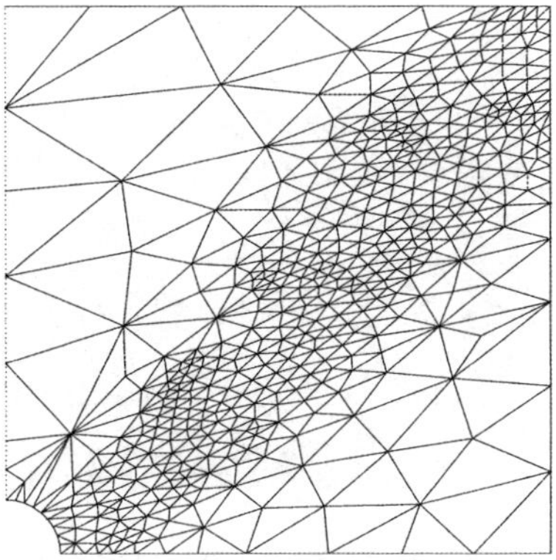

Figure 10.18. Adaptive mesh refinements for one quarter of plate from initial mesh at $\delta = 40$ mm; a) Initial mesh (mesh 1), b) First adaptive remeshing (mesh 2), c) Second adaptive remeshing (mesh 3)

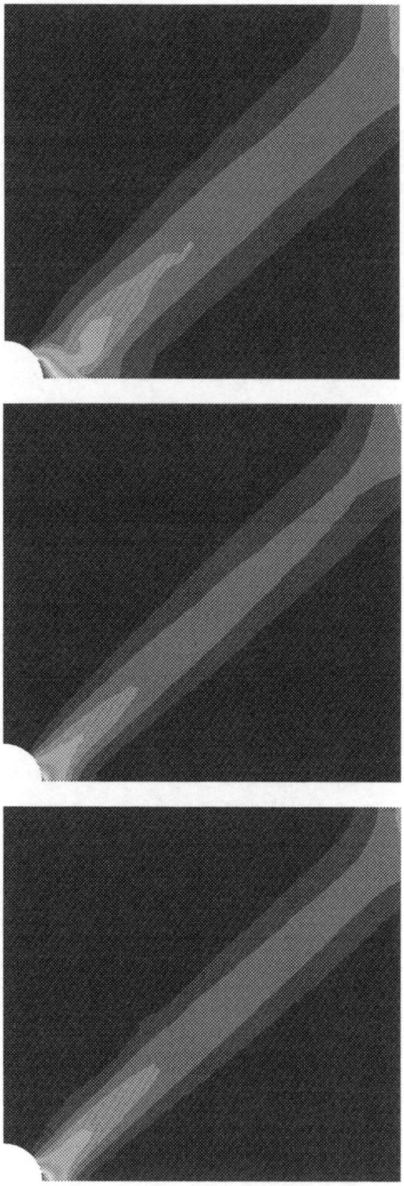

Figure 10.19. A plate with a hole; The effective plastic strain contours for initial and adapted meshes at $\delta = 40$ mm (see colour plate section at beginning of book)

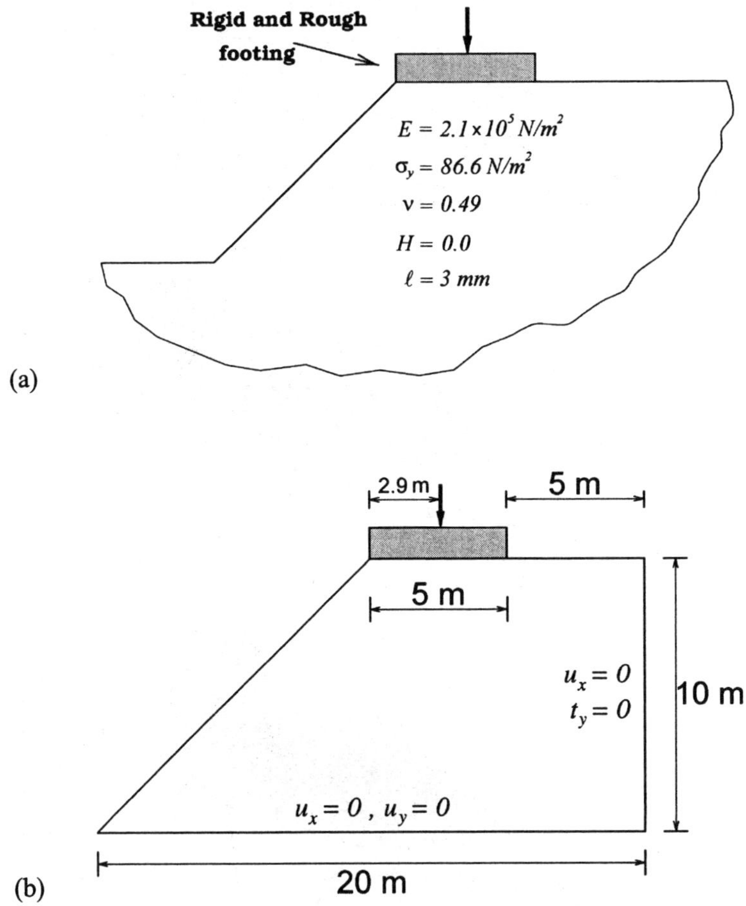

Figure 10.20. A plane strain slope; a) Problem description and elasto-plastic Cosserat material properties, b) Geometry and boundary conditions

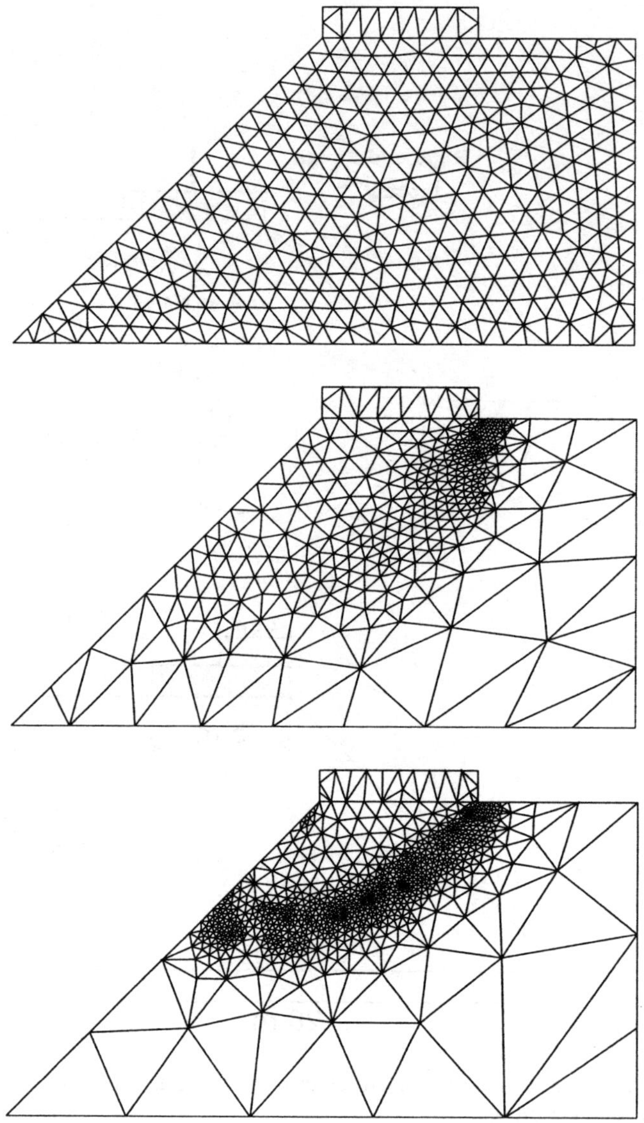

(a)

(b)

(c)

Figure 10.21. Adaptive mesh refinements for a slope from initial mesh; a) Initial mesh, b) First adaptive remeshing at $\delta = 50$ mm, c) Second adaptive remeshing at $\delta = 80$ mm

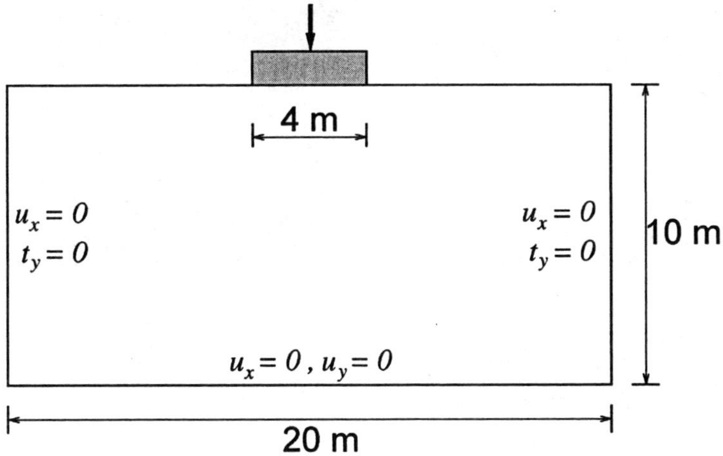

Figure 10.22. A foundation soil; Geometry and boundary conditions

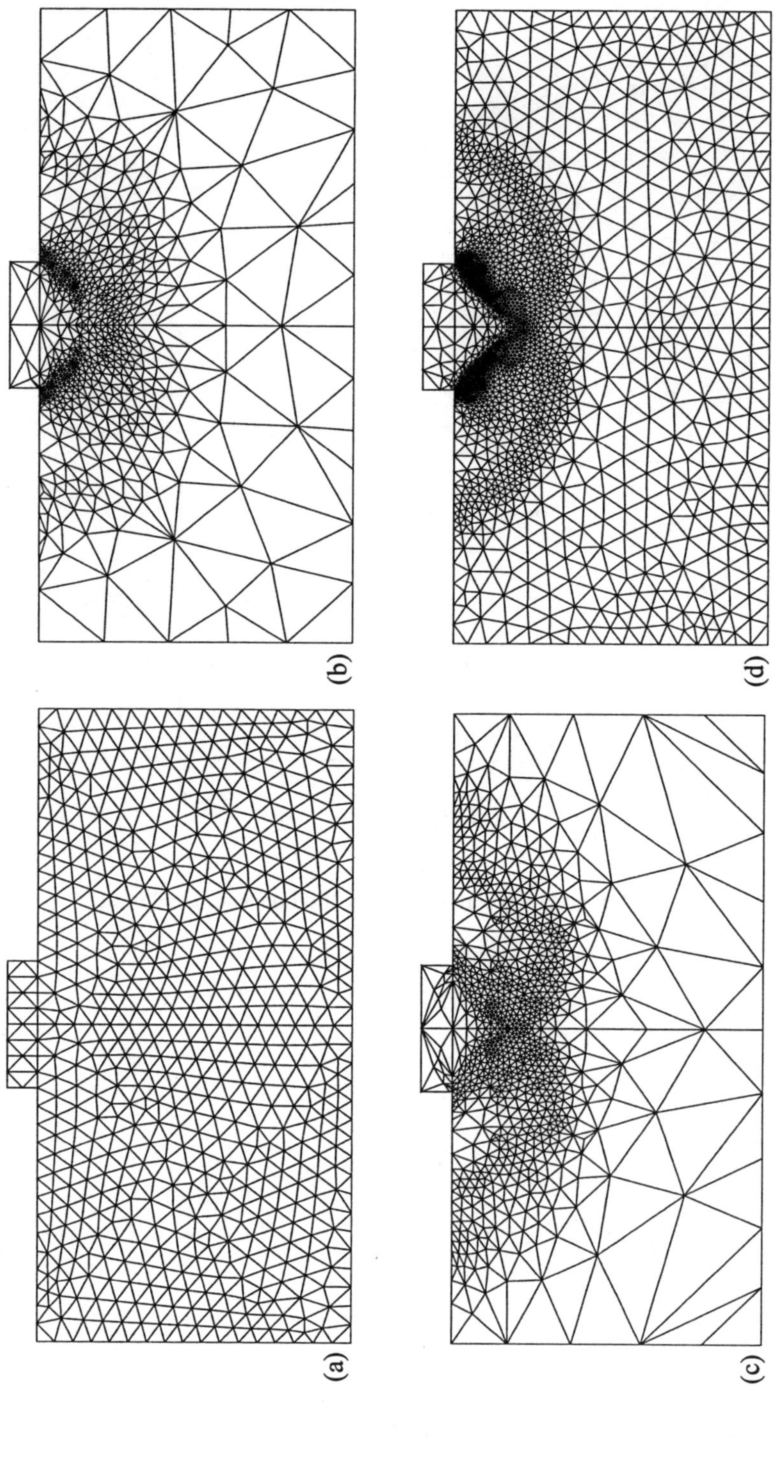

Figure 10.23. Adaptive mesh refinements for a foundation soil from initial mesh; a) Initial mesh, b) First adaptive remeshing at $\delta = 100$ mm, c) Second adaptive remeshing at $\delta = 120$ mm, d) Third adaptive remeshing at $\delta = 140$ mm

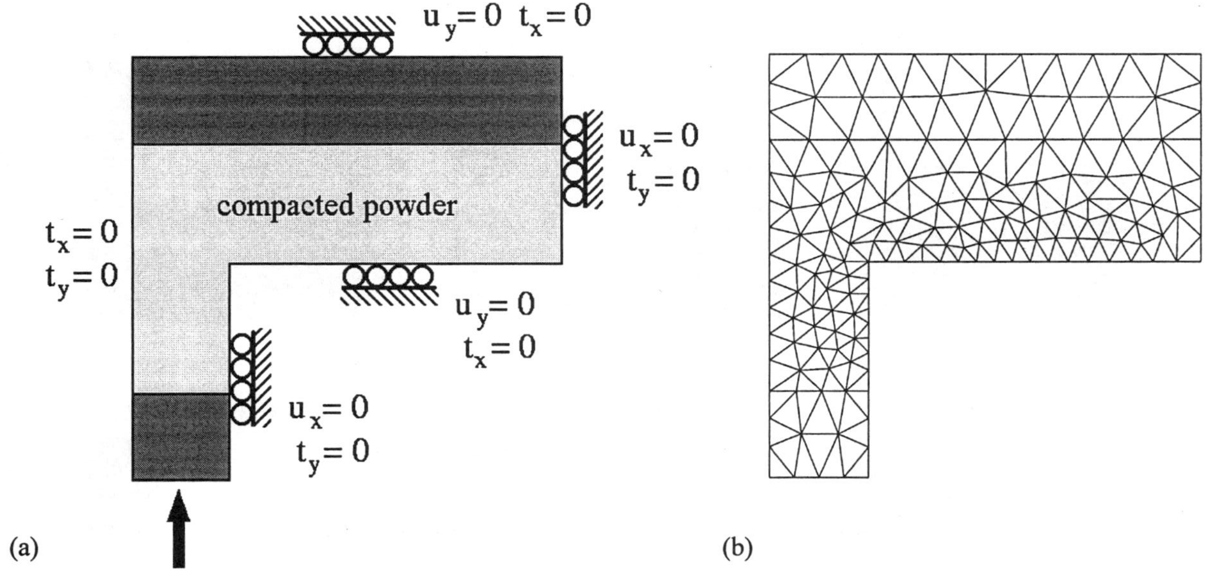

Figure 10.24. A compacted flanged component; a) Material properties and boundary conditions, b) Finite element mesh using T6C/3C elements (initial mesh 1)

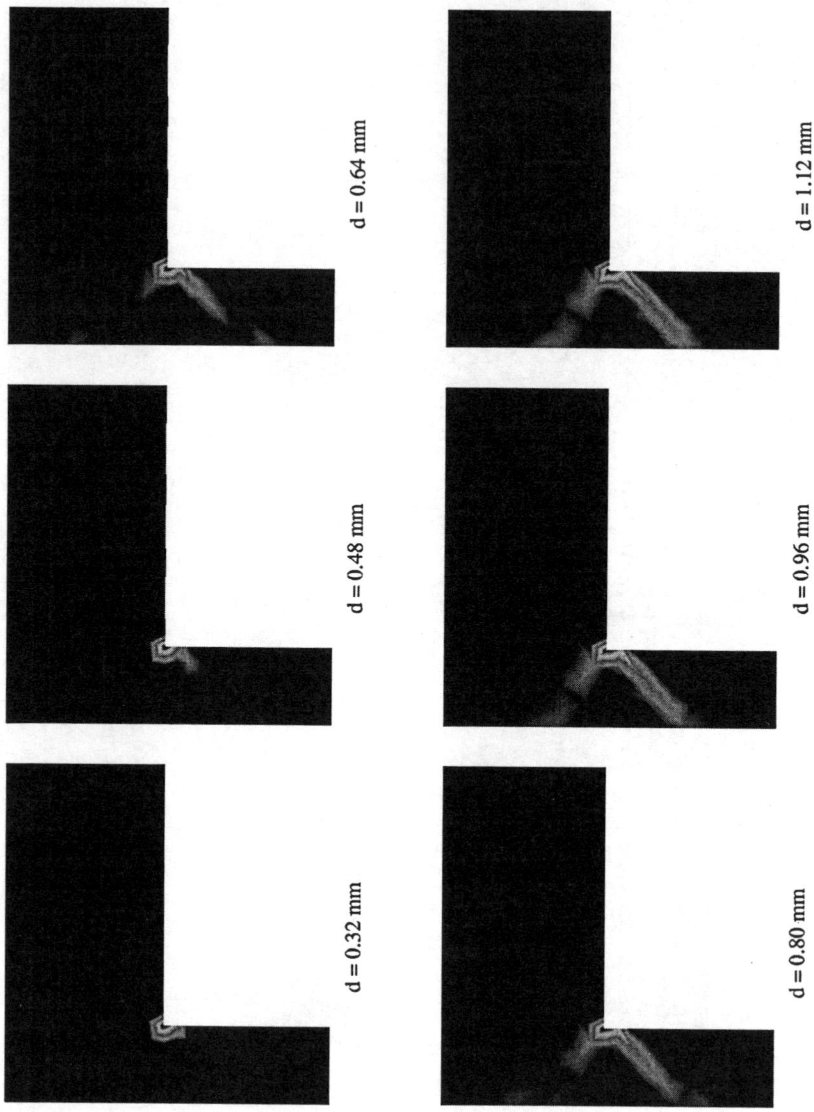

Figure 10.25. Effective plastic strain contours at different displacements using T6C/3C elements (H = -5000) (see colour plate section at beginning of book)

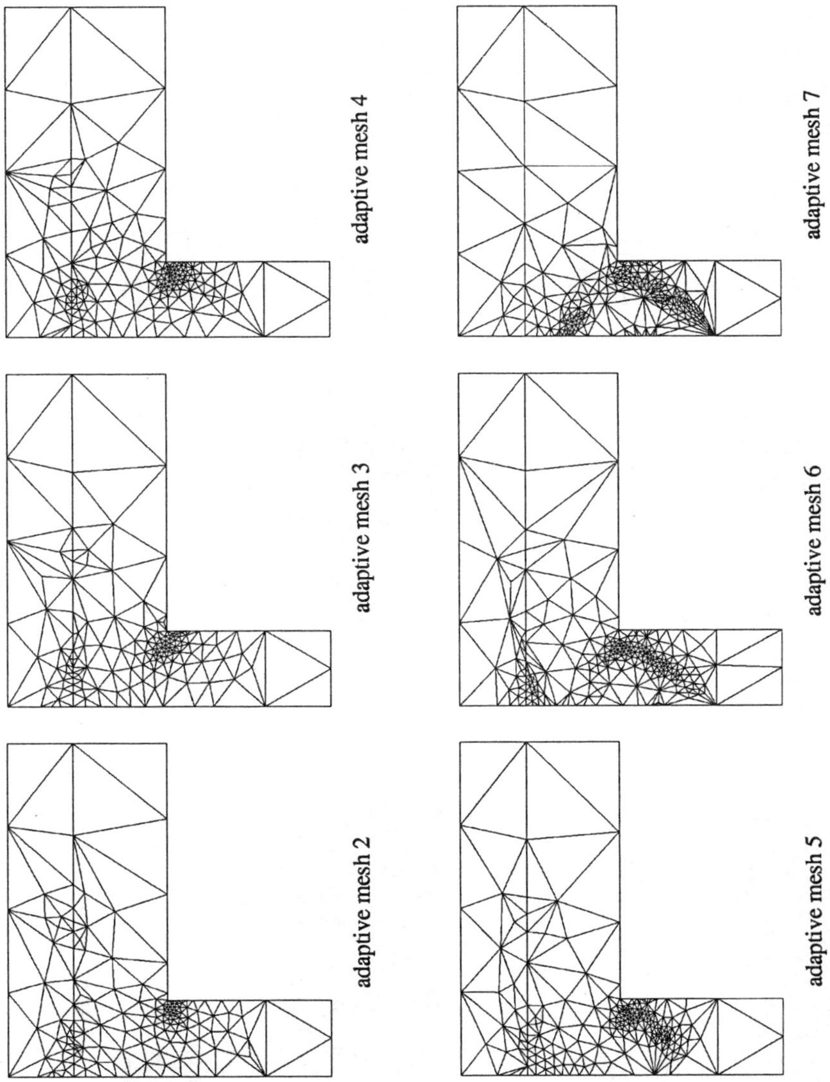

Figure 10.26. Adaptive remeshing from initial mesh 1 at different displacements (H = -5000)

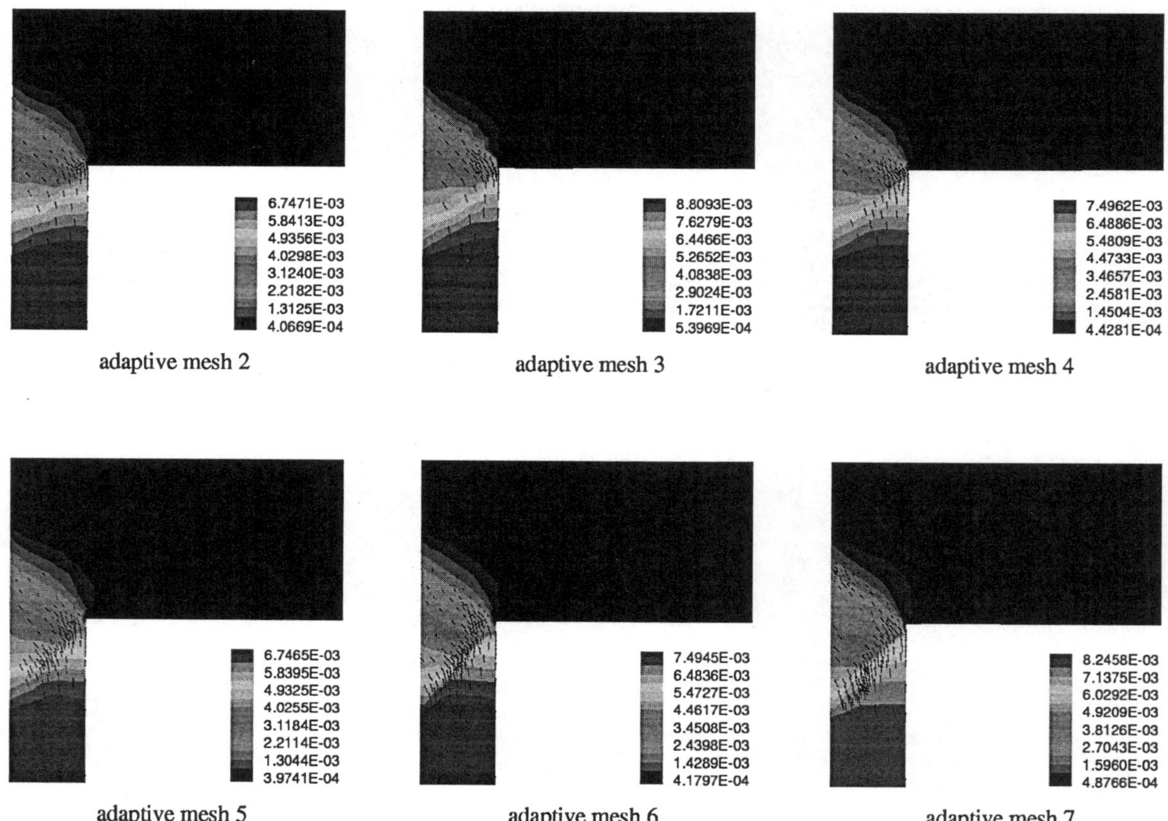

Figure 10.27. Final displacement contours along with their vectors for different adapted meshes ($H = -5000$) (see colour plate section at beginning of book)

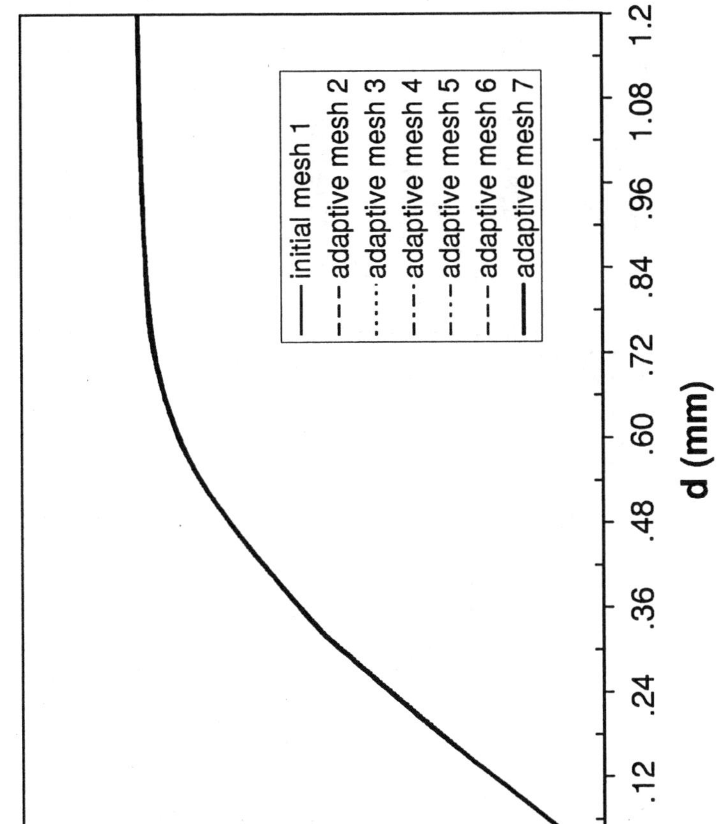

Figure 10.28. Comparison of bottom punch reaction versus prescribed displacement curves using various meshes for ideal plasticity ($H = 0$, $\nu = 0.49$)

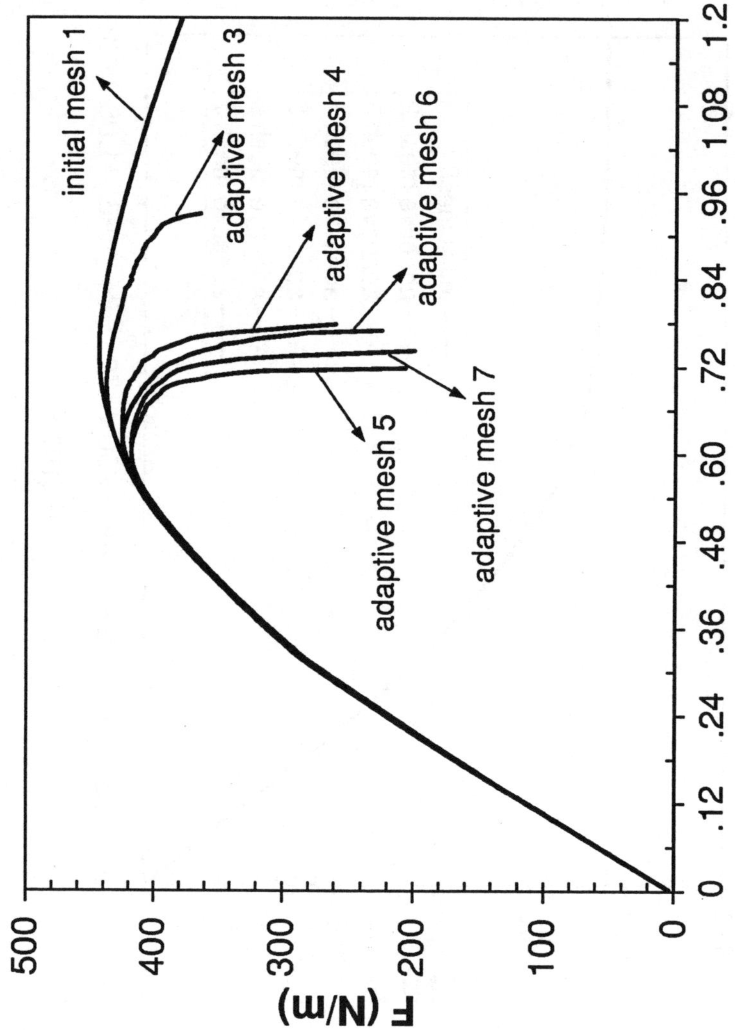

Figure 10.29. Comparison of bottom punch reaction versus prescribed displacement curves using various meshes for softening plasticity ($H = -5000$, $\nu = 0.49$)

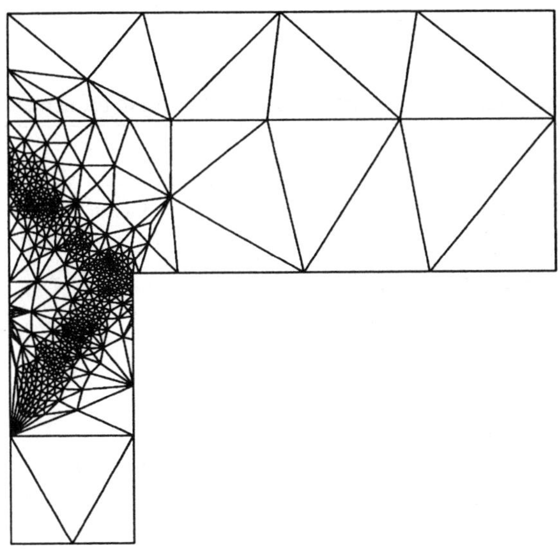

Figure 10.30. Adaptive remeshing from initial mesh 1 at d = 1.12 *mm* using $\bar{r}$ = $r/2$ for the adapted mesh 7

Table 10.1. The algorithm for adaptive mesh refinement

i) FE analysis using an initial mesh

ii) A solution including displacements and its first and second derivatives

iii) The maximum value of r in the whole domain of the old mesh, by

$$r = h_{old} \left| \partial u^h / \partial \bar{x} \right|_{max}$$

iv) The value of r in the new mesh

$$r = \beta\, r_{max}\,, \qquad 0.1 < \beta < 0.4$$

v) The new element size

$$h_{new} = \beta\, r_{max} \Big/ \left| \partial u^h / \partial \bar{x} \right|_{max}$$

vi) The elongation of element

$$s = \left\{ \left| \partial^2 u^h / \partial \bar{x}^2 \right|_{max} \Big/ \left| \partial^2 u^h / \partial \bar{x}^2 \right|_{min} \right\}^{1/2}$$

11 POWDER COMPACTION SOFTWARE

As computer simulation increasingly supports engineering design and manufacture, the requirement for a computer software environment providing an integration platform for computational engineering software increases. A key component of an integrated environment is the use of computational engineering to assist and support solutions for complex design.

As product complexity increases and economic constraints result in a demand for greater efficiency, industry must invert in new innovative ways to assist design and manufacture. As a first step in the long-term development of such a system, a computer software environment has been developed for pre- and post-processing for unstructured grid-based computational simulation. This chapter describes the powder compaction software (PCS_SUT), which is designed for pre- and post-processing for computational simulation of the process compaction of powder. Pre-processing software is used to create the model, generate an appropriate finite element grid, apply the appropriate boundary conditions, and view the total model. Post-processing provides visualization of the computed results. This chapter presents details of the environment and includes an example of the integration of application software.

11.1. Introduction

The finite element method has become a very important analysis tool in many industries as a result of the increase in computer power and the progress in numerical computation technologies. It provides a powerful procedure to mathematically model physical phenomena. The technique is numerically formulated and is effectively used on a broad range of computers. The method has increased in both popularity and functionality with the development of user-friendly pre- and post-processing software. Effective use of the method was enhanced with the introduction of low cost computer graphics systems that provided the means to both automate the generation of the mathematical model and visually display both the model and the computed results. The procedure used to automate the generation of the numerical definition of the model is named finite element pre-processing and that used to visualize the results is named post-processing.

Computer simulation methods for structural, flow and thermal analysis are well developed and have been used in design for many years. Many software packages are now available which provide an advanced capability. However, they are not designed for modeling of powder forming processes. In powder compaction simulation, there are several requirements need to be considered (Khoei and Lewis 1998, 1999, 2001, Khoei 2002, Khoei *et al.* 2002–2004). Firstly, as the compaction process involves a very large reduction in volume, the formulation adopted must be capable of representing this physical process. Secondly, as powder is a frictional and compressible material, the densification is dependent on the compaction stage and the material model needs to reflect the frictional and yielding characteristics of the powder. Thirdly, during compaction, powders exhibit strain or work hardening, the volume reduces and the material becomes harder. In this case, the yield function needs to be a moving surface dependent on the deviatoric and hydrostatic stress state. Finally, the involvement of two different materials, which have contact and relative movement in relation to each other must be considered.

A general framework for the finite element simulation of powder forming processes was presented in preceeding chapters. The behavior of the powders was assumed to be that of a rate-independent elasto-plastic material. The process was described by a large displacement finite element formulation, based on a total and updated Lagrangian formulation, for the spatial discretization. Although the raw material, a very fine powder, resembles a fluid, which could suggest that an Eulerian approach be adopted, it was observed that a Lagrangian approach has several computational advantages. In fact, compressibility of the material and, consequently, evolution of the density play a fundamental role in the process and cannot be neglected. In a Lagrangian description mass conservation can be stated locally in a closed form, thus not enlarging the number of unknowns of the discretized system of equations to be solved. However, in an Eulerian description mass conservation is locally stated through the continuity equations which have to be numerically integrated adding up to a new non-linear equation (due to convective terms) to the discretized system of equations that rule the problem.

During compaction, powders exhibit strain or work hardening, the volume reduces and the material becomes harder. In this case, the yield function was considered to be a moving surface which is dependent on the deviatoric and hydrostatic stress state. It was illustrated why the use of a yield criterion for a porous material is not suitable for loose metal powder and is more applicable to pre-compacted, sintered powder or secondary compaction processes and therefore, the adoption of a model which represents a frictional granular material is more applicable. A combination of a Mohr-Coulomb and elliptical cap model, a double-surface plasticity, and a generalized three-invariants plasticity models were utilized as a constitutive model to describe the nonlinear behavior of powder materials in the concept of the generalized plasticity formulation for the description of cyclic loading. A hardening rule was used to define the dependence of the yield surface on the degree of plastic straining. The involvement of two different materials, which have contact and relative movement in relation to each other was considered by a plasticity theory of friction, in the context of an 'interface' element formulation. The constitutive relations for friction were derived from a Coulomb friction law.

An adaptive FEM strategy into the large displacement finite element formulation of powder forming processes was utilized by employing an error estimator and applying an adaptive mesh refinement procedure. A-posteriori Zienkiewicz-Zhu estimator using the L_2 norm of strain by a recovery procedure was proposed. An *h*-refinement adaptive procedure using unstructured meshes was employed for the computation of the new element size. A set of complex powder forming simulations was demonstrated to illustrate the ability of the proposed adaptive finite element approach. The simulation of the deformation was shown as well as the distribution of stress and relative density. The predicted compaction forces at different displacements, the variation with time of the displacement, relative density and stress contours were also obtained. A good agreement was achieved between the predicted and measured density profiles. The results clearly indicated that the algorithm makes it possible to simulate the powder forming problems efficiently and automatically. It was shown that the proposed adaptive finite element approach is capable of simulating powder compaction processes in an efficient and accurate manner.

The endochronic plasticity model was developed for rate-independent and rate-dependent materials. Constitutive equations were established based on coupling between deviatoric and hydrostatic behavior to the endochronic theory. A simple procedure for calibrating endochronic material parameters with cone-cap and elliptical yield function was derived. Although the concept of yield surface was not explicitly assumed in endochronic theory, it was shown that the cone-cap plasticity and elliptical density-dependent yield surfaces, which are two most commonly used plasticity models for describing the behavior of powder material can be easily derived as special cases of the proposed endochronic theory. The infinitesimal theory of endochronic plasticity was extended to large strain range on the basis of the additive decomposition of the strain rate tensor and hypoelasticity. Constitutive equations were stated in unrotated frame of reference that greatly simplifies endochronic constitutive relations in finite plasticity and yields the efficiency of the presented algorithm by total uncoupling material and geometrical non-linearities. An implicit scheme, which is accurate, stable and amenable to consistent linearization, was developed. Although the major challenge in the integration of rate constitutive equations in large deformation analysis is to achieve incremental objectivity, it was trivially achieved in the proposed algorithm. Algorithmic modulus consistent with numerical integration of constitutive equations for endochronic theory was extracted. The efficiency of the proposed constitutive model and computational algorithms was demonstrated by several numerical examples.

In the present chapter, the powder compaction software (PCS_SUT) is demonstrated for computational simulation of the powder compaction process. It is designed in both popularity and functionality with the development of user-friendly pre- and post-processing software. Pre-processing software is used to create the model, generate an appropriate finite element grid, apply the appropriate boundary conditions, and view the total model. Post-processing provides visualization of the computed results. It is attempted to summarize work directed at exploring the concepts of an integrated software environment for simulation of powder forming processes. It reviews the capabilities that are provided by pre- and post-processing and suggests enhancements and new features that will likely be developed in the near future.

11.2. Overview of the software environment

In order to design the software environment PCS_SUT, several technical requirements have been considered. These requirements are as follows; the object-oriented programming, C++ programming language, graphic engine, exception handling, data structure, and the input and output phases.

Object-oriented programming: The object-oriented programming is one of the most important techniques in design of an integrated software environment, which models the real-world objects using software counterparts. It takes the advantage of class relationships where the objects of a certain class have similar characteristics. It also takes the advantage of *inheritance* relationships, and *multiple inheritance* relationships. The object-oriented programming gives a more natural and intuitive way to view the programming process by *modeling* the real-world objects, their behaviors and their attributes, and *creating* communication between the objects. It *encapsulates* the data (attributes) and functions (behavior) into packages called *objects;* the data and functions of an object are intimately tied together. Objects have the property of *information hiding.* This means that although objects may know how to communicate with one another across well-defined *interfaces,* objects normally are not allowed to know how other objects are implemented; in fact implementation details are hidden within the objects themselves. The object-oriented programming introduces *polymorphism,* which enables to write programs in a general fashion to handle a wide variety of existing and yet-to-be-specified related classes. Inheritance and polymorphism are effective techniques for dealing with software complexity.

C++ programming language: C++ is a powerful computer programming language. It supports all characteristics of the object-oriented programming and includes many important capabilities not involved in an object-oriented system, such as: function overloading, operator overloading, templates and exception handling. PCS_SUT has been written in C++, in order to perform different aspects of these characteristics and capabilities.

Graphic engine: Inheritance is one of the most important characteristics in object-oriented programming, which is proposed in PCS_SUT. It is developed by a set of valuable graphical features, which are independent of computer hardware. Those graphical objects perform the graphical tasks in a wide range from drawing a simple line to supporting three-dimensional tasks. PCS_SUT can simply handle all of those functions in a simple way because of the object-oriented capabilities.

Exception handling: The extensibility of C++ increases substantially the number and kinds of errors that can occur. Exception handling enables programmers to write clearer, more robust and more fault-tolerant programs. The new exception handling features of C++ enable the programmer to remove the error handling code from the 'main line' of a program's execution. This improves program readability and modifiability. With the C++ style of exception handling it is possible to catch all kinds of exceptions, to catch all exceptions of a certain type or to catch all exceptions of related types. This makes program more robust by reducing the likelihood that errors will not be caught by a program. Exception handling is provided to enable program to catch and handle errors rather than letting them occur and suffering the consequences. In PCS_SUT, three major types of exceptions have been designed. The first two types are the hardware and software exceptions, which are supported by Integrated Development Environment (IDE). The third one is a set of additional exception classes declared by PCS_SUT. These additional exceptions are used to guarantee the stability of PCS_SUT in critical situations.

Data structure: PCS_SUT needs to store and retrieve all information of a problem, such as: nodes, elements, boundaries, loadings, *etc*. Data structures are used to store those information in an easy-to-use manner. Data structures are classified as either linear or nonlinear. In a linear data structure, the elements form a sequence, or a linear list. There are two basic ways of representing such linear structures in memory. One way is to have the linear relationship between the elements represented by means of sequential memory locations, called *arrays*. Another way of storing a list in memory is to have each element in the list contain a field, called a *link* or *pointer*, which contains the address of the next element in the list. Thus, successive elements in the list need not occupy adjacent space in memory. This will make it easier to insert and delete elements in the list. There is another list structure which has been applied in PCS_SUT, called a two-way list, and can be traversed in two directions; in the usual forward direction from the beginning of the list to the end, or in the backward direction from the end of the list to the beginning. Furthermore, given the location of a node in the list, one now has immediate access to both the next node and the preceding node in the list. This means, in particular, that one is able to delete a node from the list without traversing any part of the list.

The input and output phases: PCS_SUT is designed to define the input data in two different manners. One option is based on the standard Win32 dialog box by defining the data through out a text file. In another option, the input data can be defined by using the graphical tools, and selecting the nodes, elements, boundaries and other objects, directly. Furthermore, two different options have been also designed to present the output results in PCS_SUT; text file and graphical options. The software also supports other special formats for output data, which can be used for importing the results into other commercial software.

11.3. Pre-processing – Mesh generation

In finite element pre-processing, the geometric model forms the basis for defining a finite element mesh. The main purpose of the geometric model is to display objects whose geometry is given, allowing interactive control over details, such as point of view, appearance of surfaces and lines, *etc*. The main panel, which used to define the geometric model into PCS_SUT is shown in Figure 11.1. The software can handle any number of objects and allows both separate and collective control over them. It has a simple dialog box to define the nodes, elements and boundaries of background region, as shown in Figure 11.2. At each node, the values of element size, elongation ratio and the direction of elongation are defined to produce the finite element mesh. The geometry of background region can be defined simultaneously by using the dialog box and graphical interface. Figure 11.3 shows an initial layout of a tip engineering component after loading the geometry. As the geometry is the basis for the mesh, the validity and character of the geometry have a direct impact on the meshing process. In addition, boundary conditions, element types, and material properties all have an effect on the modeling process.

The finite element pre-processing influences the types of meshing algorithms that may be used, the type of elements that may be selected, the resulting mesh density, the quality of the mesh, and even the feasibility of creating a finite element mesh. The geometric model can be used to associate the mesh with the physical attributes such as element properties, material properties, or loads and boundary conditions. These attributes are assigned to the geometry and remain constant when the mesh is modified. This implies that a direct correspondence between the geometry and the mesh must be maintained. This capability is essential in performing adaptive mesh generation.

Once the geometry model is correctly defined it must be discretized to create finite elements that contains the appropriate type of elements required for the specific analysis. This process is known as mesh generation. The development of automatic meshing algorithms has become an active area of research in recent years. Sophisticated pre-processors adjust the quality of these generated elements to enhance accurate application of the mathematical background of the FEM. There are numerous algorithms available, most of which are variants of the following basic types: mapping transformations, geometry decomposition, node insertion and connection, and progressive front technique.

<u>*Mapping transformations*</u>

Mapping techniques transform a parent region with regular grid spacing to a more arbitrarily shaped region. It works best if the region has some direct correspondence with the parent shape. For a unit square, the region being mapped should have four logical sides and four rather natural corners. This technique is used mainly for the generation of quadrilateral elements, but can be adapted to the generation of triangles in *n*-sided regions. It suffers when the transformation does not match the geometry well, and can result in elements propagating outside the region and inverting around concave corners. Thus, when mapping is employed, users often manually decompose the regions into 'rectangular-like' sub-regions before using the algorithm. This technique is employed in the development of primitives or higher order shapes that can be divided into sub-regions of roughly rectangular shapes. These more general primitives have proven useful in simplifying the required manual decompositions. Figure 11.4 shows a complex geometry of tip component and a transition primitive region. Although not truly automated, the primitives speed decomposition while allowing the user to generate completely quadrilateral mesh.

Geometry decomposition

One approach to solving the automated meshing task is to devise a procedural method to replace the manual decomposition process. This allows for automation while providing a completely quadrilateral mesh. One of the techniques in automated decompositions is based on artificial intelligence to reason about the geometry, formulate plausible dissections, and then explore the consequences of performing the dissections. Figure 11.4(b) shows an example of the type of decomposition possible. The idea of using a decomposition process to form individual elements, rather than geometric primitives has also been explored. In this technique, a geometry is continuously sectioned until the remaining portions are simple elements. The discretization level must be decided at each cut with the insertion of nodes on all boundary portions. The approach has been shown to be quite general in two-dimensions, allowing for the formation of all quadrilateral meshes with rather arbitrary refinement possible.

Node insertion and connection

A more local approach to meshing a geometric region is the node insertion and then connection approach. This algorithm requires two distinct steps; initial insertion of a set of appropriately spaced nodes, and a method of hooking these nodes together in such a manner as to form valid elements. This approach forms the bulk of the mesh generation literature, with Delauney triangulation being the most popular technique for the connection of the elements. The method has been proven very general for the formation of graded triangular meshes.

Progressive front technique

Progressive front technique forms a mesh by pushing the exterior boundary of the mesh inward until the colliding fronts intersect. This technique tends to generate better elements at the exterior boundary and poorer elements where the boundaries intersect. This technique is a more efficient process constructs directly triangles of an approximately equilateral shape and of a size specified at nodes of a background mesh, which embraces the domain and interpolates these values to any point. This sophisticated methodology has proved to be extremely useful in adaptive mesh generation and is used in PCS_SUT. Figure 11.5 shows a 2D triangular finite element mesh of an axisymmetric shaped tip component in PCS_SUT using progressive front technique.

11.4. The analysis utility environment

PCS_SUT is a software environment for the purpose of performing numerical simulation of metal powder forming processes. It can deal with two-dimensional and axisymmetric applications for static and dynamic analyses. PCS_SUT uses dynamic arrays, vectors, and maps from Standard Template Library of C++ language. All input information, such as: nodes, elements, boundaries, material properties, *etc* are stored and retrieved using the Standard Template Library. It guarantees the compatibility of the source code with other compilers supporting STL. Some features of PCS_SUT are listed in the following;

a) A Lagrangian description is used which describes the material behavior with respect to either the original domain in the case of a 'total' Lagrangian frame or the domain at the previous calculation step in the case of an 'updated' Lagrangian scheme (Figure 11.6).

b) An adaptive FEM strategy is used into the large displacement finite element formulation by employing an error estimator, adaptive mesh refinement, and data transfer operator. This consists in defining new appropriate finite element mesh within the updated, deformed geometry and interpolating (mapping) the pertinent variables from one mesh to another in order to continue the simulation. Figure 11.7 presents the designed interfaces for the adaptive strategy in PCS_SUT. In Figure 11.8, the adaptive mesh refinement is shown for a shaped tip component at two different stages.

c) A double-surface cap plasticity model, based on a combination of a convex yield surface consisting of a failure envelope and a hardening elliptical cap, is employed for the non-linear behavior of powder materials in the concept of a generalized plasticity formulation, as shown in Figure 11.9. In addition, PCS_SUT supports other yield criteria, including: Tresca, von-Mices, Mohr-Coulomb, Drucker-Prager, and Pastor-Zienkiewicz models, as presented in Figure 11.10.

d) The constitutive modeling of the frictional behavior of powder is used by Coulomb's friction law and the plasticity theory of friction is modeled in the context of an interface element formulation.

e) Both triangular and quadrilateral elements are available in the element library, including 3-noded, 6-noded and 7-noded (T6B1) triangles and 4-noded, 8-noded and 9-noded quadrilaterals. In PCS_SUT, the properties of each material are corresponded to the background region defined by the boundaries in Figure 11.2.

f) The direct implicit algorithm based on the generalized Newmark scheme is used for the time integration and an automatic time step control facility is provided (Figure 11.11).

g) For non-linear iteration, choices among fully or modified Newton-Raphson method and quasi-Newton method, using the initial stiffness method, Davidon inverse method or BFGS inverse method, are possible, as shown in Figure 11.11. For solving the linear algebraic equations, a symmetric core profile direct solver or Preconditioned Conjugate Gradient iterative solvers can be used.

11.5. Post-processing

The output resulting from finite element analysis is, to say the least, voluminous. When the finite element model has been completed and the analysis performed, it is desirable to use graphical techniques to visualize the results. Output data of a finite element analysis can be visualized in different form, such as: iso-lines, iso-bars, *xy*-plots, contour-plots, vector-plots, animation, *etc.* Post-processing is vital to allow the appropriate interpretation of the completed results of the finite element analysis. It provides the visual means to interpret the vast amounts of computed results generated. PCS_SUT visualization based post-processing capabilities take the drudgery out of interpreting the results of the FE analysis. Figure 11.12 shows the density contour in continuous color for an axisymmetric shaped tip component in PCS_SUT. It can display the nodal and elemental analysis results in a wide array of graphic formats or sorted and culled printed output.

Post-processing in PCS_SUT is designed into two distinct categories; *graphical* post-processing and *text-based* post-processing. Both forms are extremely useful in interpreting the results of FE analysis. The combined option makes the FE analysis more productive. The graphical post-processing allows to quickly locate gross behaviors of the model, such as: maximum density, maximum stress level, *etc.* The text based post-processing provides detailed numerical data often necessary to fully understand the results of FE analysis. PCS_SUT contains several unique tools that make it possible to sort and cull the large amount of nodal and elemental data returned by the FE analysis.

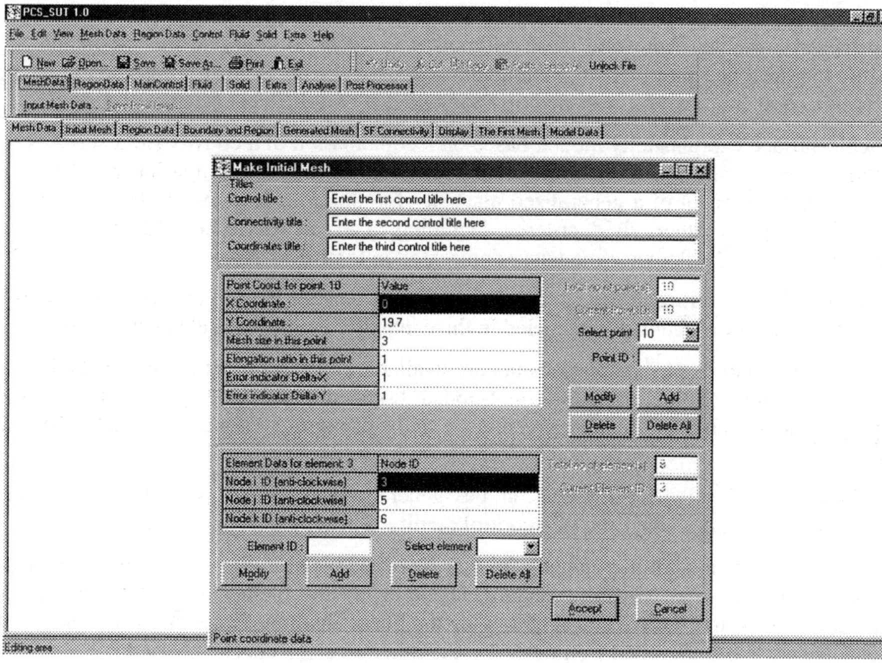

Figure 11.1. Initial layout of PCS_SUT

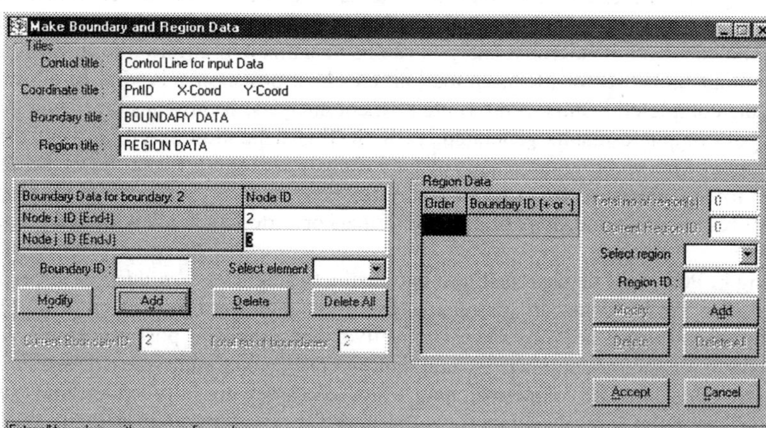

Figure 11.2. Definition of the nodes, elements and boundaries of background region in PCS_SUT

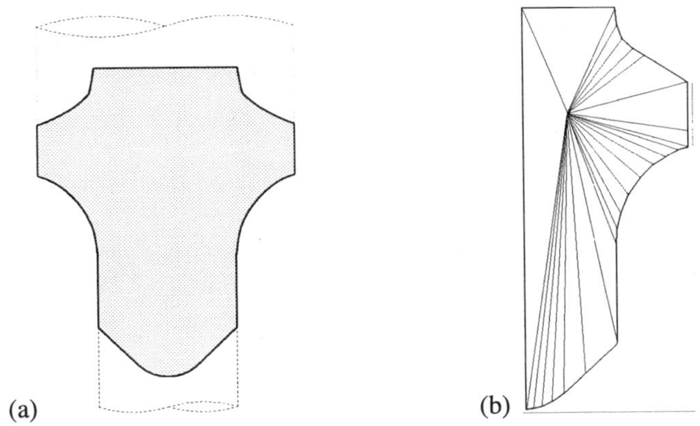

Figure 11.3. An axisymmetric shaped tip component; a) Geometry, b) Initial layout

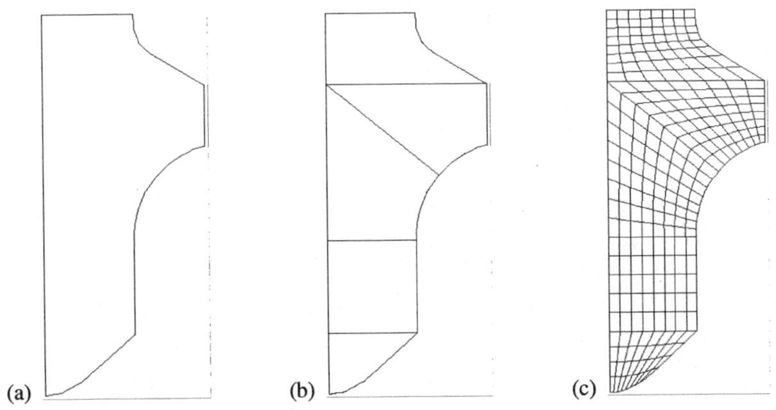

Figure 11.4. An axisymmetric tip component; a) The original body, b) The decomposed body, c) The resulting mesh

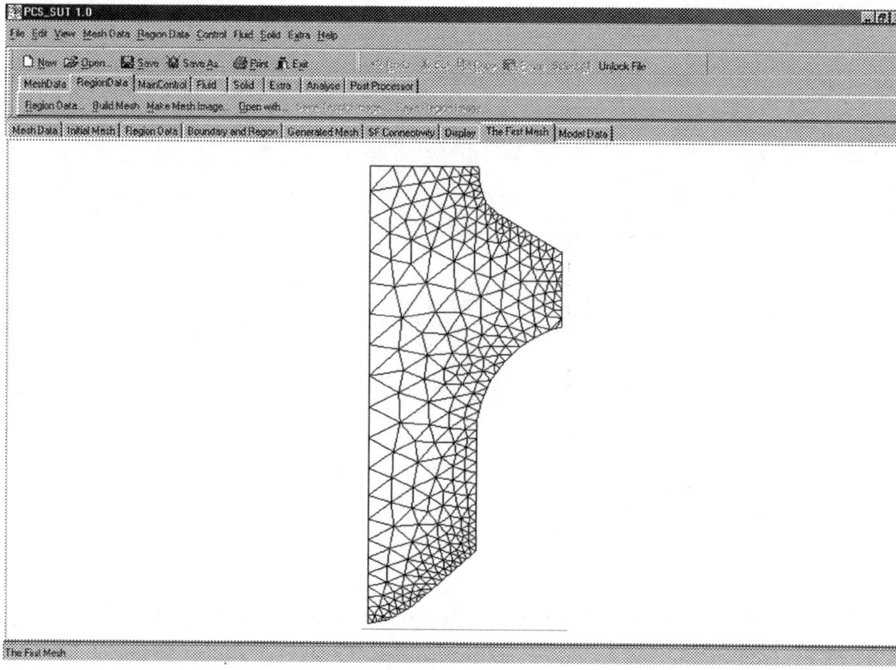

Figure 11.5. 2D triangular FE mesh of an axisymmetric shaped tip component in PCS_SUT

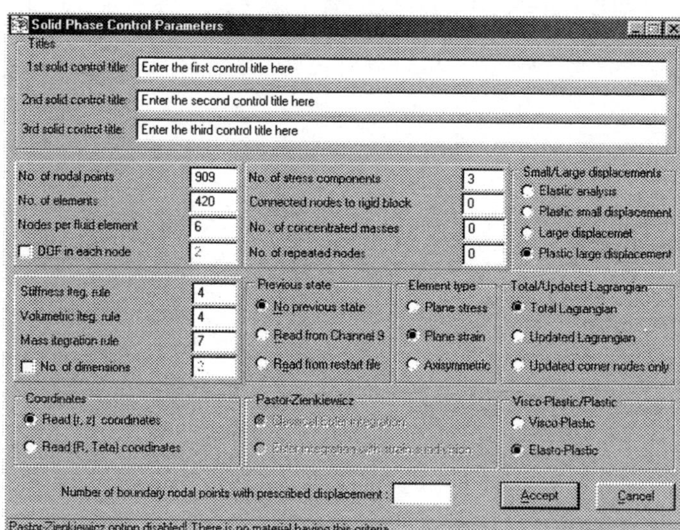

Figure 11.6. Definition of solid phase parameters in PCS_SUT

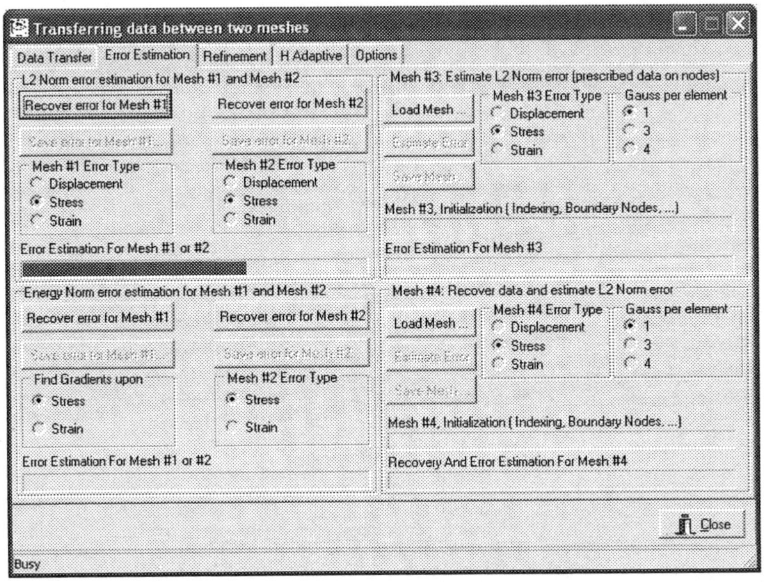

(a)

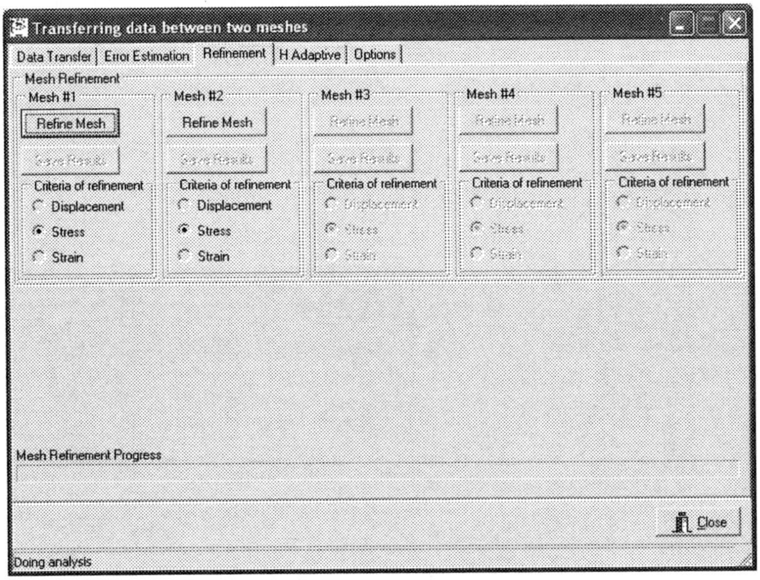

(b)

Figure 11.7. Adaptive FEM strategy in PCS_SUT; a) Error estimation, b) Adaptive mesh refinement, c) Data transfer operator

(c) 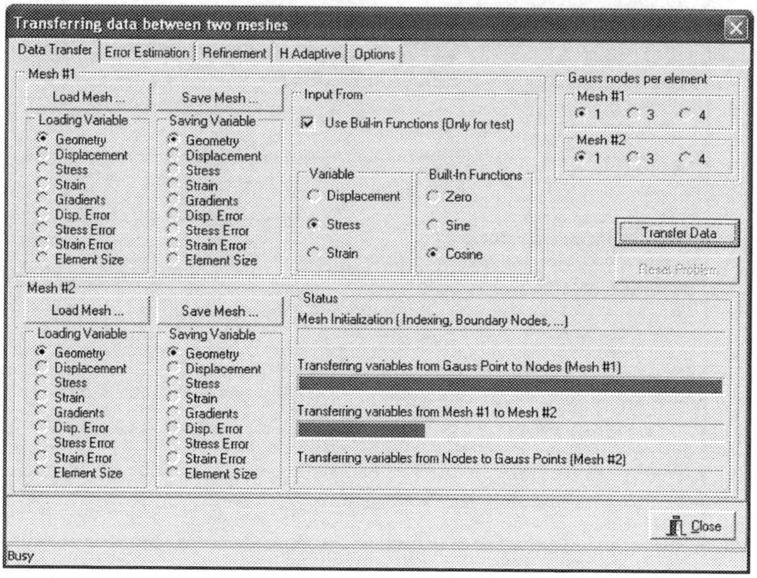

Figure 11.7. (continued)

(a) 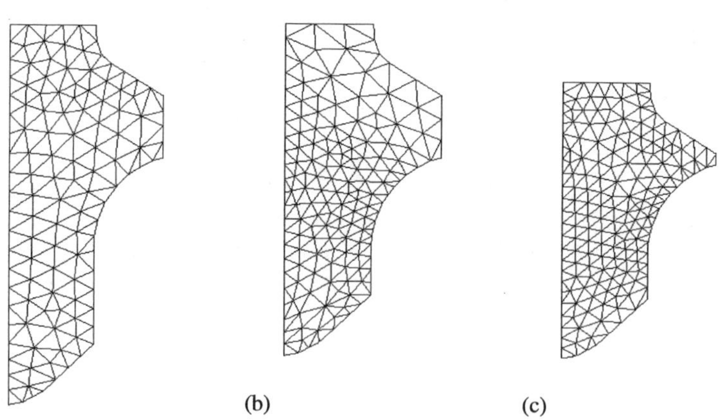

Figure 11.8. Adaptive remeshing for a shaped tip component; a) Initial mesh, b) First adaptive mesh after bottom punch movement, c) Second adaptive mesh after top punch movement

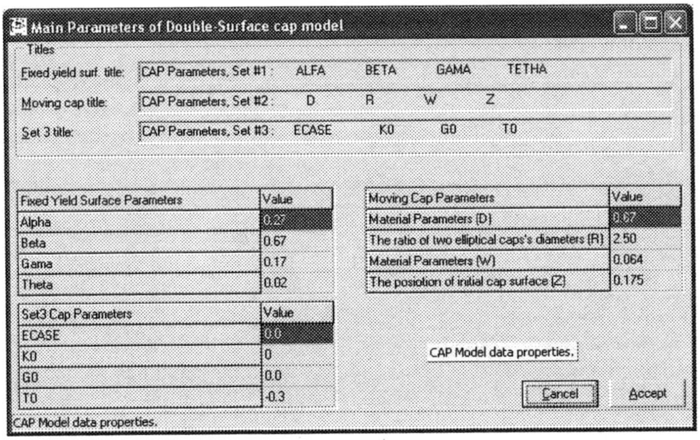

Figure 11.9. Definition of the parameters of double-surface plasticity model in PCS_SUT

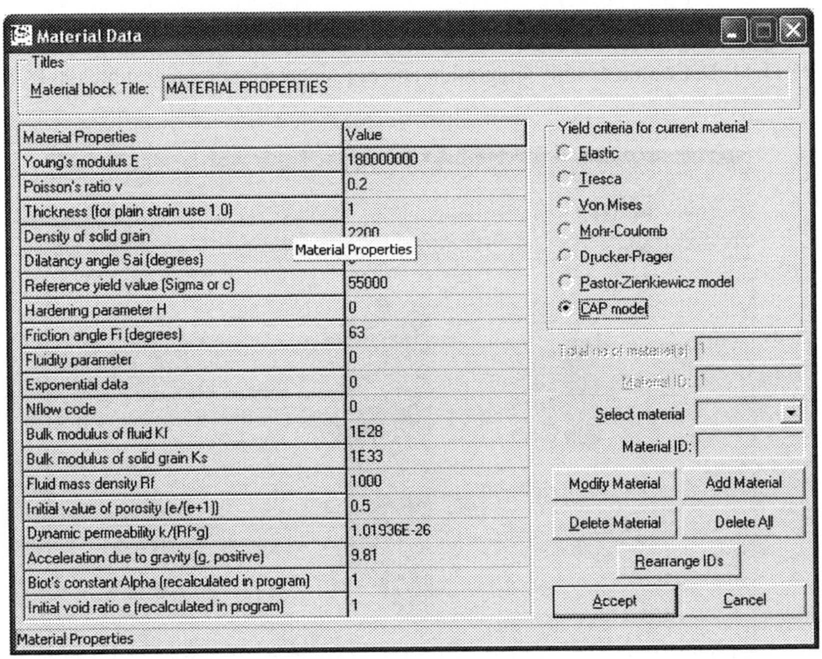

Figure 11.10. Definition of the material parameters in PCS_SUT

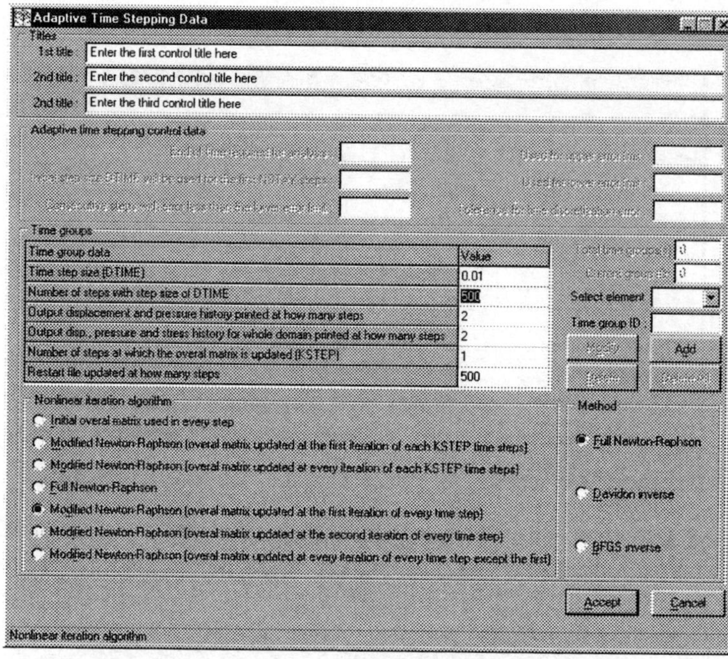

Figure 11.11. Definition of time integration and non-linear iteration algorithms in PCS_SUT

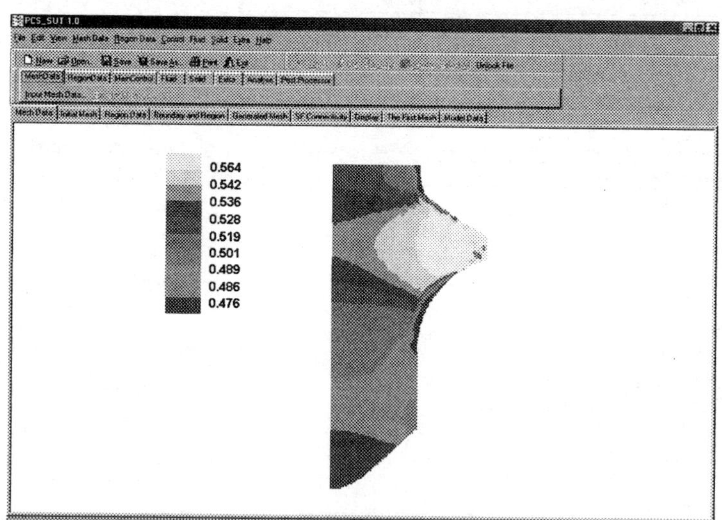

Figure 11.12. Density contour for an axisymmetric shaped tip component in PCS_SUT (see colour plate section at beginning of book)

REFERENCES

M. Abdel-Karim and N. Ohno, Kinematic hardening model suitable for ratcheting with steady state, *Int. J. Plasticity*, **16**, 225-240, 2000.

E.C. Aifantis, The physics of plastic deformation, *Int. J. Plasticity,* **3**, 221, 1987.

O. Alm, Mechanical testing of powders and powder compacts, *Scand. J. Metall.*, **12,** 302, 1983.

H. Armen and R.F. Jones, Applications of Numerical Methods to Forming Processes, *American Society of Mechanical Engineers, ASME*, ADM **28**, 1978.

P.J. Armstrong and C.O. Frederick, A mathematical representation of the multiaxial Bauschinger effect, CEGB report RD/B/N731, Berkeley, Nuclear Laboratories: Berkeley, 1966.

B. Avitzur and Y. Nakamura, Analytical determination of friction resistance as a function of normal load and geometry of surface irregularities, *Wear*, **107**, 367-383, 1986.

I. Aydin, B.J. Briscoe and K.Y. Sanliturk, The internal form of compacted ceramic components. A comparison of a finite element modeling with experiment, *Powder Tech.*, **89,** 239-254, 1996.

A.R. Azami and A.R. Khoei, A three-invariant isotropic hardening cap plasticity for numerical simulation of powder forming processes, Submitted to: *Comp. Meth. Appl. Mech. Eng.,* 2004.

I. Babuska and W.C. Rheinboldt, A posteriori error estimates for the finite element method, *Int. J. Num. Meth. Eng.*, **12**, 1597-1615, 1978.

I. Babuska and W.C. Rheinboldt, Adaptive approaches and reliability estimates in finite element analysis, *Comp. Meth. Appl. Mech. Eng.*, **17/18**, 519-540, 1979.

I. Babuska and W.C. Rheinboldt, A posteriori error analysis of finite element solutions for one-dimensional problems, *SIAM J. Num. Anal.*, **18**, 565-589, 1981.

I. Babuska, O.C. Zienkiewicz, J. Gago and E.R. de A. Oliveira, Accuracy Estimates and Adaptive Refinement in Finite Element Computations, *Wiley*, 1986.

J.A. Bailey, S.L. Haas and K.C. Nawab, Anisotropy in plastic torsion, *J. Basic Engng.,* 231, 1972.

A. Bakhshiani, A.R. Khoei and M. Mofid, An endochronic plasticity model for powder compaction processes, *J. Mater. Proc. Tech.*, **125-126**, 138-143, 2002.

A. Bakhshiani, M. Mofid, A.R. Khoei and S.L. McCabe, Finite strain simulation of thin-walled tube under torsion using endochronic theory of plasticity, *Thin-Walled Struct.,* **41**, 435-459, 2003.

A. Bakhshiani, A.R. Khoei and M. Mofid, A density-dependent endochronic plasticity for powder compaction processes, *Comput. Mechanics,* **34**, 53-66, 2004.

A. Baltov and A. Sawczuk, A rule for anisotropic hardening, *Acta Mech.,* **1**, 81-92, 1965.

D.J. Bammann, A micro-mechanically motivated model of finite deformation plasticity, *Advances in plasticity*, Proceeding of Plasticity 89, The second international symposium on plasticity and its current applications, A.S. Khan and M. Tokuda (Eds.), 283, 1989.

S. Bari and T. Hassan, Anatomy of coupled constitutive models for ratcheting simulation, *Int. J. Plasticity*, **16**, 381-409, 2000.

S. Bari and T. Hassan, An advancement in cyclic plasticity modeling for multiaxial ratcheting simulation, *Int. J. Plasticity*, **18**, 873-894, 2002.

K.J. Bathe, Finite Elements Procedures, *Prentice Hall*, New Jersey, 1996.

N. Bay and T. Wanheim, Real area of contact and friction stresses at high pressure sliding contact, *Wear*, **38**, 201-209, 1976.

Z.P. Bazant, A.M. Ansal and R.J. Krizek, Visco-plasticity of transversely isotropic clays, *J. Eng. Mech. Div. ASCE*, **105**, 549-565, 1979.

Z.P. Bazant and P.D. Bhat, Endochronic theory of inelasticity and failure of concrete. *J. Eng. Mech. Div. ASCE*, **102**, 701-722, 1976.

Z.P. Bazant and R.J. Krizek, Endochronic constitutive law for liquefaction of sand, *J. Eng. Mech. Div. ASCE*, **102**, 225-238, 1976.

G. Beer, An isoparametric joint/interface element for finite element analysis, *Int. J. Num. Meth. Eng*, **21**, 585-600, 1985.

T. Belytschko, An overview of semidiscretizatin and time integration procedures, Computational Methods for Transient Analysis, (Eds. T. Belytschko and T.J.R. Hughes), *Elsevier Science Publisher*, 1-95, 1983.

T. Belytschko, W.K. Liu and B. Moran, Nonlinear Finite Elements for Continua and Structures, *Wiley*, 2000.

T. Belytschko and M. Tabbara, *H*-adaptive finite element methods for dynamic problems with emphasis on localization, *Int. J. Num. Meth. Eng.*, **36**, 4245-4265, 1993.

T. Belytschko, B.L. Wong and E.J. Plaskacz, Fission-fusion adaptivity in finite element analysis for nonlinear dynamics of shell, *Comput. Struct.*, **33**, 1327-1333, 1985.

J.J. Bikerman, Adhesion in friction, *Wear*, **39**, 1-13, 1976.

H.H. Bleich, On uniqueness in ideally elasto-plastic problems in case of non-associated flow rules, *J. Appl. Mech. Trans., ASME*, 983-987, 1972.

A. Benallal and V. Tvergaard, Nonlocal continuum effects on bifurcation in the plane strain tension-compression test, *J. Mech. Phys. Solids*, **43**, 741-770, 1995.

R.I. Borja, Cam-clay plasticity, Part II: Implicit integration of constitutive relation based on a nonlinear elastic stress predictor, *Comp. Meth. Appl. Mech. Eng.*, **88**, 225-240, 1991.

R.I. Borja and A.P. Amies, Multiaxial cyclic plasticity model for clays, *J. Geotech. Eng.*, **120**, 1051-1070, 1994.

R.I. Borja, F.J. Chao, F.K. Montans and C.H. Lin, Non-linear ground response at Lotung LSST site, *J. Geotech. Eng.*, **125**, 187-197, 1999.

R.I. Borja and S.R. Lee, Cam-clay plasticity, Part I: Implicit integration of elastoplastic constitutive relations, *Comp. Meth. Appl. Mech. Eng.*, **78**, 49-72, 1990.

F.P. Bowden and D. Tabor, The Friction and Lubrication of Solids, Part II, *Clarendon Press*, Oxford, 1964.

J. Brandt and L. Nilsson, A constitutive model for compaction of granular media, with account for deformation induced anisotropy, *Mech. Cohes. Frict. Mater.*, **4**, 391-418, 1999.

W.A.M. Brekelmans, J.D. Janssen, A.A.F. Van de Ven and G.de With, An Eulerian approach for die compaction processes, *Int. J. Num. Meth. Eng.*, **31**, 509-524, 1991.

S.B. Brown and G. Abou-Chedid, Yield behavior of metal powder assemblages, *J. Mech. Phys. Solids*, **42**, 383-398, 1994.

S.B. Brown and G.G.A. Weber, A constitutive model for the compaction of metal powders, *Modern Development in Powder Metallurgy*, **18**, 465-476, 1988.

L.T. Campos, J.T. Oden and N. Kikuchi, A numerical analysis of a class of contact problems with friction in elastostatics, *Comp. Meth. Appl. Mech. Eng.*, **34**, 821-845, 1982.

J.C. Cavendish, D.A. Field and W.H. Frey, An approach to automatic three dimensional finite element mesh generation, *Int. J. Num. Meth. Eng.*, **21**, 329-347, 1985.

J. Cedergren, N.J. Sorensen and A. Bergmark, Three-dimensional analysis of compaction of metal powder, *Mechanics of Materials*, **34**, 43-59, 2002.

S. Cescotto, F. Frey and G. Fonder, Total and updated Lagrangian descriptions in nonlinear structural analysis: A unified approach, *Energy Methods in Finite Element Analysis*, R. Glowinski et al.(Eds.), Wiley, Chap. **15**, 283-296, 1979.

J.L. Chaboche, Modeling of ratcheting: evaluation of various approaches, *Europ. J. Mech., A/Solids*, **13**, 501-518, 1994.

J.L. Chaboche, On some modification of kinematic hardening to improve the description of ratcheting effects, *Int. J. Plasticity*, **7**, 661-678, 1994.

J.L. Chaboche, K. Dang-Wan and G. Cordier, Modelization of the strain memory effect on the cyclic hardening of 316 stainless steel, *Proc. 5th Int. Conf. on SMIRT*, Berlin, Germany, 1979.

S.K. Chan and I.S. Tuba, A finite element method for contact problems of solid bodies - Part 1. Theory and validation, *Int. J. Num. Meth. Eng.*, **13**, 615-625, 1971.

A. Chandra and S. Mukherjee, A finite element analysis of metal forming problems with an elastic-viscoplastic materials model, *Int. J. Num. Meth. Eng.*, **10**, 1613-1628, 1984.

A.B. Chaudaray and K.J. Bathe, A solution method for static and dynamic analysis of contact problems with friction, *Comput. Struct.*, **24**, 855-873, 1986.

W.F. Chen, Limit Analysis and Soil Plasticity, *Elsevier*, 1975.

W.F. Chen and G.Y. Baladi, Soil Plasticity, Theory and Implementation, *Elsevier*, Amsterdam, 1985.

J.H. Cheng, Automatic adaptive remeshing for finite element simulation of forming processes, *Int. J. Num. Meth. Eng.*, **23**, 1-8, 1986.

J.H. Cheng and N. Kikuchi, A mesh rezoning technique for finite element simulations of metal forming processes, *Int. J. Num. Meth. Eng.*, **23**, 219-228, 1986.

J.L. Chenot, F. Bay and L. Fourment, Finite element simulation of metal powder forming, *Int. J. Num. Meth. Eng.*, **30**, 1649-1674, 1990.

J.L. Chenot, R.D. Wood and O.C. Zienkiewicz, Numerical Methods in Industrial Forming Processes, *NUMIFORM 92*, *A.A. Balkema,* Rotterdam, 1992.

H. Chtourou, M. Guillot and A. Gakwaya, Modeling of the metal powder compaction process using the cap model. Part I. Experimental material characterization and validation. *Int. J. Solids Struct.*, **39**, 1059-1075, 2002.

Y. Corapcioglu and T. Uz, Constitutive equation for plastic deformation of porous materials, *Powder Tech.*, **21**, 269-274, 1978.

G.C. Cornfield and R.H. Johnson, Theoretical prediction of plastic flow in hot rolling including the effect of various temperature distributions, *J. Iron Steel Inst.*, **211**, 567-573, 1973.

E. Cosserat and F. Cosserat, Theorie des Corps Deformable, *Hermann*, Paris, 1909.

B.A. Cotter and R.S. Rivlin, Tensors associated with time dependent stress, *Quart. Appl. Math.* **13,** 177-182, 1955.

C.A. Coulomb, Essai sur une application des regles des maximis et minimis a quelquels problemesde statique relatifs, a la architecture, *Mem. Acad. Roy. Div. Sav.,* **7,** 343-387, 1776.

H. Cramer, R. Findeiss, G. Steinl and W. Wunderlich, An approach to the adaptive finite element analysis in associated and non-associated plasticity considering localization phenomena, *Comput. Meth. Appl. Mech. Eng.*, **176**, 187-202, 1999.

J.E. Crawford and P. Lindskog, Constitutive equations and their role in the modelling of the cold pressing process, *Scand. J. Metall.*, **12**, 277-281, 1983.

M.A. Crisfield, Nonlinear Finite Element Analysis of Solids and Structures, *Wiley*, Chichester, **1**, 1991.

A. Curnier, A theory of friction, *Int. J. Solids Struct.*, **20**(7), 637-647, 1984.

A. Curnier and P. Alart, Generalisation of Newton type methods to contact problems with friction, *J. de Mecanique Theorique et Appliquee*, **7**, 67-82, 1988.

Y.F. Dafalias, The plastic spin concept and a simple illustration of its role in finite plastic transformations, *Mech. Mater.*, **3**, 223-233, 1984.

Y.F. Dafalias, Bounding surface plasticity; Mathematical foundation and hypo-plasticity, *ASCE J. Eng. Mech.*, **112**, 966-987, 1986.

Y.F. Dafalias and E.P. Popov, Plastic internal variables formalism of cyclic plasticity, *J. Appl. Mech.*, **43**, 645-650, 1976.

P.R. Dawson and E.G. Thompson, Finite element analysis of steady state elasto-viscoplastic flow by the initial stress rate method, *Int. J. Num. Meth. Eng.*, **12**, 47-57, 1978.

R. de Borst, Bifurcations in finite element models with a nonassociated flow law, *Int. J. Numer. Anal. Meth. Geomech.*, **12**, 99-116, 1988.

R. de Borst, Simulation of strain localization: a reappraisal of Cosserat continuum, *Engng. Comput.*, **8**, 317-332, 1991.

R. de Borst and H.B. Muhlhaus, Gradient-dependent plasticity: formulation and algorithmic aspects, *Int. J. Num. Meth. Engng.*, **35**, 521-539, 1992.

R. de Borst, L.J. Sluys, H.B. Muhlhaus and J. Pamin, Fundamental issues in finite element analysis of localization of deformation, *Engng. Comput.*, **10**, 99-121, 1993.

A.D. De Pater and J.J. Kalker, The Mechanics of the Contact between Deformable Bodies, *Delft University Press*, 1975.

J.P. De S.R. Gago, D.W. Kelly, O.C. Zienkiewicz and I. Babuska, A posteriori error analysis and adaptive processes in the finite element method, Part II: Adaptive mesh refinement, *Int. J. Num. Meth. Eng.*, **19**, 1621-1656, 1983.

A.D. Deb, J.H. Prevost and B. Loret, Adaptive meshing for dynamic strain localization, *Comput. Meth. Appl. Mech. Eng.*, **137**, 285-306, 1996.

P. Delobelle, P. Robinet and L. Bocher, Experimental study and phenomenological modelization of ratcheting under uniaxial and bi-axial loading on an austenitic stainless steel, *Int. J. Plasticity*, **11**, 295-330, 1995.

J.E. Dennis and J. More, Quasi-Newton methods, motivation and theory, *SIAM Rev.*, **19**, 46-89, 1977.

C.S. Desai, A general basis for yield, failure and potential functions in plasticity, *Int. J. Num. Anal. Meth. Geomech.*, **4**, 361-375, 1980.

C.S. Desai and Q.S.E. Hashmi, Analysis, evaluation, and implementation of a nonassociative model for geologic materials, *Int. J. Plasticity*, **5**, 397-420, 1989.

C.S. Desai, H.V. Phan and S.C. Sture, Procedure, selection, and application of plasticity models for soil, *Int. J. Num. Anal. Meth. Geomech.*, **5**, 295-311, 1981.

C.S. Desai and H.J. Siriwardane, A concept of correction functions to account for non-associative characteristics of geologic media, *Int. J. Num. Anal. Meth. Geomech.*, **4**, 377-387, 1980.

C.S. Desai and H.J. Siriwardane, Constitutive Laws for Engineering Materials, with emphasis on geologic materials, *Prentice-Hall*, New Jersey, 1984.

J.K. Dienes, On the analysis of rotation and stress rate in deforming bodies, *Acta Mech.*, **32**, 217-232, 1979.

P. Diez and A. Huerta, A unified approach to remeshing strategies for finite element *h*-adaptivity', *Comput. Meth. Appl. Mech. Eng.*, **176**, 215-229, 1999.

F.L. DiMaggio and I.S. Sandler, Material models for granular soils, *J. Eng. Mech. Div., ASCE*, 935-950, 1971.

S.M. Doraivelu, H.L. Gegel, J.S. Gunasekera, J.C. Malas, J.T. Morgan and J.F. Thomas, A new yield function for compressible P/M materials, *Int. J. Mech. Sci.*, **26**, 527-535, 1984.

P. Doremus, C. Geindreau, A. Marttin, L. Debove, R. Lecot and M. Dao, High pressure triaxial cell for metal powder, *Powder Metall.*, **38**, 284-287, 1995.

D.C. Drucker, Coulomb friction, plasticity and limit loads, *J. Appl. Mech. Trans. ASME*, **21**, 71-74, 1954.

D.C. Drucker, R.E. Gibson and D.J. Henkel, Soil mechanics and work hardening theories of plasticity, *Trans. ASCE*,, **122**, 338-346, 1957.

D.C. Drucker and L. Palgen, On stress-strain relations suitable for cyclic and other loadings, *J. Appl. Mech.*, **48**, 479, 1981.

D.C. Drucker and W. Prager, Soil mechanics and plastic analysis on limit design, *Q. J. Appl. Math.*, **10**, 157-165, 1952.

S.F. Duffy, A unified inelastic constitutive theory for sintered powder metals, *Mech. Mater.*, **7**, 245-254, 1988.

M. Dutko, D. Peric, D.R.J. Owen, A.J.L. Crook, Z. Wei and J. Yu, Bulk forming simulation by adaptive explicit FEM, *Computational Plasticity: fundamentals and applications*, D.R.J. Owen *et al.*(Eds.), CIMNE, Barcelona, **2**, 1305-1312, 1997.

M. Dyduch, A.M. Habraken and S. Cescotto, Automatic adaptive remeshing for numerical simulations of metal forming, *Comp. Meth. Appl. Mech. Eng.*, **101**, 283-298, 1992.

M. Dyduch, S. Cescotto and A.M. Habraken, Error estimates and indicators for adaptive analysis of bulk forming, *Computational Plasticity: Fundamentals and Applications*, D.R.J. Owen *et al.*(Eds.), Pineridge Press, Swansea, **2**, 1355-1367, 1995.

A. Elgamal, Z. Yang, E. Parra and A. Ragheb, Modeling of cyclic mobility in saturated cohesionless soils, *Int. J. Plasticity*, **19**, 883-905, 2003.

A. Eterovic and K. Bathe, A hyperelastic based large strain elasto-plastic constitutive formulation with combined isotropic-kinematic hardening using the logarithmic stress and strain measures, *Comp. Meth. Appl. Mech. Eng*, **30**, 1099-1114, 1990.

R.A. Eve and B.D. Reddy, The variational formulation and solution of problems of finite strain elastoplasticiy based on the use of dissipation function, *Int. J. Num. Meth. Eng.*, **37**, 1673-1695, 1994.

X. Feaugas and C. Gaudin, Ratchetting process in the stainless steel AISI 316L at 300 K: an experimental investigation, *Int. J. Plasticity*, **20**, 643-662, 2004.

J. Fish and K. Shek, Finite deformation plasticity based on additive split of the rate of deformation and hyperelasticity, *Comp. Meth. Appl. Mech. Eng.*, **140,** 75-93, 2000.

N.A. Fleck, On the cold compaction of powders. *J. Mech. Phys. Solids*, **43,** 409-1431, 1995.

N.A. Fleck and J.W. Hutchinson, A phenomenological theory of strain gradient plasticity, *J. Mech. Phys. Solids*, **41**, 1825-1857, 1993.

N.A. Fleck, L.T. Kuhn and R.M. McMeeking, Yielding of metal powder bounded by isolated contacts. *J. Mech. Phys. Solids,* **40,** 1139, 1992.

R. Fletcher, Practical Methods of Optimization, *Wiley*, New York, **1**, 1980.

L. Fourment and J.L. Chenot, Error estimators for viscoplastic materials: Application to forming processes, *Eng. Comput.*, **12**, 469-490, 1995.

L. Fourment, M.P. Miles, F. Bay, D. Carpentier and J.L. Chenot, Adaptive strategies for the simulation of 2D and 3D forming processes, *Computational Plasticity: Fundamentals and Applications*, D.R.J. Owen *et al.*(Eds.), Pineridge Press, Swansea, **1**, 395-405, 1995.

A. Fracavilla and O.C. Zienkiewicz, A note on the numerical computation of elastic contact problems, *Int. J. Num. Meth. Eng.*, **9**, 913-924, 1975.

B. Fredriksson, Finite element solution of surface non-linearities in structural mechanics with special emphasis to contact and fracture mechanics problems, *Comp. Struct.*, **6**, 281-290, 1976.

F.J. Gallego and J.J. Anza, A mixed finite element for the elastic contact problem, *Int. J. Num. Meth. Eng.*, **28**, 1249-1264, 1989.

C.W. Gear, Numerical Initial Value Problems in Ordinary Differential Equations, *Prentice-Hall*, New Jersey, 1971.

C.J.M. Gelten and A.W.A. Konter, Application of mesh-rezoning in the updated Lagrangian method to metal forming analysis, *Numerical Methods in Industrial Forming Processes, NUMIFORM 82*, J.F.T. Pittman *et al.*(Eds.), Pineridge Press, Swansea, 511-521, 1982.

M. Geradin, S. Idelsohn and M. Hogge, Nonlinear structural dynamics via Newton and quasi-Newton methods, *Nuclear Eng. Design*, **58**, 339-348, 1980.

D.T. Gethin, R.W. Lewis and A.K. Ariffin, Modelling compaction and ejection processes in the generation of green powder compacts, *ASME Int. Mech. Eng. Congress and Exposition*, San Francisco, ADM-**216**, 27-45, 1995.

D.T. Gethin, R.W. Lewis, D.V. Tran and A.K. Ariffin, Finite element modelling of multilevel compaction of powders, *Powder Metallurgy World Congress*, Toronto, 1994.

D.T. Gethin, D.V. Tran, R.W. Lewis and A.K. Ariffin, An investigation of powder compaction processes, *Int. J. Powder Metall.*, **30**, 385-398, 1994.

J. Ghaboussi, E.L. Wilson and J. Isenberg, Finite element for rock joints and interfaces, *J. Soil Mech. Foun. Div., ASCE*, **99**, 833-848, 1973.

G.Y. Goon, P.I. Poluchin, W.P. Poluchin and B.A. Prudcowsky, The plastic deformation of metals, *Metallurgica Moscow*, 1968. (in Russian)

W.J. Gordon and C.A. Hall, Construction of curvilinear co-ordinate systems and applications to mesh generation, *Int. J. Num. Meth. Eng.*, **7**, 461-477, 1973.

A.E. Green and P.M. Naghdi, A general theory of an elastic-plastic continuum. *Archive Rational Mech. Anal.*, **18**, 251-281, 1965.

R.J. Green, A plasticity theory for porous solids, *Int. J. Mech. Sci.*, **14**, 215-224, 1972.

C. Gu, M. Kim and L. Anand, Constitutive equations for metal powders: application to powder forming processes, *Int. J. Plasticity*, **17,** 147-209, 2001.

A.L. Gurson, Continuum theory of ductile rupture by void nucleation and growth: part I - Yield criteria and flow rules for porous ductile media, *J. Eng. Mater. Tech., Trans. ASME,* **2**, 99, 1977.

H-A. Haggblad, Constitutive models for powder material, *Powder Tech.*, **76**, 127-136, 1991.

H-A. Haggblad and W-B. Li, A micro mechanical based constitutive model for finite element simulation of hot isostatic pressing of powder, *Comp. Meth. Appl. Mech. Eng.*, **128**, 191-198, 1995.

H-A. Haggblad and M. Oldenburg, Modelling and simulation of metal powder die pressing with use of explicit time integration, *Model. Simul. Mater. Sci. Eng.*, **2**, 893-911, 1994.

E. Hairer, S.P. Norsett and G. Wanner, Solving Ordinary Differential Equations I, Nonstiff Problems, *Spring-Verlag*, Berlin, 1987.

J.O. Hallquist, NIKE2D-a Vectorized implicit finite deformation finite element code for analyzing the static and dynamic response of 2D solids, *Technical report UCID-19677*, Lawrence Livermore National Laboratory, University of California, 1983.

J.O. Hallquist, G.L. Goudreau and D. Benson, Sliding interfaces with contact-impact in large-scale Lagrangian computations, *Comp. Meth. Appl. Mech. Eng.*, **51**, 107-137, 1985.

J.O. Hallquist, K. Schweizerhof and D. Stillman, Efficiency refinements of contact strategies and algorithms in explicit FE programming, *Computational Plasticity: fundamentals and applications*, D.R.J. Owen *et al.*(Eds.), Pineridge Press, Swansea, 457-481, 1992.

P. Hartley, I. Pillinger and C. Sturgess, Numerical Modelling of Material Deformation Processes, Research, Development and Applications, *Springer-Verlag*, 1992.

K. Hashiguchi and S. Tsutsumi, Shear band formation analysis in soils by the subloading surface model with tangential stress rate effect, *Int. J. Plasticity*, **19**, 1651-1677, 2003.

J.C. Heinrich, P.S. Huyakorn, A.R. Mitchell and O.C. Zienkiewicz, An 'upwind' finite element scheme for two-dimensional convective transport equation, *Int. J. Num. Meth. Eng.*, **11**, 131-143, 1977.

H.D. Hibbitt, P.V. Marcal and J.R. Rice, A finite element formulation for problems of large strain and large displacement, *Int. J. Solids Struct.*, **6**, 1069-1086, 1970.

T. Hisatsune, T. Tabata and S. Masaki, A yield criterion of porous material with anisotropy caused by geometry or distribution of pores, *J. Eng. Mat. Tech. Trans., ASME*, **113**, 425-429, 1991.

G. Hofstetter, J.C. Simo and R.L. Taylor, A modified cap model: closest point solution algorithms, *Int. J. Comp. Struct.*, **46**, 203-214, 1993.

B.P. Holownia, Balanced die method for metal powder compaction, *Powder Metall.*, **39**(3), 207-209, 1996.

S.Y. Hsu and O.H. Griffin, On stability and efficiency of numerical integration of endochronic constitutive equations, *Comp. Struct.*, **44,** 657-665, 1992.

S.Y. Hsu and O.H. Griffin, Algorithmic tangent matrix approach for mixed hardening model of endochronic plasticity, *Comp. Meth. Appl. Mech. Eng.*, **133**, 1-14, 1996.

S.Y. Hsu, S.K. Jain and O.H. Griffin, Verification of endochronic theory for nonproportional loading paths, *ASCE J. Eng. Mech.*, **117**, 110-131, 1991.

J. Huetink, Analysis of forming processes based on a combined Eulerian-Lagrangian finite element formulation, *Numerical Methods in Industrial Forming Processes, NUMIFORM 82*, J.F.T. Pittman *et al.* (Eds.), Pineridge Press, Swansea, 501-509, 1982.

J. Huetink, R. Akkerman, H.G. Mooi and G. Rekers, Aspect of Lagrangian-Eulerian formulation, *Simulation of Materials Processing: Theory, Methods and Applications, NUMIFORM 95*, S-F. Shen and P. Dawson (Eds.), A.A. Balkema, 67-76, 1995.

T.J.R. Hughes and M. Mallet, A new finite element formulation for computational fluid dynamics; The generalized streamline operator for multi-dimensional advective-diffusive systems, *Comp. Mech. Appl. Meth. Eng.*, **58**, 305-328, 1986.

T.J.R. Hughes, R.L. Taylor, J.L. Sackman, A. Curnier and W. Kanoknukulchai, A finite element method for a class of contact-impact problems, *Comp. Mech. Appl. Meth. Eng.*, **8**, 249-276, 1976.

T.J.R. Hughes and J. Winget, Finite rotation effects in numerical integration of rate constitutive equations arising in large deformation analysis, *Int. J. Num. Meth. Eng.*, **15**, 1862-1867, 1980.

A. Ibrahimbegovic and L. Chorfi, Viscoplasticity model at finite deformation with combined isotropic and kinematic hardening, *Comp. Struct.*, **77**, 509-525, 2000.

S. Im and S.N. Atluri, A study of two finite strain plasticity models: an internal time theory using Mandel's director concept and a general isotropic/kinematic hardening theory, *Int. J. Plasticity*, **3,** 163, 1987.

M.M. Iordache and K. Willam, Localized failure analysis in elastoplastic Cosserat continua, *Comput. Meth. Appl. Mech. Eng.*, **151**, 559-586, 1998.

H. Ishikawa, Constitutive model of plasticity in finite deformation, *Int. J. Plasticity,* **15**, 299-317, 1999.

W.D. Iwan, On a class of models for the yielding behaviour of continuous and composite systems, *J. Appl. Mech., Trans. ASME*, **34**, 612-617, 1967.

G. Jaumann, Geschlossenes system physikalischer und chemischer differenzialgesetze. Sitzber. Akad. Wiss. Wien (Ha) **120**, 385-530, 1911.

G. Jelenic and M. Saje, A kinematically exact space finite strain beam model - finite element formulation by generalized virtual work principle, *Comp. Meth. Appl. Mech. Eng.*, **120**, 131-161, 1995.

A.G.K. Jinka, R.W. Lewis and D.T. Gethin, Finite element simulation of powder compaction via the flow formulation, *Advances in Powder Metallurgy*, L.F. Pearl *et al.*(Eds.), Princeton, 1991.

G.C. Johnson and D.J. Bammann, A discussion of stress rates in finite deformation problems, *Int. J. Solids Struct.*, **20,** 725-737, 1984.

J-W. Ju, R.L. Taylor and L.Y. Cheng, A consistent finite element formulation of nonlinear frictional contact problems, *Numerical Methods in Engineering: Theory and Applications, NUMETA 87*, G.N. Pande and J. Middleton (Eds.), **1**, 1-13, 1987.

M.G. Katona, A general family of single-step methods for numerical time integration of structural dynamic equations, *Numerical Methods in Engineering: Theory and Applications, NUMETA 85*, J. Middleton and G.N. Pande (Eds.), A.A. Balkema, 213-225, 1985.

M.G. Katona and O.C. Zienkiewicz, A unified set of single step algorithms - Part 3: The Beta-*m* method, a generalization of the Newmark scheme, *Int. J. Num. Meth. Eng.*, **21**, 1345-1359, 1985.

D.W. Kelly, J.P.De S.R. Gago, O.C. Zienkiewicz and I. Babuska, A posteriori error analysis and adaptive processes in the finite element method: Part I - error analysis, *Int. J. Num. Meth. Eng.*, **19**, 1593-1619, 1983.

A.S. Khan and S. Huang, Continuum Theory of Plasticity, *Wiley*, New York, 1995.

A.S. Khan and M.K. Jackson, On the evolution of isotropic and kinematic hardening with finite plastic deformation; Part I: Compression/tension loading of OFHC copper cylinders, *Int. J. Plasticity*, **15**, 1265-1275, 1999.

A.R. Khoei, An integrated software environment for finite element simulation of powder compaction processes, *J. Mater. Proc. Tech.*, **130-131**, 171-177, 2002.

A.R. Khoei, PCS_SUT: A finite element software for simulation of powder forming processes, *J. Mater. Proc. Tech.*, **125-126**, 602-607, 2002.

A.R. Khoei, Numerical simulation of powder compaction processes using an inelastic finite element analysis, *Materials & Design*, **23**, 523-529, 2002.

A.R. Khoei, Application of a failure analysis in powder forming processes, *J. Mater. Proc. Tech.*, **143-144,** 46-51, 2003.

A.R. Khoei, M. Anahid and M. Mofid, An application of arbitrary Lagrangian-Eulerian method in numerical simulation of forming processes using cap plasticity model, *12th Int. Sci. Conf. Achievements in Mechanical and Materials Engineering, AMME'2003*, Gliwice, Poland, 2003.

A.R. Khoei and A.R. Azami, 3D computational modeling of powder compaction processes using a three-invariant hardening cap plasticity model, Submitted to: *Finite Elem. Analy. Des.*, 2004.

A.R. Khoei and A.R. Azami, A single cone-cap plasticity with an isotropic hardening rule for powder materials, *Int. J. Mech. Sciences,* 2005. (in press)

A.R. Khoei, A.R. Azami and S. Azizi, Computational modeling of 3D powder compaction processes, *J. Mater. Proc. Tech.*, 2005. (in press)

A.R. Khoei, A.R. Azami and S.M. Haeri, Implementation of plasticity based models in dynamic analysis of saturated-unsaturated earth and rockfill dams, *Comput. Geotechnics,* **31**, 385-410, 2004.

A.R. Khoei and S. Azizi, Numerical simulation of 3D powder compaction processes using cone-cap plasticity theory, *Materials & Design,* **26,** 137-147, 2005.

A.R. Khoei and A. Bakhshiani, A hypoelasto-plastic finite strain simulation of powder compaction processes with density dependent endochronic model, *Int. J. Solids Struct.,* **41,** 6081-6110, 2004.

A.R. Khoei and A. Bakhshiani, A constitutive model for finite deformation of endochronic plasticity in powder forming processes, *J. Mater. Proc. Tech.,* 2005. (in press)

A.R. Khoei and A. Bakhshiani, A hypoelasto-viscoplastic endochronic model of numerical simulation of shear band localization, *Finite Elem. Analy. Des.*, 2005. (in press)

A.R. Khoei, A. Bakhshiani and M. Mofid, Finite strain endochronic plasticity with reference to metal tube under torsion, *Eng. Comput.*, **20**, 248-273, 2003.

A.R. Khoei, A. Bakhshiani and M. Mofid, An implicit algorithm for hypoelasto-plastic and hypoelasto-viscoplastic endochronic theory in finite strain isotropic-kinematic-hardening model, *Int. J. Solids Struct.*, **40**, 3393-3423, 2003.

A.R. Khoei, A. Bakhshiani and M. Mofid, An endochronic plasticity model for numerical simulation of industrial powder compaction processes, *Comm. Num. Meth. Eng.*, **19**, 521-534, 2003.

A.R. Khoei, A. Bakhshiani and M. Mofid, An endochronic plasticity model for finite strain deformation of powder forming processes, *Finite Elem. Analy. Des.*, 2003.

A.R. Khoei, A. Bakhshiani and M. Mofid, Numerical simulation of powder compaction processes using an endochronic theory in finite plasticity model, *Int. J. Nonlinear Science Numer. Simul.*, **4**, 17-30, 2003.

A.R. Khoei and S.A. Gharehbaghi, Modelling of localized plastic deformation via the adaptive mesh refinement, *Int. J. Nonlinear Science Numer. Simul.*, **4**, 31-46, 2003.

A.R. Khoei and S. Iranfar, 3D numerical simulation of elasto-plastic behaviour in powder compaction process using a quasi-nonlinear technique, *J. Mater. Proc. Tech.*, **143-144,** 886-890, 2003.

A.R. Khoei and N. Jamali, On the implementation of a multi-surface kinematic hardening plasticity and its applications, *Int. J. Plasticity,* 2005. (in press)

A.R. Khoei and R.W. Lewis, Finite element modelling of large elasto-plastic deformation for the dynamic analysis of powder compaction processes, *Computational Plasticity: Fundamentals and Applications*, D.R.J. Owen *et al.* (Eds.), CIMNE, Barcelona, **2**, 1329-1334, 1997.

A.R. Khoei and R.W. Lewis, Modelling of powder compaction processes via the displacement formulation and adaptive analysis, *Numerical Methods in Industrial Forming Processes, NUMIFORM 98*, 1998.

A.R. Khoei and R.W. Lewis, Numerical simulation of elasto-plastic analysis in metal powder forming using adaptive methods, *Fourth World Congress on Computational Mechanics, WCCM IV*, 1998.

A.R. Khoei and R.W. Lewis, Finite element simulation for dynamic large elasto-plastic deformation in metal powder forming, *Finite Elem. Analy. Des.*, **30**, 335-352, 1998.

A.R. Khoei and R.W. Lewis, Adaptive finite element remeshing in a large deformation analysis of metal powder forming, *Int. J. Num. Meth. Eng.*, **45**, 801-820, 1999.

A.R. Khoei and R.W. Lewis, *H*-adaptive finite element analysis for localization phenomena with reference to metal powder forming, *Finite Elem. Anal. Des.*, **38**, 503-519, 2002.

A.R. Khoei, R.W. Lewis and O.C. Zienkiewicz, Application of the finite element method for localized failure analysis in dynamic loading, *Finite Elem. Anal. Des.*, **27**, 121-131, 1997.

A.R. Khoei, M. Mofid and A. Bakhshiani, Modeling of powder compaction process using an endochronic plasticity model, *J. Mater. Proc. Tech.*, **130-131**, 175-180, 2002.

A.R. Khoei, A. Riahi, A.R. Tabarraie and S.A. Gharehbaghi, Error estimation, adaptivity and data transfer in enriched plasticity continua to analysis of shear band localization, Submitted to: *Applied Mathematical Modelling,* 2004.

A.R. Khoei, A.R. Tabarraie and S.A. Gharehbaghi, H-adaptive mesh refinement for shear band localization in elasto-plasticity Cosserat continuum, *Comm. Nonlinear Science Numer. Simul.*, **10**, 253-286, 2005.

K.T. Kim and J.H. Cho, A densification model for mixed metal powder under cold compaction. *Int. J. Mechanical Science,* **43**, 2929-2946, 2001.

M. Kobayashi and N. Ohno, Implementation of cyclic plasticity models based on a general form of kinemetic hardening, *Int. J. Num. Meth. Eng.*, **53**, 2217-2238, 2002.

S. Kobayashi, S-I. Oh and T. Altan, Metal Forming and the Finite Element Method, *Oxford University Press*, New York, 1989.

A.S. Kravchuk, On the theory of contact problems taking account of friction on the contact surface, *J. Appl. Math. Mech. (PMM)*, **44**, 83-85, 1981.

R.D. Krieg, A practical two surface plasticity theory, *J. Appl. Mech.*, **42**, 641-646, 1975.

R.D. Krieg and S.M. Key, Implementation of a time dependent plasticity theory into structural computer programs. *Constitutive Equations in Viscoplasticity: Computational and Engineering Aspects*, ASME, **20**, 125-137, 1976.

R.D. Krieg and D.B. Krieg, Accuracies of numerical solution methods for the elastic-perfectly plastic model, *J. Pressure Vessel Technol., ASME*, **99**, 510-515, 1977.

H. Kudo and S. Matsubara, Joint examination project of validity of various numerical methods for the analysis of metal forming processes, *Metal Forming Plasticity* (Ed. H. Lippmann), Springer-Verlag, Berlin, 378, 1979.

M. Kuroda, Interpretation of behavior of metals under large plastic shear deformations: Comparison of macroscopic predictions to physically based predictions, *Int. J. Plasticity*, **15**, 1217-1236, 1999.

H.A. Kuhn and C.L. Downey, Deformation characteristics and plasticity theory of sintered powder materials, *Int. J. Powder Metall.*, **7**, 15-25, 1971.

P.V. Lade, Elasto-plastic stress-strain theory for cohesionless soil with curved yield surfaces, *Int. J. Solids Struct.*, **13**, 1019-1035, 1977.

P.V. Lade and J.M. Duncan, Elasto-plastic stress-strain theory for cohesionless soil, *J. Geotech. Eng. Div., ASCE*, **101**, 1037-1053, 1975.

P.V. Lade and K. Kim Moon, Single hardening constitutive model for soil, rock and concrete, *Int. J. Solids Structures*, **32**, 1963-1978, 1995.

C.H. Lee and S. Kobayashi, New solutions to rigid-plastic deformation problems using a matrix method, *J. Eng. Ind., Trans. ASME*, **95**, 865-873, 1973.

E. Lee, Elastic-plastic deformation at finite strains, *J. Appl. Mech.*, **36,** 1-6, 1969.

E.H. Lee, R.L. Mallett and W.H. Yang, Stress and deformation analysis of metal extrusion forming process, *Comp. Meth. Appl. Mech. Eng*., **10**, 339-353, 1977.

N.S. Lee and K.J. Bathe, Error indicators and adaptive remeshing in large deformation finite element analysis, *Finite Elem. Anal. Des*., **16**, 99-139, 1994.

T.H. Lehmann, Z.H. Guo and H.Y. Liang, The conjugacy between Cauchy stress and logarithm of the left stretch tensor, *Eur. J. Mech. A/Solids,* **10**, 395-404, 1991.

R.W. Lewis, A.G.K. Jinka and D.T. Gethin, Computer aided simulation of metal powder die compaction process, *Powder Metall. Int.*, **25**, 287-293, 1993.

R.W. Lewis and A.R. Khoei, Numerical modelling of large deformation in Metal Powder Forming, *Comp. Meth. Appl. Mech. Eng*., **159**, 291-328, 1998.

R.W. Lewis and A.R. Khoei, Numerical analysis of strain localization in metal powder forming processes, *Int. J. Num. Meth. Eng*., **52**, 489-501, 2001.

R.W. Lewis and A.R. Khoei, A plasticity model for metal powder forming processes. *Int. J. Plasticity*, **17**, 1659-1692, 2001.

R.W. Lewis and B.A. Schrefler, The Finite Element Method in the Deformation and Consolidation of Porous Media, *Wiley*, New York, 1987.

R.W. Lewis and D.V. Tran, Finite element approach to problems in particulate media with special reference to powder metal forming, *Bull. Tech. Univ. Istanbul*, **47**, 295-310, 1994.

H.C. Lin, B.J. Hsieh and R.A. Valentin, The application of endochronic plasticity theory in modeling the dynamic inelastic response of structural systems, *Nucl. Engng. Des.*, **66,** 213-221, 1981.

H.C. Lin and H.C. Wu, Strain rate effect in the endochronic theory of viscoplasticity, *J. Appl. Mech.*, **98**, 92-96, 1976.

S.H. Lo, A new mesh generation scheme for arbitrary planar domains, *Int. J. Num. Meth. Eng.*, **21**, 1403-1425, 1985.

B. Loret and J.H. Prevost, Accurate numerical solutions for Drucker-Prager elastic plastic models, *Comp. Meth. Appl. Mech. Eng.*, **54**, 258-277, 1986.

R.H. Mallett and P.V. Marcal, Finite element analysis of nonlinear structures, *J. Struct. Div. Am. Soc. Civ. Eng.*, **94**, 2081-2105, 1968.

H.C. Martin, Derivation of stiffness matrices for the analysis of large deflection and stability problems, *Proc. 1st Conf. on Matrix Methods in Struct. Mech.*, 607-715, 1966.

G. Massing, Eigenspannungen und verfestigung beim messing, *Proc. 2nd Int. Cong. Appl. Mech.*, Zurich, Switzerland, 1926.

A. Matzenmiller and R.L. Taylor, A return mapping algorithm for isotropic elastoplasticity, *Int. J. Num. Meth. Eng.*, **37**, 813-826, 1994.

D.L. McDowell, Evaluation of intersection conditions for two surface plasticity theory, *Int. J. Plasticity*, **5**, 29-50, 1989.

R.M. McMeeking, G. Jefferson and G.K. Haritos, Elastic and viscoelastic response of finite particle junctions in granular materials. *Recent Development in Computer Modeling of Powder Metallurgy Process*. Zavaliangos and Laptev (Eds.), IOS Press, 2001.

R.M. McMeeking and R.J. Rice, Finite element formulations for problems of large elastic-plastic deformation, *Int. J. Solids Struct.*, **11**, 601, 1975.

R. Michalowski and Z. Mroz, Associated and non-associated sliding rules in contact friction problems, *Arch. Mech.*, **39**, 259-276, 1978.

G.P. Mitchell and D.R.J. Owen, Numerical solutions for elastic-plastic problems, *Eng. Comp.*, **5**, 274-284, 1988.

F.J. Montans, Bounding surface plasticity model with extended Massing behavior, *Comp. Meth. Appl. Mech. Eng.*, **182**, 135-162, 2000.

F.J. Montans, Implicit algorithm for multilayer J_2 plasticity, *Comp. Meth. Appl. Mech. Eng.*, **189**, 673-700, 2000.

D.F. Moore, Principles and Applications of Tribology, *Pergamon Press*, Oxford, 1975.

K. Mori and K. Osakada, Analysis of the forming process of sintered powder metals by a rigid-plastic finite-element method, *Int. J. Mech. Sci.*, **29**, 229-238, 1987.

Y. Morimoto, T. Hayashi and T. Taikei, Mechanical behaviour of powders during compaction in a mould with variable cross sections, *Int. J. Powder Metall.*, **18**, 129-145, 1982.

P. Mosbah, D. Bouvard, J. Lauiear and E. Oue, Rheological Characterization of metal powder following various loading paths, *Powder Metallurgy World Congress*, Paris, **1**, 677-680, 1994.

Z. Mroz, On the description of anisotropic work-hardening, *J. Mech. Phys. Solids*, **15**, 163-175, 1967.

H.B. Muhlhaus and E. Ainfantis, A variational principle for gradient plasticity, *Int. J. Solids Struct.*, **28**, 845-857, 1991.

H. Murakami and H.E. Read, A second-order numerical scheme for integrating the endochronic plasticity equations, *Comput. Structures,* **31**, 663-672, 1989.

J.C. Nagtegaal and J.E. De Jong, Some computational aspects of elastic-plastic large strain analysis, *Int. J. Num. Meth. Eng.*, **17**, 15, 1981.

S.E. Navti, Numerical Simulation of Free Surface Flow During Mould Filling, *University of Wales Swansea*, C/Ph/206/96, 1996.

G.C. Nayak and O.C. Zienkiewicz, Elasto-plastic stress analysis, Generalization for various constitutive relations including strain softening, *Int. J. Num. Meth. Eng.*, **5**, 113-135, 1972.

D.J. Naylor, In Stress-Strain Laws for Soil Development in Soil Mechanics, (Edited by: R.F. Scott), *Applied Science*, London, 39-67, 1978.

D.J. Naylor, G.N. Pande, B. Simpson and R. Tabb, Finite Elements in Geotechnical Engineering, *Pineridge Press*, Swansea, 1981.

A. Needleman, Material rate dependence and mesh sensitivity on localization problems, *Comp. Meth. Appl. Mech. Eng.*, **67**, 69-86, 1988.

I. Nelson, Constitutive models for use in numerical computations, *Proc. Int. Symp. on Dynamic Methods in Soil and Rock Mechanics*, Balkema, Rotterdam, 45-97, 1978.

I. Nelson and G.Y. Baladi, Outrunning ground shock computed with different models, *J. Eng. Mech. Div., ASCE*, **103**, 377-393, 1977.

S. Nemat-Nasser, Decomposition of strain measures and their rates in finite deformation elastoplasticity, *Int. J. Solids Struct.*, **15,** 155-166, 1979.

S. Nemat-Nasser, On finite deformation elasto-plasticity, *Int. J. Solids Struct.*, **18,** 857-872, 1982.

R.E. Nickell, Nonlinear dynamics by mode superposition, *Comp. Meth. Appl. Mech. Eng.*, **7**, 107-129, 1976.

J.T. Oden, D.R. Bhandari, G. Yagewa and T.J. Chung, A new approach to the finite element formulation and solution of a class of problems in coupled thermo-elastoviscoplasticity of solids, *Nuclear Eng. Design*, **24**, 240, 1973.

J.T. Oden and J.A.C. Martins, Models and computational methods for dynamic friction phenomena, *Comp. Meth. Appl. Mech. Eng.*, **52**, 527-634, 1985.

J.T. Oden and E.B. Pires, Nonlocal and nonlinear friction laws and variational principles for contact problems in elasticity, *J. Appl. Mech.*, **50**, 67-76, 1983.

H-K. Oh and J-K. Lee, A study of the extrusion of sintered porous metal, *J. Mech. Wkg. Tech.*, **11**, 53, 1985.

N. Ohno and J.D. Wang, Kinematic hardening rules with critical state of dynamic recovery, part I: formulation and basic features for ratcheting behavior, *Int. J. Plasticity*, **9**, 375-390, 1993.

J. Oliver, S. Oller and J.C. Cante, A plasticity model for simulation of industrial powder compaction processes, *Int. J. Solids Struct.*, **33**, 3161-3178, 1996.

M. Ortiz and J.J. Quigley, Adaptive mesh refinement in strain localization problems, *Comput. Meth. Appl. Mech. Eng.*, **90**, 781-804, 1991.

M. Ortiz and E.P. Popov, Accuracy and stability of integration algorithms for elastoplastic constitutive relations, *Int. J. Num. Meth. Engrg.*, **21**, 1561- 1576, 1985.

M. Ortiz and J.C. Simo, An analysis of a new class of integration algorithms for elastoplastic constitutive relations, *Int. J. Num. Meth. Engrg.*, **23**, 353- 366, 1986.

K. Osakada, J. Nakano and K. Mori, Finite element method for rigid-plastic analysis of metal forming-formulation for finite deformation, *Int. J. Mech. Sci.*, **24**, 459-468, 1982.

D.R.J. Owen and E. Hinton, Finite Elements in Plasticity: Theory and Practice, *Pineridge Press*, Swansea, 1980.

M. Oyane, S. Shima and Y. Kono, Theory of plasticity for porous metals, *Bull. JSME*, **16**, 1254, 1973.

M. Oyane, S. Shima and T. Tabata, Consideration of basic equations, and their application, in the forming of metal powders and porous metals, *J. Mech. Wkg. Tech.*, **1**, 325, 1978.

M. Oyane and T. Tabata, Slip-line field theory and upper-bound theory for porous materials, *J. Jap. Soc. Tech. Plast.*, **15**, 43, 1974.

W.F. Pan, Endochronic simulation for finite viscoplastic deformation, *Int. J. Plasticity*, **13**, 571-586, 1997.

W.F. Pan and C.H. Chern, Endochronic description of viscoplastic behavior of material under multiaxial loading, *Int. J. Solids Struct.*, **34**, 2131-2160, 1997.

W.F. Pan, W.J. Chiang and C.K. Wang, Endochronic analysis for rate-dependent elasto-plastic deformation, *Int. J. Solids Struct.*, **36**, 3215-3237, 1999.

W.F. Pan, T.H. Lee and W.C. Yeh, Endochronic analysis for finite elastoplastic deformation and application to metal tube under torsion and metal rectangular block under biaxial compression, *Int. J. Plasticity*, **12,** 1287-1316, 1996.

H. Parisch, A consistent tangent stiffness matrix for three-dimensional contact analysis, *Int. J. Num. Meth. Eng.*, **28**, 1803-1812, 1989.

M. Pastor, J. Peraire and O.C. Zienkiewicz, Adaptive remeshing for shear band localization problem, *Arch. Appl. Mech.*, **61**, 30-39, 1991.

M. Pastor, C. Rubio, P. Mira, J. Peraire, J.P. Vilotte and O.C. Zienkiewicz, Numerical analysis of localization, *Numerical Models in Geomechanics*, G.N. Pande and S. Pietruszczak (Eds.), A.A. Balkema, 339-348, 1992.

J. Peraire, M. Vahdati, K. Morgan and O.C. Zienkiewicz, Adaptive remeshing for compressible flow computations, *J. Comp. Phys.*, **72**, 449-466, 1987.

A. Perez-Foguet, A. Rodriguez-Ferran, A. Huerta, Consistent tangent matrices for density-dependent finite plasticity models, *Int. J. Num. Analyt. Meth. Geomech.*, **25,** 1045-1075, 2001.

A. Perez-Foguet, A. Rodriguez-Ferran and A. Huerta, Efficient and accurate approach for powder compaction problems, *Computational Mechanics,* (2003) in press.

D. Peric and D.R.J. Owen, Computational model for 3-D contact problems with friction based on the penalty method, *Int. J. Num. Meth. Eng.*, **35**, 1289-1309, 1992.

D. Peric, J. Yu and D.R.J. Owen, On error estimates and adaptivity in elastoplastic solids: Applications to the numerical simulation of strain localization in classical and Cosserat continua, *Int. J. Num. Meth. Eng.*, **37**, 1351-1379, 1994.

H. Petersson and E.P. Popov, Constitutive equations for generalized loadings, *J. Eng. Mech. Div., ASCE,* **104**, 611, 1977.

G. Pijaudier-Cabot and A. Benallal, Strain localization and bifurcation in a nonlocal continuum, *Int. J. Solids Structures,* **30**, 1761-1771, 1993.

J.F.T. Pittman, O.C. Zienkiewicz, R.D. Wood and J.M. Alexander, Numerical Analysis of Forming Processes, *Wiley*, 1984.

S. Pietruszczak and D.F.E. Stolle, Deformation of strain softening materials, Part I: Objectivity of finite element solutions based on conventional strain softening formulations, *Comp. Geotech.*, **1**, 99-115, 1985.

S. Pietruszczak and D.F.E. Stolle, Deformation of strain softening materials, Part II: Modelling of strain softening response, *Comp. Geotech.*, **4**, 109-123, 1987.

J.P. Ponthot, Efficient mesh management in Eulerian-Lagrangian method for large deformations analysis, *Numerical Methods in Industrial Forming Processes*, E.G. Thompson *et al.*(Eds.), A.A. Balkema, 203-210, 1989.

J.P. Ponthot, Unified stress update algorithms for the numerical simulation of large deformation elasto-plastic and elasto-viscoplastic processes, *Int. J. Plasticity*, **18,** 91-126, 2002.

J.H. Prevost, Plasticity theory for soil stress-strain behavior, *J. Eng. Mech. Div., ASCE,* **104**, 1177-1194, 1978.

J.H. Prevost, Two surface versus multi-surface plasticity theories: a critical assessment, *Int. J. Num. Anal. Meth. Geomech.*, **6**, 323-338, 1982.

J.H. Prevost, A simple plasticity theory for frictional cohesionless soils, *Soil Dyn. Earthq. Eng.*, **4**, 9-17, 1985.

R.S. Ransing, I. Cameron, N.P. Lavery, A.R. Khoei, R.W. Lewis and D.T. Gethin, Powder compaction modeling and material characterization, *Proc Int. Workshop Modeling Metal Powder Forming Processes*, Grenoble, France, 189-201, 1997.

R.S. Ransing, D.T. Gethin, A.R. Khoei, P. Mosbah and R.W. Lewis, Powder compaction modelling via the discrete and finite element method, *Mater. Design,* **21,** 263-269, 2000.

R.S. Ransing, R.W. Lewis, D.T. Gethin, A.J.L. Crook and M. Dutko, Compaction of granular materials using deformable discrete elements, *Sixth Int. Conf. Numer. Meth. Indust. Forming Processes, NUMIFORM 98*, 1998.

M.M. Rashid, Incremental kinematics for finite element applications, *Int. J. Num. Meth. Eng.*, **36**, 3937-3956, 1993.

K.W. Reed and S. Atluri, Constitutive modeling and computational implementation for finite strain plasticity, *Int. J. Plasticity*, **1**, 63-87, 1985.

W.D. Reinhardt and R.N. Dubey, Application of objective rates in mechanical modeling of solids, *ASME J. Appl. Mech.*, **63**, 692-698, 1996.

J.R. Rice, The localization of plastic deformation, *Theoretical and Applied Mechanics*, W.T. Koiter (Ed.), North-Holland, Amsterdam, 207-220, 1977.

M.D. Riero and J.M. Prado, Modelling of cold compaction process of metal powders, *Powder Metallurgy World Congress*, Paris, **1**, 685-688, 1994.

R.S. Rivlin, Some comments on the endochronic theory of plasticity, *Int. J. Solids Struct.* **17,** 231-248, 1981.

T. Rodic and D.R.J. Owen, A plasticity theory of friction and joint elements, *Computational Plasticity II: Models, Software and Applications*, D.R.J. Owen *et al.*(Eds.), Pineridge Press, Swansea, 1043-1062, 1989.

A. Rodriguez-Ferran, P. Pegon and A. Huerta, Two stress update algorithms for large strains: Accuracy analysis and numerical implementation, *Int. J. Num. Meth. Eng.*, **40**, 4363-4404, 1997.

A. Rodriguez-Ferran, A. Perez-Foguet and A. Huerta, Arbitrary Lagrangian-Eulerian (ALE) formulation for hyperelastoplasticity, *Int. J. Num. Meth. Eng.*, (2003) in press.

K.H. Roscoe and J.B. Burland, On the generalized stress-strain behaviour of wet clay, *Engineering Plasticity*, J. Heyman and F.A. Leckie (Eds.), Cambridge University Press, England, 535-609, 1968.

K.H. Roscoe, A.N. Schofield and A. Thurairjah, Yielding of clays in state wetter than critical, *Geotechnique*, **13**, 211-240, 1963.

K. Runneson, N.S. Ottossen and D. Peric, Discontinuous bifurcation of elasto-plastic solution at plane stress and plane strain, *Int. J. Plasticity*, **7**, 99-121, 1991.

A. Samuelsson and N.E. Wiberg, Finite element adaptivity in dynamics and elastoplasticity, *The finite element method in the 1990's, O.C. Zienkiewicz's anniversary*, E. Onate *et al.*(Eds.), CIMNE and Springer-Verlag, Barcelona, 152-162, 1991.

I.S. Sandler, On the uniqueness and stability of endochronic theories of material behavior. *J. Appl. Mech.* **45,** 263-266, 1978.

I.S. Sandler and M.L. Baron, Recent development in the constitutive modelling of geological materials, *Proc. 3rd Int. Conf. Numerical Methods in Geomechanics*, Aachen, Germany, 363-376, 1979.

I.S. Sandler, F.L. DiMaggio and G.Y. Baladi, Generalized cap model for geological materials, *J. Geotech. Eng. Div., ASCE*, **102**, 683-699, 1976.

I.S. Sandler and D. Rubin, An algorithm and a modular subroutine for the cap model, *Int. J. Num. Anal. Meth. Geomech.*, **3**, 173-186, 1979.

M. Sava, Constitutive equations for cold pressed iron powder - formulation and experimental identification, *Int. Workshop on Modelling of Metal Powder Forming Processes*, Grenoble, France, 159-167, 1997.

J.C.J. Schellekens and R.de Borst, On the integration of interface elements, *Int. J. Num. Meth. Eng.*, **36**, 43-66, 1993.

J.C.J. Schellekens and R.de Borst, The application of interface elements and enriched or rate-dependent continua to micro-mechanical analyses of fracture in composites, *Comput. Mech.*, **14**, 68-83, 1994.

A. Schofield and P. Wroth, Critical State Soil Mechanics, *McGraw-Hill*, London, 1968.

H.L. Schreyer, R.F. Kulak and J.M. Kramer, Accurate numerical solutions for elastic-plastic models, *J. Pressure Vessel Technol., ASME*, **101**, 226-234, 1979.

G.S. Sekhon and J.L. Chenot, Numerical simulation of continuous chip formulation during non-steady orthogonal cutting, *Eng. Comput.*, **10**, 31-48, 1993.

S.N. Shah, C.H. Lee and S. Kobayashi, Compression of tall, circular, solid cylinders between parallel flat dies, *Proc. Int. Conf. Prod. Eng.*, Tokyo, 295-300, 1974.

S-F. Shen and P.R. Dawson, Simulation of Materials Processing: Theory, Methods and Applications, *NUMIFORM 95, A.A. Balkema,* Rotterdam, 1995.

W.M. Shen, T. Kimura, K. Takita and K. Hosono, Numerical simulation of powder transfer and compaction based on continuum model. *Simulation of Materials Processing: Theory, Methods and Applications* (Ed. K. Mori), 1027-1032, 2001.

S. Shima and M. Oyane, Plasticity theory for porous metals, *Int. J. Mech. Sci.*, **18**, 285-291, 1976.

S. Shima, T. Tabata, M. Oyane and T. Kawakami, Upper bound theory for deformation of porous materials, *Mem. Fac. Eng. Kyoto Univ.*, **38**, 117, 1976.

J.C. Simo, A framework for finite strain elastoplasticity based on maximum plastic dissipation and the multiplicative decomposition. Part I: Continuum formulation, *Comp. Meth. Appl. Mech. Eng.*, **66**, 199-219, 1988.

J.C. Simo, A framework for finite strain elastoplasticity based on maximum plastic dissipation and the multiplicative decomposition. Part II: Computational aspects, *Comp. Meth. Appl. Mech. Eng.*, **68,** 1-31, 1988.

J.C. Simo and T.J.R. Hughes, General return mapping algorithms for rate-independent plasticity, *Constitutive Laws for Engineering Materials: Theory and Applications*, Elsevier, **1,** 221-232, 1987.

J.C. Simo and T.J.R. Hughes, Computational Inelasticity, *Springer-Verlag*, New York, 1998.

J.C. Simo, J.W. Ju, K.S. Pister and R.L. Taylor, An assessment of the cap model: Consistent return algorithms and rate-dependent extensions, *J. Eng. Mech. Div., ASCE,* **114**, 191-218, 1988.

J.C. Simo and T.A. Laursen, An augmented Lagrangian treatment of contact problems involving friction, *Comp. Struct.*, **42**(1), 97-116, 1992.

J.C. Simo, J. Oliver and F. Armero, An analysis of strong discontinuities induced by strain-softening in rate-independent inelastic solids, *Comput. Mech.* **12**, 227-249, 1993.

J.C. Simo and M. Ortiz, A unified approach to finite deformation elastoplastic analysis based on the use of hyperelastic constitutive equations, *Comp. Meth. Appl. Mech. Engrg.*, **49**, 221-245, 1985.

J.C. Simo and R.L. Taylor, Consistent tangent operator for rate-independent elastoplasticity, *Comp. Meth. Appl. Mech. Eng.*, **48,** 101-118, 1985.

J.C. Simo and R.L. Taylor, A consistent return mapping algorithm for plane strain elastoplasticity, *Int. J. Num. Meth. Engrg.,* 22, 649-670, 1986.

J.C. Simo, P. Wriggers and R.L. Taylor, A perturbed Lagrangian formulation for the finite element solution of contact problems, *Comp. Meth. Appl. Mech. Eng.*, **51**, 163-180, 1985.

S.W. Sloan, Substepping schemes for the numerical integration of elastoplastic stress-strain relations, *Int. J. Num. Meth. Eng.*, **24**, 893-911, 1987.

S.W. Sloan and M.F. Randolph, Numerical prediction of collapse loads using finite element methods, *Int. J. Num. Anal. Geomech.*, **6**, 47-76, 1982.

A. Sluzalec, Yield function in simulation of the powder metal forming, *Computational Plasticity II: models, software and applications*, D.R.J. Owen et al.(Eds.), Pineridge Press, Swansea, 995-1005, 1989.

R. Sowerby and E. Chu, Rotations, stress rate and strain measures in homogeneous deformation processes, *Int. J. Solids Struct.*, **20,** 1037-1048, 1984.

N.P. Suh, A yield criterion for plastic, frictional, work--hardening granular materials, *Int. J. Powder Metall.*, **5**, 69-77, 1969.

X-K. Sun and K-T. Kim, Simulation of cold die compaction densification behaviour of iron and copper powders by cam-clay model, *Powder Metall.*, **40**, 193-195, 1997.

W. Swift, Length changes in metals under torsion overstrain, *Engineering*, **163**, 253-257, 1947.

T. Tabata, S. Masaki and K. Kamata, Determination of the coefficient of friction between metal powder and die wall in compaction, *Int. J. Plasticity*, **21**, 773-779, 1980.

D. Tabor, Friction - The present state of our understanding, *J. Lubr. Technol.*, **183**, 169-179, 1981.

R.L. Taylor and P. Papadopoulos, On a patch test for contact problems in two dimensions, *Computational Methods in Nonlinear Mechanics*, P. Wriggers and W. Wagner (Eds.), Springer-Verlag, Berlin, 690-702, 1991.

R.M. Thomas and I. Gladwell, Variable-order variable-step algorithms for second-order systems, Part 1: The method, *Int. J. Num. Meth. Eng.*, **26**, 39-53, 1988.

T.Y. Thomas, Plastic Flow and Fracture in Solids, *Academic Press*, New York, 1961.

E.G. Thompson and H.M. Berman, Steady-state analysis of elasto-viscoplastic flow during rolling, *Numerical Analysis of Forming Processes*, J.F.T. Pittman *et al.*(Eds.), Wiley, Chap. **9**, 269-283, 1984.

J.F. Thompson, F.C. Thomas and C.W. Mastin, Automatic numerical generation of body fitted curvilinear coordinate system for field containing any number of arbitrary two-dimensional bodies, *J. Comp. Phys.*, **15**, 299-319, 1974.

D.V. Tran, R.W. Lewis, D.T. Gethin and A.K. Ariffin, Numerical modelling of powder compaction processes: Displacement based finite element method, *Powder Metall.*, **36**, 257-266, 1993.

K.J. Trassoras, M. Tony and B.L. Ferguson, Modelling of powder compaction using the finite element method, *Proc. Powder Metallurgy Conf. Exhibition VI*, 85-104, 1989.

N. Triantafyllidis and E. Ainfantis, A gradient approach to localization of deformation. I. Hyperelastic materials, *J. Elasticity*, **16**, 225-237, 1986.

C. Truesdell, Hypo-elasticity, *J. Rat. Mech. Anal.*, **4**, 83-133, 1955.

C.D. Turner, A micromechanical model of multiaxial powder densification, *Powder Metallurgy World Congress*, Paris, 713-716, 1994.

M.J. Turner, E.H. Dill, H.C. Martin and R.J. Melosh, Large deflection analysis of complex structures subjected to heating and external loads, *J. Aero Space Sci.*, **27**, 97-106, 1960.

K.C. Valanis, A theory of viscoplasticity without a yield surface, Part I: general theory, *Arch. Mech.*, **23**, 517-534, 1971.

K.C. Valanis, A theory of viscoplasticity without yield surface, Part II: Application to mechanical behavior of metals. *Arch. Mech.*, **23**, 535-551, 1971.

K.C. Valanis, On the formulation of endochronic theory of plasticity, *Arch. Mech.*, **27**, 857-872, 1975.

K.C. Valanis, Fundamental consequences of a new intrinsic time measure: plasticity as a limit of endochronic theory, *Archs Mech.*, **32,** 171-191, 1980.

K.C. Valanis, Continuum formulation of plasticity, *ASME J. Engng. Mater. Tech.*, **106**, 367-375, 1984.

K.C. Valanis and J. Fan, A numerical algorithm for endochronic plasticity and comparison with experiment, *Comp. Struct.*, **19**, 717-724, 1984.

K.C. Valanis and C.F. Lee, Endochronic theory of cyclic plasticity with applications, *J. Appl. Mech.,* **51**, 367-374, 1984.

K.C. Valanis and H.E. Read, An endochronic plasticity theory for concrete, *Mech. Mater.*, **5**, 277-295, 1986.

K.C. Valanis and J.F. Peters, An endochronic plasticity theory with shear–volumetric coupling, *Int. J. Num. Anal. Meth. Geomech.*, **15,** 77-102, 1991.

H.A.M. Van Eekelen, Isotropic yield surfaces in three dimensions for use in soil mechanics, *Int. J. Num. Anal. Meth. Geomech.*, **4,** 89-101, 1980.

P.A. Vermeer, A double hardening model for sand, *Geotechnique*, **28**, 413-433, 1978.

P.A. Vermeer, *Formulation and analysis of sand deformation problems*. Ph.D. Thesis, Department of Civil Engineering, Delft University, The Netherlands, 1980.

W.M. Wang, L.J. Sluys and R. de Borst, Viscoplasticity for instabilities due to strain softening and strain-rate softening, *Int. J. Num. Meth. Eng.*, **40**, 3839-3864, 1997.

O. Watanbe and S.N. Atluri, A new endochronic approach to computational elastoplasticity: example of a cyclically loaded cracked plate, *ASME, J. Appl. Mech.*, **52,** 857-864, 1985.

O. Watanbe and S.N. Atluri, Internal time, general internal variable and multi-yield surface theories of plasticity and creep: A unification of concepts, *Int. J. Plasticity*, **2**, 37, 1986.

T.J. Watson and J.A. Wert, On the development of constitutive relations for metallic powders, *Metallurgical Trans.*, **24**, 1993-2071, 1993.

J.L. Wilkins, Calculation of elastic-plastic flow. *Methods of Computational Physics*, Academic Press, New York, **8**, 1964.

W.L. Wood, A unified set of single step algorithms - Part 2: Theory, *Int. J. Num. Meth. Eng*., **20**, 2303-2309, 1984.

P. Wriggers and J.C. Simo, A note on tangent stiffness for fully nonlinear contact problems, *Comm. Appl. Num. Meth.*, **1**, 199-203, 1985.

P. Wriggers, J.C. Simo and R.L. Taylor, Penalty and augmented Lagrangian formulations for contact problems, *Numerical Methods in Engineering: Theory and Applications, NUMETA 85*, J. Middleton and G.N. Pande (Eds.), A.A. Balkema, 97-106, 1985.

H.C. Wu and C.C. Ho, An investigation of transient creep by means of endochronic viscoplasticity and experiment, *ASME J. Eng. Mat. Tech.*, **117**, 260-267, 1995.

H.C. Wu, J.K. Lu and W.F. Pan, Endochronic equations for finite plastic deformation and application to metal tube under torsion, *Int. J. Plasticity*, **32**, 1079-1097, 1995.

H.C. Wu, Z.K. Wang, and M.R. Aboutorabi, Endochronic modeling of sand in true triaxial test, *J. Eng. Mech., ASCE*, **111**, 1257-1276, 1985.

H.C. Wu, Z. Xu and P. Wang, Torsion text of aluminum in the large strain rang, *Int. J. Plasticity,* **13**, 873-892, 1998.

H.C. Wu and M.C. Yip, Strain rate and strain-rate history effects on the dynamic behavior of metallic materials, *Int. J. Solids Struct.*, **16**, 515-536, 1980.

H. Xiao, O.T. Bruhns and A. Meyers, Logarithmic strain, logarithmic spin and logarithmic rate, *Acta Mech.*, **124**, 89-105, 1997.

H. Xiao, O.T. Bruhns and A. Meyers, On objective corotational rates and their defining spin tensors. *Int. J. Solids Struct.*, **35**, 4001-4014, 1998.

H. Xiao, O.T. Bruhns and A. Meyers, Large strain responses of elastic-prefect plasticity and kinematic hardening plasticity with the logarithmic rate: Swift effect in torsion, *Int. J. Plasticity,* **17**, 211-235, 2001.

Y.M. Xie, N. Bicanic and O.C. Zienkiewicz, A note on local error estimators for adaptive time stepping in dynamic analysis, *University of Wales Swansea*, CR/622/88, 1988.

Y. Yamada, N. Yishimura and T. Sakurai, Plastic stress-strain matrix and its application for the solution of elasto-plastic problems by the finite element method, *Int. J. Mech. Sci.*, **10**, 343-354, 1968.

M.A. Yerry and M.S. Shepard, Automatic three-dimensional mesh generation by the modified-octree technique, *Int. J. Num. Meth. Eng*., **20**, 1965-1990, 1984.

P.J. Yoder and B.G. Whirley, On the numerical implementation of elastoplastic models. *J. Appl. Mech., ASME*, **51**, 283-228, 1984.

F. Yoshida, A constitutive model of cyclic plasticity, *Int. J. Plasticity*, **16**, 359-380, 2000.

S. Zaremba, Sur une forme perfectionee de la theorie de la relaxation, *Bull. Intl. Acad. Sci.* Cracovie, 594-614, 1903.

O.C. Zienkiewicz, Flow formulation for numerical solution of forming processes, *Numerical Analysis of Forming Processes*, J.F.T. Pittman *et al.*(Eds.), Wiley, Chap. **1**, 1-44, 1984.

O.C. Zienkiewicz and P.N. Godbole, Flow of plastic and viscoplastic soilds with special reference to extrusion and forming processes, *Int. J. Num. Meth. Eng.*, **8**, 3-16, 1974.

O.C. Zienkiewicz and G.C. Huang, Adaptive modelling of transient coupled metal forming processes, *Numerical Methods in Industrial Forming Processes, NUMIFORM 89*, E.G. Thompson *et al.*(Eds.), A.A. Balkema, 3-10, 1989.

O.C. Zienkiewicz, G.C. Huang and Y.C. Liu, Adaptive FEM computation of forming processes, Application to porous and non-porous materials, *Int. J. Num. Meth. Eng.*, **30**, 1527-1553, 1990.

O.C. Zienkiewicz, M. Huang and M. Pastor, Localization problems in plasticity using finite elements with adaptive remeshing, *Int. J. Num. Anal. Meth. Geomech.*, **19**, 127-148, 1995.

O.C. Zienkiewicz, C. Humpheson and R.W. Lewis, Associated and non-associated visco-plasticity and plasticity in soil mechanics, *Geotechnique*, **25**, 671-689, 1975.

O.C. Zienkiewicz, Y.C. Liu and G.C. Huang, Error estimation and adaptivity in flow formulation for forming problems, *Int. J. Num. Meth. Eng.*, **25**, 23-42, 1988.

O.C. Zienkiewicz, Y.C. Liu and G.C. Huang, Error estimates and convergence rates for various incompressible elements, *Int. J. Num. Meth. Eng.*, **28**, 2191-2202, 1989.

O.C. Zienkiewicz and Z. Mroz, Generalised plasticity formulation and applications to geomechanics, *Mechanics of Materials*, C.S. Desai and R.H. Gallagher (Eds.), Wiley, 665-679, 1984.

O.C. Zienkiewicz, E. Onate and J.C. Heinrich, Plastic flow in metal forming, *ASME*, 107-120, 1978.

O.C. Zienkiewicz, E. Onate and J.C. Heinrich, A general formulation for coupled thermal flow of metals using finite elements, *Int. J. Num. Meth. Eng.*, **17**, 1497-1514, 1981.

O.C. Zienkiewicz and G.N. Pande, Some useful forms of isotropic yield surfaces for soil and rock mechanics, *Finite Element in Geomechanics* (Edited Gudehus), John Wiley, London, 179-190, 1977.

O.C. Zienkiewicz, M. Pastor and M. Huang, Softening, localization and adaptive remeshing. Capture of discontinuous solutions, *Comp. Mech.*, **17**, 98-106, 1995.

O.C. Zienkiewicz and D.V. Phillips, An automatic mesh generation scheme for plane and curved surfaces by isoparametric co-ordinates, *Int. J. Num. Meth. Eng.*, **3**, 519-528, 1971.

O.C. Zienkiewicz, S. Qu, R.L. Taylor and S. Nakazawa, The patch test for mixed formulation, *Int. J. Num. Meth. Eng.*, **23**, 1873-1883, 1986.

O.C. Zienkiewicz and R.L. Taylor, The Finite Element Method, *McGraw-Hill*, London, **1**, 1989.

O.C. Zienkiewicz and R.L. Taylor, The Finite Element Method, *McGraw-Hill*, London, **2**, 1991.

O.C. Zienkiewicz, S. Toyoshima, Y.C. Liu and J.Z. Zhu, Flow formulation for numerical solution of forming processes II, Some new directions, *Numerical Methods in Industrial Forming Processes, NUMIFORM 86*, K. Mattiasson *et al.*(Eds.), 1986.

O.C. Zienkiewicz, S. Valliappan and I.P. King, Elasto-plastic solutions of engineering problems. Initial-stress, finite element approach, *Int. J. Num. Meth. Eng.*, **1**, 75-100, 1969.

O.C. Zienkiewicz, W.L. Wood and R.L. Taylor, An alternative single-step algorithm for dynamic problems, *Earthq. Eng. Struct. Dyn.*, **8**, 31-40, 1980.

O.C. Zienkiewicz, W.L. Wood, N.W. Hine and R.L. Taylor, A unified set of single step algorithms - Part 1: General formulation and applications, *Int. J. Num. Meth. Eng.*, **20**, 1529-1552, 1984.

O.C. Zienkiewicz and J.Z. Zhu, A simple error estimator and adaptive procedure for practical engineering analysis, *Int. J. Num. Meth. Eng.*, **24**, 337-357, 1987.

O.C. Zienkiewicz and J.Z. Zhu, Adaptivity and mesh generation, *Int. J. Num. Meth. Eng.*, **32**, 783-810, 1991.

O.C. Zienkiewicz and J.Z. Zhu, The superconvergent patch recovery and a posteriori error estimates, Part 1: the recovery technique, *Int. J. Num. Meth. Eng.*, **33**, 1331-1364, 1992.

O.C. Zienkiewicz and J.Z. Zhu, The superconvergent patch recovery and a posteriori error estimates, Part 2: error estimates and adaptivity, *Int. J. Num. Meth. Eng.*, **33**, 1365 –1382, 1992.

AUTHOR INDEX

SUBJECT INDEX